STUDENT SOLUTIONS GUIDE

to accompany

Introduction to
ORGANIC CHEMISTRY

second edition

by William H. Brown

DAVID R. BENSON
South Dakota State University

BRENT IVERSON & SHEILA IVERSON
University of Texas

WILEY

JOHN WILEY & SONS, INC.

To order books or for customer service please, call 1(800)-CALL-WILEY (225-5945).

ISBN 0-470-00405-3

Printed in the United States of America
10 9 8 7 6 5 4

Preface

This Student Solutions Guide is a companion to *Introduction to Organic Chemistry*, second edition by William H. Brown. All of the problems from the text are reproduced in this guide and are followed by detailed stepwise solutions. This solutions guide was reviewed for accuracy by Paul Silver and by William H. Brown. If you have any comments or questions, please direct them to Professor David Benson, Department of Chemistry, University of Kansas, Lawrence, KS 66045. E-mail: drb@ukans.edu.

David R. Benson
University of Kansas
July 1999

TABLE OF CONTENTS

1   Covalent Bonding.................................1

2   Acids and Bases.................................20

3   Alkanes and Cycloalkanes.................28

4   Chirality...44

5   Alkenes and Alkynes.......................58

6   Reactions of Alkenes.......................70

7   Alkyl Halides..................................85

8   Alcohols, Ethers and Thiols.............97

9   Benzene and Its Derivatives............116

10   Amines..139

11   Aldehydes and Ketones....................150

12   Carboxylic Acids.............................172

13   Derivatives of Carboxylic Acids........183

14   Enolate Anions................................201

15   Organic Polymer Chemistry..............216

16   Carbohydrates.................................222

17   Lipids..236

18   Amino Acids and Proteins.................247

19   Nucleic Acids.................................261

20   The Organic Chemistry of Metabolism..............271

21   Nuclear Magnetic Resonance Spectroscopy.......278

22   Infrared Spectroscopy.......................296

CHAPTER 1
Solutions to Problems

Problem 1.1 Write and compare the ground-state electron configurations for the elements in each set.
(a) Carbon and silicon

$$C \text{ (6 electrons): } 1s^2 2s^2 2p^2$$
$$Si \text{ (14 electrons): } 1s^2 2s^2 2p^6 3s^2 3p^2$$

Both carbon and silicon have four electrons in their highest (valence) shells.

(b) Oxygen and sulfur

$$O \text{ (8 electrons): } 1s^2 2s^2 2p^4$$
$$S \text{ (16 electrons): } 1s^2 2s^2 2p^6 3s^2 3p^4$$

Both oxygen and sulfur have six electrons in their highest (valence) shells.

(c) Nitrogen and phosphorus

$$N \text{ (7 electrons): } 1s^2 2s^2 2p^3$$
$$P \text{ (15 electrons): } 1s^2 2s^2 2p^6 3s^2 3p^3$$

Both nitrogen and phosphorus have five electrons in their highest (valence) shells.

Problem 1.2 Show that the following obey the octet rule:
(a) Sulfur forms S^{2-}.

$$S \text{ (16 electrons): } 1s^2 2s^2 2p^6 3s^2 3p^4$$

$$S^{2-} \text{ (18 electrons): } 1s^2 2s^2 2p^6 3s^2 3p^6$$

(b) Magnesium forms Mg^{2+}.

$$Mg \text{ (12 electrons): } 1s^2 2s^2 2p^6 3s^2$$

$$Mg^{2+} \text{ (10 electrons): } 1s^2 2s^2 2p^6$$

Problem 1.3 Judging from their relative positions in the Periodic Table, which element in each set has the larger electronegativity?
(a) Lithium or potassium

In general, electronegativity increases from left to right across a row and from bottom to top of a column in the Periodic Table. Lithium is higher up on the table and, thus, more electronegative than potassium.

(b) Nitrogen or phosphorus

Nitrogen is higher up on the table and, thus, more electronegative than phosphorus.

(c) Carbon or silicon

Carbon is higher up on the table and, thus, more electronegative than silicon.

Problem 1.4 Classify each bond as nonpolar covalent, polar covalent, or ionic.
(a) S-H   (b) P-H   (c) C-F   (d) C-Cl

Bond	Differences in electronegativity	Type of bond
S-H	2.5 - 2.1 = 0.4	nonpolar covalent
P-H	2.1 - 2.1 = 0.0	nonpolar covalent
C-F	4.0 - 2.5 = 1.5	polar covalent
C-Cl	3.0 - 2.5 = 0.5	polar covalent

Problem 1.5 Using the symbols δ- and δ+, indicate the direction of polarity in these polar covalent bonds.
(a) C-N

$$\overset{\delta+}{C}-\overset{\delta-}{N}$$

Nitrogen is more electronegative than carbon.

(b) N-O

$$\overset{\delta+}{N}-\overset{\delta-}{O}$$

Oxygen is more electronegative than nitrogen.

(c) C-Cl

$$\overset{\delta+}{C}-\overset{\delta-}{Cl}$$

Chlorine is more electronegative than carbon.

Problem 1.6 Draw Lewis structures for these molecules showing all valence electrons.
(a) C_2H_6 (b) CS_2 (c) HCN

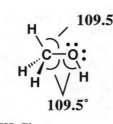

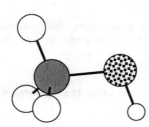

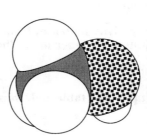

Problem 1.7 Draw Lewis structures for these ions, and show which atom in each bears the formal charge.
(a) $CH_3NH_3^+$ (b) CO_3^{2-} (c) OH^-
    Methyl ammonium ion   Carbonate ion   Hydroxide ion

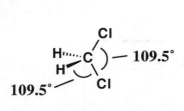

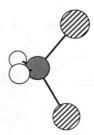

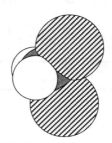

Problem 1.8 Predict all bond angles for these molecules.

(a) CH_3OH

(b) CH_2Cl_2

(c) H_2CO_3 (Carbonic Acid)

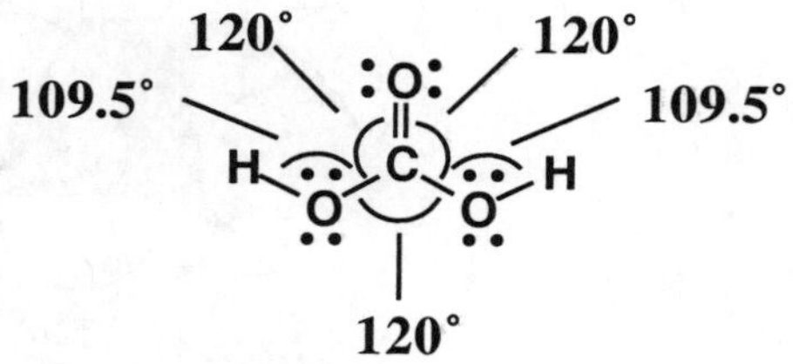

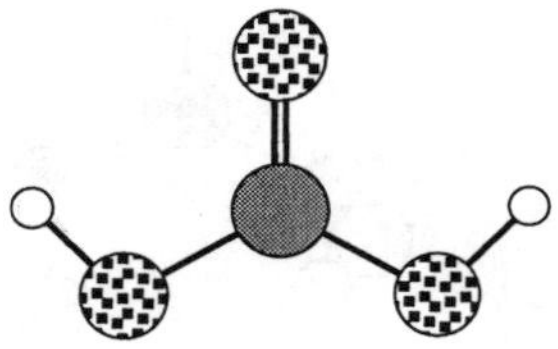

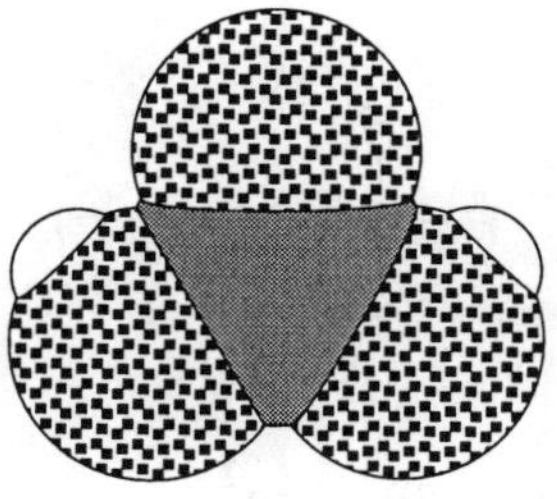

Problem 1.9 Both carbon dioxide, CO_2, and ozone, O_3, are triatomic molecules. Account for the fact that carbon dioxide is a nonpolar molecule whereas ozone is a polar molecule.

The difference lies in the fact that carbon dioxide is a linear molecule whereas ozone is bent. Oxygen is more electronegative than carbon and therefore both C=O bonds in CO_2 have significant dipoles. However, because CO_2 is a linear molecule, the dipoles cancel one another and therefore CO_2 is not dipolar. In contrast to carbon dioxide, ozone cannot be adequately represented by a single Lewis structure. Two structures can be drawn for it. In each, the central oxygen atom bears a full positive charge and one of the terminal oxygen atoms bears a full negative charge. Because the O-O-O angle is close to 120°, these dipoles do not cancel and the ozone molecule has a non-zero dipole moment.

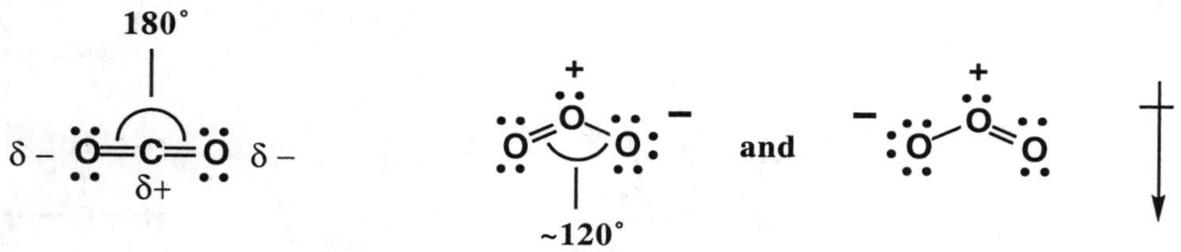

Problem 1.10 Which sets are pairs of contributing structures?

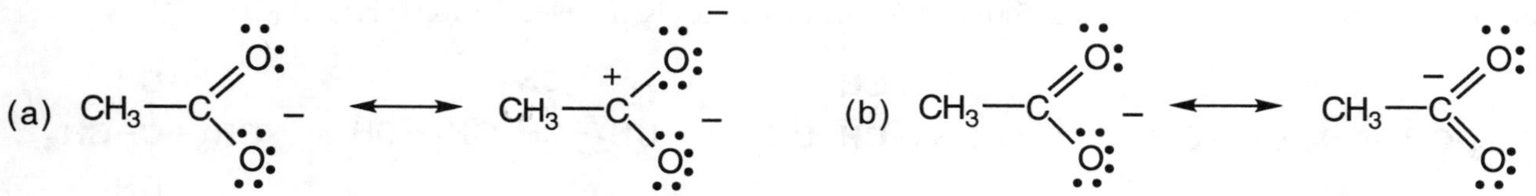

(a) Contributing Structures. They differ only in the distribution of valence electrons.
(b) Not a set of contributing structures. The octet rule is violated for the carboxylate carbon atom in the structure on the right (it has ten valence electrons).

Problem 1.11 Use curved arrows to show the redistribution of electrons in converting contributing structure (a) to (b), and then (b) to (c).

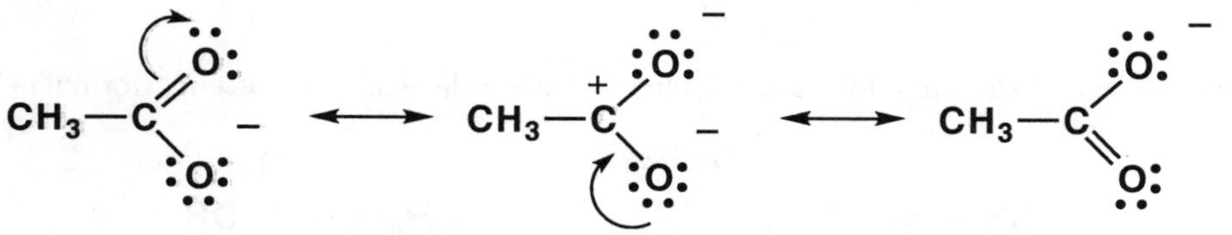

Problem 1.12 Describe the bonding in these molecules in terms of the atomic orbitals involved and predict all bond angles.

(a)   H–C–C=C–H (with H atoms shown)

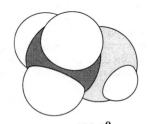

H–C–C=C–H with sp^3 and sp^2 labels

H–C–C=C–H with $\sigma_{sp^3\text{-}1s}$, $\sigma_{sp^2\text{-}sp^2}$, $\sigma_{sp^3\text{-}sp^2}$, $\sigma_{sp^2\text{-}1s}$, $\pi_{2p\text{-}2p}$

109.5° 120° H–C–C=C–H

(b)   H–C–N–H (with H atoms, lone pair on N)

H–C–N–H with sp^3 label

H–C–N–H with $\sigma_{sp^3\text{-}1s}$, $\sigma_{sp^3\text{-}sp^3}$, $\sigma_{sp^3\text{-}1s}$

109.5° H–C–N–H

Problem 1.13 Write condensed structural formulas for the four alcohols of molecular formula $C_4H_{10}O$.

$CH_3\text{-}CH_2\text{-}CH_2\text{-}CH_2\text{-}OH$ $CH_3\text{-}CH_2\text{-}\underset{\overset{|}{OH}}{CH}\text{-}CH_3$ $CH_3\text{-}\underset{\overset{|}{CH_3}}{CH}\text{-}CH_2\text{-}OH$ $CH_3\text{-}\underset{\overset{|}{CH_3}}{\overset{\overset{OH}{|}}{C}}\text{-}CH_3$

Problem 1.14 Write condensed structural formulas for the three ketones of molecular formula $C_5H_{10}O$.

$CH_3\text{-}CH_2\text{-}CH_2\text{-}\overset{\overset{O}{\|}}{C}\text{-}CH_3$ $CH_3\text{-}\underset{\overset{|}{CH_3}}{CH}\text{-}\overset{\overset{O}{\|}}{C}\text{-}CH_3$ $CH_3\text{-}CH_2\text{-}\overset{\overset{O}{\|}}{C}\text{-}CH_2\text{-}CH_3$

Problem 1.15 Write condensed structural formulas for the two carboxylic acids of molecular formula $C_4H_8O_2$.

$CH_3\text{-}CH_2\text{-}CH_2\text{-}\overset{\overset{O}{\|}}{C}\text{-}OH$ $CH_3\text{-}\underset{\overset{|}{CH_3}}{CH}\text{-}\overset{\overset{O}{\|}}{C}\text{-}OH$

<u>Problem 1.16</u> Draw condensed structural formulas for the three secondary amines of molecular formula $C_4H_{11}N$.

$$CH_3-CH_2-\overset{\cdot\cdot}{\underset{H}{N}}-CH_2-CH_3 \qquad CH_3-CH_2-CH_2-\overset{\cdot\cdot}{\underset{H}{N}}-CH_3 \qquad CH_3-\underset{\underset{H}{|}}{\overset{\overset{CH_3}{|}}{CH}}-\overset{\cdot\cdot}{N}-CH_3$$

Electronic Structure of Atoms

<u>Problem 1.17</u> Write the ground-state electron configuration for each atom. After each is given its atomic number.
(a) Sodium (11)

Na (11 electrons) $1s^22s^22p^63s^1$

(b) Magnesium (12)

Mg (12 electrons) $1s^22s^22p^63s^2$

(c) Oxygen (8)

O (8 electrons) $1s^22s^22p^4$

(d) Nitrogen (7)

N (7 electrons) $1s^22s^22p^3$

<u>Problem 1.18</u> Write the ground-state electron configuration for each atom.
(a) Potassium

K (19 electrons) $1s^22s^22p^63s^23p^64s^1$

(b) Aluminum

Al (13 electrons) $1s^22s^22p^63s^23p^1$

(c) Phosphorus

P (15 electrons) $1s^22s^22p^63s^23p^3$

(d) Argon

Ar (18 electrons) $1s^22s^22p^63s^23p^6$

<u>Problem 1.19</u> Which element has the ground-state electron configuration of:
(a) $1s^22s^22p^63s^23p^4$

Sulfur (16) has this ground-state electron configuration.

(b) $1s^22s^22p^4$

Oxygen (8) has this ground-state electron configuration.

<u>Problem 1.20</u> Which element does not have the ground-state electron configuration $1s^22s^22p^63s^23p^6$?
(a) S^{2-} (b) Cl^- (c) Ar   (d) Ca^{2+} (e) K

Potassium (K) has one more electron than all of the other atoms. Its configuration is $1s^22s^22p^63s^23p^64s^1$. The potassium ion (K^+) has the same electron configuration as the other atoms.

<u>Problem 1.21</u> Define valence shell and valence electron.

The valence shell is the outermost occupied shell of an atom, namely the highest numbered shell. A valence electron is an electron in the valence shell.

<u>Problem 1.22</u> How many electrons are in the valence shell of each atom?
(a) Carbon

With a ground-state electron configuration of $1s^2 2s^2 2p^2$, there are four electrons in the valence shell of carbon.
(b) Nitrogen

With a ground-state electron configuration of $1s^2 2s^2 2p^3$, there are five electrons in the valence shell of nitrogen.

(c) Chlorine

With a ground-state electron configuration of $1s^2 2s^2 2p^6 3s^2 3p^5$, there are seven electrons in the valence shell of chlorine.

(d) Aluminum

With a ground-state electron configuration of $1s^2 2s^2 2p^6 3s^2 3p^1$, there are three electrons in the valence shell of aluminum.

(e) Oxygen

With a ground-state electron configuration of $1s^2 2s^2 2p^4$, there are six electrons in the valence shell of oxygen.

Lewis Structures
<u>Problem 1.23</u> Judging from their relative positions in the Periodic Table, which atom in each set is more electronegative?
(a) Carbon or nitrogen

In general, electronegativity increases from left to right across a row and from bottom to top of a column in the Periodic Table. Nitrogen is farther to the right on the table and, thus, more electronegative than carbon.

(b) Chlorine or bromine

Chlorine is higher up on the table and, thus, more electronegative than bromine.

(c) Oxygen or sulfur

Oxygen is higher up on the table and, thus, more electronegative than sulfur.

<u>Problem 1.24</u> Which compounds have nonpolar covalent bonds, which have polar covalent bonds, and which have ionic bonds?
(a) LiF   (b) CH$_3$F   (c) MgCl$_2$ (d) HCl

Using the rule that an ionic bond is formed between atoms with an electronegativity difference of 1.9 or greater, and that polar covalent bonds have electronegativity differences between 0.5 and 1.8, the following table can be constructed:

Bond	Differences in electronegativity	Type of bond
Li-F	4.0 - 1.0 = 3.0	ionic
C-H	2.5 - 2.1 = 0.4	nonpolar covalent
C-F	4.0 - 2.5 = 1.5	polar covalent
Mg-Cl	3.0 - 1.2 = 1.8	polar covalent
H-Cl	3.0 - 2.1 = 0.9	polar covalent

<u>Problem 1.25</u> Using the symbols δ- and δ+, indicate the direction of polarity, if any, in each covalent bond.

$$\overset{\delta+ \quad \delta-}{\text{C-Cl}}$$

(a) C-Cl   **C-Cl**

Chlorine is more electronegative than carbon.

(b) S-H $\overset{\delta-\quad\delta+}{\text{S-H}}$

Sulfur is more electronegative than hydrogen.

(c) C-S

Carbon and sulfur have the same electronegativities, so there is no polarity in a C-S bond.

(d) P-H

Phosphorus and hydrogen have the same electronegativities, so there is no polarity in a P-H bond.

Problem 1.26 Write Lewis structures for these compounds. Show all valence electrons. None of them contains a ring of atoms.

(a) H_2O_2
Hydrogen peroxide

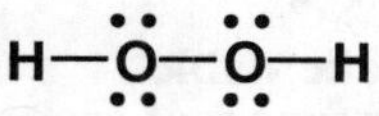

(b) N_2H_4
Hydrazine

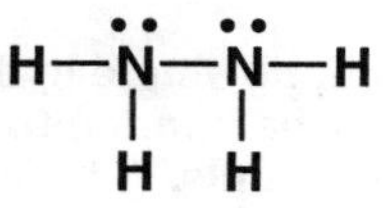

(c) CH_3OH
Methanol

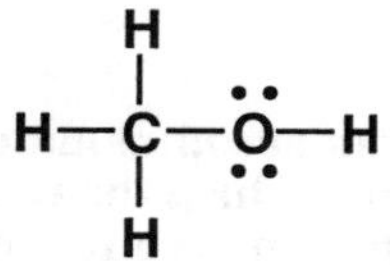

(d) CH_3SH
Methanethiol

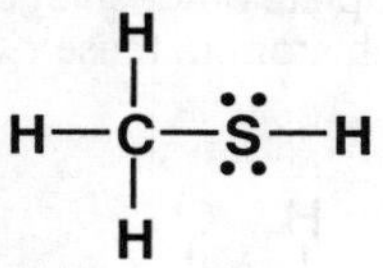

(e) CH_3NH_2
Methylamine

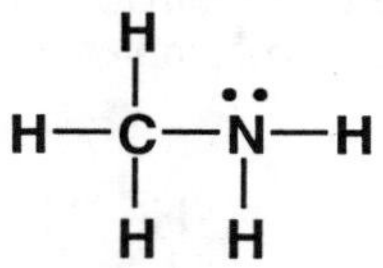

(f) CH_3Cl
Chloromethane

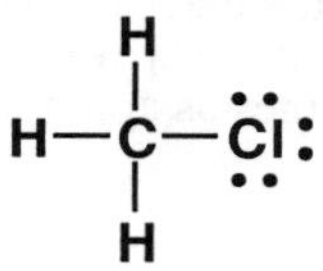

(g) CH_3OCH_3
Dimethyl ether

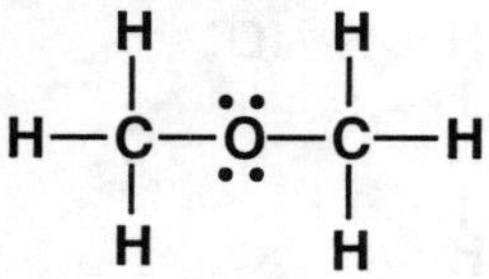

(h) C_2H_6
Ethane

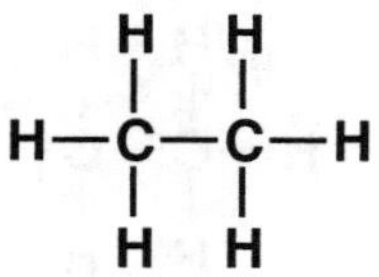

(i) C_2H_4
Ethylene

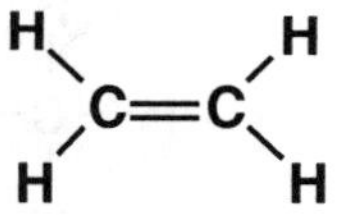

(j) C_2H_2
Acetylene

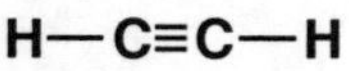

(k) CO_2
Carbon dioxide

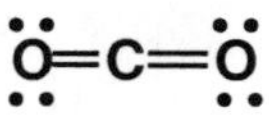

(l) CH_2O
Formaldehyde

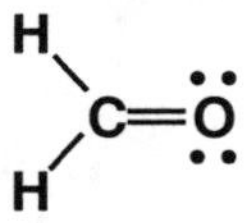

(m) CH_3COCH_3
Acetone

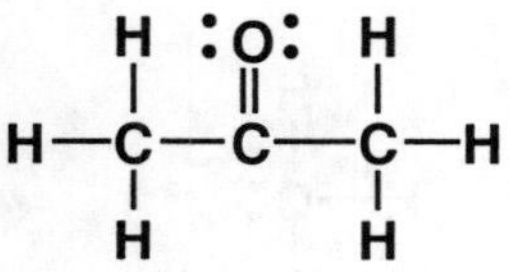

(n) H_2CO_3
Carbonic acid

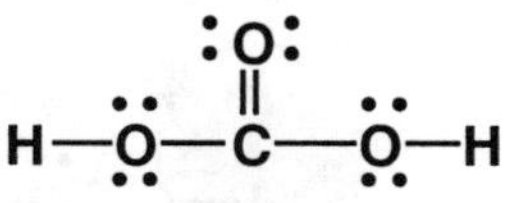

(o) CH_3CO_2H
Acetic acid

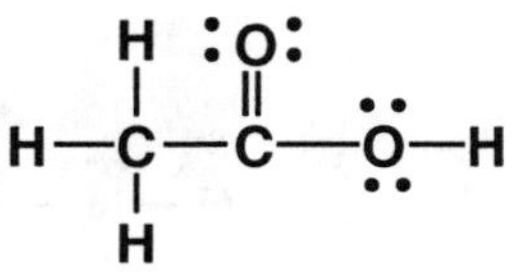

Problem 1.27 Write Lewis structures for these ions. Be certain to show all valence electrons and all formal charges.

(a) HCO_3^-
   Bicarbonate ion

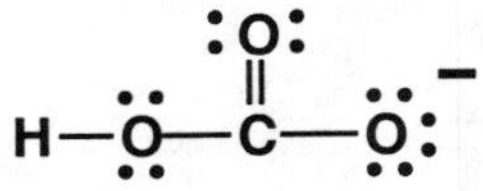

(b) $CO_3{}^{2-}$
   Carbonate ion

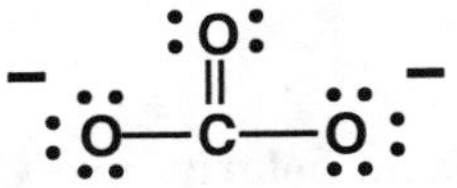

$CH_3CO_2^-$
   Acetate ion

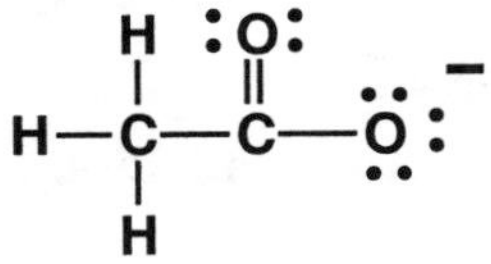

(d) Cl^-
   Chloride ion

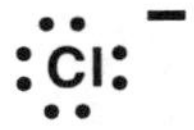

Problem 1.28 Why are the following molecular formulas impossible?
(a) CH_5

Carbon atoms can accommodate only four bonds, and each hydrogen atom can accommodate only one bond. Thus, there is no way for a stable bonding arrangement to be created that utilizes one carbon atom and five hydrogen atoms.

(b) C_2H_7

Because hydrogen atoms can accommodate only one bond each, no single hydrogen atom can make stable bonds to both carbon atoms. Thus, the two carbon atoms must be bonded to each other. This means that each of the bonded carbon atoms can accommodate only three more bonds. Therefore, only six hydrogen atoms can be bonded to the carbon atoms, not seven hydrogen atoms.

Problem 1.29 Following the rule that each atom of carbon, oxygen, and nitrogen reacts to achieve a complete outer shell of eight valence electrons, add unshared pairs of electrons as necessary to complete the valence shell of each atom in these ions. Then, assign formal charges as appropriate.

(a)   (b) 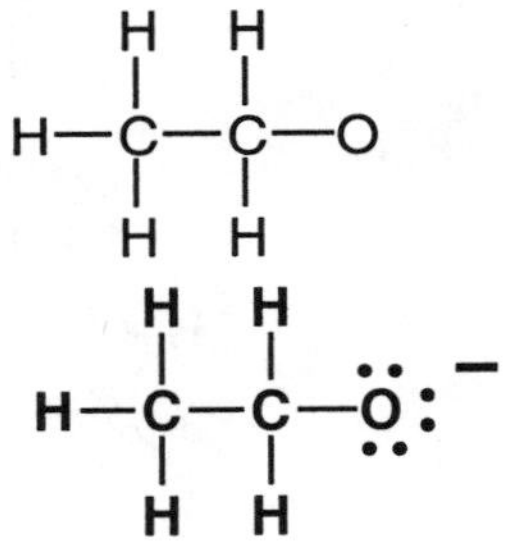   (c) 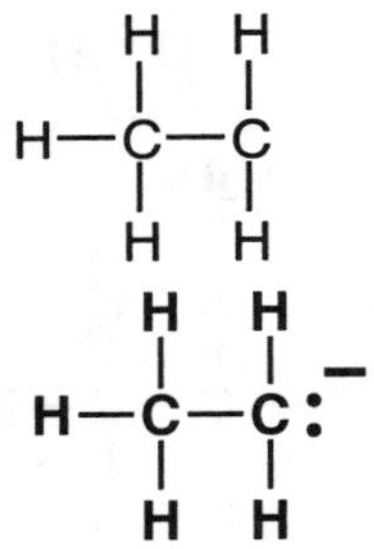   (d)

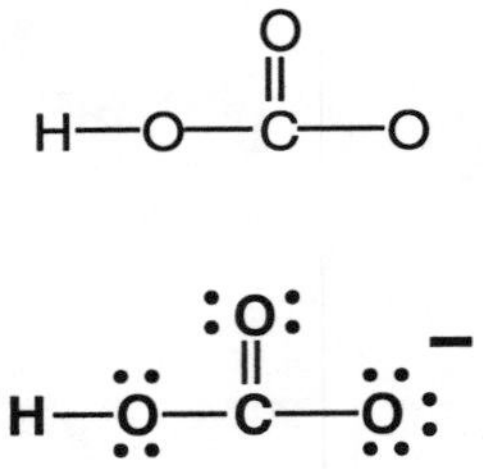

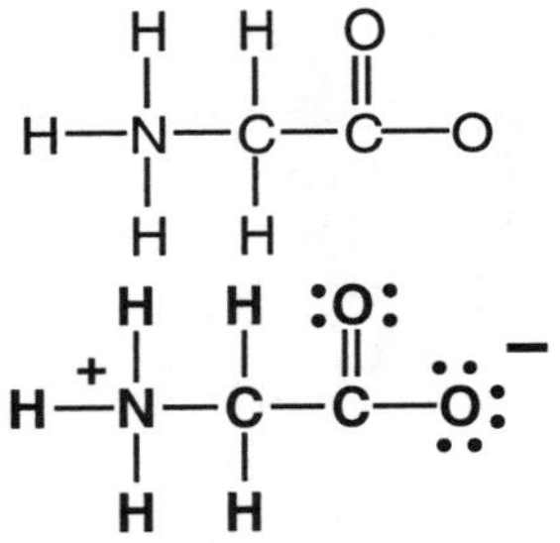

Problem 1.30 Following are several Lewis structures showing all valence electrons. Assign formal charges in each structure as appropriate

(a)   (b) 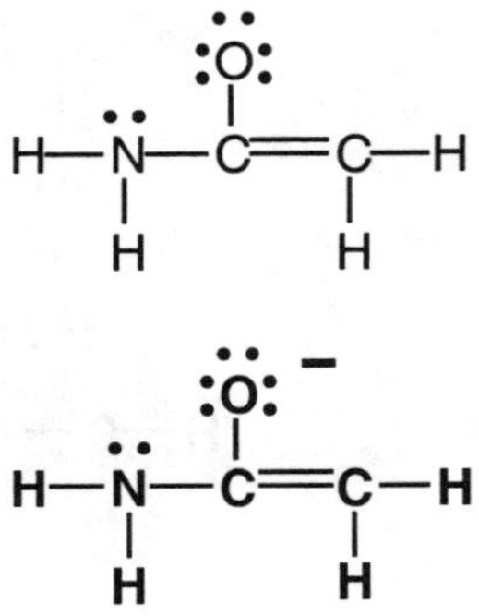   (c) 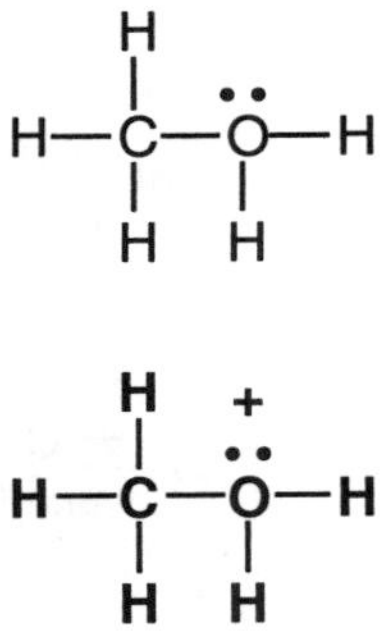   (d)

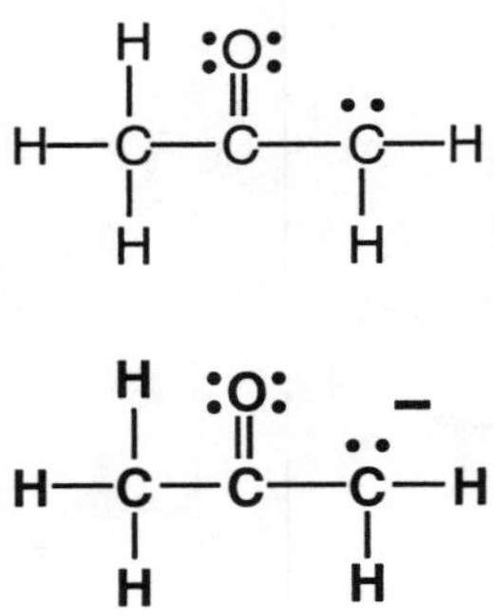

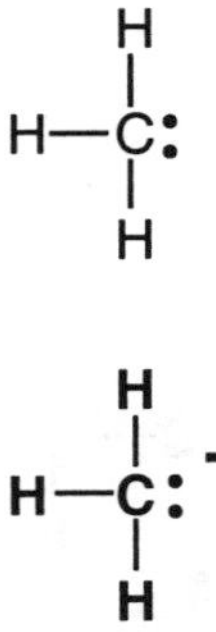

Problem 1.31 Each compound contains both ionic and covalent bonds. Draw a Lewis structure for each and show by charges which bonds are ionic and by dashes which bonds are covalent.

(a) NaOH

(b) NaHCO$_3$

(c) NH$_4$Cl

(d) CH$_3$CO$_2$Na

(e) CH$_3$ONa

Polarity of Covalent Bonds

Problem 1.32 Which statement is true about electronegativity?

(a) Electronegativity increases from left to right in a period of the Periodic Table.
(b) Electronegativity increases from top to bottom in a group of the Periodic Table.
(c) Hydrogen, the element with the lowest atomic number, has the smallest electronegativity.
(d) The higher the atomic number of an element, the greater its electronegativity.

Electronegativity increases from left to right across a row and from bottom to top of a column in the Periodic Table. Thus, statement (a) is true, but (b), (c), and (d) are false.

Problem 1.33 Why does fluorine, the element in the upper right corner of the Periodic Table, have the largest electronegativity of any element?

Electronegativity increases with increasing positive charge on the nucleus and with decreasing distance of the valence electrons from the nucleus. Fluorine is that element for which these two parameters lead to maximum electronegativity.

Problem 1.34 Arrange the single covalent bonds within each set in order of increasing polarity.

(a) C-H, O-H, N-H
  C-H < N-H < O-H
  (0.4) (0.9) (1.4)

(b) C-H, C-Cl, C-I
  C-I < C-H < C-Cl
  (0) (0.4) (0.5)

(c) C-S, C-O, C-N
  C-S < C-N < C-O
  (0) (0.5) (1.0)

(d) C-Li, C-Hg, C-Mg
  C-Hg < C-Mg < C-Li
  (0.6) (1.3) (1.5)

The difference in electronegativities is shown in parentheses under each bond.

Problem 1.35 Using the values of electronegativity given in Table 1.5, predict which indicated bond in each set is more polar and, using the symbols δ+ and δ-, show the direction of its polarity.

(a) CH$_3$-OH or CH$_3$O-H

(b) H-NH$_2$ or CH$_3$-NH$_2$ -

(c) CH$_3$-SH or CH$_3$S-H

(d) CH$_3$-F or H-F

  δ– δ+
CH$_3$O-H

  δ+ δ–
H-NH$_2$

  δ– δ+
CH$_3$S-H

  δ+ δ–
H-F

Problem 1.36 Identify the most polar bond in each molecule.

(a) HSCH$_2$CH$_2$OH

The O-H bond

(b) CHCl$_2$F

The C-F bond

(c) HOCH$_2$CH$_2$NH$_2$

The O-H bond

Problem 1.37 Predict whether the carbon-metal bond in these organometallic compounds is nonpolar covalent, polar covalent, or ionic. For each polar covalent bond, show its direction of polarity using the symbols δ+ and δ-.

(a)
$$\delta-$$
$$CH_2CH_3$$
$$\overset{\delta-}{CH_3CH_2}-\overset{|\,\delta+}{Pb}-\overset{\delta-}{CH_2CH_3}$$
$$\underset{\delta-}{CH_2CH_3}$$
Tetraethyllead

(b)
$$\overset{\delta-}{CH_3}-\overset{\delta+}{Mg}-\overset{\delta-}{Cl}$$
Methylmagnesium chloride

(c)
$$\overset{\delta-}{CH_3}-\overset{\delta+}{Hg}-\overset{\delta-}{CH_3}$$
Dimethylmercury

All of the above carbon-metal bonds are polar covalent.

Bond Angles and Shapes of Molecules
Problem 1.38 Use the VSEPR model to predict bond angles about each highlighted atom.

Approximate bond angles as predicted by the valence-shell electron-pair repulsion model are shown.

Problem 1.39 Use the VSEPR model to predict bond angles about each atom of carbon, nitrogen, and oxygen in these molecules. Hint: first add unshared pairs of electrons as necessary to complete the valence shell of each atom and then make your predictions of bond angles.

Problem 1.40 Silicon is immediately below carbon in the Periodic Table. Predict the C-Si-C bond angle in tetramethylsilane, $(CH_3)_4Si$.

Silicon is in Group 4 of the Periodic Table and, like carbon, has four valence electrons. In tetramethylsilane, $(CH_3)_4Si$, silicon is surrounded by four regions of electron density. Therefore, predict all C-Si-C bond angles to be 109.5º, so the molecule is tetrahedral around Si.

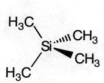 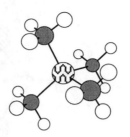 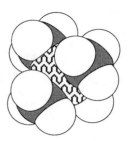

Polar and Nonpolar Molecules

<u>Problem 1.41</u> Draw a three-dimensional representation for each molecule Indicate which molecules are polar and the direction of the polarity.

(a) CH₃F

  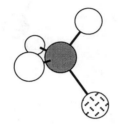 

(b) CH₂Cl₂

  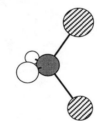 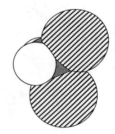

(c) CHCl₃

  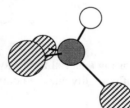 

(d) CCl₄

No dipole

 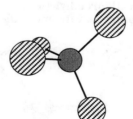 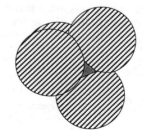

(e) $CH_2=CCl_2$

(f) $CH_2=CHCl$

(g) CH_3CN

(h) $(CH_3)_2C=O$

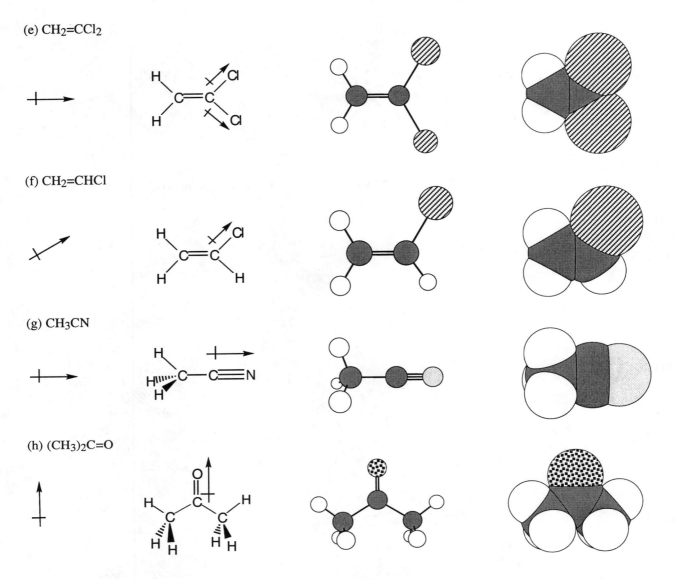

<u>Problem 1.42</u> Tetrafluoroethylene, C_2F_4, is the starting material for the synthesis of the polymer poly(tetrafluoroethylene), commonly known as Teflon®. Molecules of tetrafluoroethylene are nonpolar. Propose a structural formula for the compound.

Fluorine can form only one bond to another atom and carbon requires four bonds. Therefore, each fluorine atom in tetrafluoroethylene must be bonded to a carbon atom, and the carbon atoms are joined together in a carbon-carbon double bond. Thus, tetrafluoroethylene is a planar molecule. Even though each C-F bond is highly polar, the dipoles cancel one another and the molecule has a dipole moment of zero.

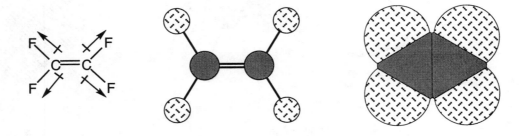

Problem 1.43 The two chlorofluorocarbons (CFCs) most widely used as heat transfer media for refrigeration systems were trichlorofluoromethane, CCl_3F, and dichlorodifluoromethane, CCl_2F_2. Draw a three-dimensional representation of each molecule, and indicate the direction of its polarity.

CCl_3F

CCl_2F_2

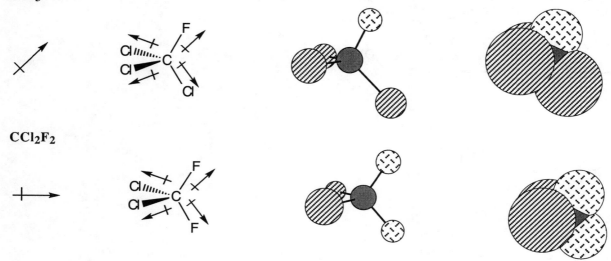

Resonance and Contributing Structures

Problem 1.44   Which of these statements are true about resonance contributing structures?

(a) All contributing structures must have the same number of valence electrons.
(b) All contributing structures must have the same arrangement of atoms.
(c) All atoms in a contributing structure must have complete valence shells.
(d) All bond angles in sets of contributing structures must be the same.

For sets of contributing structures, electrons (usually π electrons or lone pair electrons) move, but the atomic nuclei maintain the same arrangement in space. Thus, statements (b) and (d) are true. In addition, the total number of electrons, valence and inner shell electrons, in each contributing structure must be the same, so statement (a) is also true. However, the movement of electrons often leaves one or more atoms without a filled valence shell in a given contributing structure, so statement (c) is false.

Problem 1.45 Draw the contributing structure indicated by the curved arrow(s) and assign formal charges as appropriate.

(a)

(b)

(c)

<u>Problem 1.46</u> Using the VSEPR model, predict the bond angles about the carbon atom in each pair of contributing structures in Problem 1.45. In what way do the bond angles change from one contributing structure to the other?

As shown on the following structural formulas, carbon atoms that are cations are trigonal planar, exhibiting bond angles of 120°, just like the neutral carbon atoms that are sp² hybridized. Therefore, bond angles do not change from one contributing structure to the other.

(a)

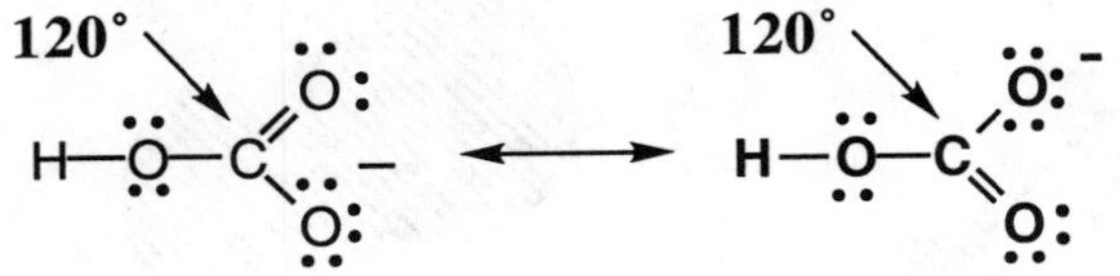

(b)

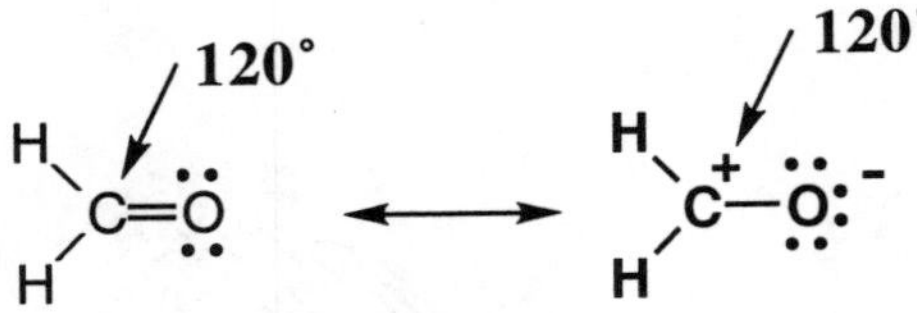

(c)

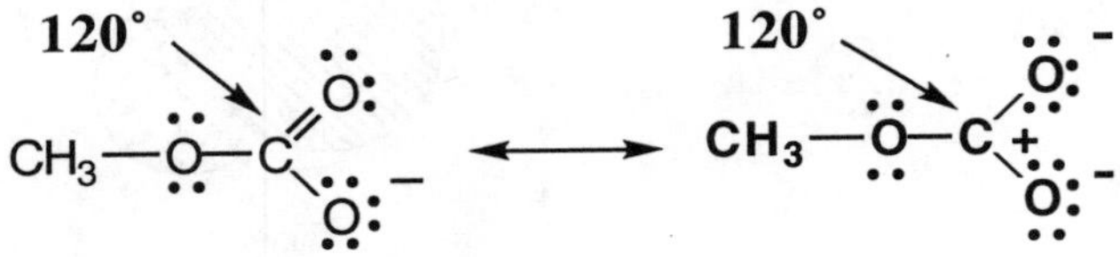

<u>Problem 1.47</u> State the orbital hybridization of each highlighted atom.

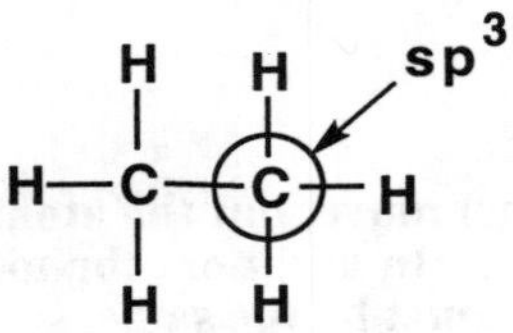

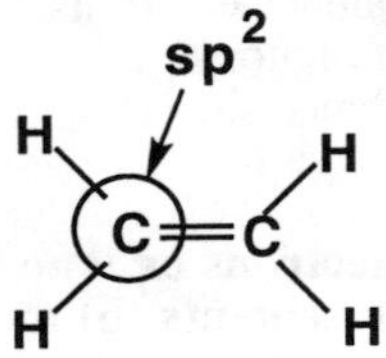

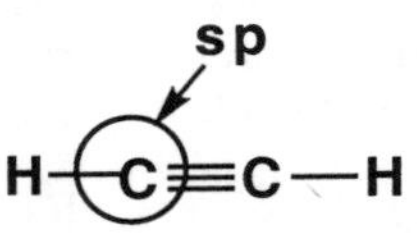

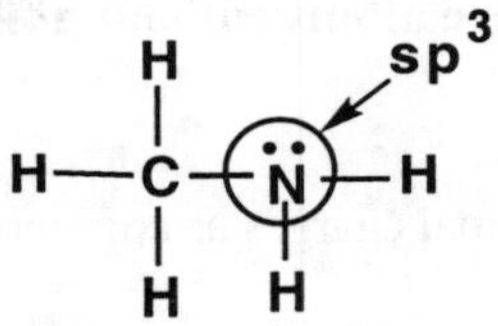

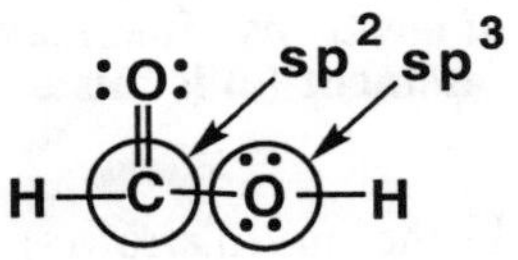

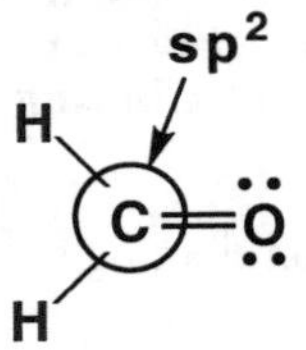

<u>Problem 1.48</u> Describe each highlighted bond in terms of the overlap of atomic orbitals.

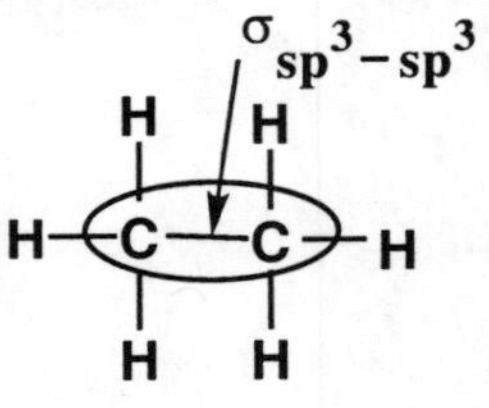

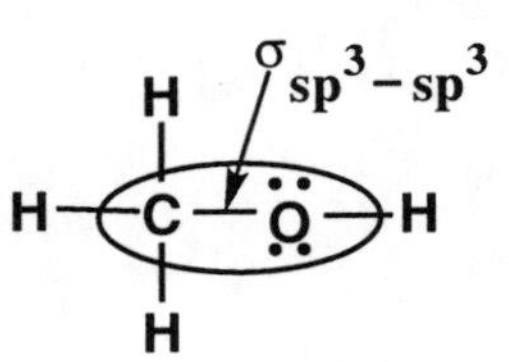

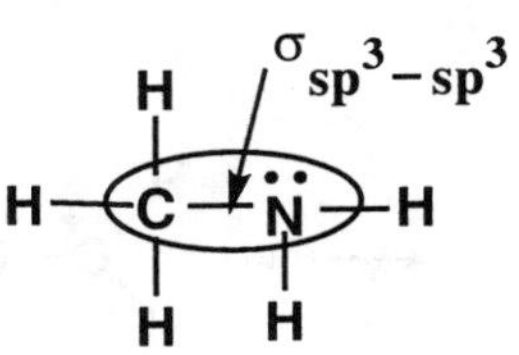

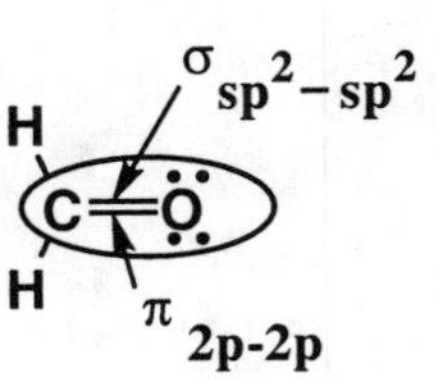

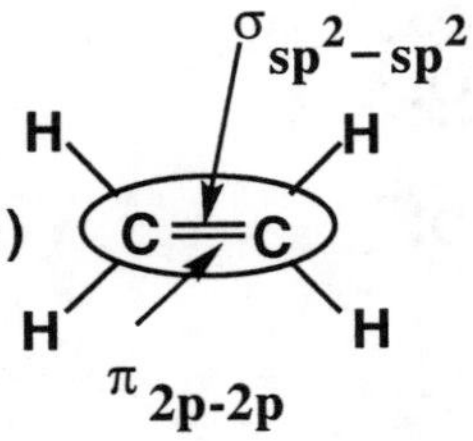

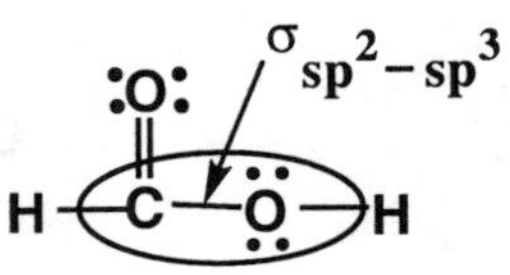

<u>Problem 1.49</u> Following is a structural formula and a space-filling model of benzene, C_6H_6.

(a) Predict each H-C-C and each C-C-C bond angle on benzene.

Each carbon atom in benzene has three regions of electron density around it, so according to the VSEPR model, the carbon atoms are trigonal planar. Predict each H-C-C and C-C-C bond angle to be 120°.

(b) State the hybridization of each carbon atom in benzene.

Each carbon atom is sp^2 hybridized because each one makes three σ bonds and one π bond.

(c) Predict the shape of a benzene molecule.

Since all of the carbon atoms in the ring are sp^2 hybridized and thus trigonal planar, predict carbon atoms in benzene to form a flat hexagon in shape, with the hydrogen atoms in the same plane as the carbon atoms.

<u>Problem 1.50</u> In Chapter 6, we study a group of organic cations called carbocations. Following is the structure of one such carbocation, the *tert*-butyl cation.

(a) How many electrons are in the valence shell of the carbon bearing the positive charge?

The carbon bearing the positive charge has six valence electrons.

(b) Predict the bond angles about this carbon.

The central carbon atom in the *tert*-butyl cation has three regions of electron density around it, so according to the VSEPR model, the central carbon is trigonal planar. Predict the C-C-C bond angle to be 120°.

(c) Given the bond angles you predicted in (b), what hybridization do you predict for this carbon?

The central carbon atom is sp^2 hybridized because it makes three σ bonds and has an empty p orbital.

<u>Problem 1.51</u> We will also study the isopropyl cation, $(CH_3)_2CH^+$.

(a) Write a Lewis structure for this cation. Use a plus sign to show the location of the positive charge.

(b) How many electrons are in the valence shell of the carbon bearing the positive charge?

The carbon bearing the positive charge has six valence electrons.

(c) Use the VSEPR model to predict all bond angles about the carbon bearing the positive charge.

The central carbon atom in the isopropyl cation has three regions of electron density around it, so according to the VSEPR model, the central carbon is trigonal planar. Predict the C-C-C and C-C-H bond angles to be 120°.

(d) Describe the hybridization of each carbon in this cation.

The central carbon atom is sp^2 hybridized because it makes three σ bonds and has an empty p orbital.

Functional Groups

Problem 1.52 Draw Lewis structures for these functional groups. Be certain to show all valence electrons on each.

(a) Carbonyl group   (b) Carboxyl group   (c) Hydroxyl group   (d) Primary amino group

Problem 1.53 Draw the structural formula for a compound of molecular formula

(a) C_2H_6O that is an alcohol   (b) C_3H_6O that is an aldehyde   (c) C_3H_6O that is a ketone

(d) $C_3H_6O_2$ that is a carboxylic acid   (e) $C_4H_{11}N$ that is a tertiary amine

Problem 1.54 Draw condensed structural formulas for all compounds of molecular formula C_4H_8O that contain:

(a) A carbonyl group (there are two aldehydes and one ketone).

(b) A carbon-carbon double bond and a hydroxyl group (there are eight).

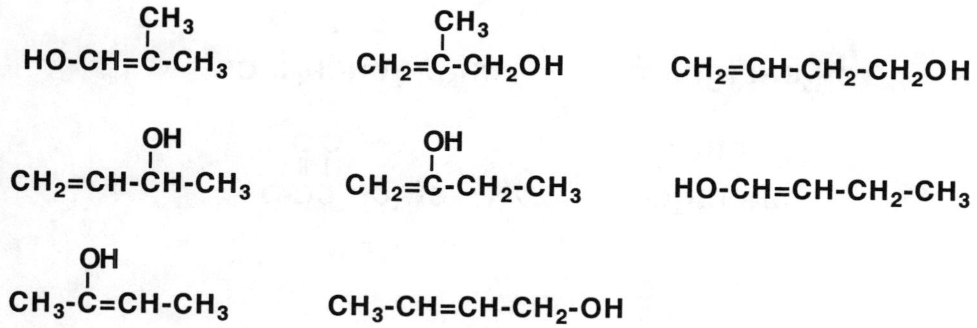

$$CH_3$$
HO-CH=C-CH_3

$$CH_3$$
CH_2=C-CH_2OH

CH_2=CH-CH_2-CH_2OH

$$OH$$
CH_2=CH-CH-CH_3

$$OH$$
CH_2=C-CH_2-CH_3

HO-CH=CH-CH_2-CH_3

$$OH$$
CH_3-C=CH-CH_3

CH_3-CH=CH-CH_2-OH

Problem 1.55 Draw structural formulas for:

(a) The eight alcohols of molecular formula $C_5H_{12}O$

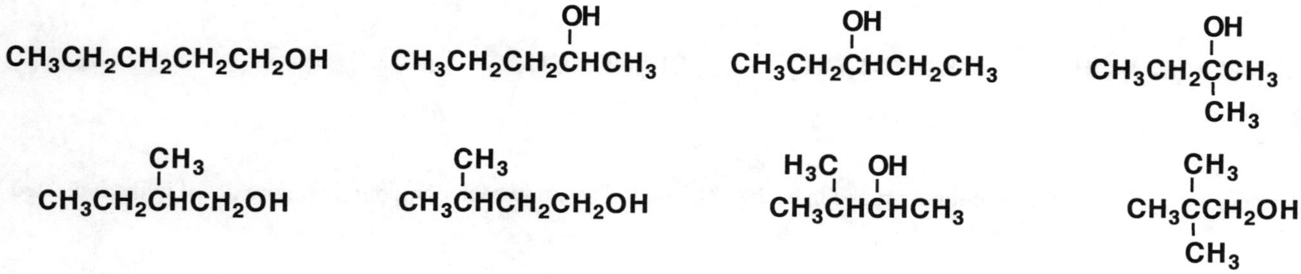

CH_3CH_2CH_2CH_2CH_2OH

$$OH$$
CH_3CH_2CH_2CHCH_3

$$OH$$
CH_3CH_2CHCH_2CH_3

$$OH$$
CH_3CH_2CCH_3
$$CH_3$$

$$CH_3$$
CH_3CH_2CHCH_2OH

$$CH_3$$
CH_3CHCH_2CH_2OH

$$H_3C \quad OH$$
CH_3CHCHCH_3

$$CH_3$$
CH_3CCH_2OH
$$CH_3$$

(b) The eight aldehydes of molecular formula $C_6H_{12}O$

$$O$$
CH_3CH_2CH_2CH_2CH_2C-H

$$CH_3 \quad O$$
CH_3CHCH_2CH_2C-H

$$CH_3 \quad O$$
CH_3CH_2CHCH_2C-H

$$H_3C \quad O$$
CH_3CH_2CH_2CHC-H

$$CH_3 \quad O$$
CH_3CCH_2C-H
$$CH_3$$

$$H_3C \quad O$$
CH_3CH_2CC-H
$$H_3C$$

$$H_3C \quad O$$
CH_3CHCHC-H
$$CH_3$$

$$O$$
CH_3CH_2CHC-H
$$CH_2CH_3$$

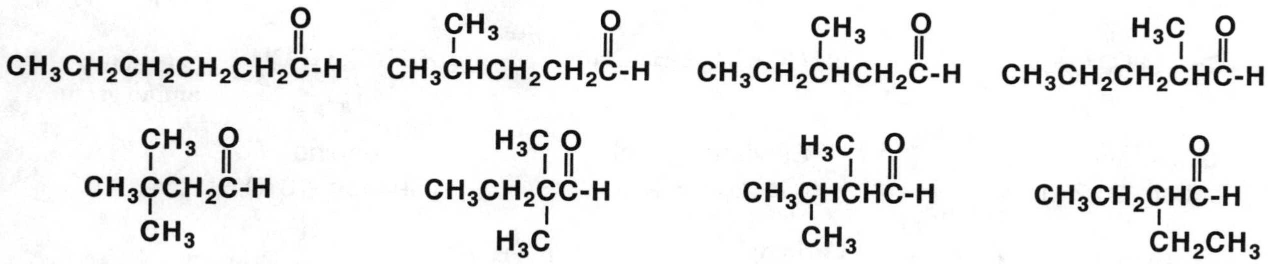

(c) The six ketones of molecular formula $C_6H_{12}O$

$$O$$
CH_3CH_2CH_2CH_2CCH_3

$$O$$
CH_3CH_2CH_2CCH_2CH_3

$$O$$
CH_3CHCCH_2CH_3
$$CH_3$$

$$O$$
CH_3CCHCH_2CH_3
$$CH_3$$

$$O$$
CH_3CCH_2CHCH_3
$$CH_3$$

$$OCH_3$$
CH_3CCCH_3
$$CH_3$$

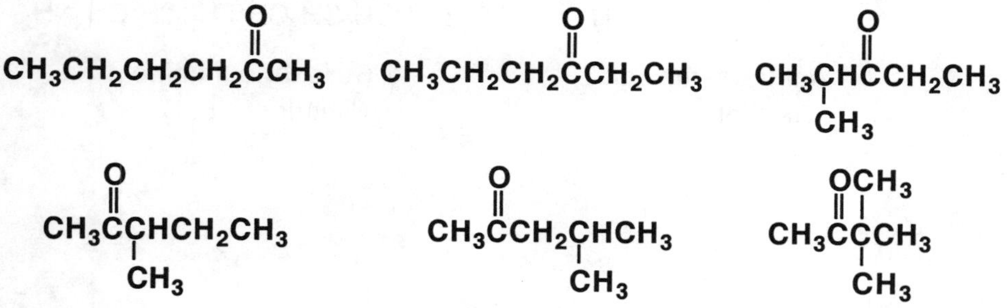

(d) The eight carboxylic acids of molecular formula $C_6H_{12}O_2$

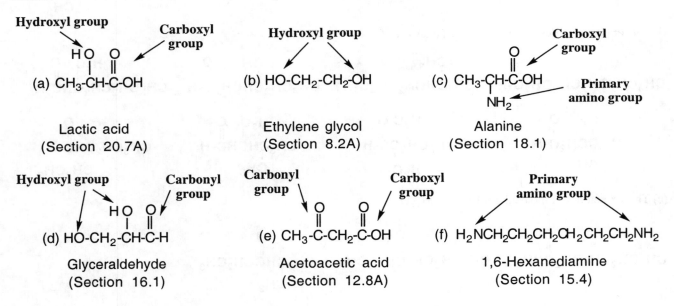

(e) The three tertiary amines of molecular formula $C_5H_{13}N$

Problem 1.56 Identify the functional groups in each compound. We study each compound in more detail in the indicated section.

Hydroxyl group   **Carboxyl group**

           HO O
(a) $CH_3-CH-C-OH$

Lactic acid
(Section 20.7A)

Hydroxyl group

(b) $HO-CH_2-CH_2-OH$

Ethylene glycol
(Section 8.2A)

Carboxyl group

           O
(c) $CH_3-CH-C-OH$
           |
           NH_2 **Primary amino group**

Alanine
(Section 18.1)

Hydroxyl group   **Carbonyl group**

           HO O
(d) $HO-CH_2-CH-C-H$

Glyceraldehyde
(Section 16.1)

Carbonyl group   **Carboxyl group**

           O   O
(e) CH_3-C-CH_2-C-OH

Acetoacetic acid
(Section 12.8A)

Primary amino group

(f) $H_2NCH_2CH_2CH_2CH_2CH_2CH_2NH_2$

1,6-Hexanediamine
(Section 15.4)

Interactive Questions
Problem 1.57 Measure the carbon-carbon and carbon-oxygen bond lengths in these five molecules and record your results in the appropriate table. What is the correlation between bond length and type of bond (single, double, triple)?

As shown in the table, single bonds are longer than double bonds, which in turn are longer than triple bonds.

Compound	Name	Measured C-C Bond Length (Å)
Ethane	CH_3-CH_3	1.531 Å
Ethylene	$CH_2=CH_2$	1.399 Å
Acetylene	$CH\equiv CH$	1.212 Å

Compound	Name	Measured C-O Bond Length (Å)
Methanol	CH_3OH	1.427 Å
Formaldehyde	CH_2O	1.207 Å

Problem 1.58 Measure all bond angles in each molecule and compare your results with the values predicted by the valence-shell electron-pair repulsion model.

Note that unshared pairs are not shown on the Chem3D models!

(a) CH_3CH_3 (b) $CH_2=CH_2$ (c) CH_3OH (d) CH_3OCH_3

(e) $CH_2=O$ (f) $(CH_3)_2C=O$ (g) CH_3NH_2 (h) $CH_2=NH$

(i) H_2CO_3 (j) HCO_2H (k) CH_3CO_2H

Reminder: four regions of electron density = 109.5°, three regions = 120°, and two regions = 180°.

Problem 1.59 Measure the carbon-oxygen bond lengths in acetic acid, CH_3CO_2H, and acetate ion, $CH_3CO_2^-$. How do you account for any differences between them?

For acetic acid, C=O is 1.208° and C-O is 1.345Å. For acetate ion, each C-O bond has the same length, 1.251Å. Remember that the bond orders for the carbon-oxygen single and double bonds in acetic acid are 1.00 and 2.00 respectively. In acetate ion, the C-O bonds each have a bond order of 1.50, as predicted from resonance theory. As you can see from the Chem3D structures, the C-O bond lengths of the acetate ion are nearly midway between the C-O single and double bond lengths of carboxylic acid.

Problem 1.60 Identify the functional group(s) in each molecule. Do not be concerned with names; we discuss how to name organic compounds in subsequent chapters.

(a) 1-Butanol   (b) 2-Butanamine   (c) 2-Butanone
(d) 2-Hydroxybutanoic acid   (e) Acetoacetic acid   (f) Acrylic acid
(g) Alanine   (h) Allyl alcohol   (i) Dihydroxyacetone
(j) Ethanolamine   (k) Propanal   (l) Isoprene

(a) hydroxyl group; this compound is an alcohol
(b) a 1° amino group
(c) carbonyl group of a ketone
(d) an hydroxyl group and a carboxyl group
(e) carbonyl group of a ketone and a carboxyl group
(f) a carbon-carbon double bond and a carboxyl group
(g) a 1° amino group and a carboxyl group
(h) a carbon-carbon double bond and an hydroxyl group
(i) a carbonyl group and two hydroxyl groups
(j) a 1° amino group and an hydroxyl group
(k) a carbonyl group of a ketone
(l) two carbon-carbon double bonds

CHAPTER 2
Solutions to Problems

Problem 2.1 Write each acid-base reaction as a proton-ransfer reaction. Label which reactant is the acid and which the base; which product is the conjugate base of the original acid and which the conjugate acid of the original base. Use curved arrows to show the flow of electrons in each reaction.

(a) $CH_3S\text{-}H$ + OH^- $\longrightarrow$ $CH_3\text{-}S^-$ + H_2O

acid   base   conjugate   conjugate
                                         base   acid

(b) $CH_3O\text{-}H$ + NH_2^- $\longrightarrow$ CH_3O^- + NH_3

acid   base   conjugate   conjugate
                                         base   acid

Problem 2.2 For each value of K_a, calculate the corresponding value of pK_a. Which compound is the stronger acid?
(a) Acetic acid, $K_a = 1.74 \times 10^{-5}$ (b) Water, $K_a = 2.00 \times 10^{-16}$

The pK_a is equal to $-\log_{10}K_a$. The pK_a of acetic acid is 4.76 and the pK_a of water is 15.7. Acetic acid, with the smaller pK_a value, is the stronger acid.

Problem 2.3 Predict the position of equilibrium for each acid-base reaction. See Table 2.1 for pK_a values of each acid.

(a) CH_3NH_2 + CH_3CO_2H $\rightleftharpoons$ $CH_3NH_3^+$ + $CH_3CO_2^-$
       Methylamine   Acetic acid   Methylammonium   Acetate
                                                        ion   ion

Acetic acid is a much stronger acid than the methylammonium ion; equilibrium lies far to the right.

CH_3NH_2 + CH_3CO_2H $\rightleftharpoons$ $CH_3NH_3^+$ + $CH_3CO_2^-$

pK_a 4.76 pK_a 9.64

(stronger base)   (stronger acid)   (weaker acid)   (weaker base)

(b) $CH_3CH_2O^-$ + NH_3 $\rightleftharpoons$ CH_3CH_2OH + NH_2^-
       Ethoxide   Ammonia   Ethanol   Amide
         ion   ion

Ethanol is a much stronger acid than ammonia; equilibrium lies very far to the left.

$CH_3CH_2O^-$ + NH_3 $\rightleftharpoons$ CH_3CH_2OH + NH_2^-

pK_a 33 pK_a 15.9

(weaker base)   (weaker acid)   (stronger acid)   (stronger base)

Problem 2.4 Write an equation for the reaction between each Lewis acid-base pair, showing electron flow by means of curved arrows. Hint: Aluminum is in Group IIIA of the Periodic Table, just under boron. Aluminum in $AlCl_3$ has only six electrons in its valence shell and has an incomplete octet.

(a) Cl^- + $AlCl_3$ $\longrightarrow$

(b) CH_3Cl + $AlCl_3$ $\longrightarrow$

Brønsted-Lowry Acids and Bases

Problem 2.5 Complete a net ionic equation for each proton-transfer reaction using curved arrows to show the flow of electron pairs in each reaction. In addition, write Lewis structures for all starting materials and products. Label the original acid and its conjugate base; label the original base and its conjugate acid. If you are uncertain about which substance in each equation is the proton donor, refer to Table 2.1 for the pK_a values of proton acids.

(a) NH_3 + HCl $\longrightarrow$

(b) $CH_3CH_2O^-$ + HCl $\longrightarrow$

(c) HCO_3^- + OH^- $\longrightarrow$

(d) $CH_3CO_2^-$ + NH_4^+ $\longrightarrow$

(e) NH_4^+ + HO^- ⟶

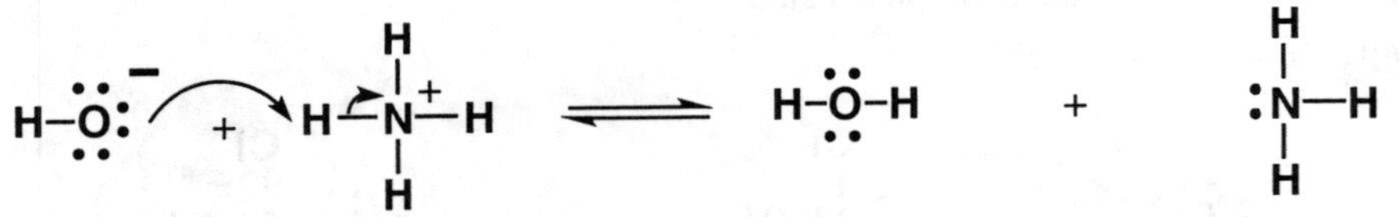

base   acid   conjugate acid   conjugate base

(f) $CH_3CO_2^-$ + $CH_3NH_3^+$ ⟶

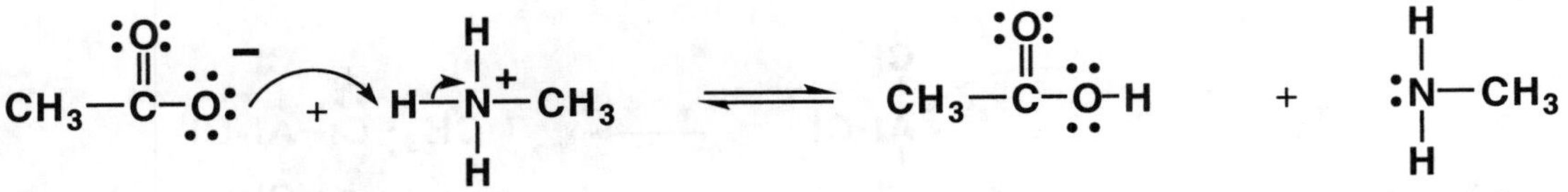

base   acid   conjugate acid   conjugate base

(g) $CH_3CH_2O^-$ + NH_4^+ ⟶

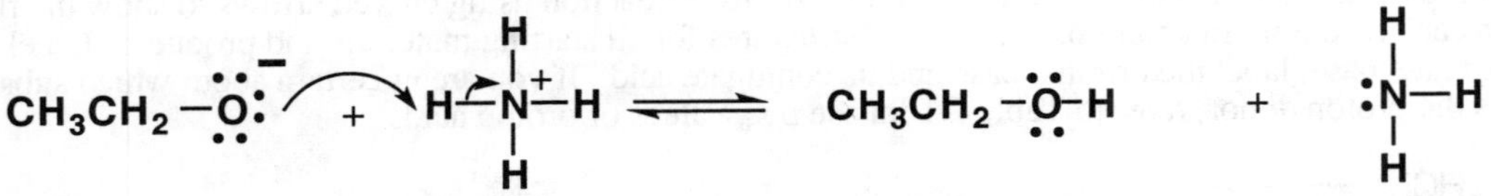

base   acid   conjugate acid   conjugate base

(h) $CH_3NH_3^+$ + OH^- ⟶

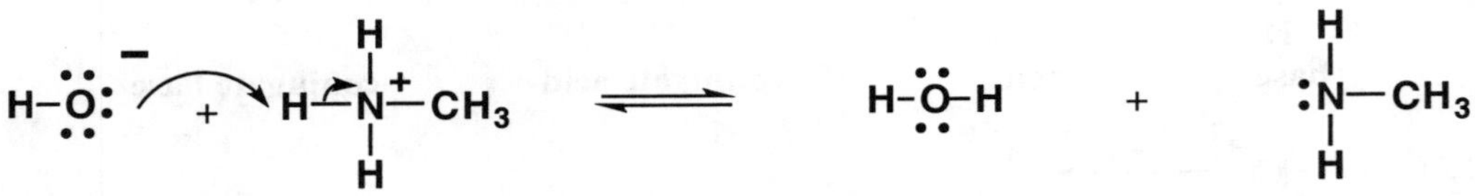

base   acid   conjugate acid   conjugate base

<u>Problem 2.6</u> Each of these molecules and ions can function as a base. Complete the Lewis structure of each base, and write the structural formula of the conjugate acid formed by its reaction with HCl.

(a) CH_3CH_2OH (b) 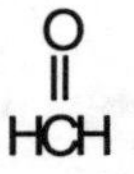 $\overset{O}{\overset{\|}{H-C-H}}$ (c) $(CH_3)_2NH$ (d) HCO_3^-

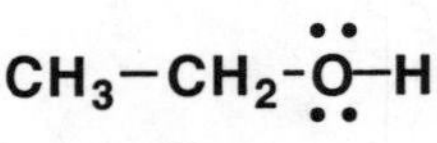

Base   Base   Base   Base

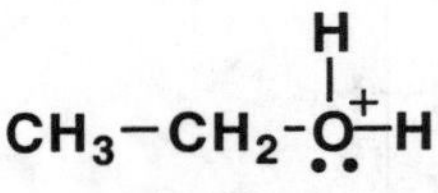

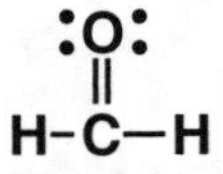

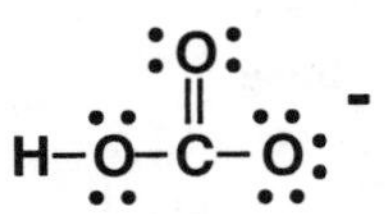

Conjugate acid   Conjugate acid   Conjugate acid   Conjugate acid

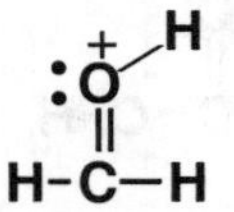

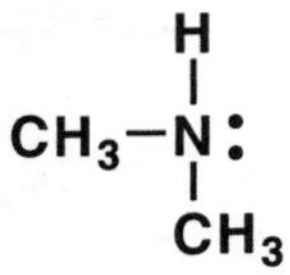

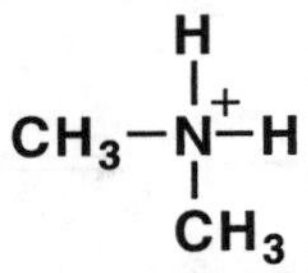

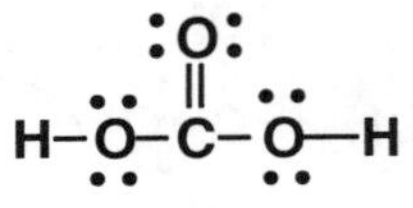

Problem 2.7 Offer an explanation for the following observations.
(a) H_3O^+ is a stronger acid than NH_4^+.

Oxygen is more electronegative than nitrogen. It is less able to bear a formal positive charge.

(b) Nitric acid, HNO_3, is a stronger acid than nitrous acid, HNO_2 (pK_a 3.7).

For the nitrate anion (NO_3^-), three identical resonance structures can be drawn. Only two resonance structures can be drawn for the nitrite anion (NO_2^-). Thus, the negative charge is more delocalized in the nitrate anion.

(c) Ethanol, CH_3CH_2OH, and water have approximately the same acidity.

Deprotonation of ethanol and water gives ethoxide and hydroxide anions, respectively. In both cases, the negative charge is on an oxygen atom which cannot be stabilized by resonance.

(d) Trichloroacetic acid, CCl_3CO_2H (pK_a 0.64), is a stronger acid than acetic acid, CH_3CO_2H (pK_a 4.74).

The carbon atom of a carbonyl group bears a partial positive charge due to the electronegativity difference between carbon and oxygen. To compensate, this carbon inductively withdraws electron density from the OH group of a carboxylic acid, thereby weakening the O-H bond. This effect (together with resonance stabilization of the resultant anion) causes carboxylic acids to be stronger acids than alcohols. The three chlorine atoms in trichloroacetic acid cause the attached carbon atom to have a partial positive charge. To compensate, the carbon atom of the Cl_3C group pulls electron density from the carbonyl carbon, which withdraws even more electron density from the OH group than occurs in acetic acid. As a result, the O-H bond in trichloroacetic acid is even weaker than the O-H bond in acetic acid.

(e) Trifluoroacetic acid, CF_3CO_2H (pK_a 0.23), is a stronger acid than trichloroacetic acid, CCl_3CO_2H (pK_a 0.64).

The effect is the same as in (d), but the fluorine atoms are more electronegative than chlorine and withdraw even more electron density from the attached carbon atom than occurs in trichloroacetic acid.

Problem 2.8 As we shall see in Chapter 14, hydrogens on a carbon adjacent to a carbonyl group are far more acidic than those not adjacent to a carbonyl group. The highlighted H in propanone, for example, is more acidic than the highlighted H in ethane. Account for the greater acidity of propanone in terms of (a) the inductive effect, and (b) the resonance effect.

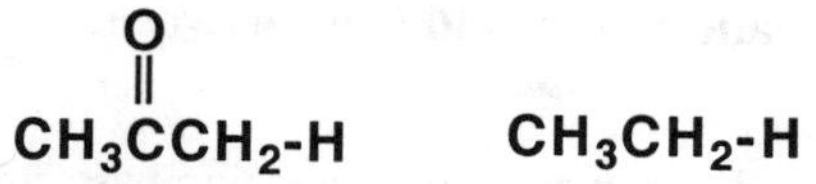

$$O$$
$$\|$$
$$CH_3CCH_2\text{-}H \qquad CH_3CH_2\text{-}H$$

Propanone   Ethane
$pK_a = 22$ $pK_a = 51$

(a) The difference in electron density between oxygen and carbon results in the carbon atom having a partial positive charge. To compensate, this carbon inductively pulls electron density from atoms directly attached to it. Thus, the C-H bonds in propanone are weakened and more easily ionized than a C-H bond in ethane.

(b) Following deprotonation, the negative charge can be stabilized via resonance with the carbonyl group:

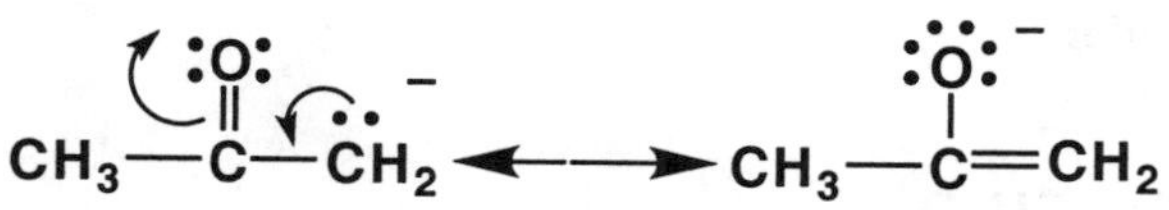

Quantitative Measure of Acid Strength
Problem 2.9 Which has the larger numerical value:
(a) The pK_a of a strong acid or the pK_a of a weak acid?

The weaker acid will have the larger pK_a.

(b) The K_a of a strong acid or the K_a of a weak acid?

The stronger acid will have the larger K_a.

Problem 2.10 In each pair, select the stronger acid:

The stronger acid (underlined in each answer) is the one with the lower pK_a or the larger K_a.
(a) Pyruvic acid (pK_a 2.49) or lactic acid (pK_a 3.85)

(b) Citric acid (pK_{a1} 3.08) or phosphoric acid (pK_{a1} 2.10)

(c) Nicotinic acid (niacin, K_a 1.4 x 10^{-5}) or acetylsalicylic acid (aspirin, K_a 3.3 x 10^{-4})

(d) Phenol (K_a 1.12 x 10^{-10}) or acetic acid (K_a 1.74 x 10^{-5})

Problem 2.11 Arrange the compounds in each set in order of increasing acid strength. Consult Table 2.1 for pK_a values of each acid.

(a) CH_3CH_2OH $HOCO^-$ C_6H_5COH
           Ethanol   Bicarbonate ion   Benzoic acid

pK_a: 15.9   10.33   4.19

The compounds are already in order of increasing acid strength. Ethanol is the weakest acid, benzoic acid is the strongest acid, and bicarbonate ion is in between.

(b) $HOCOH$ CH_3COH HCl
         Carbonic acid   Acetic acid   Hydrogen chloride

pK_a: 6.36   4.76   -7

Again, the compounds are already in order of increasing acid strength. Carbonic acid is the weakest acid, hydrogen chloride is the strongest acid, and acetic acid is in between.

Problem 2.12 Arrange the compounds in each set in order of increasing base strength. Consult Table 2.1 for pK_a values for the conjugate acid of each base. (Hint: The stronger the acid, the weaker its conjugate base, and vice versa).

The weaker the conjugate acid (higher pK_a), the stronger the base.

(a) NH_3 $HOCO^-$ $CH_3CH_2O^-$

           9.24   6.34   15.9   pK_a of conjugate acid
Base strength increases in the order:

$HOCO^-$ < NH_3 < $CH_3CH_2O^-$

(b) OH^- $HOCO^-$ CH_3CO^-

           15.7   6.34   4.76   pK_a of conjugate acid

Base strength increases in the order:

$$CH_3\overset{\displaystyle O}{\overset{\|}{C}}O^- \quad < \quad HO\overset{\displaystyle O}{\overset{\|}{C}}O^- \quad < \quad OH^-$$

(c) H_2O NH_3 $CH_3\overset{\displaystyle O}{\overset{\|}{C}}O^-$

      -1.74   9.24   4.76 pK_a of conjugate acid

Base strength increases in the order:

$$H_2O \quad < \quad CH_3\overset{\displaystyle O}{\overset{\|}{C}}O^- \quad < \quad NH_3$$

(d) NH_2^- $CH_3\overset{\displaystyle O}{\overset{\|}{C}}O^-$ OH^-

      33   4.76   15.7 pK_a of conjugate acid

Base strength increases in the order:

$$CH_3\overset{\displaystyle O}{\overset{\|}{C}}O^- \quad < \quad OH^- \quad < \quad NH_2^-$$

The Position of Equilibrium in Acid-Base Reactions

Problem 2.13 Unless under pressure, carbonic acid in aqueous solution breaks down into carbon dioxide and water, and carbon dioxide is evolved as bubbles of gas. Write an equation for the conversion of carbonic acid to carbon dioxide and water.

$$HO\overset{\displaystyle O}{\overset{\|}{C}}OH \quad \longrightarrow \quad H_2O \quad + \quad CO_2\uparrow$$

Problem 2.14 Will carbon dioxide be evolved when sodium bicarbonate is added to an aqueous solution of:
(a) H_2SO_4 (b) CH_3CH_2OH (c) NH_4Cl?

 In order for carbon dioxide to be evolved, the bicarbonate ion must be protonated to give carbonic acid (Problem 2.13). The pK_a of carbonic acid is 6.36. The pK_a's for sulfuric acid, ethanol and ammonium chloride are -5.2, 15.9, and 9.24, respectively. Thus, sulfuric acid is the only acid strong enough to protonate sodium bicarbonate and evolve carbon dioxide.

Problem 2.15 Acetic acid, CH_3CO_2H, is a weak organic acid, pK_a 4.76. Write equations for the equilibrium reactions of acetic acid with each base. Which equilibria lie considerably toward the left? Which lie considerably toward the right?

(a) CH_3CO_2H + HCO_3^- $\rightleftharpoons$ $CH_3CO_2^-$ + H_2CO_3
        pK_a 4.76 pK_a 6.36

(b) CH_3CO_2H + NH_3 $\rightleftharpoons$ $CH_3CO_2^-$ + NH_4^+
        pK_a 4.76 pK_a 9.24

(c) CH_3CO_2H + H_2O $\rightleftharpoons$ $CH_3CO_2^-$ + H_3O^+
        pK_a 4.76 pK_a -1.74

(d) CH_3CO_2H + HO^- $\rightleftharpoons$ $CH_3CO_2^-$ + H_2O
        pK$_a$ 4.76   pK$_a$ 15.7

Based on the pK$_a$ values shown, reactions (a), (b), and (d) have equilibria that lie considerably to the right, while reaction (c) has an equilibrium that lies considerably to the left.

Problem 2.16 Alcohols are weak organic acids, pK$_a$ 16-18. The pK$_a$ of ethanol, CH_3CH_2OH, is 15.9. Write equations for the equilibrium reactions of ethanol with each base. Which equilibria lie considerably toward the right? Which lie considerably toward the left?

(a) CH_3CH_2OH + HCO_3^- $\rightleftharpoons$ $CH_3CH_2O^-$ + H_2CO_3
        pK$_a$ 15.9   pK$_a$ 6.36

(b) CH_3CH_2OH + OH^- $\rightleftharpoons$ $CH_3CH_2O^-$ + H_2O
        pK$_a$ 15.9   pK$_a$ 15.7

(c) CH_3CH_2OH + NH_2^- $\rightleftharpoons$ $CH_3CH_2O^-$ + NH_3
        pK$_a$ 15.9   pK$_a$ 33

(d) CH_3CH_2OH + NH_3 $\rightleftharpoons$ $CH_3CH_2O^-$ + NH_4^+
        pK$_a$ 15.9   pK$_a$ 9.24

Based on the pK$_a$ values shown, reaction (c) has an equilibrium that lies considerably to the right, while reactions (a) and (d) have equilibria that lie considerably to the left. Reaction (b) has an equilibrium that lies somewhat to the left.

Problem 2.17 Benzoic acid, $C_6H_5CO_2H$, is insoluble in water, but its sodium salt, $C_6H_5CO_2^-Na^+$, is quite soluble in water. Will benzoic acid dissolve in:

(a) Aqueous sodium hydroxide?   (b) Aqueous sodium bicarbonate?   (c) Aqueous sodium carbonate?

The pK$_a$ of benzoic acid is 4.19. The pK$_a$ values for the conjugate acids of sodium hydroxide, sodium bicarbonate, and sodium carbonate are 15.7, 6.36, and 10.33, respectively. Thus, equilibrium will favor reaction of benzoic acid with all three of these bases to give the soluble $C_6H_5CO_2^-Na^+$. Therefore, benzoic acid will dissolve in aqueous solutions of all three bases.

Problem 2.18 Phenol, C_6H_5OH, is only slightly soluble in water, but its sodium salt, $C_6H_5O^-Na^+$, is quite soluble in water. Will phenol dissolve in:

(a) Aqueous NaOH?   (b) Aqueous $NaHCO_3$?   (c) Aqueous Na_2CO_3?

The pK$_a$ of phenol is 9.95. The pK$_a$ values for the conjugate acids of sodium hydroxide, sodium bicarbonate, and sodium carbonate are 15.7, 6.36, and 10.33, respectively. Thus, equilibrium will favor reaction of phenol with only sodium hydroxide and sodium carbonate to give the soluble $C_6H_5O^-$ Na^+. Phenol will dissolve in aqueous solutions of these two bases. Sodium bicarbonate is not a strong enough base to deprotonate phenol, so phenol will not dissolve in an aqueous solution of sodium bicarbonate.

Problem 2.19 For an acid-base reaction, one way to indicate the predominant species at equilibrium is to say that the reaction arrow points to the acid with the higher value of pK_a. For example:

$$NH_4^+ \; + \; H_2O \quad \longleftarrow \quad NH_3 \; + \; H_3O^+$$
$$pK_a \; 9.24 \qquad\qquad\qquad\qquad pK_a \; -1.74$$

$$NH_4^+ \; + \; OH^- \quad \longrightarrow \quad NH_3 \; + \; H_2O$$
$$pK_a \; 9.24 \qquad\qquad\qquad\qquad pK_a \; 15.7$$

Explain why this rule works.

In acid-base reactions, the position of equilibrium favors reaction of the stronger acid and stronger base to give the weaker acid and weaker base. The acid with the higher pK_a is the weaker acid, so the arrow will point toward it.

Lewis Acids and Bases
Problem 2.20 Complete equations for these reactions between Lewis acid-Lewis base pairs. Label which starting material is the Lewis acid and which is the Lewis base, and use a curved arrow to show the flow of the electron pair in each reaction. In solving these problems, it is essential that you show all valence electrons for the atoms participating directly in each reaction.

CHAPTER 3
Solutions to Problems

Problem 3.1 Do the structural formulas in each set represent the same compound or constitutional isomers?

(a) 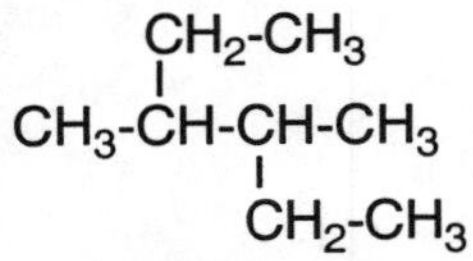
$$CH_3\text{-}CH\text{-}CH\text{-}CH_3$$
with $CH_2\text{-}CH_3$ above and $CH_2\text{-}CH_3$ below

and

$$CH_3\text{-}CH_2\text{-}CH\text{-}CH_2\text{-}CH\text{-}CH_3$$
with CH_3 and CH_3 above

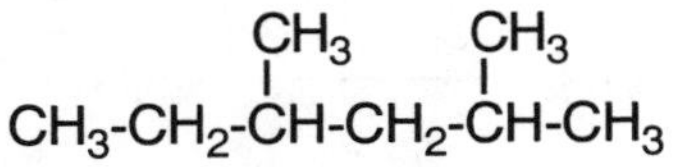

These molecules are constitutional isomers. Each has six carbons in the longest chain. The first has one-carbon branches on carbons 3 and 4 of the chain; the second has one-carbon branches on carbons 2 and 4 of the chain.

(b) 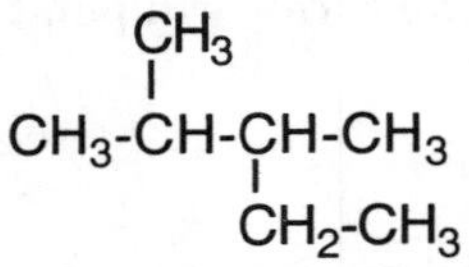
$$CH_3\text{-}CH\text{-}CH\text{-}CH_3$$
with CH_3 above and $CH_2\text{-}CH_3$ below

and

$$CH_3\text{-}CH\text{-}CH\text{-}CH_2\text{-}CH_3$$
with CH_3 above and CH_3 below

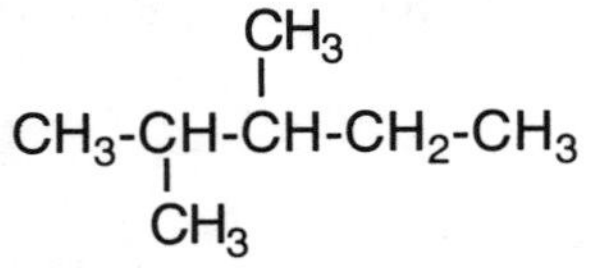

These molecules are identical. Each has five carbons in the longest chain, and one-carbon branches on carbons 2 and 3 of the chain.

Problem 3.2 Draw structural formulas for the three constitutional isomers of molecular formula C_5H_{12}.

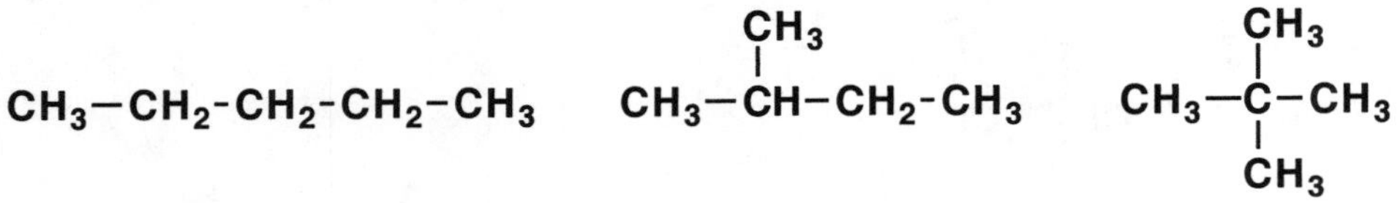

$$CH_3-CH_2-CH_2-CH_2-CH_3 \qquad CH_3-\underset{\underset{CH_3}{|}}{CH}-CH_2-CH_3 \qquad CH_3-\underset{\underset{CH_3}{|}}{\overset{\overset{CH_3}{|}}{C}}-CH_3$$

Problem 3.3 Write IUPAC names for these alkanes.

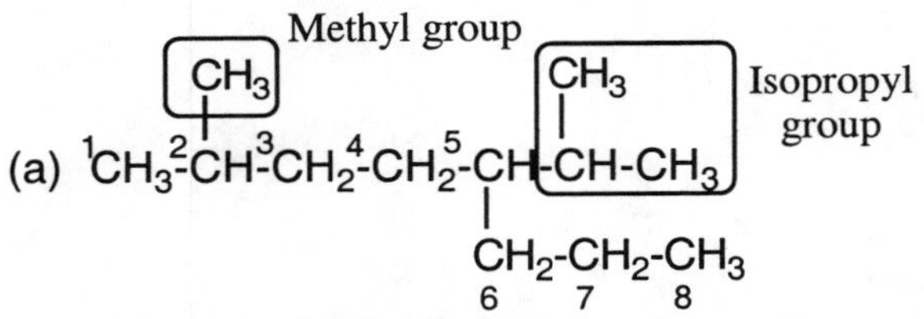

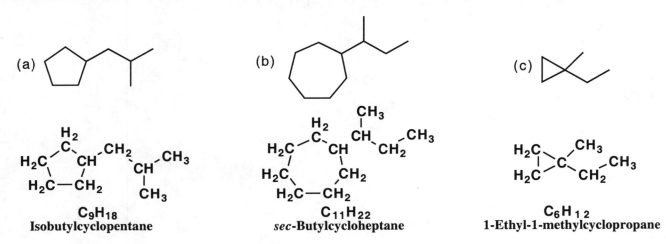

(a) Methyl group [CH_3]   Isopropyl group
$$^1CH_3\text{-}^2CH\text{-}^3CH_2\text{-}^4CH_2\text{-}^5CH\text{-}CH\text{-}CH_3$$
with [CH_3] at carbon 2, [CH_3] at the CH–CH, and
$$CH_2\text{-}CH_2\text{-}CH_3$$
$$\quad\; 6 \quad\;\; 7 \quad\;\; 8$$
5-Isopropyl-2-methyloctane

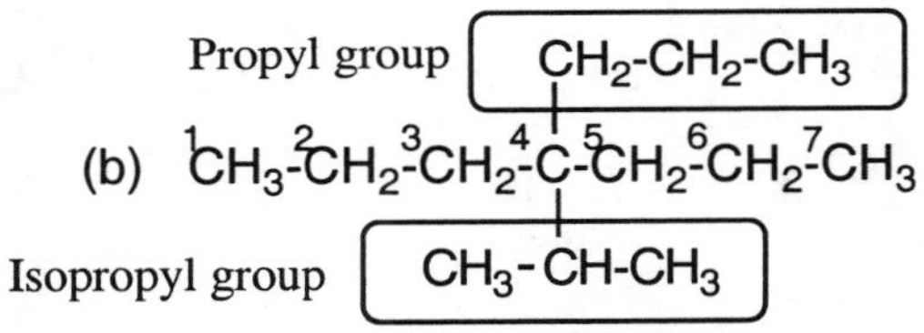

(b) Propyl group [$CH_2\text{-}CH_2\text{-}CH_3$]
$$^1CH_3\text{-}^2CH_2\text{-}^3CH_2\text{-}^4C\text{-}^5CH_2\text{-}^6CH_2\text{-}^7CH_3$$
Isopropyl group [$CH_3\text{-}CH\text{-}CH_3$]
4-Isopropyl-4-propylheptane

Problem 3.4 Write the molecular formula and IUPAC name for each cycloalkane.

(a)

C_9H_{18}
Isobutylcyclopentane

(b)

$C_{11}H_{22}$
***sec*-Butylcycloheptane**

(c)

C_6H_{12}
1-Ethyl-1-methylcyclopropane

<u>Problem 3.5</u> Combine the proper prefix, infix and suffix and write the IUPAC name for each compound.

(a) CH₃-C-CH₃
with O double bonded to the central C

(b) CH₃-CH₂-CH₂-CH₂-C-H
with O double bonded to C

(c) [cyclopentane ring]=O

(d) [cycloheptene ring]

Propanone   **Pentanal**   **Cyclopentanone**   **Cycloheptene**

<u>Problem 3.6</u>
(a) Draw Newman projections for two staggered and two eclipsed conformations of 1,2-dichloroethane.

Following are three eclipsed conformations of 1,2-dichoroethane drawn as Newman projections. The eclipsed conformation on top has the highest energy because the chlorine atoms are closest to each other.

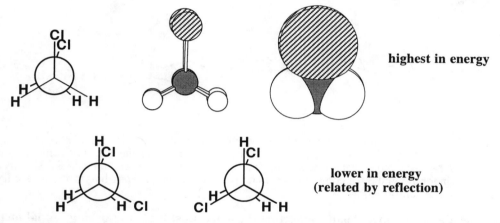

highest in energy

lower in energy
(related by reflection)

Following are three staggered conformations of 1,2-dichoroethane drawn as Newman projections. These staggered conformations are all lower in energy than the eclipsed conformations above. The staggered conformation on top has the lowest overall energy, because the chlorine atoms are farthest away from each other.

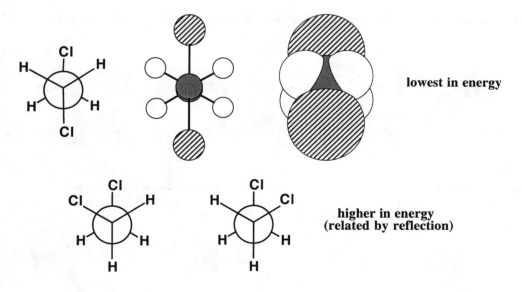

lowest in energy

higher in energy
(related by reflection)

<u>Problem 3.7</u> Following is a chair conformation of cyclohexane with carbon atoms numbered 1 through 6.

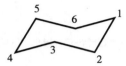

(a) Draw hydrogen atoms that are above the plane of the ring on carbons 1 and 2 and below the plane of the ring on
     carbon 4.
(b) Which of these hydrogens are equatorial? Which are axial?
(c) Draw the other chair conformation. Now, which hydrogens are equatorial? Which are axial?
     Which are above the plane of the ring and which are below it?

**Axial and equatorial hydrogens are labelled (a) and (e), respectively. In both structures, the hydrogens
attached to carbons 1 and 2 are above the plane of the ring, while the hydrogen attached to carbon 4 is below
the plane of the ring.**

<u>Problem 3.8</u> The conformational equilibria for methyl, ethyl, and isopropylcyclohexane are all about 95% in favor of the
equatorial conformation, but the conformational equilibrium for *tert*-butylcyclohexane is almost completely on the
equatorial side. Explain by using molecular models and using the molecular models on the CD-ROM, why the
conformational equilibria for the first three compounds are comparable, but that for *tert*-butylcyclohexane lies considerably
farther toward the equatorial conformation.

**Rotation is possible about the single bond connecting the axial substituent to the ring. Axial methyl, ethyl
and isopropyl groups can assume a conformation in which a hydrogen creates the axial-axial interactions.
With a *tert*-butyl substituent, however, a bulkier -CH₃ group creates the axial-axial interaction. Because of the
increased steric strain (nonbonded interactions) created by the axial *tert*-butyl group, the potential energy of
the axial conformation is considerably greater than that for the equatorial conformation.**

**As shown here, an axial isopropyl group can adopt a conformation with only a minimal 1,3-diaxial
interaction:**

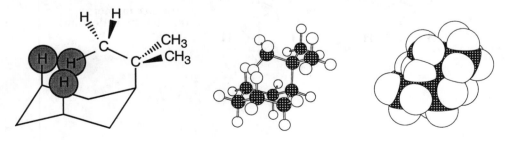

On the other hand, an axial *tert*-butyl group leads to a very severe 1,3-diaxial interaction:

Problem 3.9 Which cycloalkanes show cis-trans isomerism? For each that does, draw both isomers.

(a) 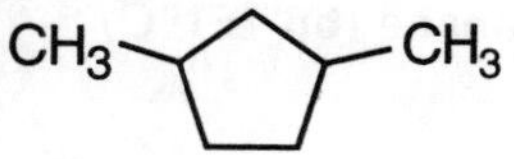 CH₃ — CH₃

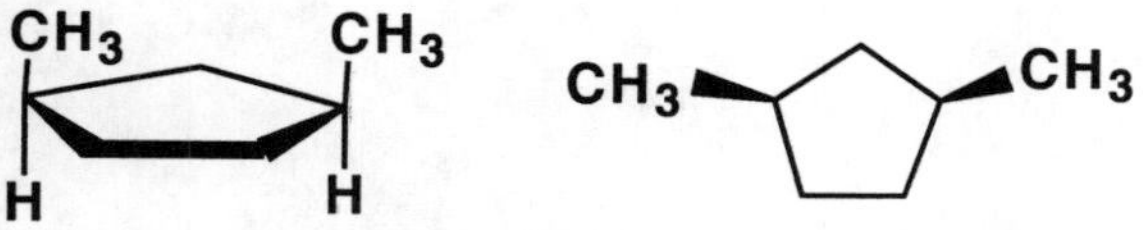

cis-1,3-Dimethylcyclopentane 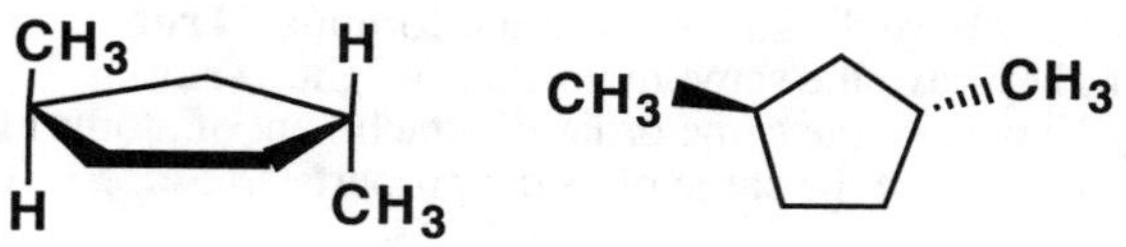   **trans-1,3-Dimethylcyclopentane**

(b) 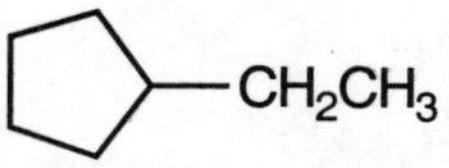 — CH₂CH₃

Ethylcyclopentane does not show cis-trans isomerism.

(c) 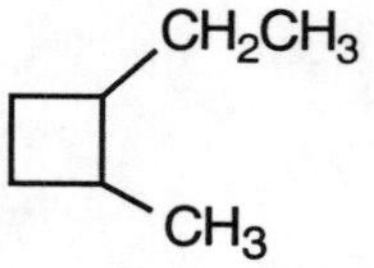
CH₂CH₃

CH₃

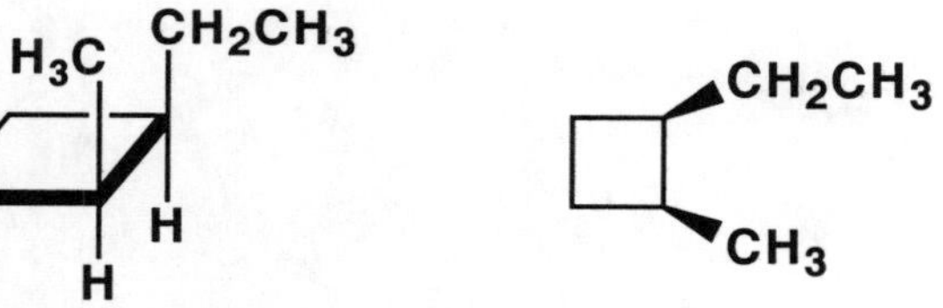

cis-1-Ethyl-2-methylcyclobutane 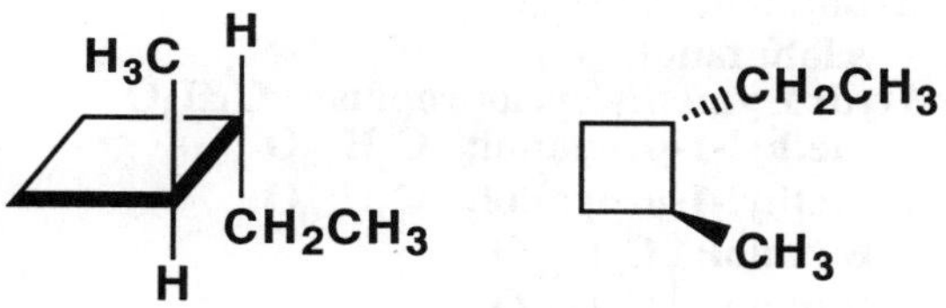   **trans-1-Ethyl-2-methylcyclobutane**

Problem 3.10 Following is a planar hexagon representation for one isomer of 1,2,4-trimethylcyclohexane. Draw alternative chair conformations of this compound, and state which chair conformation is the more stable.

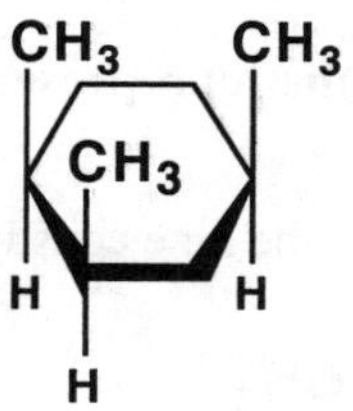

Following are alternative chair conformations for this isomer of 1,2,4-trimethylcyclohexane. The chair conformation on the right is the more stable because it has only one axial methyl group.

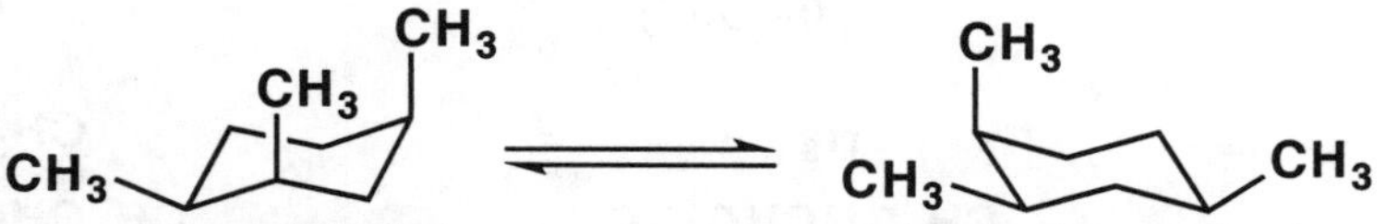

less stable chair   **more stable chair**
(two methyl groups axial)   **(one methyl group axial)**

Problem 3.11 Arrange the alkanes in each set in order of increasing boiling point.
(a) 2-Methylbutane, 2,2-dimethylpropane, and pentane

2,2-Dimethylpropane (bp 9.5ºC), 2-methylbutane (bp 29ºC), pentane (bp 36ºC)

(b) 3,3-Dimethylheptane, 2,2,4-trimethylhexane, and nonane

2,2,4-Trimethylhexane (bp 112°C), 3,3-dimethylheptane (bp 137°C), nonane (bp 151°C)

Constitutional Isomerism
Problem 3.12 Which statements are true about constitutional isomers?
(a) They have the same molecular formula. **True**
(b) They have the same molecular weight. **True**
(c) They have the same order of attachment of atoms. **False**
(d) They have the same physical properties. **False**

Problem 3.13 Which structural formulas represent identical compounds and which represent constitutional isomers?

(a) CH$_3$CH$_2$CHCH$_3$ (b) ◇—OH   (c) HOCH$_2$—◁   (d) CH$_3$CHCH$_3$
             | |
             OH   CH$_2$OH

                    CH$_3$ CH$_2$CH$_3$ CH$_3$
                     | | |
(e) HOCH$_2$CHCH$_3$ (f) CH$_3$CH$_2$CH$_2$CH$_2$OH   (g) CH$_3$CHOH   (h) CH$_3$CCH$_3$
                                                                                              |
                                                                                              OH

Following are names and molecular formulas of each

(a) **2-Butanol; C$_4$H$_{10}$O**
(b) **Cyclobutanol; C$_4$H$_8$O**
(c) **Hydroxymethylcyclopropane; C$_4$H$_8$O**
(d) **2-Methyl-1-propanol; C$_4$H$_{10}$O**
(e) **2-Methyl-1-propanol; C$_4$H$_{10}$O**
(f) **1-Butanol; C$_4$H$_{10}$O**
(g) **2-Butanol; C$_4$H$_{10}$O**
(h) **2-Methyl-2-propanol; C$_4$H$_{10}$O**

The following represent identical compounds: (a),(g)   (d),(e)

The following four compounds represent one set of constitutional isomers: 1-butanol (f), 2-butanol (a)(g), 2-methyl-1-propanol (d)(e), 2-methyl-2-propanol (h).

Cyclobutanol (b) and hydroxymethylcyclopropane (c) represent a second set of constitutional isomers.

Problem 3.14 Name and draw structural formulas for the nine constitutional isomers of molecular formula C$_7$H$_{16}$.

                                                              CH$_3$ CH$_3$
                                                               | |
CH$_3$CH$_2$CH$_2$CH$_2$CH$_2$CH$_2$CH$_3$ CH$_3$CHCH$_2$CH$_2$CH$_2$CH$_3$ CH$_3$CH$_2$CHCH$_2$CH$_2$CH$_3$

            Heptane   **2-Methylhexane**   **3-Methylhexane**
            (bp 94.8)   **(bp 90.0)**   **(bp 92.0)**

        CH$_3$ CH$_3$ CH$_3$
         | | |
CH$_3$CCH$_2$CH$_2$CH$_3$ CH$_3$CHCHCH$_2$CH$_3$ CH$_3$CHCH$_2$CHCH$_3$
         | | |
        CH$_3$ CH$_3$ CH$_3$

2,2-Dimethylpentane   **2,3-Dimethylpentane**   **2,4-Dimethylpentane**
      (bp 79.2)   **(bp 89.8)**   **(bp 80.5)**

$$CH_3$$
$$|$$
$$CH_3CH_2CCH_2CH_3$$
$$|$$
$$CH_3$$

3,3-Dimethylpentane
(bp 86.1)

$$CH_2CH_3$$
$$|$$
$$CH_3CH_2CHCH_2CH_3$$

3-Ethylpentane
(bp 93.5)

$$H_3C \quad CH_3$$
$$| \quad \quad |$$
$$CH_3CHCCH_3$$
$$|$$
$$CH_3$$

2,2,3-Trimethylbutane
(bp 80.9)

<u>Problem 3.15</u> Tell whether the compounds in each set are constitutional isomers.

(a) CH_3-CH_2-OH and CH_3-O-CH_3

(b) $CH_3-\overset{\displaystyle O}{\overset{\|}{C}}-CH_3$ and $CH_3-CH_2-\overset{\displaystyle O}{\overset{\|}{C}}-H$

(c) $CH_3-\overset{\displaystyle O}{\overset{\|}{C}}-O-CH_3$ and $CH_3-CH_2-\overset{\displaystyle O}{\overset{\|}{C}}-OH$

(d) $CH_3-\overset{\displaystyle OH}{\overset{|}{C}}H-CH_2-CH_3$ and $CH_3-\overset{\displaystyle O}{\overset{\|}{C}}-CH_2-CH_3$

(e) [pentagon] and $CH_3-CH_2-CH_2-CH_2-CH_3$

(f) [pentagon] and $CH_2{=}CH-CH_2-CH_2-CH_3$

Sets (a), (b), (c), and (f) contain constitutional isomers; sets (d) and (e) do not.

<u>Problem 3.16</u> Draw structural formulas for:
(a) The four alcohols of molecular formula $C_4H_{10}O$.

$$CH_3-CH_2-CH_2-CH_2-OH \quad CH_3-CH_2-\overset{\displaystyle OH}{\overset{|}{C}}H-CH_3 \quad CH_3-\overset{\displaystyle CH_3}{\overset{|}{C}}H-CH_2-OH \quad CH_3-\overset{\displaystyle CH_3}{\underset{\displaystyle CH_3}{\overset{|}{\underset{|}{C}}}}-OH$$

(b) The two aldehydes of molecular formula C_4H_8O.

$$CH_3-CH_2-CH_2-\overset{\displaystyle O}{\overset{\|}{C}}-H \quad\quad CH_3-\underset{\displaystyle CH_3}{\overset{|}{C}}H-\overset{\displaystyle O}{\overset{\|}{C}}-H$$

(c) The one ketone of molecular formula C_4H_8O.

$$CH_3-CH_2-\overset{\displaystyle O}{\overset{\|}{C}}-CH_3$$

(d) The three ketones of molecular formula $C_5H_{10}O$.

$$CH_3-CH_2-\overset{\displaystyle O}{\overset{\|}{C}}-CH_2-CH_3 \quad CH_3-CH_2-CH_2-\overset{\displaystyle O}{\overset{\|}{C}}-CH_3 \quad CH_3-\underset{\displaystyle CH_3}{\overset{|}{C}}H-\overset{\displaystyle O}{\overset{\|}{C}}-CH_3$$

(e) The four carboxylic acids of molecular formula $C_5H_{10}O_2$.

$$CH_3-\underset{\displaystyle CH_3}{\overset{|}{C}}H-CH_2-\overset{\displaystyle O}{\overset{\|}{C}}-OH \quad CH_3-CH_2-CH_2-CH_2-\overset{\displaystyle O}{\overset{\|}{C}}-OH \quad CH_3-\underset{\displaystyle CH_3}{\overset{|}{\underset{|}{C}}}-\overset{\displaystyle O}{\overset{\|}{C}}-OH \quad CH_3-CH_2-\underset{\displaystyle CH_3}{\overset{|}{C}}H-\overset{\displaystyle O}{\overset{\|}{C}}-OH$$

Nomenclature of Alkanes and Cycloalkanes

Problem 3.17 Write IUPAC names for these alkanes and cycloalkanes.

(a) $CH_3CHCH_2CH_2CH_3$
            |
           CH_3

2-Methylpentane
(Isohexane)

(b) $CH_3CHCH_2CH_2CHCH_3$
             | |
          CH_3 CH_3

2,5-Dimethylhexane

(c) $CH_3(CH_2)_4CHCH_2CH_3$
                 |
              CH_2CH_3

3-Ethyloctane

(d) $(CH_3)_2CHC(CH_3)_3$

2,2,3-Trimethylbutane

(e) 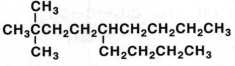

Isobutylcyclopentane

(f) 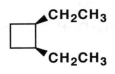

1-*tert*-Butyl-2,4-dimethylcyclohexane

Problem 3.18 Write structural formulas for these alkanes.

(a) 2,2,4-Trimethylhexane

     CH_3 CH_3
     | |
$CH_3CCH_2CHCH_2CH_3$
     |
     CH_3

(b) 2,2-Dimethylpropane

     CH_3
     |
CH_3CCH_3
     |
     CH_3

(c) 3-Ethyl-2,4,5-trimethyloctane

   CH_3CH_2 CH_3
     | |
$CH_3CHCHCHCHCH_2CH_2CH_3$
     | |
     CH_3 CH_3

(d) 5-Butyl-2,2-dimethylnonane

    CH_3
    |
$CH_3CCH_2CH_2CHCH_2CH_2CH_2CH_3$
    | |
    CH_3 $CH_2CH_2CH_2CH_3$

(e) 4-Isopropyloctane

$CH_3CH_2CH_2CHCH_2CH_2CH_2CH_3$
              |
          CH_3CHCH_3

(f) 3,3-Dimethylpentane

       CH_3
       |
$CH_3CH_2CCH_2CH_3$
       |
       CH_3

(g) *trans*-1,3-Dimethylcyclopentane

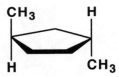

(h) *cis*-1,2-Diethylcyclobutane

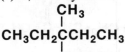

Problem 3.19 Explain why each is an incorrect IUPAC name. Write the correct IUPAC name for the intended compound.

(a) 1,3-Dimethylbutane

    CH_3
    |
$CH_3CHCH_2CH_2CH_3$ **The longest chain is pentane. The IUPAC name is 2-methylpentane.**

(b) 4-Methylpentane

    CH_3
    |
$CH_3CHCH_2CH_2CH_3$ **The pentane is numbered incorrectly. The IUPAC name is 2-methylpentane.**

(c) 2,2-Diethylbutane

      CH_3
      |
$CH_3CH_2CCH_2CH_3$
      |
      CH_2CH_3 **The longest chain is pentane. The IUPAC name is 3-ethyl-3-methylpentane.**

(d) 2-Ethyl-3-methylpentane

$$CH_3$$
$$|$$
$$CH_3CHCHCH_2CH_3$$
$$|$$
$$CH_2CH_3$$

The longest chain is hexane. The IUPAC name is 3,4-dimethylhexane.

(e) 2-Propylpentane

$$CH_3$$
$$|$$
$$CH_3CH_2CH_2CHCH_2CH_2CH_3$$

The longest chain is heptane. The IUPAC name is 4-methylheptane.

(f) 2,2-Diethylheptane

$$CH_3$$
$$|$$
$$CH_3CH_2CCH_2CH_2CH_2CH_2CH_3$$
$$|$$
$$CH_2CH_3$$

The longest chain is octane. The IUPAC name is 3-ethyl-3-methyloctane.

(g) 2,2-Dimethylcyclopropane

The ring is numbered incorrectly. The IUPAC name is 1,1-dimethylcyclopropane.

(h) 1-Ethyl-5-methylcyclohexane

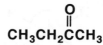

The name is numbered incorrectly. The IUPAC name is
1-ethyl-3-methylcyclohexane.

The IUPAC System of Nomenclature
Problem 3.20 Draw a structural formula for each compound.

(a) Ethanol   (b) Ethanal   (c) Ethanoic acid   (d) Butanone

$$CH_3CH_2OH$$

$$\overset{O}{\overset{||}{CH_3C\text{-}H}}$$

$$\overset{O}{\overset{||}{CH_3C\text{-}OH}}$$

$$\overset{O}{\overset{||}{CH_3CH_2CCH_3}}$$

(e) Butanal   (f) Butanoic Acid   (g) Propanal   (h) Cyclopropanol

$$\overset{O}{\overset{||}{CH_3CH_2CH_2C\text{-}H}}$$

$$\overset{O}{\overset{||}{CH_3CH_2CH_2C\text{-}OH}}$$

$$\overset{O}{\overset{||}{CH_3CH_2C\text{-}H}}$$

(i) Cyclopentanol   (j) Cyclopentene   (k) Cyclopentanone

Problem 3.21 Write the IUPAC name for each compound.

(a) $CH_3\overset{\displaystyle O}{\overset{\|}{C}}CH_3$

Propanone

(b) $CH_3(CH_2)_3\overset{\displaystyle O}{\overset{\|}{C}}\text{-H}$

Pentanal

(c) $CH_3(CH_2)_8\overset{\displaystyle O}{\overset{\|}{C}}\text{-OH}$

Decanoic acid

(d)

Cyclohexene

(e) =O

Cyclohexanone

(f) —OH

Cyclobutanol

Conformations of Alkanes and Cycloalkanes

Problem 3.22 How many different staggered conformations are there for 2-methylpropane? How many different eclipsed conformations are there?

Looking down any of the carbon-carbon bonds, there is one staggered and one eclipsed conformation of 2-methylpropane.

$$\underset{\text{2-Methylpropane}}{CH_3\overset{\displaystyle \overset{CH_3}{|}}{C}HCH_3}$$

Staggered

Eclipsed

Problem 3.23 Looking along the bond between carbons 2 and 3 of butane, there are two different staggered conformations and two different eclipsed conformations. Draw Newman projections of each and arrange them in order from most stable conformation to least stable conformation.

The two staggered conformations are more stable than the eclipsed conformations, with the staggered anti conformation being the most stable. Of the eclipsed conformations, the one with the methyl groups eclipsing each other is the least stable.

Staggered

Eclipsed

Most Stable ◄———————————————————— **Least Stable**

Problem 3.24 Demonstrate, using molecular models on the CD-ROM, that in cyclohexane, an equatorial substituent is almost equidistant from the axial and equatorial groups on each adjacent carbon.

We use chlorocyclohexane as an example. The best way to see the relationship between the equatorial chlorine atom and the hydrogen atoms on an adjacent carbon is to draw a Newman projection along one of the carbon-carbon bonds. As can be seen, the equatorial chlorine atom on the front carbon is in between and, thus, equidistant from the axial and equatorial hydrogen atoms on the rear carbon.

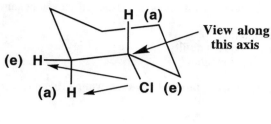

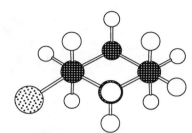

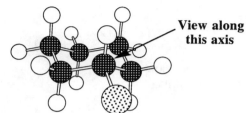

Cis-Trans Isomerism in Cycloalkanes

Problem 3.25   Name and draw structural formulas for the cis and trans isomers of 1,2-dimethylcyclopropane.

cis-1,2-Dimethylcyclopropane   *trans*-1,2-Dimethylcyclopropane

Problem 3.26   Name and draw structural formulas for all cycloalkanes of molecular formula C_5H_{10}. Be certain to include cis-trans isomers as well as constitutional isomers.

Cyclopentane   **Methylcyclobutane**   **Ethylcyclopropane**

     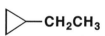

cis-1,2-Dimethylcyclopropane   *trans*-1,2-Dimethylcyclopropane   **1,1-Dimethylcyclopropane**

Problem 3.27   Using a planar pentagon representation for the cyclopentane ring, draw structural formulas for the cis and trans isomers of:
(a) 1,2-Dimethylcyclopentane   (b) 1,3-Dimethylcyclopentane

          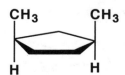   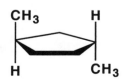

cis-1,2-Dimethyl-
cyclopentane   *trans*-1,2-Dimethyl-
cyclopentane   *cis*-1,3-Dimethyl-
cyclopentane   *trans*-1,3-Dimethyl-
cyclopentane

Problem 3.28 Draw the alternative chair conformations for the cis and trans isomers of 1,2-dimethylcyclohexane, 1,3-dimethylcyclohexane, and 1,4-dimethylcyclohexane.
(a) Indicate by a label whether each methyl group is axial or equatorial.
(b) For which isomer(s) are the alternative chair conformations of equal stability?
(c) For which isomer(s) is one chair conformation more stable than the other?

Cis and trans isomers are drawn as pairs. The more stable chair is labeled in cases where there is a difference.

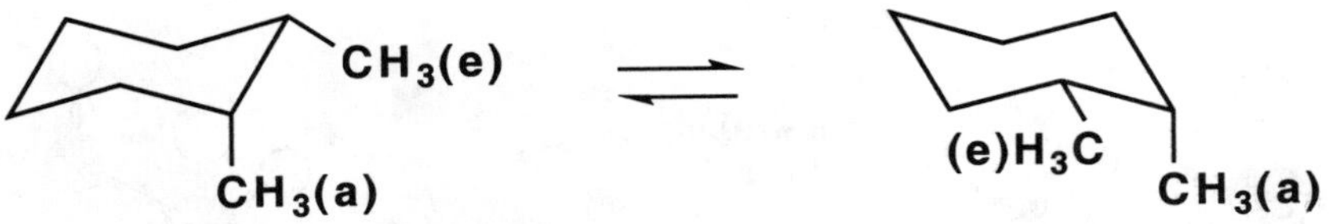

$CH_3(e)$ $⇌$ $(e)H_3C$ $CH_3(a)$

cis-1,2-Dimethylcyclohexane
(chairs of equal stability)

more stable chair

$CH_3(e)$
$CH_3(e)$ $⇌$ $CH_3(a)$
$CH_3(a)$

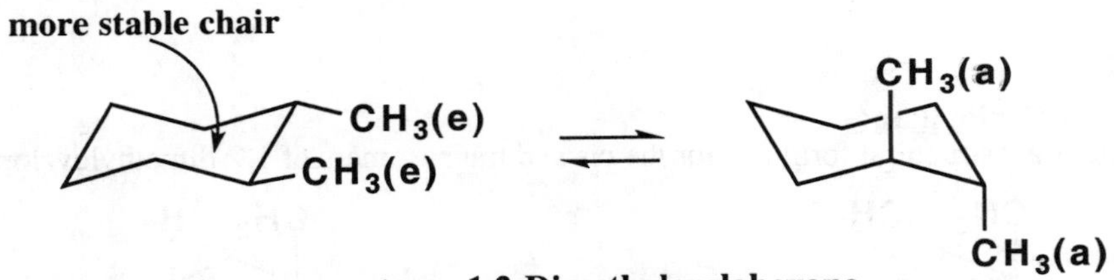

trans-1,2-Dimethylcyclohexane

more stable chair

$CH_3(e)$ $⇌$
$(e)H_3C$ $(a)H_3C$ $CH_3(a)$

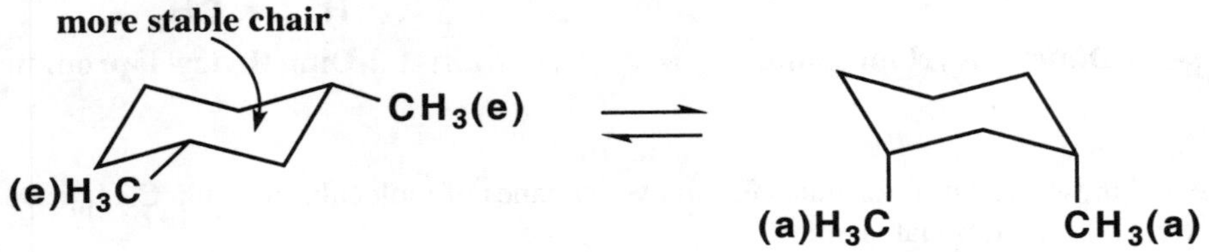

cis-1,3-Dimethylcyclohexane

$CH_3(a)$
$CH_3(e)$ $⇌$ $(e)H_3C$
                                        $CH_3(a)$

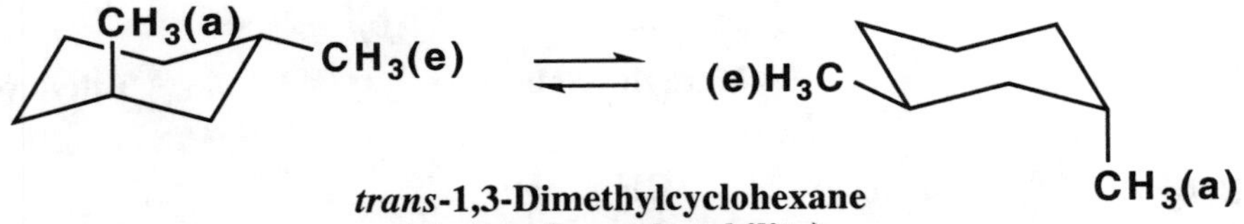

trans-1,3-Dimethylcyclohexane
(chairs of equal stability)

$(e)H_3C$ $⇌$
$CH_3(a)$ $(a)H_3C$ $CH_3(e)$

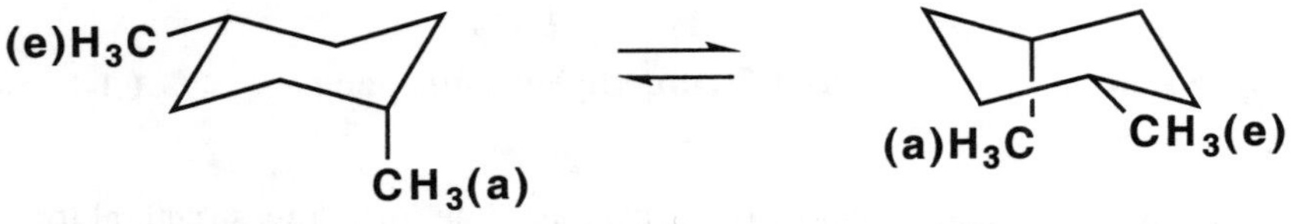

cis-1,4-Dimethylcyclohexane
(chairs of equal stability)

more stable chair

$(e)H_3C$ $⇌$ $CH_3(a)$
        $CH_3(e)$ $(a)H_3C$

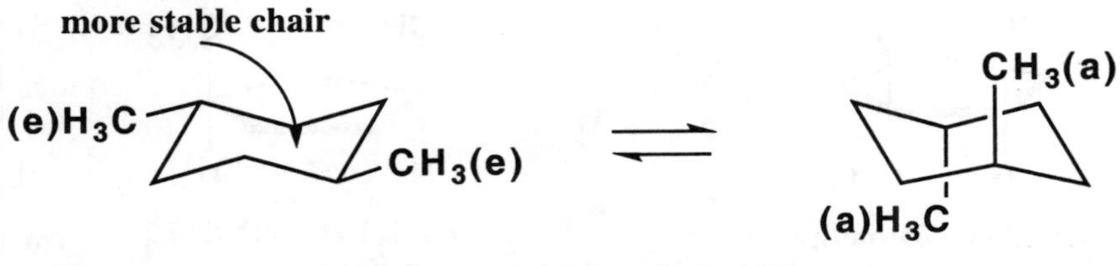

trans-1,4-Dimethylcyclohexane

Problem 3.29 Use your answers from problem 3.28 to complete the table showing correlations between cis, trans and axial, equatorial for the disubstituted derivatives of cyclohexane.

These relationships are summarized in the following table.

Position of Substitution	cis	trans
1,4	a,e or e,a	e,e or a,a
1,3	e,e or a,a	a,e or e,a
1,2	a,e or e,a	e,e or a,a

Problem 3.30 There are four cis-trans isomers of 2-isopropyl-5-methylcyclohexanol.

2-Isopropyl-5-methylcyclohexanol

(a) Using a planar hexagon representation for the cyclohexane ring, draw structural formulas for these four isomers.

Following are planar hexagon representations for the four cis-trans isomers. In each, the isopropyl group is shown by the symbol R. One way to arrive at these structural formulas is to take one group as a reference and then arrange the other two groups in relation to it. In these drawings, -OH is taken as the reference and placed above the plane of the ring. Once -OH is fixed, there are only two possible arrangements for the isopropyl group on carbon 2; either cis or trans to -OH. Similarly, there are only two possible arrangements for the methyl group on carbon 5; either cis or trans to -OH. Note that, even if you take another substituent as a reference, and even if you put the reference below the plane of the ring, there are still only four cis-trans isomers for this compound.

cis-2-Isopropyl-*cis*-5-methyl-cyclohexanol

cis-2-Isopropyl-*trans*-5-methyl-cyclohexanol

trans-2-Isopropyl-*cis*-5-methyl-cyclohexanol

trans-2-Isopropyl-*trans*-5-methyl-cyclohexanol

(b) Draw the more stable chair conformation for each of your answers in part (a).

cis-2-Isopropyl-*cis*-5-methylcyclohexanol

cis-2-Isopropyl-*trans*-5-methylcyclohexanol

trans-2-Isopropyl-*cis*-5-methylcyclohexanol

trans-2-Isopropyl-*trans*-5-methylcyclohexanol

(c) Of the four cis-trans isomers, which is the most stable? If you answered this part correctly, you picked the isomer found in nature and given the name menthol

Menthol is *trans*-2-isopropyl-*cis*-5-methylcyclohexanol. This isomer is the most stable because all substituents are in equatorial positions.

<u>Problem 3.31</u> Draw alternative chair conformations for each substituted cyclohexane and state which chair is the more stable.

(a)

(chairs of equal stability)

(b)

more stable chair

(c)

more stable chair

(d)

more stable chair

<u>Problem 3.32</u> What kinds of conformations do the six-membered rings exhibit in adamantane?

Adamantane

In adamantane, the cyclohexane rings all have chair conformations.

Problem 3.33 Glucose (Section 16.2B) contains a six-membered ring. In the more stable chair conformation of this molecule, all substituents on the ring are equatorial. Draw this more stable chair conformation.

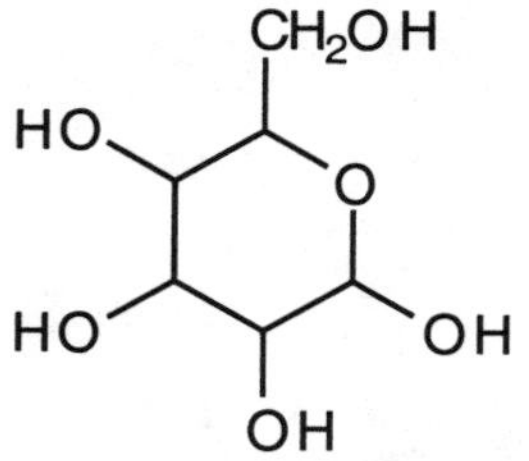

Glucose

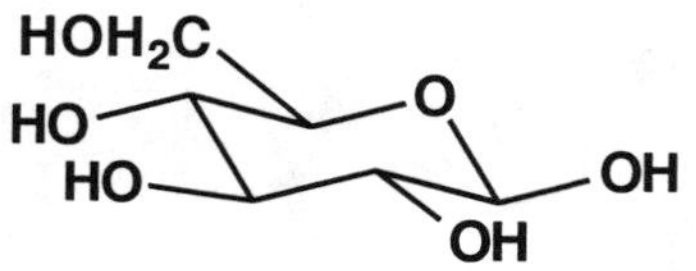

Problem 3.34 Following is a structural formula and ball-and-stick model of cholestanol. The only difference between this compound and cholesterol (Section 17.4A) is that cholesterol has a carbon-carbon double bond in ring B.

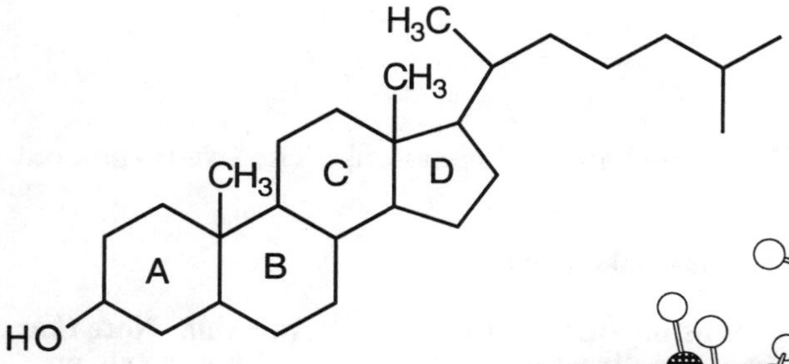

Cholestanol

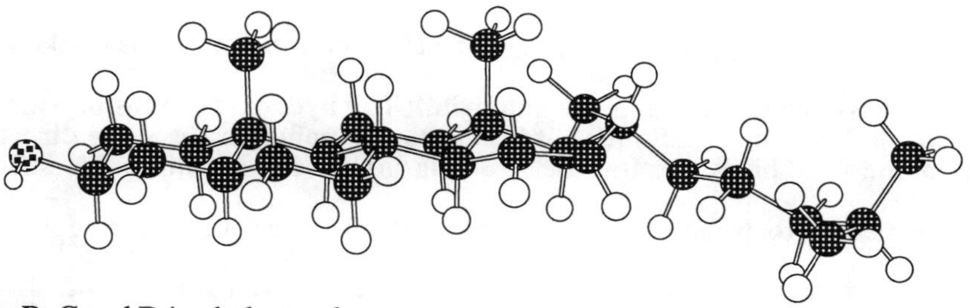

(a) Describe the conformation of rings A, B, C and D in cholestanol.

The six-membered rings A, B and C are all in chair conformations. The five-membered ring D is in an envelope conformation.

(b) Is the hydroxyl group on ring A axial or equatorial?

The hydroxyl group is equatorial.

(c) Consider the methyl group at the junction of rings A/B. Is it axial or equatorial to ring A? Is it axial or equatorial to ring B?

The methyl group at the A/B ring junction is axial with respect to both rings.

(d) Is the methyl group at the junction of rings C/D axial or equatorial to ring C? .

This methyl group is axial to ring C.

Problem 3.35 Following is a structural formula of cholic acid (Section 17.4A), a component of human bile whose function is to aid in the absorption and digestion of dietary fats.

Cholic Acid

(a) What is the conformation of ring A? of ring B? of ring C? of ring D?

As in cholestanol, all of the six-membered rings (A, B and C) are in chair conformations. The five-membered ring is in an envelope conformation.

(b) There are hydroxyl groups on rings A, B and C. Tell whether each is axial or equatorial.

The hydroxyl group on ring A is equatorial. The hydroxyl groups on rings B and C are both axial. Note that the hydroxyl on ring A, although it is equatorial, points in the same direction as the two axial hydroxyls on rings B and C. This is because the A/B ring juncture in cholic acid is different than in cholestanol.

(c) Is the methyl group at the junction of rings A/B axial or equatorial to ring A? Is it axial or equatorial to ring B?

With respect to ring A, this methyl group is an equatorial substituent. However, it is axial to ring B.

(d) Is the methyl group at the junction of rings C/D axial or equatorial to ring C?

As in cholestanol, this methyl group is axial to ring C.

Physical Properties of Akanes and Cycloalkanes
Problem 3.36 In Problem 3.14, you drew structural formulas for all constitutional isomers of molecular formula C_7H_{16}. Predict which isomer has the lowest boiling point and which has the highest boiling point.

Names and boiling points of these isomers are given in the solution to Problem 3.14. The isomer with the lowest boiling point is 2,2-dimethylpentane, bp 79.2°C. The isomer with the highest boiling point is heptane, bp 94.8°C.

Problem 3.37 What generalizations can you make about the densities of alkanes relative to that of water?

(1) All alkanes are less dense than water; (2) As alkane molecular weight increases, density increases; (3) Constitutional isomers have similar densities.

Problem 3.38 What unbranched alkane has about the same boiling point as water (see Table 3.4)? Calculate the molecular weight of this alkane, and compare it with that of water.

Heptane, C_7H_{16}, has a boiling point of 98.4°C and a molecular weight of 100. Its molecular weight is approximately 5.5 times that of water. Although considerably lower in molecular weight, water molecules are held together by the relatively strong forces of hydrogen bonding, while the much heavier heptane molecules are held together only by relatively weak dispersion forces.

Reactions of Alkanes

<u>Problem 3.39</u> Write balanced equations for combustion of each hydrocarbon. Assume that each is converted completely to carbon dioxide and water.

(a) Hexane

$$2 \; CH_3(CH_2)_4CH_3 \;\; + \;\; 19 \; O_2 \;\; \longrightarrow \;\; 12 \; CO_2 \;\; + \;\; 14 \; H_2O$$

(b) Cyclohexane

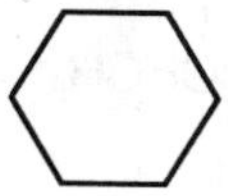

$$+ \;\; 9 \; O_2 \;\; \longrightarrow \;\; 6 \; CO_2 \;\; + \;\; 6 \; H_2O$$

(c) 2-Methylpentane

$$2 \;\; CH_3\underset{\underset{CH_3}{|}}{CH}CH_2CH_2CH_3 \;\; + \;\; 19 \; O_2 \;\; \longrightarrow \;\; 12 \; CO_2 \;\; + \;\; 14 \; H_2O$$

<u>Problem 3.40</u> Following are heats of combustion of methane and propane. On a gram-for-gram basis, which of these hydrocarbons is the better source of heat energy?

Hydrocarbon	Component of	Heat of Combustion [kcal/mol (kJ/mol)]
CH_4	natural gas	-212 (-886)
$CH_3CH_2CH_3$	LPG	-530 (-2220)

On a gram-per-gram basis, methane is the better source of heat energy.

Hydrocarbon	Molecular Weight	Heat of Combustion (kcal/mol)	Heat of Combustion (kcal/gram)
methane	16.04	-212	-13.3
propane	44.09	-531	-12.0

CHAPTER 4
Solutions to Problems

<u>Problem 4.1</u> Each molecule has one stereocenter. Draw stereorepresentations for the enantiomers of each.

The stereocenters are labeled with an asterisk.

(a)

OH
|
−CHCH₃

(b) CH₃CHCHCH₃ with OH above and CH₃ below

<u>Problem 4.2</u> Assign priorities to the groups in each set.

(a) -CH₂OH and -CH₂CH₂OH

The -CH₂OH group has higher priority because the FIRST point of difference is the underlined O atom of -CH₂OH that takes priority over the underlined C atom of -CH₂CH₂OH.

(b) -CH₂OH and -CH=CH₂

$$\underset{1}{-CH}=\underset{2}{CH_2}\qquad\text{is treated as}\qquad -\overset{C}{\underset{H}{\underset{|}{C}}}\overset{C}{\underset{H}{\underset{|}{C}}}-H$$

Nevertheless, the FIRST point of difference is the underlined O atom of -CH₂OH that takes priority over any of the atoms attached to carbon 1 of -CH=CH₂. Thus, the -CH₂OH group takes priority over the -CH=CH₂ group.

(c) -CH₂OH and -C(CH₃)₃

The -CH₂OH group has higher priority because the underlined O atom attached to carbon 1 takes priority over the underlined C atoms attached to carbon 1 of -C(CH₃)₃, even though there are three of them.

<u>Problem 4.3</u> Assign an R or S configuration to each stereocenter and give each molecule an IUPAC name.

The drawings to the right of each molecule show the order of priority, the perspective from which to view the molecule, and the R,S designation for the configuration.

(a)

view from this perspective.

If you view from the perspective shown, this is what you see.

(b)

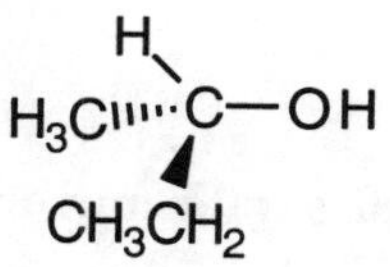

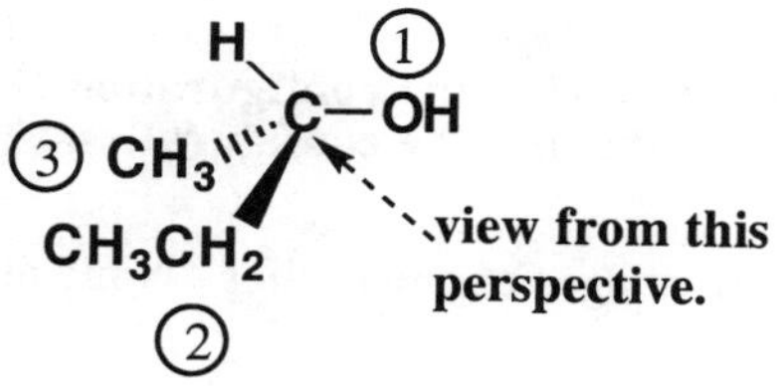

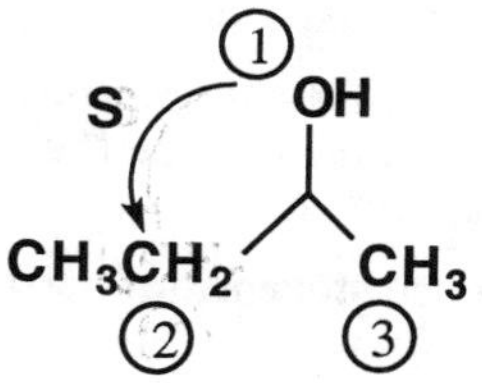

(c)

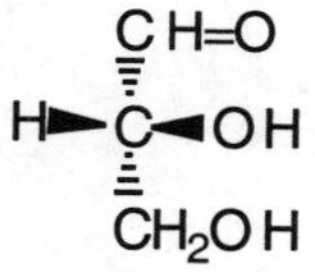

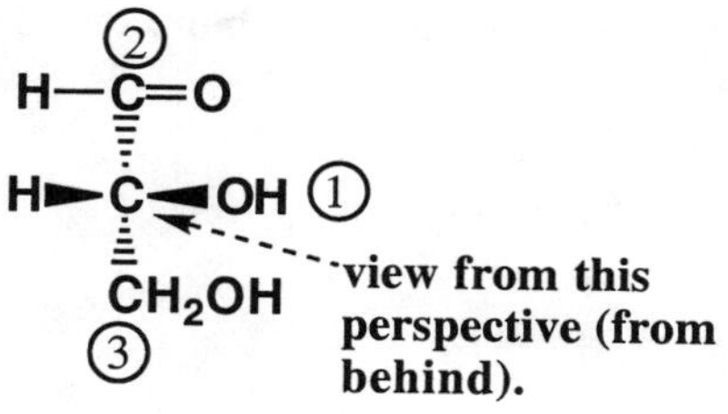

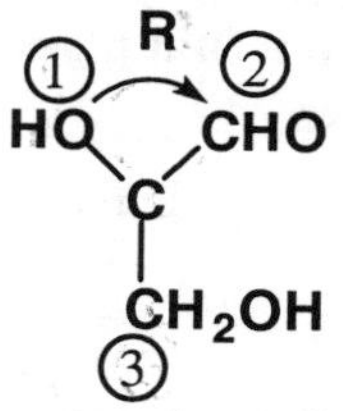

Problem 4.4 Following are stereorepresentations for the four stereoisomers of 3-chloro-2-butanol.

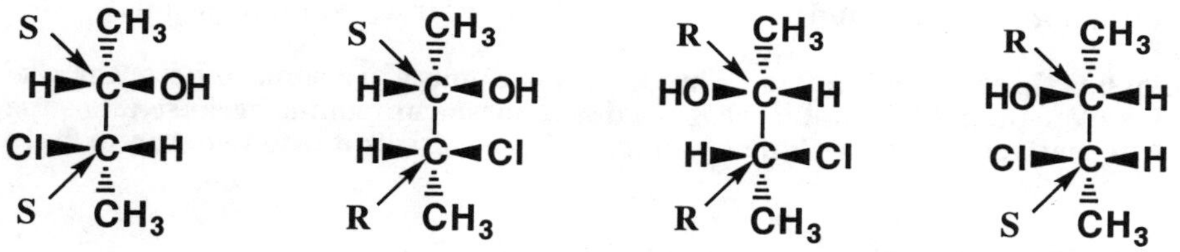

(1)   (2)   (3)   (4)

The configuration of each tetrahedral stereocenter is labeled on the structures. This labeling often helps when trying to establish stereochemical relationships between molecules.

(a) Which are pairs of enantiomers?

Enantiomers are stereoisomers that are mirror images of each other. The pairs of enantiomers are structures (1) and (3) (S,S and R,R) as well as structures (2) and (4) (S,R and R,S).

(b) Which compounds are diastereomers?

Diastereomers are stereoisomers that are not mirror images of each other. Pairs of diastereomers are (1)/(2), (1)/(4), (2)/(3) and (3)/(4).

Problem 4.5 Following are four Newman projection formulas for tartaric acid.

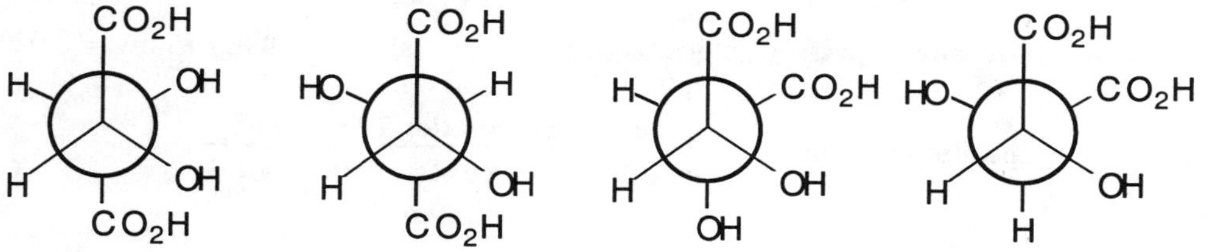

(1)   (2)   (3)   (4)

(a) Which represent the same compound?

Compounds (1) and (4) are the same compound having the configuration (2R,3R).
Compounds (2) and (3) are the same compound having the configuration (2R,3S).
(b) Which represent enantiomers?

Enantiomers are stereoisomers that are mirror images of each other. None of these molecules is the mirror image of any other.

(c) Which is the meso compound?

The meso compound of tartaric acid has the (2R,3S) configuration. Thus, compound (2)/(3) is the meso compound. This can be seen by rotating the rear carbon in structure 2 by 180°. The fact that each group on the front carbon is eclipsed by an identical group on the back carbon after this rotation shows that the molecule has an internal plane of symmetry.

Problem 4.6 How many stereoisomers exist for 1,3-cyclopentanediol?

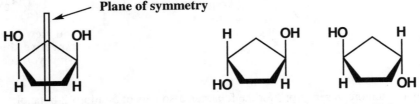

Plane of symmetry

cis-1,3-Cyclopentanediol   *trans*-1,3-Cyclopentanediol
(achiral, a meso compound)   (a pair of enantiomers)

1,3-Cyclopentanediol has three stereoisomers. The two trans isomers are enantiomers; the cis isomer is a meso compound. *Cis*-1,3-cyclopentanediol can be recognized as a meso compound because it is superposable upon its mirror image. Alternatively, it has a plane of symmetry that bisects it into two mirror halves.

Problem 4.7 How many stereoisomers exist for 1,4-cyclohexanediol?

1,4-Cyclohexanediol can exist as a pair of cis-trans isomers. Each is achiral because of a plane of symmetry that bisects each molecule into two mirror halves. In the figure below, the plane of symmetry in each molecule is in the plane of the paper.

trans-1,4-Cyclohexanediol   *cis*-1,4-Cyclohexanediol

Problem 4.8 The specific rotation of progesterone, a female sex hormone (Table 17.3), is +172°, measured at 20°C. Calculate the observed rotation for a solution prepared by dissolving 300 mg of progesterone in 15 mL of dioxane and placing it in a sample tube 1.00 dm long.

The concentration of progesterone, expressed in grams per milliliter is: 300 mg/15 mL = 0.020 g/mL

$$\text{specific rotation} = \frac{\text{observed rotation (degrees)}}{\text{length (dm) x concentration (g/mL)}}$$

Rearranging this formula to solve for observed rotation gives:

observed rotation (degrees) = specific rotation x length (dm) x concentration (g/mL)

Plugging in the experimental values gives the final answer.

observed rotation (degrees) = +172° x 1.00 dm x 0.020 g/mL = +3.4°

Chirality

Problem 4.9 Which of these objects are chiral (assume there is no label or other identifying mark)?

(a) Pair of scissors   (b) Tennis ball   (c) Paper clip   (d) Beaker

(e) The swirl created in water as it drains out of a sink or bathtub.

An object is chiral if it is not superposable upon its mirror image. Objects that have a plane of symmetry are not chiral. The tennis ball, paper clip and beaker, (b), (c), and (d), are all achiral because that all have a plane of symmetry and are all superposable on their mirror images. The pair of scissors and drain swirl, (a) and (e), are chiral because they do not have a plane of symmetry and they are not superposable on their mirror images.

Problem 4.10 Think about the helical coil of a telephone cord or the spiral binding on a notebook and suppose that you view the spiral from one end and find that it is a left-handed twist. If you view the same spiral from the other end, is it a right-handed twist, or a left-handed twist from that end as well?

A helical coil has the same handedness viewed from either end.

Problem 4.11 Next time you have the opportunity to view a collection of augers or other sea shells that have a helical twist, study the chirality of their twists. Do you find an equal number of left-handed and right-handed augers or, for example, do they all have the same handedness? What about the handedness of augers compared with other spiral shells?

This question was just meant to make you think about chirality in nature, but if you do know the answer please share it with your class.

Problem 4.12 One reason we can be sure that sp^3-hybridized carbon atoms are tetrahedral is the number of stereoisomers that can exist for different organic compounds.

(a) How many stereoisomers are possible for $CHCl_3$, CH_2Cl_2, and $CHBrClF$ if the four bonds to carbon have a tetrahedral geometry?

Both tetrahedral $CHCl_3$ and tetrahedral CH_2Cl_2 are achiral, so there is only one stereoisomer of either.

On the other hand, tetrahedral $CHBrClF$ is chiral, so there are two stereoisomers possible.

(b) How many stereoisomers are possible for each of these compounds if the four bonds to the carbon have a square planar geometry?

Even with a square planar geometry (the H and three Cl atoms are in the same plane as the C atom), there is only one stereoisomer possible for CHCl₃.

$$\overset{\displaystyle H}{\underset{\displaystyle Cl}{Cl-C-Cl}}$$

There are two possible stereoisomers of CH₂Cl₂, one with the Cl atoms adjacent to each other, and another with the Cl atoms opposite to each other.

$$\overset{\displaystyle H}{\underset{\displaystyle Cl}{H-C-Cl}} \qquad\qquad \overset{\displaystyle H}{\underset{\displaystyle H}{Cl-C-Cl}}$$

There are three possible stereoisomers of a square planar CHBrClF as shown.

$$\overset{\displaystyle Br}{\underset{\displaystyle F}{H-C-Cl}} \qquad \overset{\displaystyle Cl}{\underset{\displaystyle F}{H-C-Br}} \qquad \overset{\displaystyle Br}{\underset{\displaystyle Cl}{H-C-F}}$$

Enantiomers

Problem 4.13 Which compounds contain stereocenters?

(a) 2-Chloropentane

$$CH_3 \overset{*}{\underset{\displaystyle H}{\overset{\displaystyle Cl}{-C-}}} CH_2CH_2CH_3$$

(b) 3-Chloropentane

$$CH_3CH_2 \underset{\displaystyle H}{\overset{\displaystyle Cl}{-C-}} CH_2CH_3$$

(c) 3-Chloro-1-pentene

$$CH_2{=}CH \overset{*}{\underset{\displaystyle H}{\overset{\displaystyle Cl}{-C-}}} CH_2CH_3$$

(d) 1,2-Dichloropropane

$$ClCH_2 \overset{*}{\underset{\displaystyle H}{\overset{\displaystyle Cl}{-C-}}} CH_3$$

To be a tetrahedral stereocenter, a carbon atom must have four different groups attached. Thus, compounds (a), (c), and (d) have stereocenters (marked with an asterisk).

Problem 4.14 Using only C, H, and O, write structural formulas for the lowest-molecular weight chiral

(a) Alkane

$$CH_3CH_2CH_2 \overset{\displaystyle H}{\underset{\displaystyle CH_3}{-\overset{*}{C}-}} CH_2CH_3 \qquad \text{or} \qquad CH_3 \overset{\displaystyle H}{\underset{\displaystyle CH_3}{-C-}} \overset{\displaystyle H}{\underset{\displaystyle CH_3}{\overset{*}{C}-}} CH_2CH_3$$

3-Methylhexane   **2,3-Dimethylpentane**

(b) Alcohol

$$HO-\overset{\overset{\displaystyle H}{|}}{\underset{\underset{\displaystyle CH_3}{|}}{C^*}}-CH_2CH_3$$

2-Butanol

(c) Aldehyde

$$H\overset{\overset{\displaystyle O}{\|}}{C}-\overset{\overset{\displaystyle H}{|}}{\underset{\underset{\displaystyle CH_3}{|}}{C^*}}-CH_2CH_3$$

2-Methylbutanal

(d) Ketone

$$CH_3CH_2-\overset{\overset{\displaystyle H}{|}}{\underset{\underset{\displaystyle CH_3}{|}}{C^*}}-\overset{\overset{\displaystyle O}{\|}}{C}-CH_3$$

3-Methyl-2-pentanone

(e) Carboxylic acid

$$CH_3CH_2-\overset{\overset{\displaystyle H}{|}}{\underset{\underset{\displaystyle CH_3}{|}}{C^*}}-\overset{\overset{\displaystyle O}{\|}}{C}-OH$$

2-Methylbutanoic acid

<u>Problem 4.15</u> Which alcohols of molecular formula $C_5H_{12}O$ are chiral?

$$CH_3CH_2CH_2-\overset{\overset{\displaystyle H}{|}}{\underset{\underset{\displaystyle CH_3}{|}}{C^*}}-OH$$

2-Pentanol

$$CH_3CH_2-\overset{\overset{\displaystyle H}{|}}{\underset{\underset{\displaystyle CH_3}{|}}{C^*}}-CH_2OH$$

2-Methyl-1-butanol

$$(CH_3)_2CH-\overset{\overset{\displaystyle H}{|}}{\underset{\underset{\displaystyle CH_3}{|}}{C^*}}-OH$$

3-Methyl-2-butanol

There are five achiral alcohols with this formula. Their structures are shown in the answer to Problem 1.55a.

<u>Problem 4.16</u> Which carboxylic acids of molecular formula $C_6H_{12}O_2$ are chiral?

$$CH_3CH_2CH_2-\overset{\overset{\displaystyle H}{|}}{\underset{\underset{\displaystyle CH_3}{|}}{C^*}}-CO_2H$$

2-Methylpentanoic acid

$$CH_3CH_2-\overset{\overset{\displaystyle H}{|}}{\underset{\underset{\displaystyle CH_3}{|}}{C^*}}-CH_2CO_2H$$

3-Methylpentanoic acid

$$(CH_3)_2CH-\overset{\overset{\displaystyle H}{|}}{\underset{\underset{\displaystyle CH_3}{|}}{C^*}}-CO_2H$$

2,3-Dimethylbutanoic acid

There are five achiral carboxylic acids with this formula. Their structures are shown in the answer to Problem 1.55d.

<u>Problem 4.17</u> Write the structural formula of an alcohol of molecular formula $C_6H_{14}O$ that contains two stereocenters.

$$CH_3CH_2-\overset{\overset{\displaystyle H}{|}}{\underset{\underset{\displaystyle H_3C}{|}}{C^*}}-\overset{\overset{\displaystyle OH}{|}}{\underset{\underset{\displaystyle H}{|}}{C^*}}-CH_3$$

3-Methyl-2-pentanol

Problem 4.18 Draw mirror images for these molecules:

The mirror images are shown in bold.

(a)

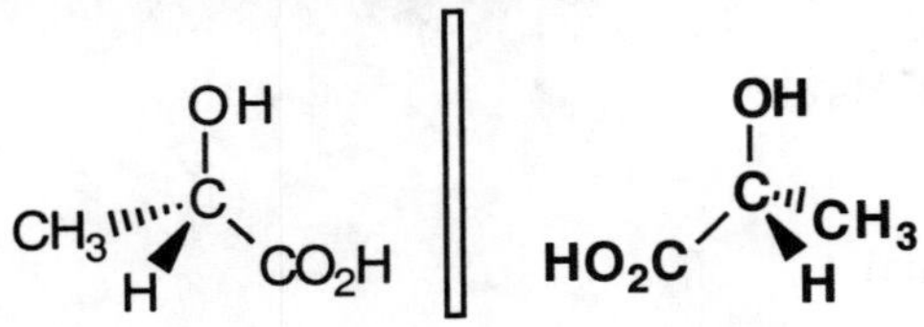

(b)

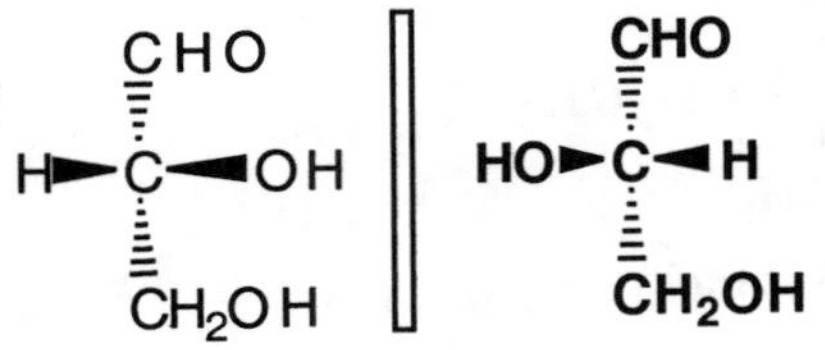

(c)

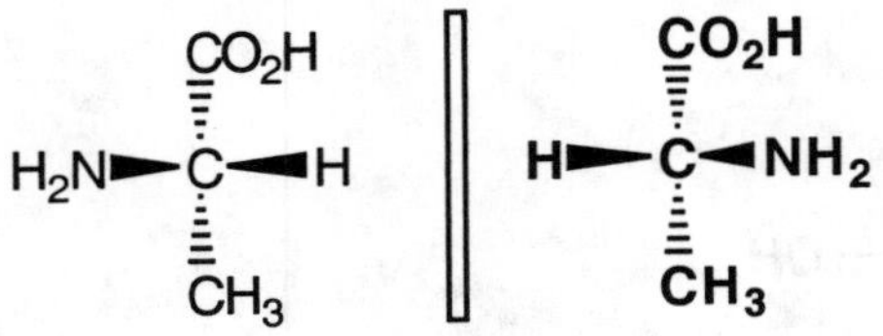

(d)

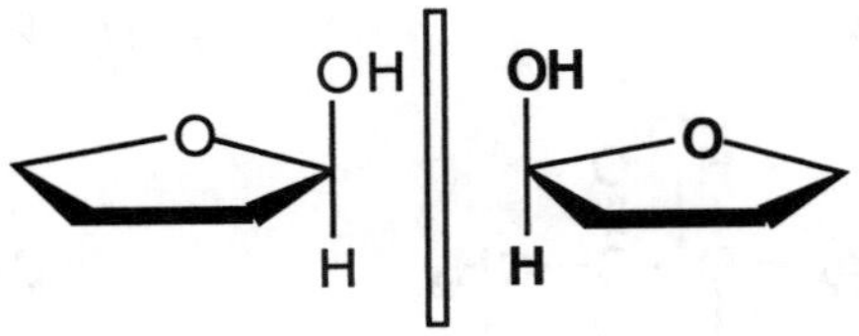

(e)

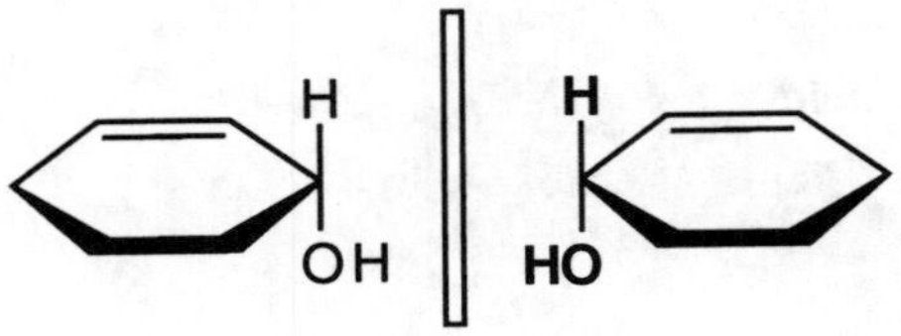

(f)

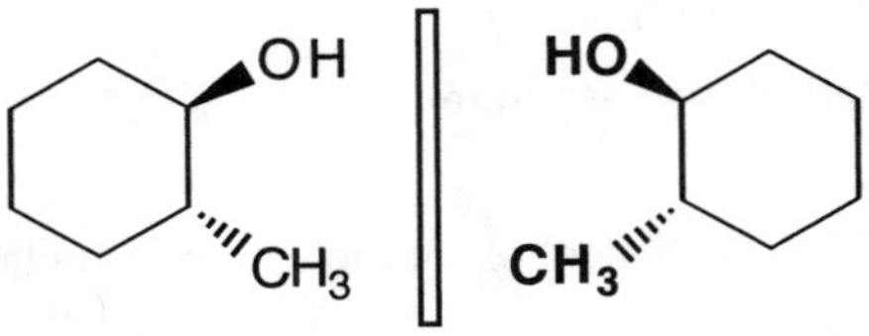

(g)

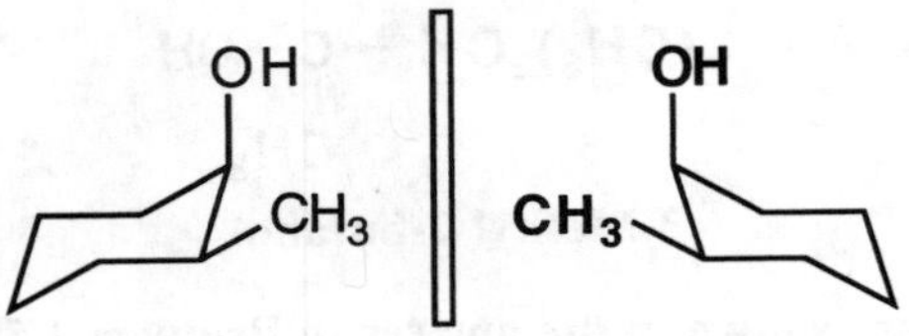

(h)

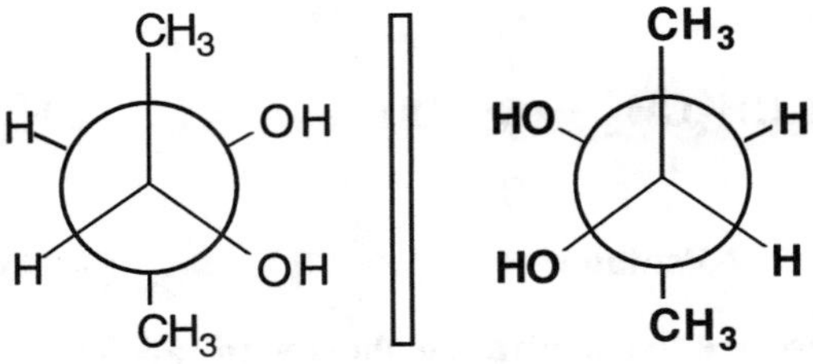

Problem 4.19 Following are four stereorepresentations for lactic acid. Take (a) as a reference structure. Which stereorepresentations are identical with (a) and which are mirror images of (a)?

(a) (b) 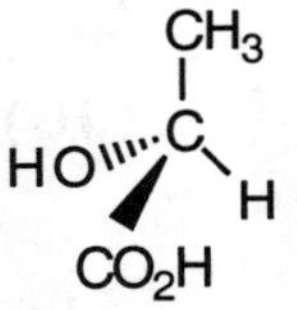 (c) 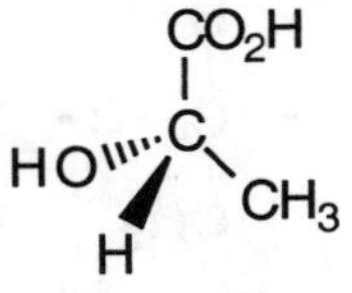 (d)

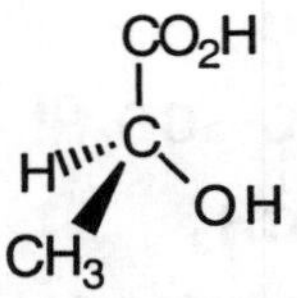

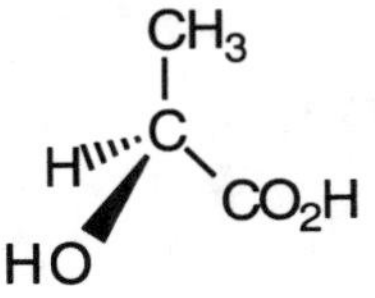

All stereorepresentations have the (S)-configuration, so they are identical.

Problem 4.20 Label each stereocenter in these molecules with an asterisk. Note that not all contain stereocenters.

(a) $CH_3-\underset{\underset{OH}{|}}{\overset{\overset{CH_3}{|}}{C}}-CH{=}CH_2$

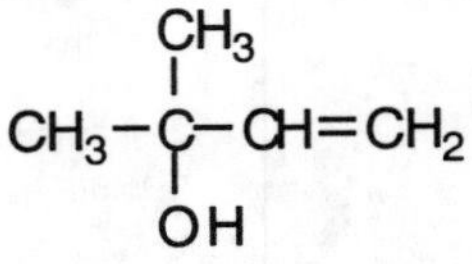

(b) $H-\underset{\underset{CH_3}{|}}{\overset{\overset{CO_2H}{|}}{C}}-OH$

(c) $CH_3-\underset{\underset{NH_2}{|}}{CH}-CH-CO_2H$

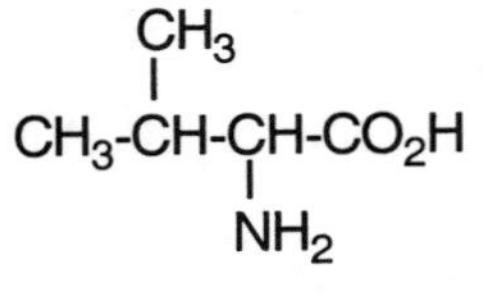

No stereocenters

$H-\overset{*}{\underset{\underset{CH_3}{|}}{\overset{\overset{CO_2H}{|}}{C}}}-OH$

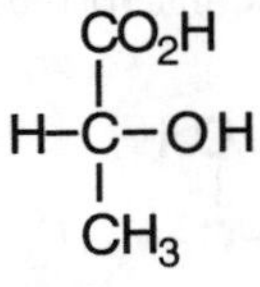

$CH_3-\underset{\underset{NH_2}{|}}{CH}-\overset{*}{CH}-CO_2H$

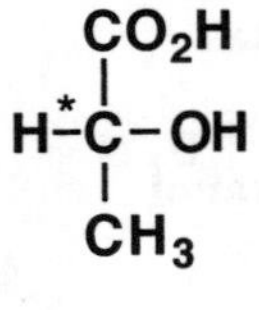

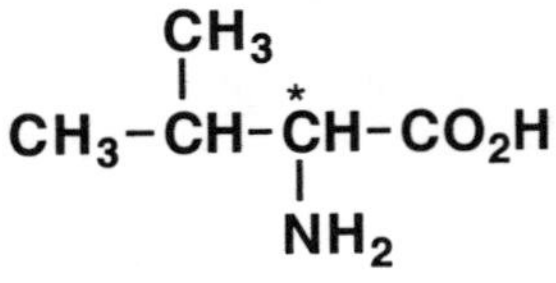

(d)

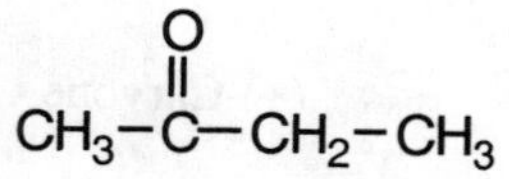

$$CH_3-\overset{\overset{\displaystyle O}{\|}}{C}-CH_2-CH_3$$

No stereocenters

(e)

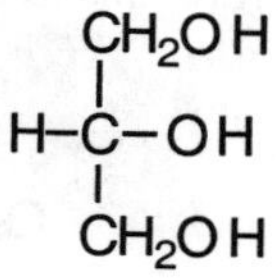

$$H-\overset{\overset{\displaystyle CH_2OH}{|}}{\underset{\underset{\displaystyle CH_2OH}{|}}{C}}-OH$$

No stereocenters

(f) $CH_3-CH_2-CH-CH=CH_2$ (OH above CH)

$$CH_3-CH_2-\overset{*}{\underset{\underset{\displaystyle }{}}{CH}}-CH=CH_2$$ (OH above CH)

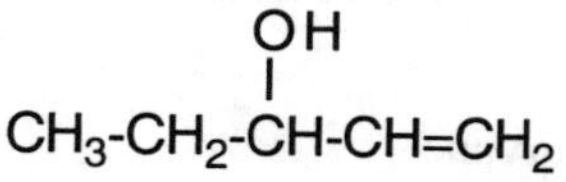

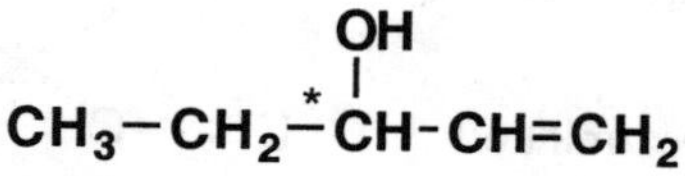

(g)

$$HO-\overset{\overset{\displaystyle CH_2-CO_2H}{|}}{\underset{\underset{\displaystyle CH_2-CO_2H}{|}}{C}}-CO_2H$$

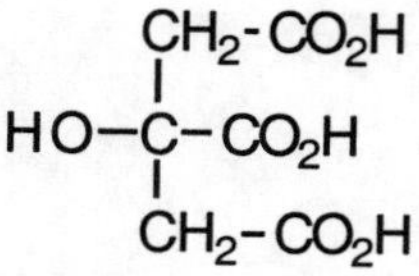

No stereocenters

Designation of Configuration: The R-S Convention
Problem 4.21 Assign priorities to the groups in each set.

The groups are ranked from highest to lowest under each problem. Remember that priority is assigned at the first point of difference.

(a) -H -CH₃ -OH -CH₂OH

-OH > -CH₂OH > -CH₃ > -H

(b) -CH₂CH=CH₂ -CH=CH₂ -CH₃ -CH₂CO₂H

-CH=CH₂ > -CH₂CO₂H > -CH₂CH=CH₂ > -CH₃

(c) -CH₃ -H -CO₂⁻ -NH₃⁺

-NH₃⁺ > -CO₂⁻ > -CH₃ > -H

(d) -CH₃ -CH₂SH -NH₃⁺ -CO₂⁻

-NH₃⁺ > -CH₂SH > -CO₂⁻ > -CH₃

Problem 4.22 Which molecules have R configurations?

(a)   CH₃ / Br / H / CH₂OH

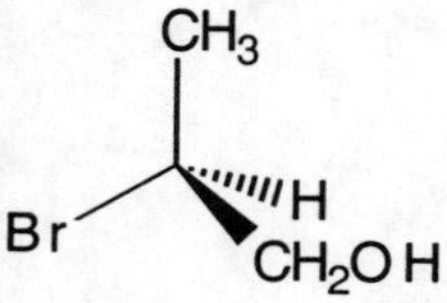

(b)   H / HOCH₂ / CH₃ / Br

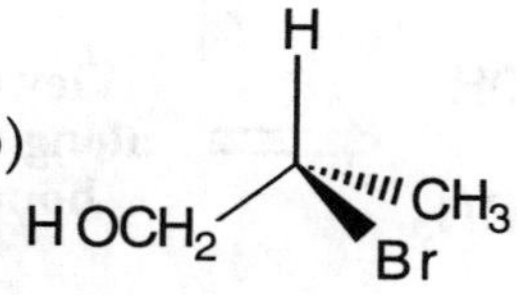

(c)   CH₂OH / H₃C / Br / H

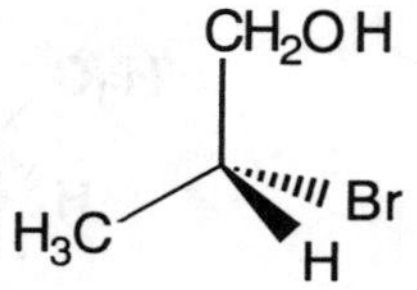

(d)   Br / H / CH₂OH / CH₃

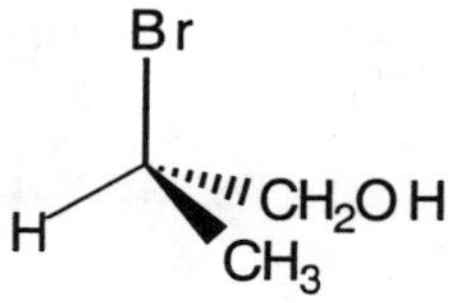

Molecules (c) and (d) have an R configuration. Group priorities are shown below for molecule (c)

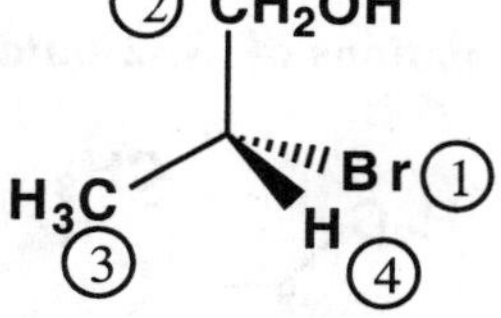

② CH₂OH

H₃C ③   ////Br ①

H ④

Problem 4.23 Following are structural formulas for the enantiomers of carvone. Each has a distinctive odor characteristic of the source from which it can be isolated. Assign R or S configurations to each enantiomer.

Following are R-S designations for each enantiomer.

(R)-(-)-Carvone   (S)-(+)-Carvone

Problem 4.24 Following is a staggered conformation for one of the stereoisomers of 2-butanol.

(a) Is this (R)-2-butanol or (S)-2-butanol?

The structure drawn is (S)-2-butanol.

(b) Draw a Newman projection for this enantiomer, viewed along the bond between carbons 2 and 3.

(c) Draw a Newman projection for one more staggered conformation of this molecule. Which of your conformations is the more stable? Assume that -OH and -CH₃ are comparable in size.

There are actually two more staggered conformations of (S)-2-butanol.

More stable

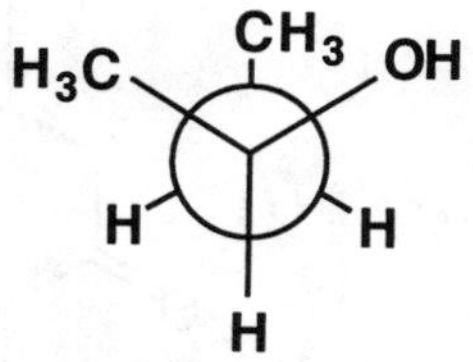

Less stable

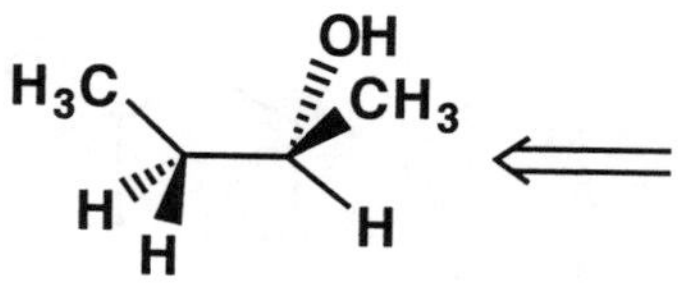

Viewed along this bond

Assuming that -OH and -CH_3 are the same size, then the structure drawn in part (b) and the upper structure shown in part (c) are of equal stability. These are more stable than the lower structure shown in part (c). This lower structure is less stable because both the -OH and -CH_3 groups are adjacent to the -CH_3 group.

<u>Molecules With Two Or More Stereocenters</u>

<u>Problem 4.25</u> For centuries, Chinese herbal medicine has used extracts of *Ephedra sinica* to treat asthma. Phytochemical investigation of this plant resulted in isolation of ephedrine, a very potent dilator of the air passages of the lungs. The naturally occurring stereoisomer is levorotatory and has the following structure. Assign R or S configuration to each stereocenter.

Ephedrine
$[\alpha]^{21}_{D} = -41°$

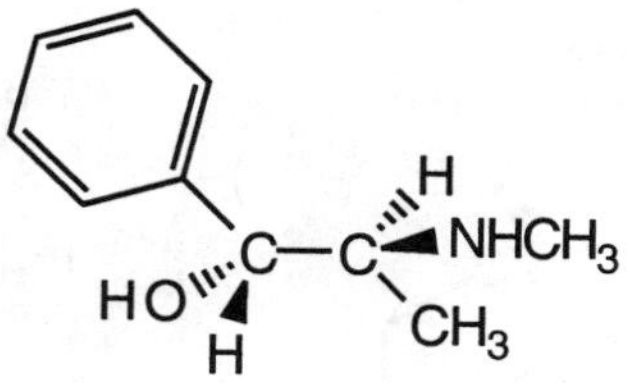

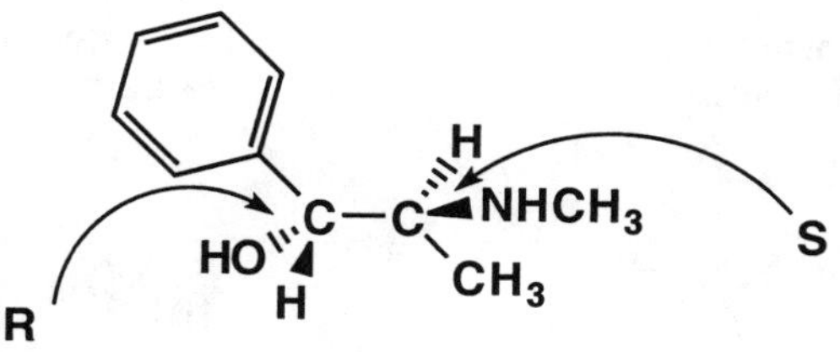

(1R,2S)-(-)-Ephedrine

<u>Problem 4.26</u> The specific rotation of naturally occurring ephedrine, shown in Problem 4.25, is -41°. What is the specific rotation of its enantiomer?

The specific rotations of enantiomers are equal in magnitude, but of opposite sign. Thus, the specific rotation of the enantiomer of ephedrine is +41°.

<u>Problem 4.27</u> Label each stereocenter in these molecules with an asterisk. How many stereoisomers are possible for each molecule?

(a) $CH_3CH(OH)—CH(OH)-CO_2H$

(b) CH_2-CO_2H | CH-CO_2H | $HO—CH$-CO_2H

(c) 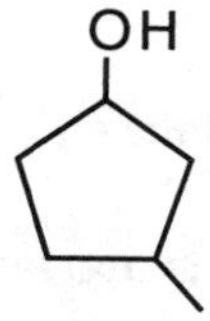

CH_3-C*H(OH)-C*H(OH)-CO_2H

4 stereoisomers (two pairs of enantiomers)

CH_2CO_2H | * $CH\cdot CO_2H$ | HO-*$CH\cdot CO_2H$

4 stereoisomers (two pairs of enantiomers)

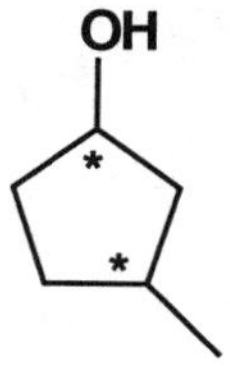

4 stereoisomers (two pairs of enantiomers)

(d)

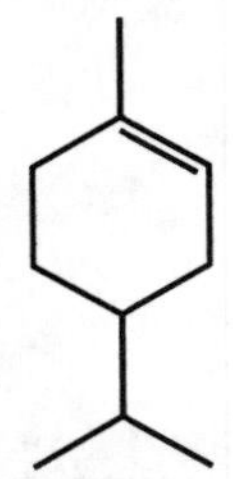

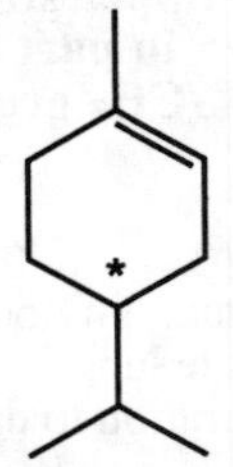

(e)

(f)

2 stereoisomers
(a pair of enantiomers)

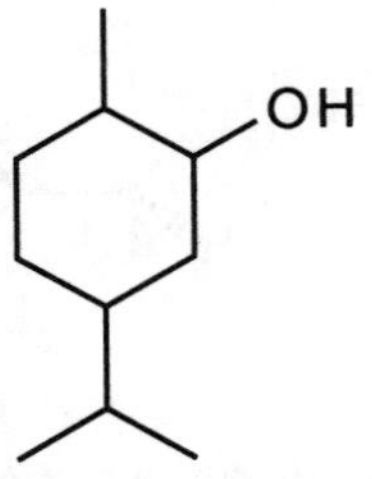

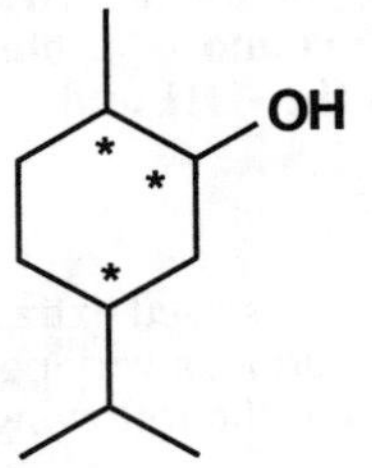

8 stereoisomers
(4 pairs of enantiomers)

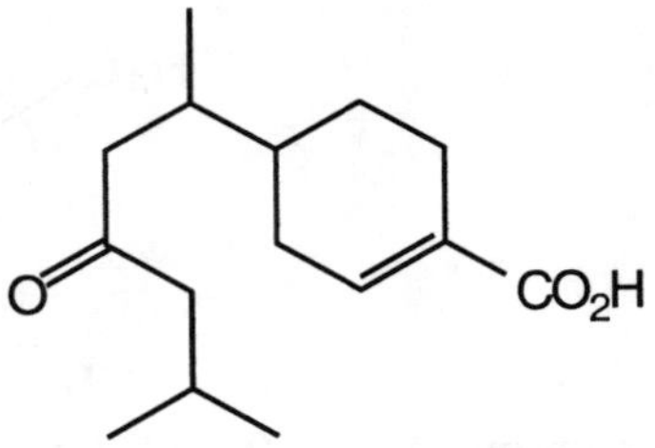

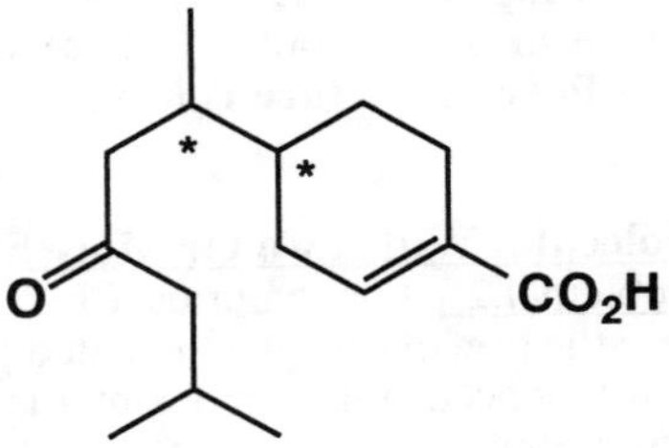

4 stereoisomers
(2 pairs of enantiomers)

(g)

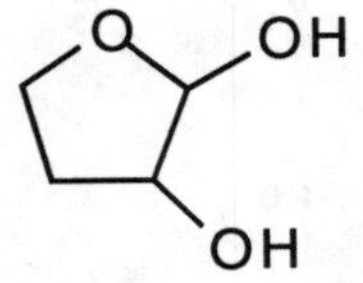

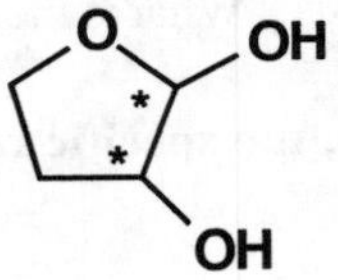

(h)

4 stereoisomers
(two pairs of enantiomers)

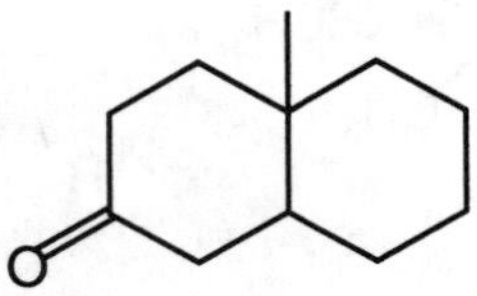

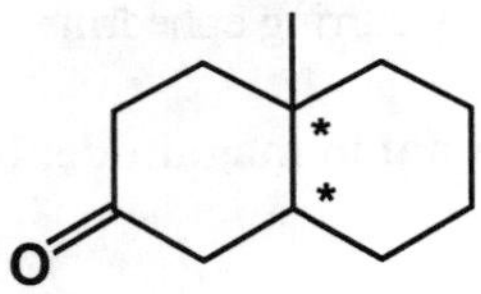

4 stereoisomers
(two pairs of enantiomers)

Problem 4.28 Label the eight stereocenters in cholesterol. How many stereoisomers are possible for a molecule of this structural formula?

Each stereocenter is marked with an asterisk.

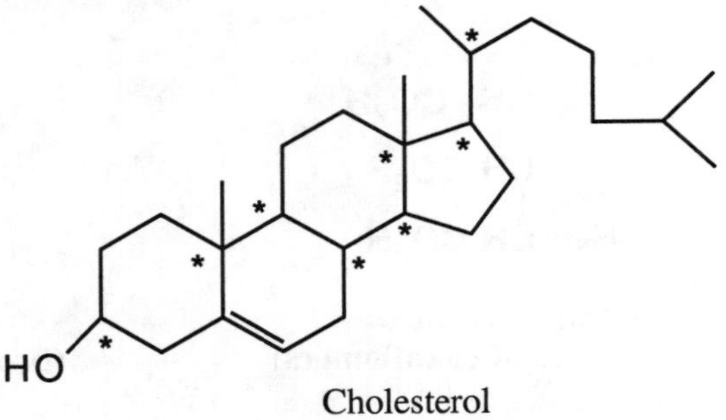

HO

Cholesterol

There are a total of $2^8 = 256$ stereoisomers possible for a molecule of this structural formula.

Problem 4.29 Label the four stereocenters in amoxicillin, which belongs to the family of semisynthetic penicillins.

Each stereocenter is marked with an asterisk.

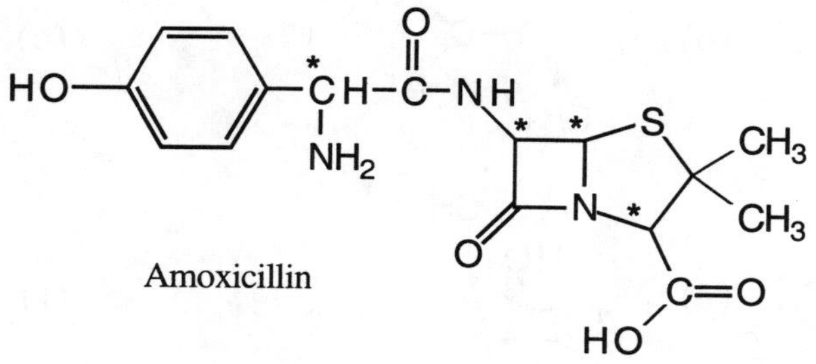

Amoxicillin

Problem 4.30 Label all stereocenters in the antihistamine terfenadine (Seldane). Terfenadine received FDA approval in May 1985 and by year's end had become the top-selling antihistamine in the United States. It provides relief from allergic disorders, but, unlike so many of the earlier antihistamines, does not cause drowsiness. Use of terfenadine in combination with certain other medications, however, has been linked to certain cardiac disorders and, in 1997, terfenadine was withdrawn from the market.

The single stereocenter is marked with an asterisk.

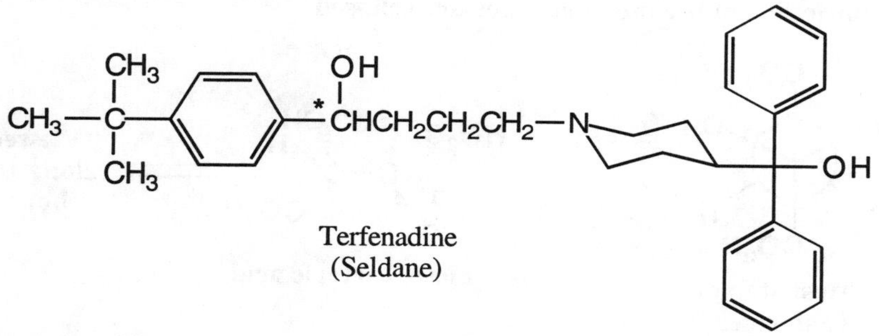

Terfenadine
(Seldane)

Problem 4.31 Label all stereocenters in loratadine (Claritin), now the top-selling antihistamine in the U.S. How many stereoisomers are possible for this compound?

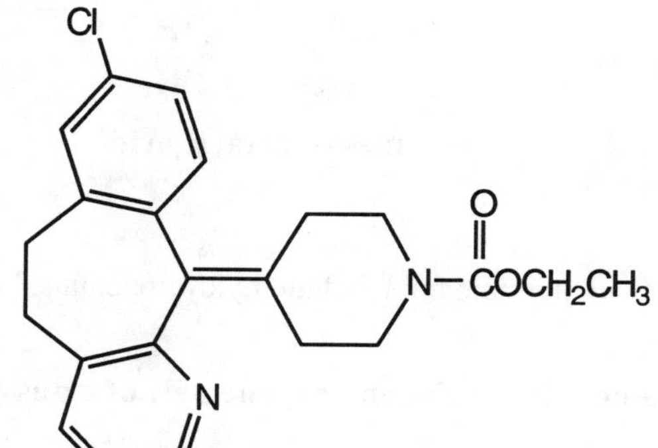

Loratadine has no stereocenters. It is an achiral molecule.

Problem 4.32 Which of these structural formulas represent meso compounds?

(a)

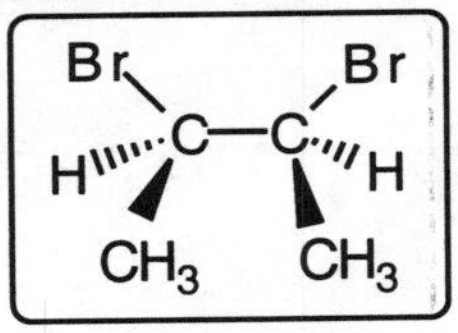

(b)

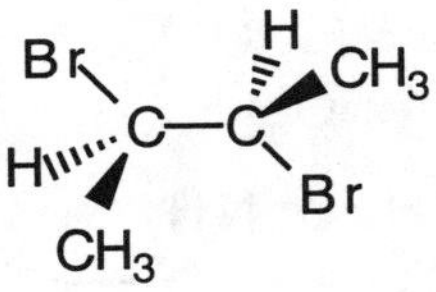

(c)

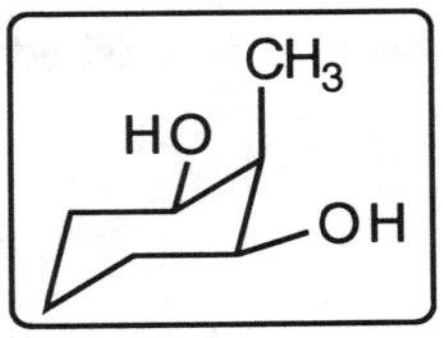

(d)

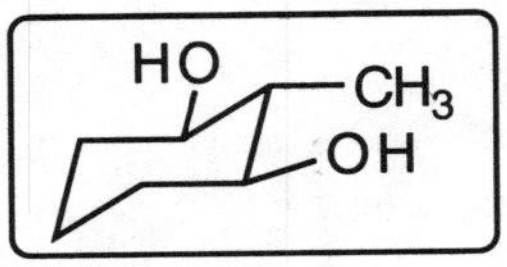

(e)

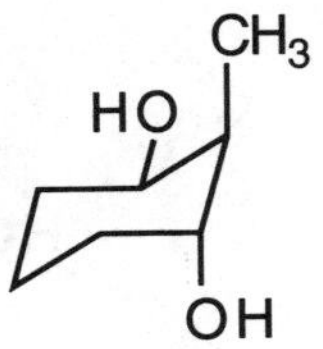

(f)

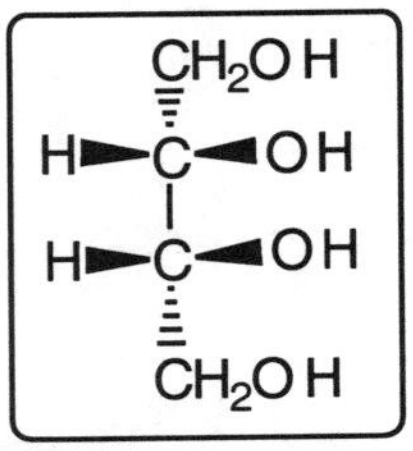

Each meso compound is circled. A meso compound has an internal plane of symmetry.

Problem 4.33 Draw a Newman projection, viewed along the bond between carbons 2 and 3, for both the most stable and the least stable conformations of meso-tartaric acid.

The carboxylic acid groups are the largest groups. In the most stable conformer, the carboxylic acid groups are anti to each other. In the least stable conformer, they are eclipsed.

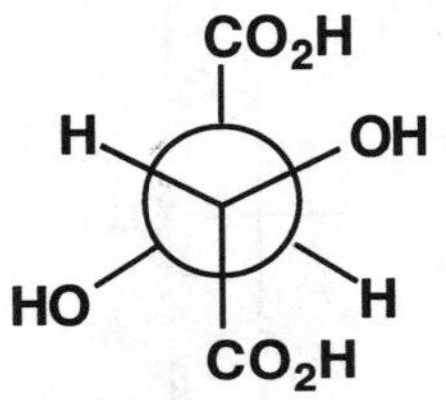

Most stable conformer

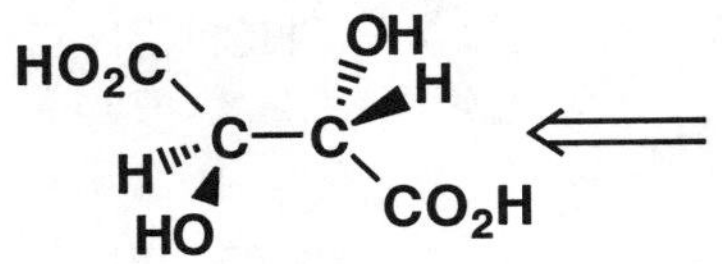

meso-Tartaric acid

Viewed along this bond

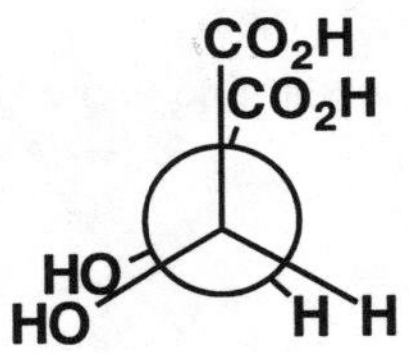

Least stable conformer

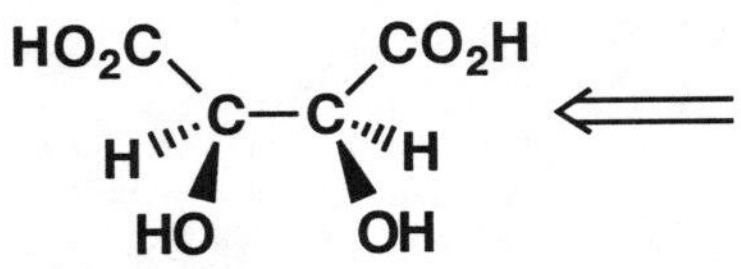

meso-Tartaric acid

Viewed along this bond

Problem 4.34 How many stereoisomers are possible for 1,3-dimethylcyclopentane? Which are pairs of enantiomers? Which are meso compounds?

There are three stereoisomers of 1,3-dimethylcyclopentane, one pair of trans enantiomers and a cis meso compound.

Plane of symmetry

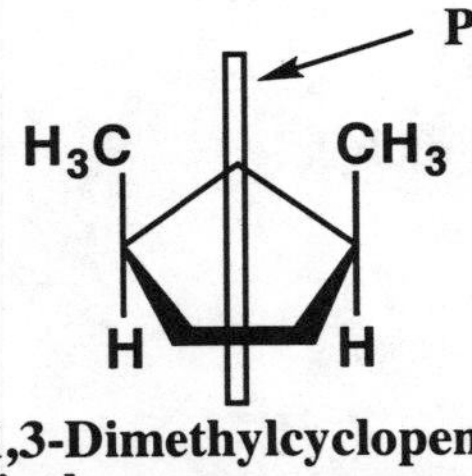

cis-1,3-Dimethylcyclopentane
(achiral, a meso compound)

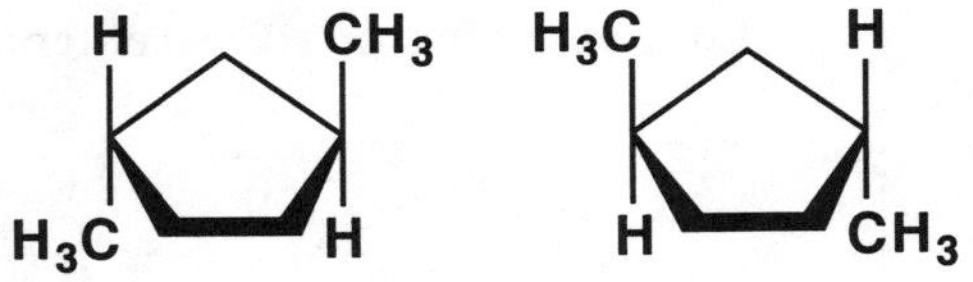

trans-1,3-Dimethylcyclopentane
(a pair of enantiomers)

Problem 4.35 In answer to Problem 3.33 you were asked to draw the more stable chair conformation of glucose, a molecule in which all groups on the six-membered ring are equatorial. Here is a drawing of that conformation.

(a) Identify all stereocenters in this molecule.

(b) How many stereoisomers are possible?

There are 2^5 = 32 stereoisomers.

(c) How many pairs of enantiomers are possible?

There are 16 pairs of enantiomers.

(a) What is the configuration (R or S) at carbons 1 and 5 in the stereoisomer shown?

The configuration at carbon 1 is R, and the configuration at carbon 5 is also R.

CHAPTER 5
Solutions to Problems

Problem 5.1 Write the IUPAC name of each hydrocarbon.

(a) $CH_3CH_2\overset{\overset{\displaystyle CH_3}{|}}{\underset{\underset{\displaystyle CH_3}{|}}{C}}CH{=}CH_2$

(b) $(CH_3)_2C{=}C(CH_3)_2$

(c) $CH_3\overset{\overset{\displaystyle CH_3}{|}}{\underset{\underset{\displaystyle CH_3}{|}}{C}}C{\equiv}CH$

3,3-Dimethyl-1-pentene   **2,3-Dimethyl-2-butene**   **3,3-dimethyl-1-butyne**

Problem 5.2 Write the common name of each hydrocarbon.

(a) $CH_3\overset{\overset{\displaystyle CH_3}{|}}{C}HC{\equiv}C\overset{\overset{\displaystyle CH_3}{|}}{C}HCH_3$

(b) 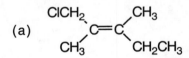$-C{\equiv}CH$

(c) $HC{\equiv}CCH_2CH_2CH_2CH_3$

Diisopropylacetylene   **Cyclohexylacetylene**   **Butylacetylene**

Problem 5.3 Which alkenes show cis-trans isomerism? For each alkene that does, draw the trans isomer.

(a) 2-Pentene   (b) 2-Methyl-2-pentene   c) 3-Methyl-2-pentene

$$\underset{H}{\overset{CH_3CH_2}{\diagdown}}C{=}C\underset{CH_3}{\overset{H}{\diagup}}$$

**No cis-trans isomers
since there are two methyl
groups on one end of
the double bond.**

$$\underset{CH_3}{\overset{CH_3CH_2}{\diagdown}}C{=}C\underset{CH_3}{\overset{H}{\diagup}}$$

trans-**2-Pentene**   *trans*-**3-Methyl-2-pentene**

Problem 5.4 Name each alkene and specify its configuration by the E,Z system.

(a) $$\underset{CH_3}{\overset{ClCH_2}{\diagdown}}C{=}C\underset{CH_2CH_3}{\overset{CH_3}{\diagup}}$$

(b) $$\underset{Br}{\overset{Cl}{\diagdown}}C{=}C\underset{CH_3}{\overset{H}{\diagup}}$$

(E)-1-Chloro-2,3-dimethyl-2-pentene   **(Z)-1-Bromo-1-chloropropene**

(c) $$\underset{CH_3}{\overset{CH_3CH_2CH_2}{\diagdown}}C{=}C\underset{CH(CH_3)_2}{\overset{CH_3}{\diagup}}$$

(E)-2,3,4-Trimethyl-3-heptene

Problem 5.5 Write the IUPAC name for each cycloalkene.

(a)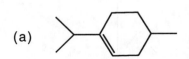

(b)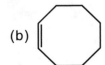

(c) $\langle$ $-C(CH_3)_3$

1-Isopropyl-4-methylcyclohexene   **Cyclooctene**   **4-*tert*-Butylcyclohexene**

Problem 5.6 Draw structural formulas for the other two cis-trans isomers of 2,4-heptadiene.

cis,trans-2,4-Heptadiene   **cis,cis-2,4-Heptadiene**

Problem 5.7 The sex pheromone from the silkworm is (10E,12Z)-10,12-hexadecadiene-1-ol. Draw a structural formula for this compound.

(10E,12Z)-10,12-Hexadecadiene-1-ol

Structure of Alkenes and Alkynes
Problem 5.8 Predict all bond angles about each highlighted carbon atom. To make these predictions, use the valence shell electron-pair repulsion (VSEPR) model (Section 1.3).

Problem 5.9 For each circled carbon atom in Problem 5.8, identify which atomic orbitals are used to form each sigma bond and which are used to form each pi bond.

Each bond is labeled sigma or pi, and the orbitals overlapping to form each bond are shown.

Problem 5.10 Below is the structure of 1,2-propadiene (allene). In it, the plane created by H-C-H of carbon 1 is perpendicular to that created by H-C-H of carbon 3.

(a) State the orbital hydridization of each carbon in allene.

(b) Account for the molecular geometry of allene in terms of the orbital overlap model.

Each terminal carbon atom is sp² hybridized and thus has three sp² orbitals oriented 120° apart. Two of these sp² hybrid orbitals form sigma bonds with hydrogen atoms while the third forms a sigma bond with the central carbon atom. One carbon 2p orbital is left over to form a π bond with the central carbon. The central carbon, being sp hybridized, has two sp orbitals oriented 180° apart. Each of these forms a sigma bond with a terminal carbon atom. Two carbon 2p orbitals are left over. Because these two p orbitals are at right angles to one another, the p orbitals on the terminal carbon atoms must also be at right angles to one another to permit orbital overlap.

Nomenclature of alkenes and alkynes
Problem 5.11 Draw a structural formula for each compound.
(a) trans-2-Methyl-3-hexene   (b) 2-Methyl-3-hexyne   (c) 2-Methyl-1-butene

(d) 3-Ethyl-3-methyl-1-pentyne   (e) 2,3-Dimethyl-2-butene   (f) cis-2-Pentene

(g) (Z)-1-Chloropropene   (h) 3-Methylcyclohexene   (i) 1-Isopropyl-4-methylcyclohexene

(j) (6E)-2,6-Dimethyl-2,6-octadiene   (k) Allylcyclopropane   (l) Diethylacetylene

(m) 2-Chloropropene

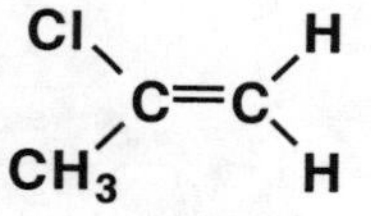

(n) Tetrachloroethylene

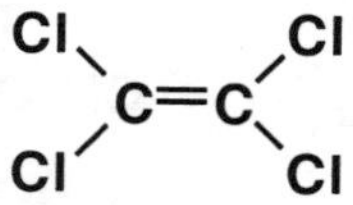

<u>Problem 5.12</u> Name these compounds.

(a) $CH_2=C$ with $(CH_2)_4CH_3$ and $CH_2CH(CH_3)_2$

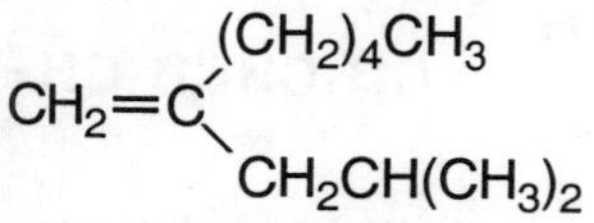

2-Isobutyl-1-heptene

(b)

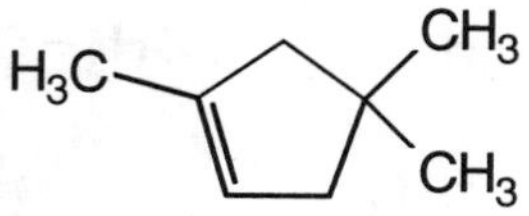

1,4,4-Trimethylcyclopentene

(c)

1,3-Cyclopentadiene

(d) $(CH_3)_2CHCH=C(CH_3)_2$

2,4-Dimethyl-2-pentene

(e) $CH_3(CH_2)_5C\equiv CH$

1-Octyne
(Hexylacetylene)

(f) $(CH_3)_2CHC\equiv CC(CH_3)_3$

2,2,5-Trimethyl-3-hexyne
(*tert*-Butylisopropylacetylene)

<u>Problem 5.13</u> Which alkenes exist as pairs of cis-trans isomers? For each alkene that does, draw the trans isomer.

For an alkene to exist as a pair of cis-trans isomers, both carbon atoms of the double bond must have two different substituents. Thus, (b), (c), and (e) exist as a pair of cis-trans isomers. The trans isomers for these alkenes are drawn under their respective condensed molecular formulas.

(a) $CH_2=CHBr$

(b) $CH_3CH=CHBr$

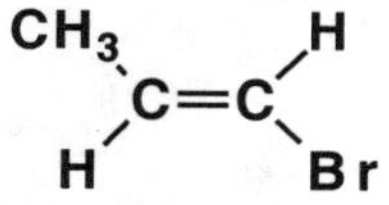

(c) $BrCH=CHBr$

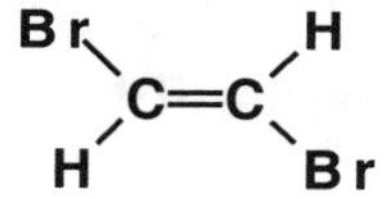

(d) $(CH_3)_2C=CHCH_3$

(e) $(CH_3)_2CHCH=CHCH_3$

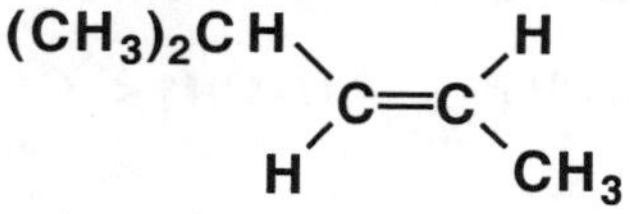

<u>Problem 5.14</u> Name and draw structural formulas for all alkenes of molecular formula C_5H_{10}. As you draw these alkenes, remember the cis and trans isomers are different compounds and must be counted separately.

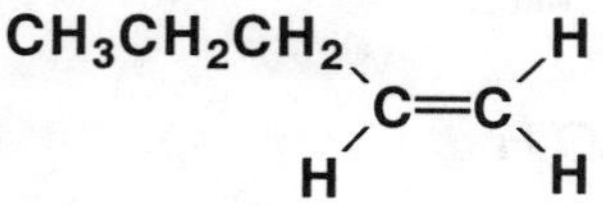

1-Pentene

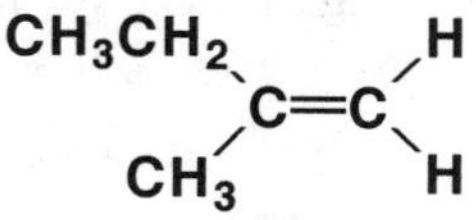

2-Methyl-1-butene

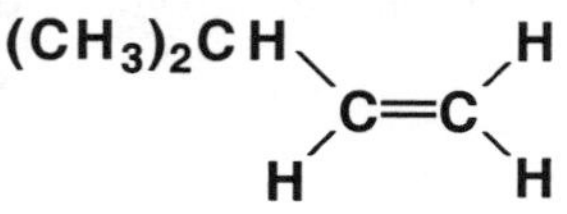

3-Methyl-1-butene

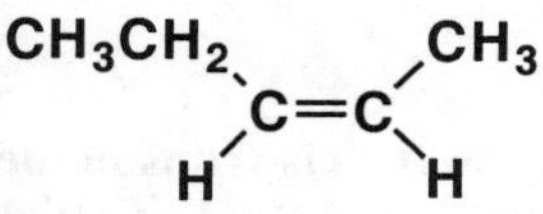

cis-**2-Pentene**

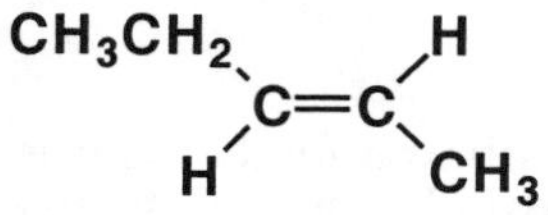

trans-**2-Pentene**

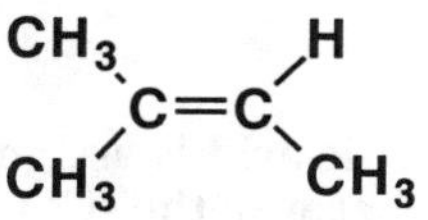

2-Methyl-2-butene

Problem 5.15 Name and draw structural formulas for all alkenes of molecular formula C_6H_{12} that have these carbon skeletons. Remember cis and trans isomers!

(a)

C–C–C–C–C with C on second carbon

$$CH_3$$
$$|$$
$$CH_2=CCH_2CH_2CH_3$$

2-Methyl-1-pentene

$$CH_3$$
$$|$$
$$CH_3C=CHCH_2CH_3$$

2-Methyl-2-pentene

$$CH_3-CH$$ $$C=C$$ with H and CH_3

***trans*-4-Methyl-2-pentene**

$$CH_3-CH$$ $$C=C$$ with CH_3, H, H

***cis*-4-Methyl-2-pentene**

$$CH_3$$
$$|$$
$$CH_3CHCH_2CH=CH_2$$

4-Methyl-1-pentene

(b)

C–C–C–C with C, C

$$CH_3$$
$$|$$
$$CH_2=CCHCH_3$$
$$|$$
$$CH_3$$

2,3-Dimethyl-1-butene

$$CH_3$$
$$|$$
$$CH_3C=CCH_3$$
$$|$$
$$CH_3$$

2,3-Dimethyl-2-butene

(c)

C–C–C–C with C, C

$$CH_3$$
$$|$$
$$CH_3CCH=CH_2$$
$$|$$
$$CH_3$$

3,3-Dimethyl-1-butene

Problem 5.16 Arrange the groups in each set in order of increasing priority (Review Section 4.3).

(a) -CH_3 -Br -CH_2CH_3

 -CH_3 < -CH_2CH_3 < -Br

(b) -OCH_3 -CH(CH_3)_2 -CH=CH_2

 -CH(CH_3)_2 < -CH=CH_2 < -OCH_3

(c) -CH_2OH -CO_2H -OH

 -CH_2OH < -CO_2H < -OH

Problem 5.17 Assign an E or Z configuration and a cis or trans configuration to these carboxylic acids, each of which is an intermediate in the tricarboxylic acid cycle. Following each is given its common name.

(a)

H, CO_2H / C=C / HO_2C, H

Fumaric acid

E (trans)

(b)

HO_2C, CO_2H / C=C / H, CH_2CO_2H

Aconitic acid

Z (trans)

Note that aconitic acid is assigned the Z configuration, but is trans according to the cis-trans nomenclature rules. This is because the -CO_2H group outranks the -CH_2CO_2H group according to the E-Z rules and is on the same side of the double bond as the higher ranking -CO_2H group on the other carbon atom of the double bond. However, it is the main chain groups that count with the cis-trans rules and the -CH_2CO_2H group is part of the main chain.

Problem 5.18 Four stereoisomers exist for 3-penten-2-ol.

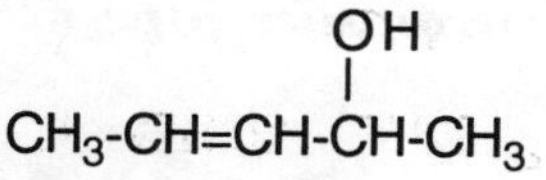

$$\underset{\text{3-Penten-2-ol}}{\overset{\overset{\displaystyle OH}{|}}{CH_3\text{-}CH=CH\text{-}CH\text{-}CH_3}}$$

(can be either E or Z)

OH   (can be either R or S)

$$CH_3\text{-}CH=CH\text{-}\underset{*}{CH}\text{-}CH_3$$

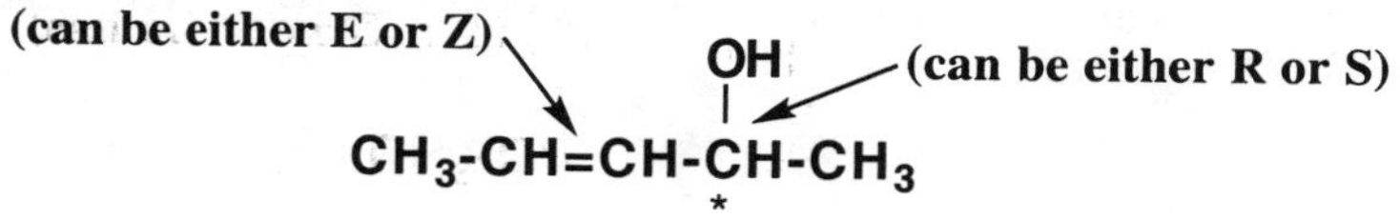

(a) Explain how these four stereoisomers arise.

The double bond can exist in two forms, E and Z (or cis- and trans-). There is also a stereocenter in the molecule (the atom marked with an asterisk in the structure above). The groups attached to this atom can be arranged to give either an R or an S configuration. Thus, four stereoisomers are possible: (E)(R), (E)(S), (Z)(R) and (Z)(S).

(b) Draw the stereoisomer having the E configuration about the carbon-carbon double bond and the R configuration at the stereocenter.

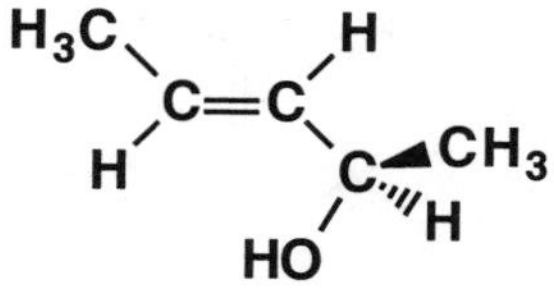

Problem 5.19 Draw the structural formula for at least one bromoalkene of molecular formula C_5H_9Br that shows:

(a) Neither E,Z isomerism nor enantiomerism.

Draw any structural formula in which there is no tetrahedral stereocenter and in which at least one carbon of the double bond has two identical atoms or groups of atoms on it. For example:

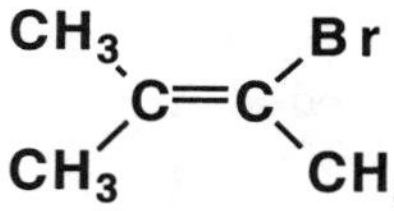

2-Bromo-3-methyl-2-butene

(b) E,Z isomerism but not enantiomerism.

Each carbon of the double bond must have two different atoms or groups of atoms on it, and there can be no tetrahedral stereocenter. For example:

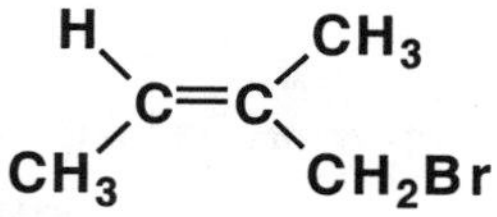

(Z)-1-Bromo-2-methyl-2-butene

(c) Enantiomerism but not E,Z isomerism.

One carbon of the double bond must have two identical atoms or groups of atoms on it, and there must be a tetrahedral stereocenter. For example:

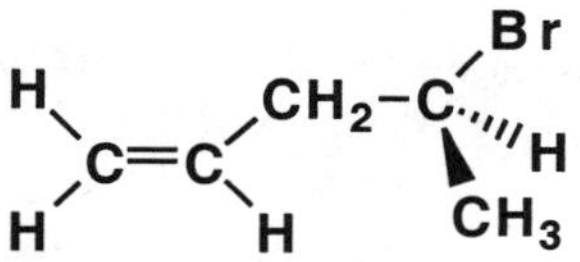

(S)-4-Bromo-1-pentene

(d) Enantiomerism and E,Z isomerism.

There is only one constitutional isomer that shows E-Z isomerism and has a tetrahedral stereocenter. Drawn is the (4R),(Z) isomer:

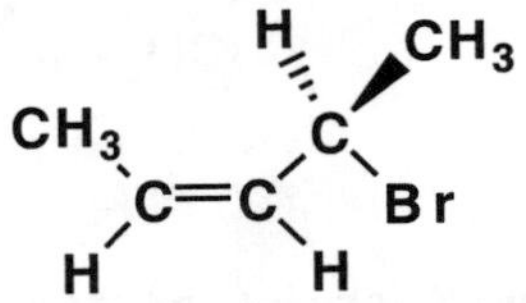

(R)-4-Bromo-(Z)-2-pentene

Problem 5.20 For each molecule that shows cis-trans isomerism, draw the cis isomer.

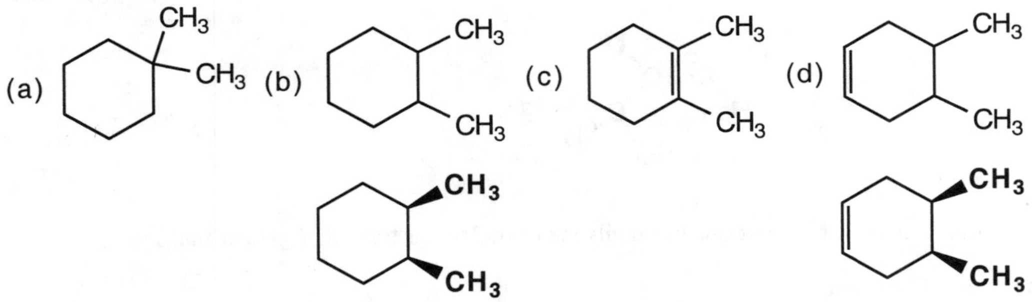

(a)   (b)   (c)   (d)

Only (b) and (d) show cis-trans isomerism. In (a), the two methyl groups are attached to the same carbon atom. In (c), the two methyl groups are attached to the double bond that must remain cis due to strain considerations in a six-member ring.

Problem 5.21 Draw structural formulas for all compounds of molecular formula C_5H_{10} that are:

(a) Alkenes that do not show cis-trans isomerism.

Four alkenes of molecular formula C_5H_{10} do not show cis-trans isomerism.

$$CH_2=CHCH_2CH_2CH_3$$

1-Pentene

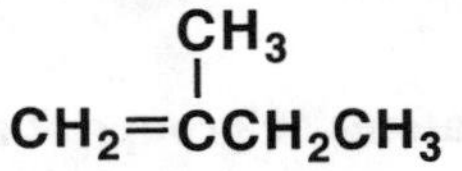

$$CH_2=\overset{\overset{\displaystyle CH_3}{|}}{C}CH_2CH_3$$

2-Methyl-1-butene

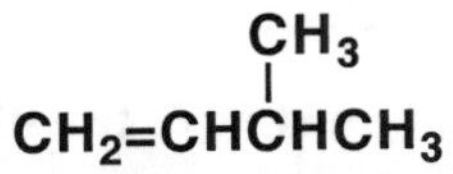

$$CH_2=CH\overset{\overset{\displaystyle CH_3}{|}}{C}HCH_3$$

3-Methyl-1-butene

$$CH_3\overset{\overset{\displaystyle CH_3}{|}}{C}=CHCH_3$$

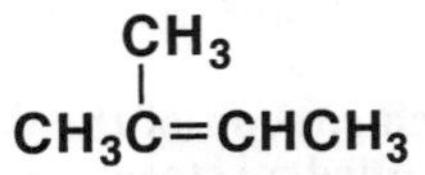

2-Methyl-2-butene

(b) Alkenes that do show cis-trans isomerism.

One alkene of molecular formula C_5H_{10} shows *cis-trans* isomerism.

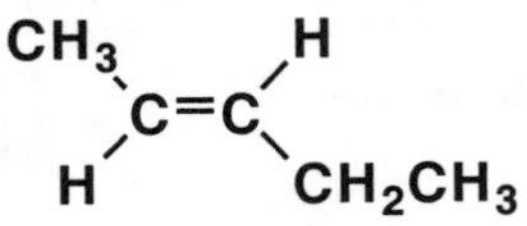

trans-2-Pentene 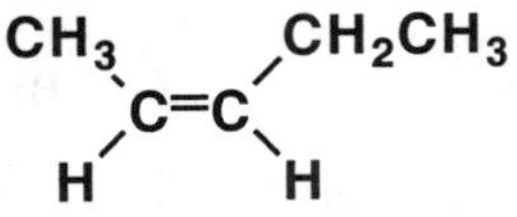   **cis-2-Pentene**

(c) Cycloalkanes that do not show cis-trans isomerism.

Four cycloalkanes of molecular formula C_5H_{10} do not show cis-trans isomerism.

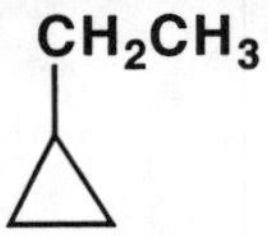

Ethylcyclopropane 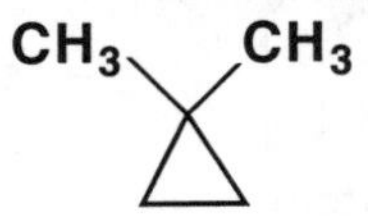   **1,1-Dimethylcyclopropane** 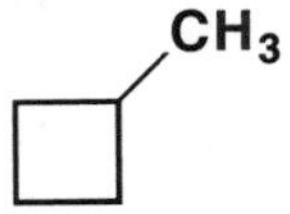   **Methylcyclobutane**    **Cyclopentane**

(d) Cycloalkanes that do show cis-trans isomerism.

Only one cycloalkane of molecular formula C₅H₁₀ shows cis-trans isomerism.

cis-1,2-Dimethylcyclopropane

trans-1,2-Dimethylcyclopropane

Problem 5.22 β-Ocimene, a triene found in the fragrance of cotton blossoms and several essential oils, has the IUPAC name (3Z)-3,7-dimethyl-1,3,6-octatriene. Draw a structural formula of β-ocimene.

β-Ocimene
(3Z)-3,7-Dimethyl-1,3,6-octatriene

Terpenes
Problem 5.23 Show how the carbon skeleton of farnesol can be coiled and cross-linked to give the carbon skeleton of caryophyllene (Figure 5.3).

The answer to this problem can best be visualized by numbering the atoms along the farnesol chain. Notice how, according to this analysis, the cyclobutane ring of caryophyllene would be formed by cross-linking farnesol from C11 to C1 and from C10 to C2. The structures of caryophyllene and an uncoiled farnesol are drawn for comparison.

Uncoiled farnesol   **Caryophyllene**   **Coiled farnesol**

Problem 5.24 Show that the structural formula of Vitamin A (Section 5.3G) can be divided into four isoprene units joined by head-to-tail linkages and cross-linked at one point to form the six-membered ring.

isoprene chain cross-linked here

Vitamin A

Problem 5.25 Following is the structural formula of lycopene, a deep red compound that is partially responsible for the red color of ripe fruits, especially tomatoes. Approximately 20 mg of lycopene can be isolated from 1 kg of ripe tomatoes (a) Show that lycopene is a terpene; that its carbon skeleton can be divided into two sets of four isoprene units with the units in each set joined head-to-tail.

The following structural formula shows the eight isoprene units of lycopene.

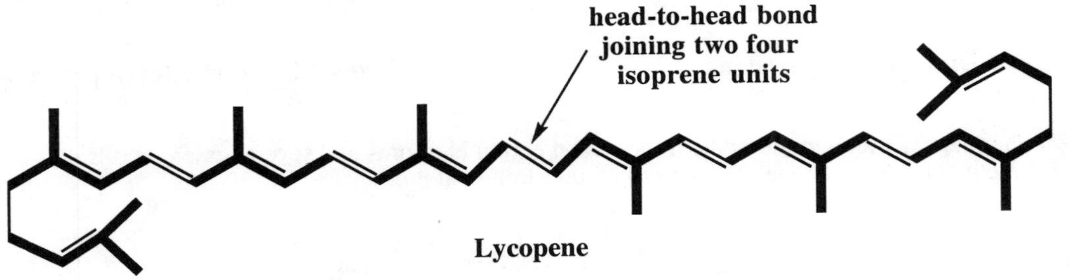

Lycopene

(b) How many of the carbon-carbon double bonds in lycopene have the possibility for cis-trans isomerism? Lycopene is the all trans isomer.

The double bonds on the two ends of the molecule cannot show cis-trans isomerism. The other 11 double bonds can show cis-trans isomerism.

Problem 5.26 The structural formula of β-carotene, precursor to vitamin A, is given in Section 17.6A. As you might suspect, it was first isolated from carrots. Dilute solutions of β-carotene are yellow, hence, its use as a food coloring. In plants, it is almost always present in combination with chlorophyll to assist in the harvesting of the energy of sunlight. As tree leaves die in the fall, the green of their chlorophyll molecules is replaced by the yellow and reds of carotene and carotene-related molecules. Compare the carbon skeletons of β-carotene and lycopene. What are the similarities? What are the differences?

The main structural difference between β-carotene and lycopene is that β-carotene has six-member rings on the ends, not an open chain. On the other hand, both β-carotene and lycopene can be divided into two sets of four isoprene units as shown below. All of the double bonds are trans in both molecules.

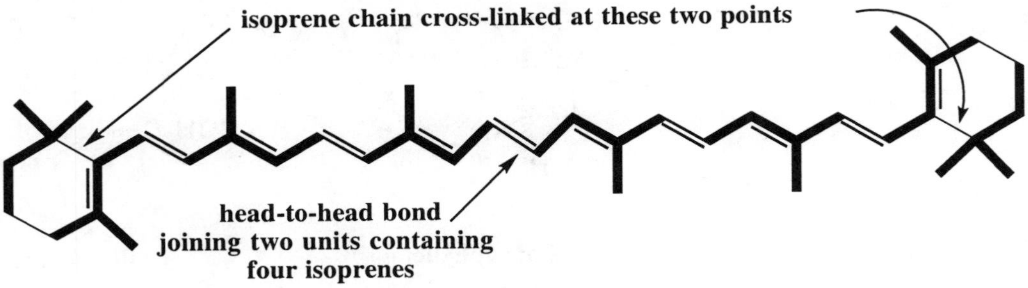

Problem 5.27 α-Santonin, isolated from the flower heads of certain species of Artemisia, is an anthelmintic; that is, a drug used to rid the body of worms (helminths). It has been estimated that over one-third of the world's population is infested with these parasites.

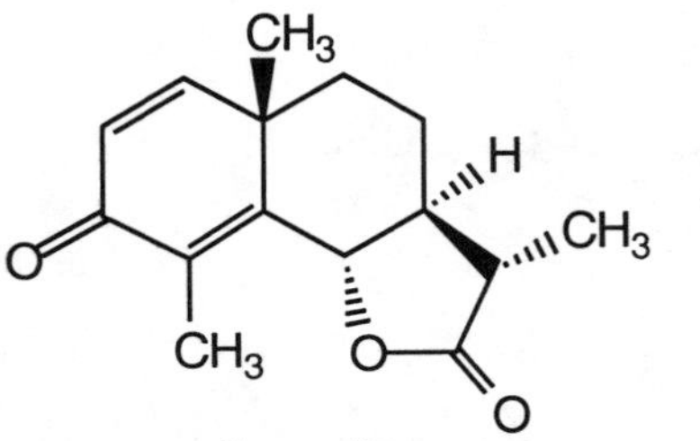

Santonin

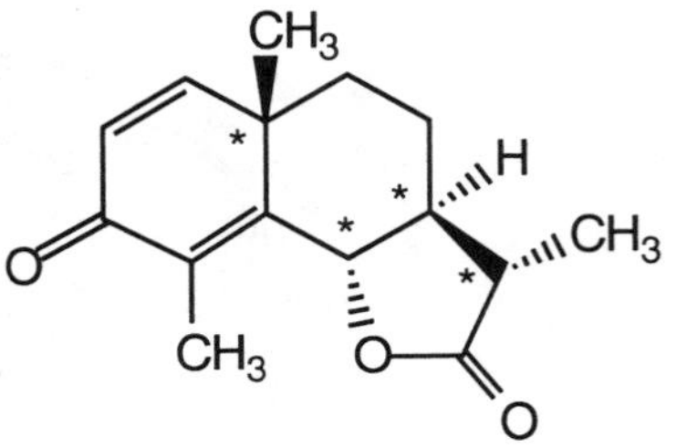

(a) Mark all stereocenters in santonin. How many stereoisomers are possible for it?

Santonin has four stereocenters, indicated with asterisks in the structure on the right. Thus, $2^4 = 16$ stereoisomers are possible.

(b) Locate the three isoprene units in santonin and show how the carbon skeleton of farnesol might be coiled and then cross-linked to give santonin. Two different coiling patterns of the carbon skeleton of farnesol can lead to santonin. Try to find them both.

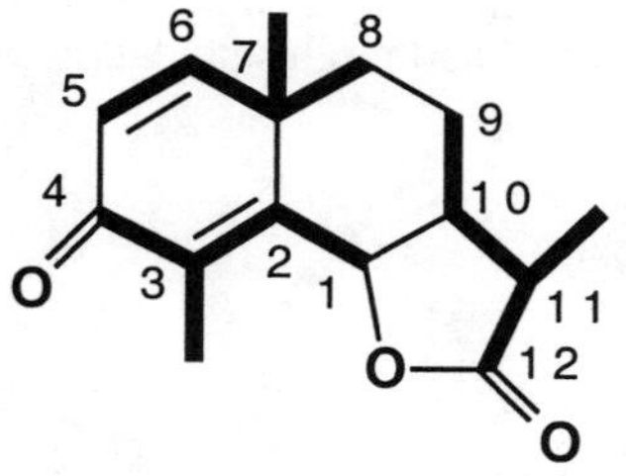

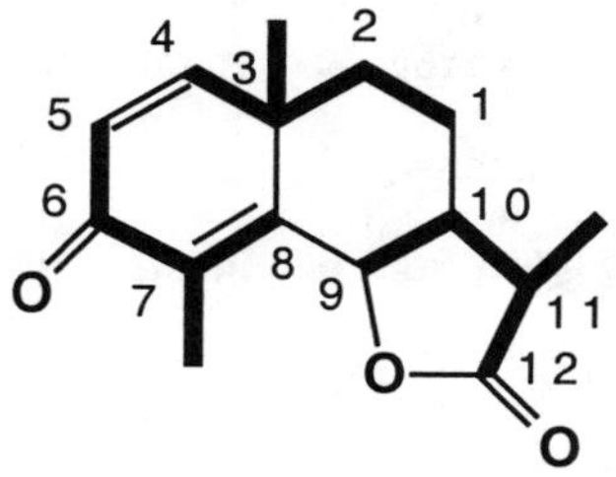

Problem 5.28 In many parts of South America, extracts of the leaves and twigs of *Montanoa tomentosa* are used as a contraceptive, to stimulate menstruation, to facilitate labor, and as an abortifacient. Phytochemical investigations of this plant have resulted in isolation of a very potent fertility-regulating compound called zoapatanol.

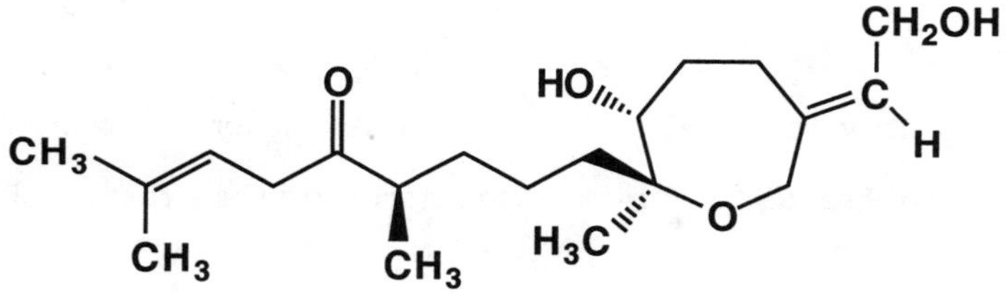

Zoapatanol

(a) Show that the carbon skeleton of zoapatanol can be divided into four isoprene units bonded head-to-tail and then cross-linked in one point along the chain.

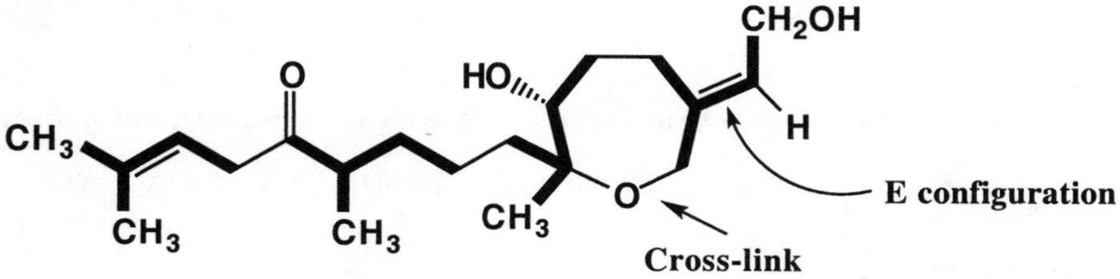

(b) Specify the configuration about the carbon-carbon double bond attached to the seven-member ring according to the E,Z system.

The double bond in question has the E configuration, because the hydroxymethyl group is on the side of the double bond opposite the higher priority carbon atom that is linked to the ether oxygen.

(c) How many cis-trans isomers are possible for zoapatanol? Consider the possibilities for cis-trans isomerism both in cyclic compounds and about carbon-carbon double bonds.

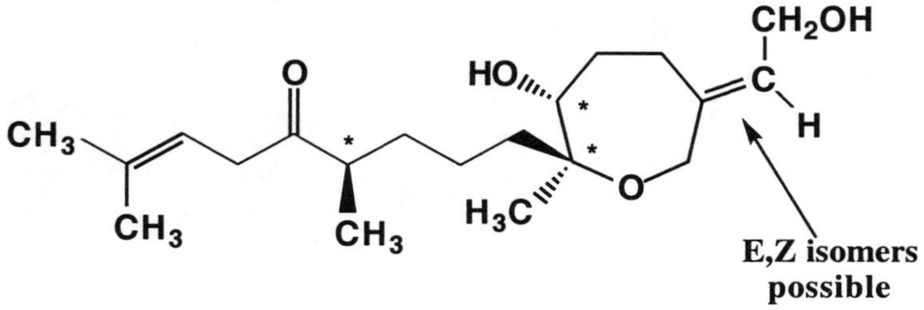

The stereocenters in zoapatanol are indicated in the structure above with asterisks. The double bond attached to the ring can be either E or Z. The other double bond has two methyl groups on one carbon atom, so it has no E,Z isomers. The -OH and -CH₃ groups on the ring can be cis (as occurs in zoapatanol) or trans. Possible combinations are E/cis, E/trans, Z/cis and Z/trans. Note that each of these isomers is chiral and, thus, can exist in two enantiomeric forms.

Problem 5.29 Following is the structural formula of warburganal, a crystalline solid isolated from the plant *Warbugia ugandensis, Canallaceae*. An important use of warburganal is its antifeeding activity against the African army worm. In addition, it acts as a plant growth regulator and has cytotoxic, antimicrobial and molluscicidal properties.

(a) Show that warburganal is a terpene.

The isoprene units are highlighted on the structure in the center.

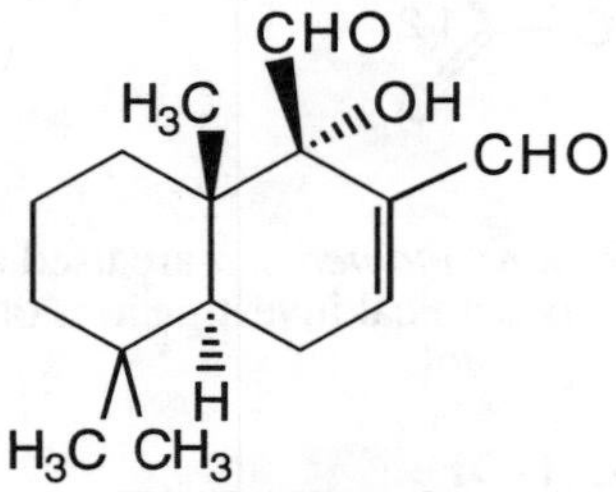

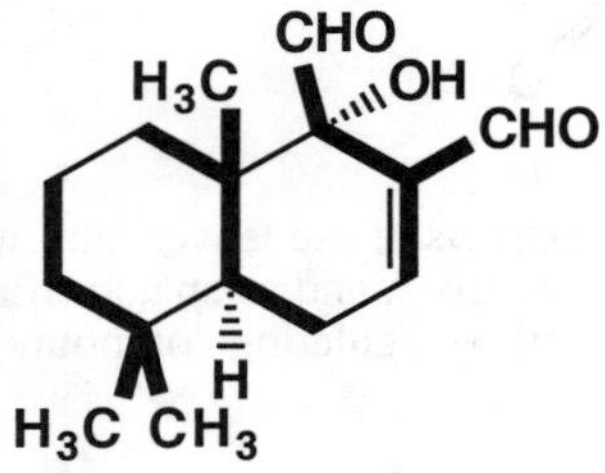

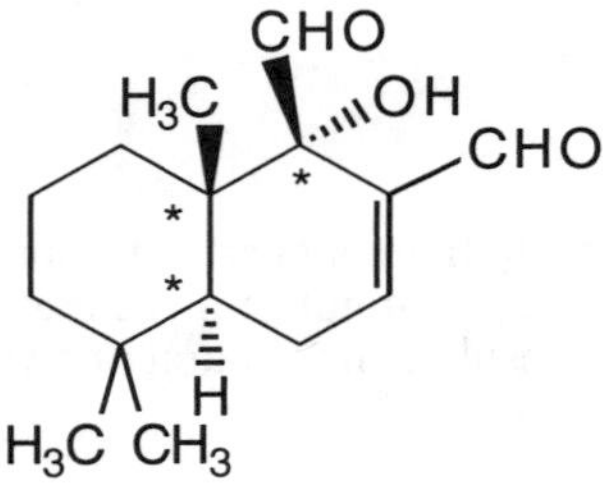

(b) Label each stereocenter and specify the number of stereoisomers possible for a molecule of this structure.

The molecule contains three stereocenters, indicated with asterisks in the structure shown above on the right. Thus, $2^3 = 8$ stereoisomers are possible.

Problem 5.30 Pyrethrin II and pyrethrosin are natural products isolated from plants of the chrysanthemum family. Pyrethrin II is a natural insecticide and is marketed as such.

(a) Label all tetrahedral stereocenters in each molecule and all carbon-carbon double bonds about which cis-trans isomerism is possible.

The stereocenters and carbon-carbon double bonds with possible cis-trans isomers are indicated on the structures.

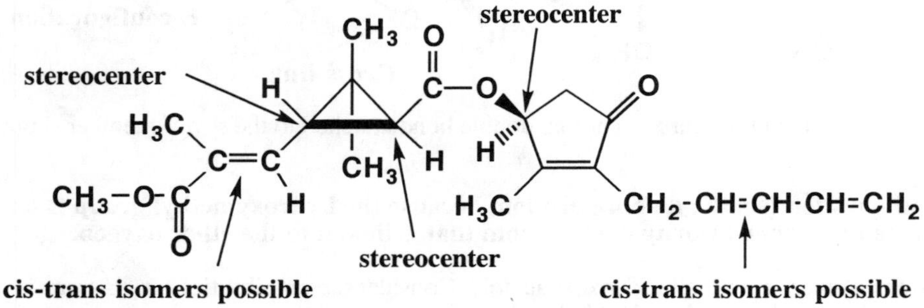

Pyrethrin ll

Pyrethrosin

(b) State the number of stereoisomers possible for each molecule.

For pyrethrin II, there are 2 alkenes with possible cis-trans isomers and there are three stereocenters. Thus, 2^2 x 2^3 = 32 stereoisomers of this molecule are possible.

For pyrethrosin, there are five stereocenters and one alkene with possible cis-trans isomers. Recall that a ten-member ring is large enough to allow the trans alkene configuration. There are a total of 2^5 x 2 = 64 possible stereoisomers of this molecule.

(c) Show that the ring system of pyrethrosin is composed of three isoprene units.

The isoprene units are highlighted on the structure.

Problem 5.31 Show that the carbon skeletons of the three terpenes drawn in the Chemical Connections box "Terpenoids of the Cotton Plant" can be divided into three isoprene units bonded head-to-tail and then cross-linked at appropriate carbons.

Spathulenol   Gossonorol   β-Bisabolol

CHAPTER 6
Solutions to Problems

Problem 6.1 Name and draw a structural formula for the product of each alkene addition reaction.

(a) $CH_3-CH=CH_2$ + HI $\longrightarrow$ $CH_3\overset{|}{C}HCH_3$
2-Iodopropane

(b) ⬡=CH_2 + HI $\longrightarrow$ (ring with CH_3 and I)
1-Iodo-1-methylcyclohexane

Problem 6.2 Arrange these carbocations in order of increasing stability.

(a) (ring)$\overset{+}{\;}-CH_3$ (b) (ring)$\overset{+}{\;}-CH_3$ (c) (ring)$-\overset{+}{C}H_2$

The order of increasing stability of carbocations is methyl < primary < secondary < tertiary.

(c) (ring)$-\overset{+}{C}H_2$ (b) (ring)$\overset{+}{\;}-CH_3$ (a) (ring)$\overset{+}{\;}-CH_3$

primary carbocation   **secondary carbocation**   **tertiary carbocation**

Problem 6.3 Propose a mechanism for the addition of HI to 1-methylcyclohexene to give 1-iodo-1-methylcyclohexane. Which step in your mechanism is rate limiting?

Step 1: (reaction scheme) $+$ H—I: $\xrightarrow[\text{limiting step}]{\text{Slow, rate-}}$ (carbocation)$-CH_3$ + :I:⁻

Step 2: (carbocation)$-CH_3$ + :I:⁻ $\longrightarrow$ (product ring with I and CH_3)

Problem 6.4 Draw a structural formula for the product of each alkene hydration reaction.

(a) $CH_3\overset{\overset{\displaystyle CH_3}{|}}{C}=CHCH_3$ + H_2O $\xrightarrow{H_2SO_4}$ $CH_3\overset{\overset{\displaystyle CH_3}{|}}{\underset{\underset{\displaystyle OH}{|}}{C}}CH_2CH_3$
2-Methyl-2-butanol

(b) $CH_2{=}\overset{\overset{\displaystyle CH_3}{|}}{C}CH_2CH_3$ + H_2O $\xrightarrow{H_2SO_4}$ $CH_3\overset{\overset{\displaystyle CH_3}{|}}{\underset{\underset{\displaystyle OH}{|}}{C}}CH_2CH_3$
2-Methyl-2-butanol

Problem 6.5 Propose a mechanism for the acid-catalyzed hydration of 1-methylcyclohexene to give 1-methylcyclohexanol. Which step in your mechanism is the rate-limiting step?

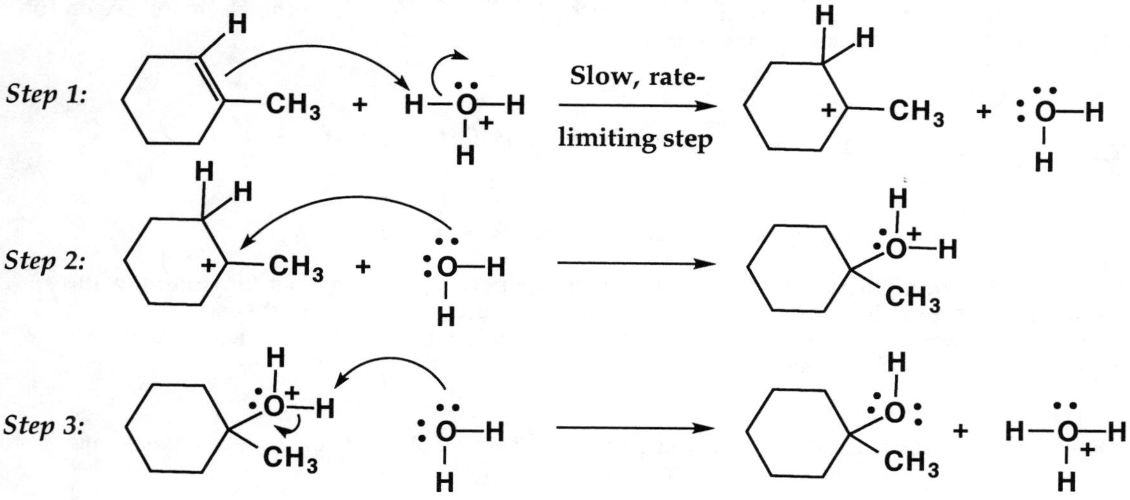

Step 1:

Slow, rate-limiting step

Step 2:

Step 3:

Problem 6.6 Complete these reactions.

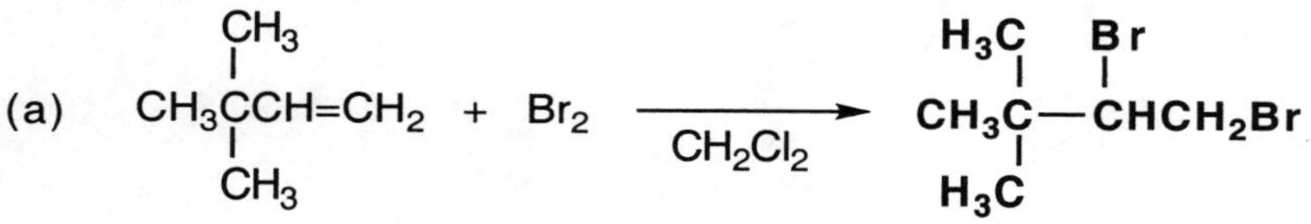

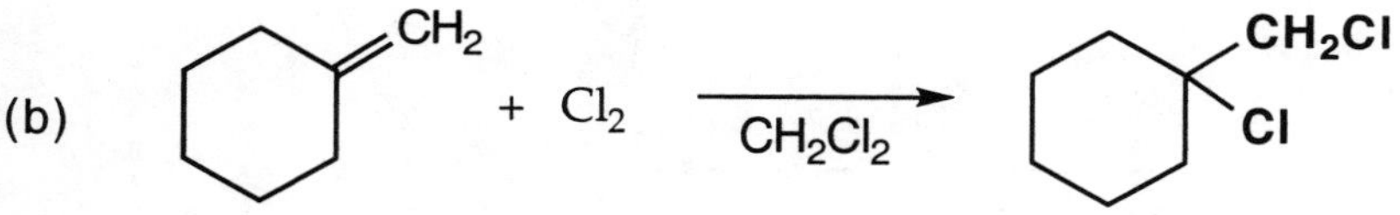

(a)

$$CH_3CCH=CH_2 + Br_2 \xrightarrow{CH_2Cl_2} CH_3C-CHCH_2Br$$

(b)

$$+ Cl_2 \xrightarrow{CH_2Cl_2}$$

Problem 6.7 Balance each half-reaction to show that each transformation involves a reduction.

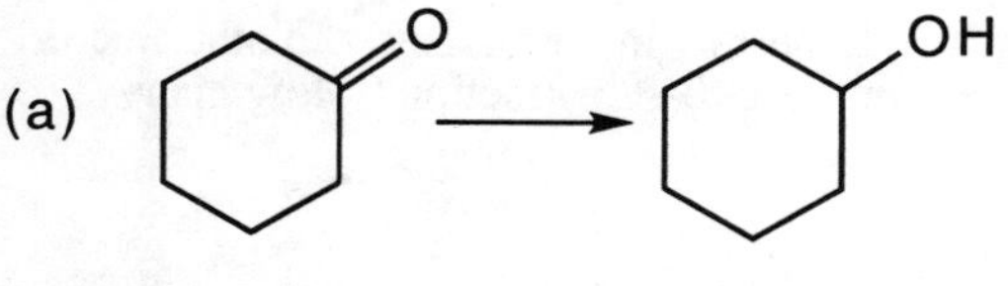

(a)

Two hydrogens are required to produce the product alcohol from the ketone. Therefore, the balanced half-reaction needs two protons and two electrons (for charge balance) on the left-hand side. Because the electrons are on the left-hand side of the equation, the reaction is a two-electron reduction.

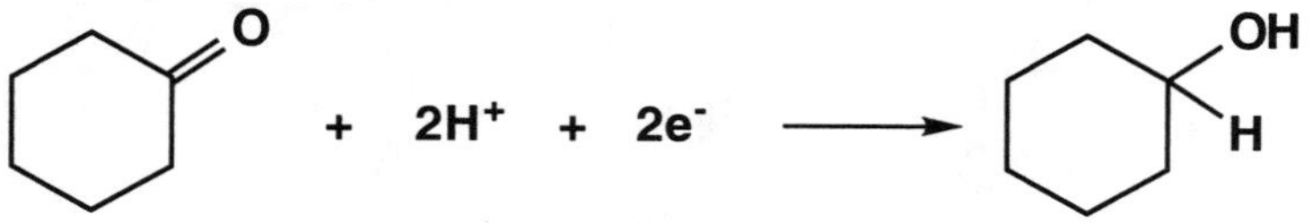

$$+ \ 2H^+ \ + \ 2e^- \longrightarrow$$

(b) $CH_3CH_2C\text{-}OH \longrightarrow CH_3CH_2CH_2OH$

Two hydrogens are required to produce the product alcohol from the carboxylic acid. Therefore, the balanced half-reaction needs two protons and two electrons (for charge balance) on the left-hand side. Additionally, the product alcohol has one less oxygen atom than the carboxylic acid starting material, so there must be an H_2O

molecule added to the right side of the equation to balance the oxygen atoms. This H_2O molecule has two more hydrogens that must be balanced by adding two more protons and electrons to the left-hand side of the equation, giving a total of four protons and four electrons on the left-hand side. Since the electrons are on the left-hand side of the equation, the reaction is a four-electron reduction.

$$\overset{\overset{\text{O}}{\|}}{\text{CH}_3\text{CH}_2\text{COH}} \ + \ 4\text{H}^+ \ + \ 4\text{e}^- \ \longrightarrow \ \text{CH}_3\text{CH}_2\text{CH}_2\text{OH} \ + \ \text{H}_2\text{O}$$

Problem 6.8 Describe the differences between a transition state and a reaction intermediate.

A transition state is the point on a reaction coordinate in which the potential energy is a maximum, while an intermediate is a potential energy minimum between two transition states. The potential energy of an intermediate is often higher than the energy of either reactants or products, so many intermediates are highly reactive and, thus, cannot be observed directly.

Problem 6.9 Sketch a potential energy diagram for a one-step reaction that is very slow and only slightly exothermic. How many transition states are present in this reaction? How many intermediates?

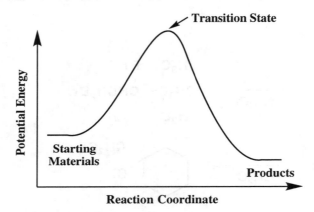

The reaction described by this potential energy diagram has one transition state and no intermediates.

Problem 6.10 Sketch a potential energy diagram for a two-step reaction that is endothermic in the first step, exothermic in the second step, and exothermic overall. How many transition states are present in this two-step reaction? How many intermediates?

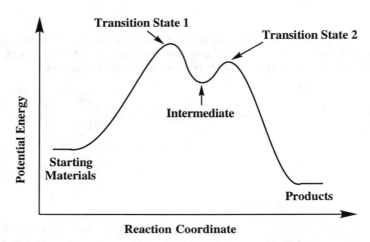

The reaction described by this potential energy diagram has two transition states and one intermediate.

Electrophilic Additions

Problem 6.11 From each pair, select the more stable carbocation.

The more stable carbocation in each pair is circled.

(a) $CH_3CH_2\overset{+}{C}H_2$ or $\boxed{CH_3\overset{+}{C}HCH_3}$ (b) $CH_3\underset{+}{C}HCHCH_3$ (with CH_3 on middle C)   or $\boxed{CH_3\underset{+}{C}CH_2CH_3}$ (with CH_3 on C)

Primary   Secondary   Secondary   Tertiary

Cation stability decreases in the order tertiary > secondary > primary > methyl.

Problem 6.12 From each pair, select the more stable carbocation.

The more stable carbocation in each pair is circled.

(a) [cyclohexyl ring with + and —CH₃, Secondary] or [boxed: cyclohexyl ring with + CH₃, Tertiary] (b) [boxed: cyclohexyl ring with + and —CH₃, Secondary] or [cyclohexyl ring with —CH₂⁺, Primary]

Secondary   Tertiary   Secondary   Primary

Problem 6.13 Draw structural formulas for the isomeric carbocation intermediates formed by reaction of each alkene with HCl. Label each carbocation primary, secondary, or tertiary, and state which of the isomeric carbocations is formed more readily.

(a) $CH_3CH_2\underset{\underset{\textstyle CH_3}{|}}{C}=CHCH_3$ ⟶ $CH_3CH_2\underset{+}{\overset{\overset{\textstyle CH_3}{|}}{C}}CH_2CH_3$ + $CH_3CH_2\underset{+}{\overset{\overset{\textstyle CH_3}{|}}{C}H}CHCH_3$

                          Tertiary   **Secondary**
                         (more stable)   **(less stable)**

(b) $CH_3CH_2CH=CHCH_3$ ⟶ $CH_3CH_2\underset{+}{C}HCH_2CH_3$ + $CH_3CH_2CH_2\underset{+}{C}HCH_3$

Both secondary carbocations
(of equal stability)

(c) [cyclopentene ring with CH₃] ⟶ [cyclopentane ring with + and —CH₃]   [cyclopentane ring with —CH₃ and +]

                    Tertiary   **Secondary**
                  (more stable)   **(less stable)**

(d) [cyclohexane ring =CH₂] ⟶ [cyclohexane ring —$\overset{+}{C}H_2$]   [cyclohexane ring with + and —CH₃]

                   Primary   **Tertiary**
                 (less stable)   **(more stable)**

Problem 6.14 From each pair of compounds, select the one that reacts more rapidly with HI. Draw the structural formula of the major product formed in each case, and explain the basis for your ranking.

Reaction rates in these electrophilic additions are a function of the stability of the carbocation intermediate formed in the rate-limiting step. The more stable the carbocation intermediate (tertiary > secondary > primary), the greater the rate of the reaction.

(a) 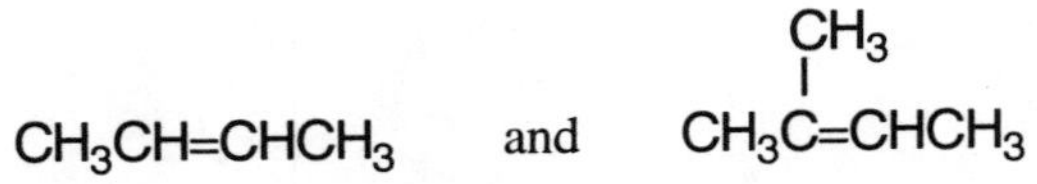 $CH_3CH=CHCH_3$ and $CH_3\overset{\overset{CH_3}{|}}{C}=CHCH_3$

$CH_3CH=CHCH_3 \longrightarrow CH_3CH_2\overset{+}{C}HCH_3 \longrightarrow CH_3CH_2\overset{\overset{I}{|}}{C}HCH_3$

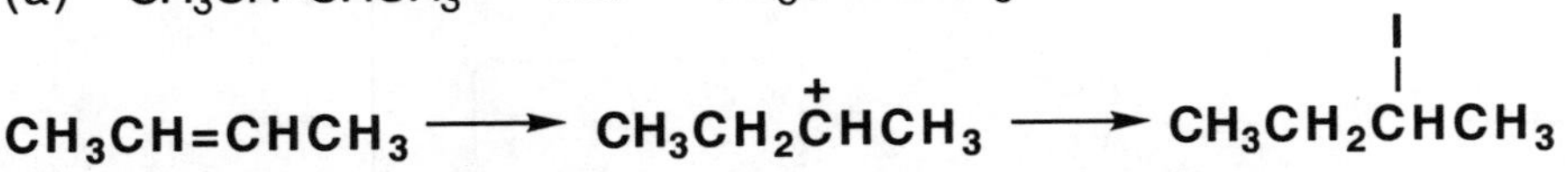

2-Butene   **a secondary**   **2-Iodobutane**
                         carbocation   **(sec-Butyl iodide)**

$CH_3\overset{\overset{CH_3}{|}}{C}=CHCH_3 \longrightarrow CH_3\overset{\overset{CH_3}{|}}{\underset{+}{C}}CH_2CH_3 \longrightarrow CH_3\overset{\overset{CH_3}{|}}{\underset{I}{C}}CH_2CH_3$

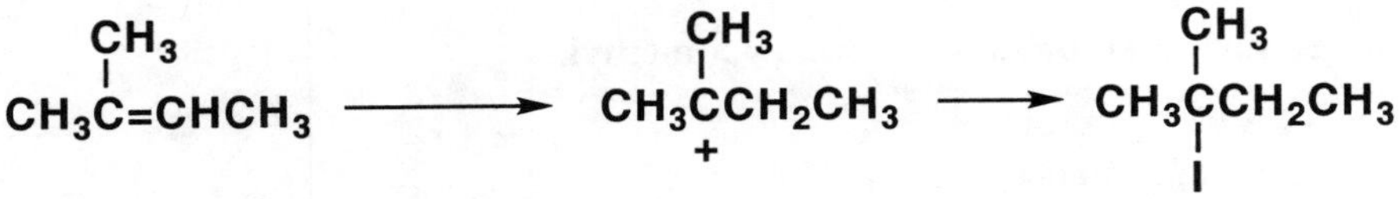

2-Methyl-2-butene   **a tertiary**   **2-Iodo-2-methylbutane**
                              carbocation   **(major product)**

The reaction of 2-methyl-2-butene involves formation of a tertiary carbocation, so 2-methyl-2-butene reacts faster with HI.

(b) 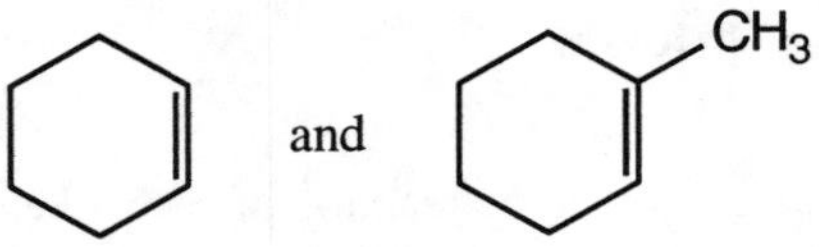 and

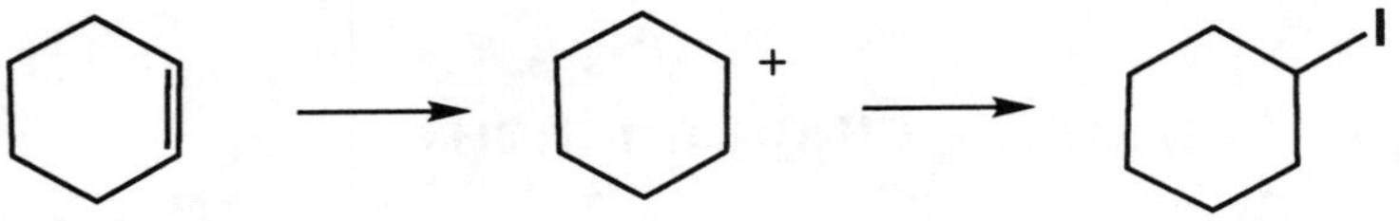

Cyclohexene   **a secondary**   **Iodocyclohexane**
                        carbocation

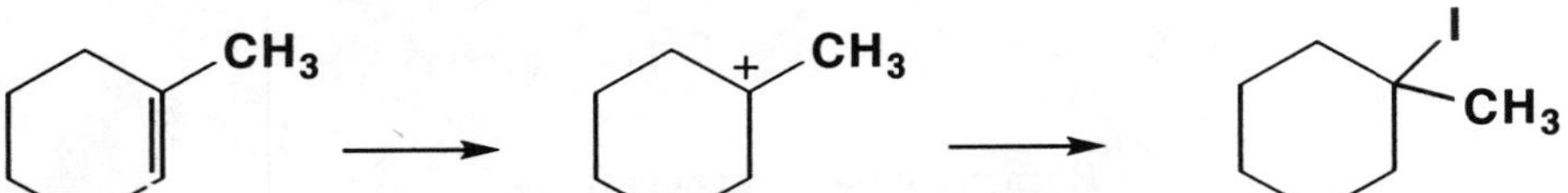

1-Methylcyclohexene   **a tertiary**   **1-Iodo-1-methylcyclohexane**
                                carbocation   **(major product)**

The reaction of 1-methylcyclohexene involves formation of a tertiary carbocation, so 1-methylcyclohexene reacts faster with HI.

Problem 6.15 Complete these equations.

(a)

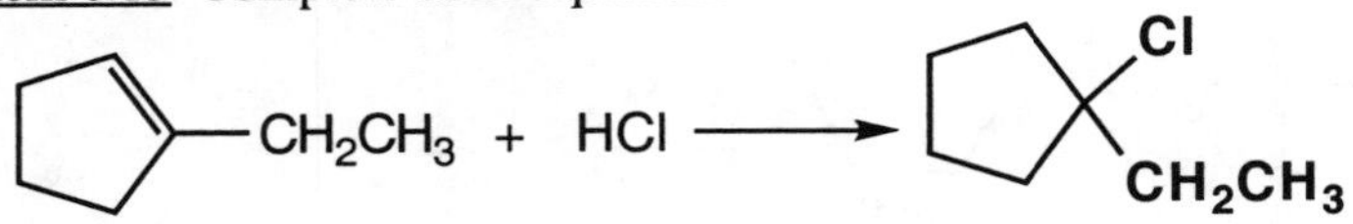

(b)

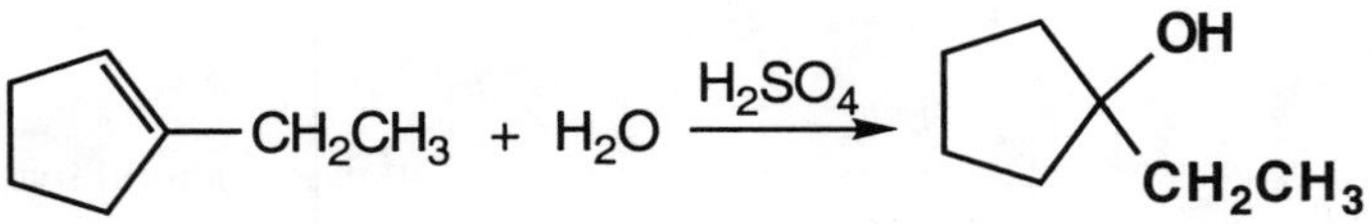

(c) 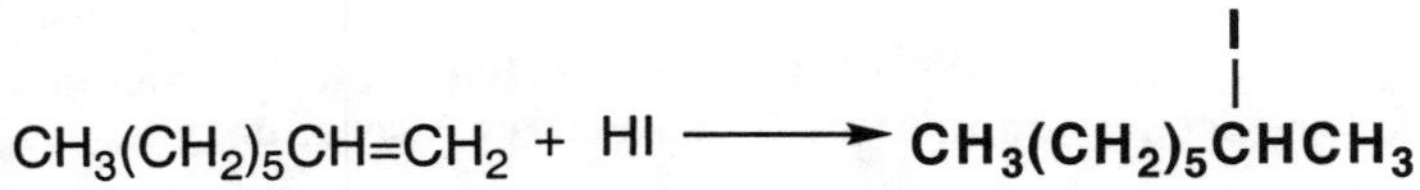 $CH_3(CH_2)_5CH=CH_2 + HI \longrightarrow CH_3(CH_2)_5\overset{\overset{I}{|}}{C}HCH_3$

(d)

(e) $CH_3CH=CHCH_2CH_3$ + H_2O $\xrightarrow{H_2SO_4}$ $CH_3\overset{\overset{OH}{|}}{C}HCH_2CH_2CH_3$ + $CH_3CH_2\overset{\overset{OH}{|}}{C}HCH_2CH_3$

Similar amounts of both products will be formed because they are both derived from a secondary carbocation.

(f) $CH_2=CHCH_2CH_2CH_3$ + H_2O $\xrightarrow{H_2SO_4}$ $CH_3\overset{\overset{OH}{|}}{C}HCH_2CH_2CH_3$

Problem 6.16 Reaction of 2-methyl-2-pentene with each reagent is regioselective. Draw a structural formula for the product of each reaction, and account for the observed regioselectivity.

(a) HI

$CH_3\overset{\overset{CH_3}{|}}{\underset{\underset{I}{|}}{C}}CH_2CH_2CH_3$

(b) H_2O in the presence of H_2SO_4

$CH_3\overset{\overset{CH_3}{|}}{\underset{\underset{OH}{|}}{C}}CH_2CH_2CH_3$

The first step in each reaction is protonation of the alkene to generate a carbocation. The observed regioselectivity arises because the reaction will occur so as to produce the more stable carbocation, which in this case is tertiary.

Problem 6.17 Addition of bromine and chlorine to cycloalkenes is stereoselective. Predict the stereochemistry of the product formed in each reaction and account for your predicted stereoselectivity.

(a) 1-Methylcyclohexene + Br_2

The product has the two bromine atoms trans to each other due to the anti addition stereoselectivity observed with these reactions.

(b) 1,2-Dimethylcyclopentene + Cl_2

The product has the two chlorine atoms trans to each other due to the anti addition stereoselectivity observed with these reactions.

Problem 6.18 Draw a structural formula for an alkene with the indicated molecular formula that gives the compound shown as the major product. Note that more than one alkene may give the same compound as the major product.

(a) C_5H_{10} + H_2O $\xrightarrow{\text{H}_2\text{SO}_4}$

$$CH_3CCH_2CH_3$$ with CH_3 and OH substituents

$$CH_3C=CHCH_3 \quad \text{or} \quad CH_2=CCH_2CH_3$$ (each with CH_3 substituent)

(b) C_5H_{10} + Br_2 $\longrightarrow$

$$CH_3CHCHCH_2$$ with CH_3 substituent and $Br\,Br$

$$CH_3CHCH=CH_2$$ with CH_3 substituent

(c) C_7H_{12} + HCl $\longrightarrow$

cyclohexane ring with CH_3 and Cl

cyclohexene ring with $-CH_3$ or   cyclohexane ring with $=CH_2$

Problem 6.19 Draw the structural formula for an alkene of molecular formula C_5H_{10} that reacts with Br_2 to give each product.

(a) $CH_3C-CHCH_3$ with CH_3 substituent and Br Br

(b) $CH_2CCH_2CH_3$ with CH_3 substituent and Br Br

(c) $CH_2CHCH_2CH_2CH_3$ with Br Br

$CH_3C=CHCH_3$ with CH_3 substituent

$CH_2=CCH_2CH_3$ with CH_3 substituent

$CH_2=CHCH_2CH_2CH_3$

Problem 6.20 Draw the structural formula for a cycloalkene of molecular formula C_6H_{10} that reacts with Cl_2 to give each compound.

(a) cyclohexane with Cl and Cl
(b) cyclopentane with Cl, Cl and $-CH_3$
(c) cyclopentane with H_3C, Cl and Cl
(d) cyclopentane with Cl and CH_2Cl

cyclohexene

cyclopentene with $-CH_3$

cyclopentene with CH_3

cyclopentane with $=CH_2$

Problem 6.21 Draw the structural formula for an alkene of molecular formula C_5H_{10} that reacts with HCl to give the indicated chloroalkane as the major product.

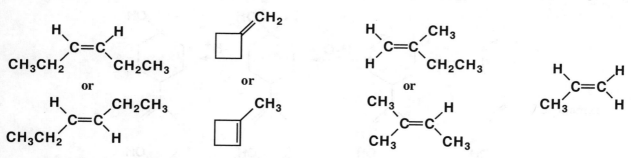

For part (a), either alkene shown gives 2-chloro-2-methylbutane as the major product as predicted by
Markovnikov's rule. For part (b), 2-methyl-2-butene would not work because the predominant product
predicted by Markovnikov's rule would be 2-chloro-2-methylbutane (part (a)), not the desired 2-chloro-3-
methylbutane. For part (c), *cis* or *trans* 2-butene would not be good choices because 3-chloropentane would be
formed in essentially the same amount as the desired product 2-chloropentane.

Problem 6.22 Draw the structural formula of an alkene that undergoes acid-catalyzed hydration to give the indicated
alcohol as the major product. More than one alkene may give each compound as the major product.

(a) 3-Hexanol   (b) 1-Methylcyclobutanol   (c) 2-Methyl-2-butanol   (d) 2-Propanol

Note that in part (a), using 2-hexene would give appreciable amounts of 2-hexanol along with the desired 3-
hexanol.

Problem 6.23 Draw the structural formula of an alkene that undergoes acid-catalyzed hydration to give each alcohol as
the major product. More than one alkene may give each compound as the major product.

(a) Cyclohexanol   (b) 1,2-Dimethylcyclopentanol   (c) 1-Methylcyclohexanol

(d) 1-Isopropyl-4-methylcyclohexanol

Problem 6.24 Terpin is prepared commercially by the acid-catalyzed hydration of limonene (Figure 5.3).

Limonene   Terpin

(a) Propose a structural formula for terpin and a mechanism for its formation.

Add water to each double bond by protonation of each double bond to give 3⁰ carbocations. Reaction of each carbocation with water and loss of the proton gives terpin. Note that the reactions do not necessarily proceed in the order shown. Both alkenes will react at similar rates because, in each case, a tertiary carbocation is formed.

(b) How many cis,trans isomers are possible for the structural formula you propose?

There are two cis-trans isomers of terpin.

(c) Terpin hydrate, the isomer of terpin in which the one-carbon and three-carbon substituents are cis to each other, is used as an expectorant in cough medicines. Draw alternative chair conformations for terpin hydrate and state which of the two is the more stable.

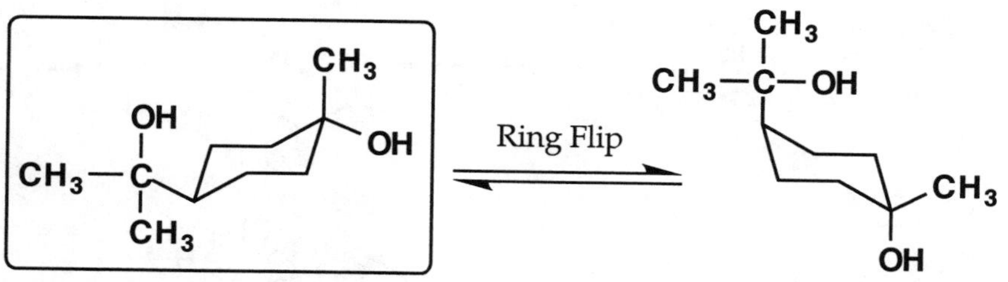

The circled conformation is more stable because the bulky three-carbon substituent (about the same size as a *tert*-butyl group) is equatorial.

Problem 6.25 Treatment of 2-methylpropene with methanol in the presence of a sulfuric acid catalyst gives *tert*-butyl methyl ether. Propose a mechanism for formation of this ether.

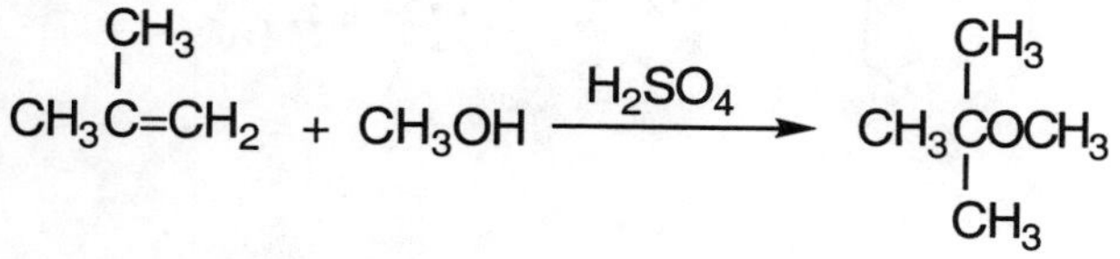

Reaction of the alkene with a proton donor in the first step gives a tertiary carbocation intermediate. Reaction of this intermediate with the oxygen atom of methanol yields a protonated ether. Transfer of the proton to a methanol molecule (an efficient process because the solvent is present in large excess) then gives the final product.

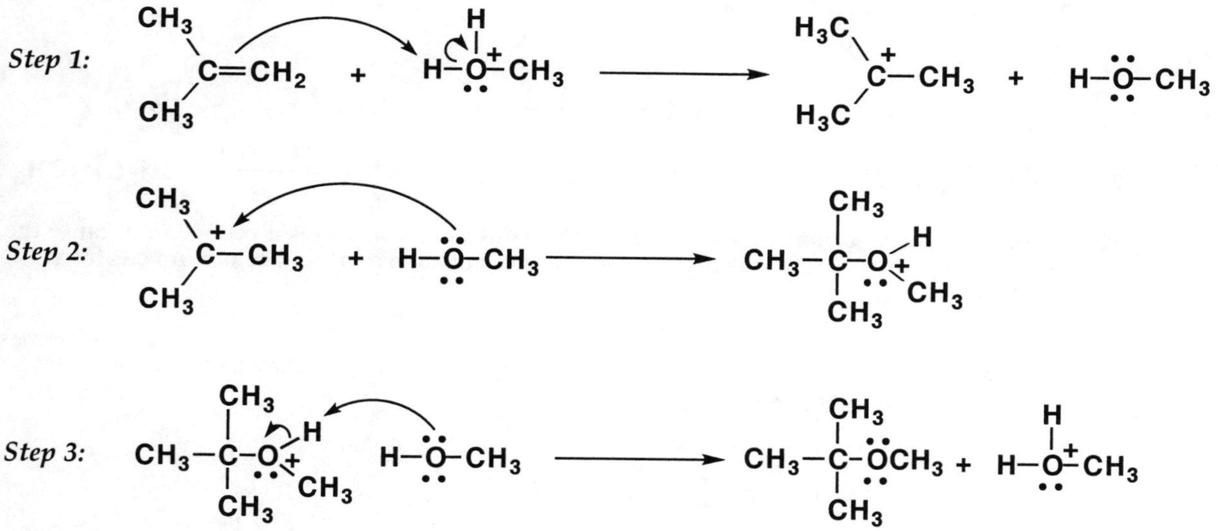

Step 1:

Step 2:

Step 3:

Problem 6.26 Treatment of 1-methylcyclohexene with methanol in the presence of a sulfuric acid catalyst gives a compound of molecular formula $C_8H_{16}O$. Propose a structural formula for this compound and a mechanism for its formation.

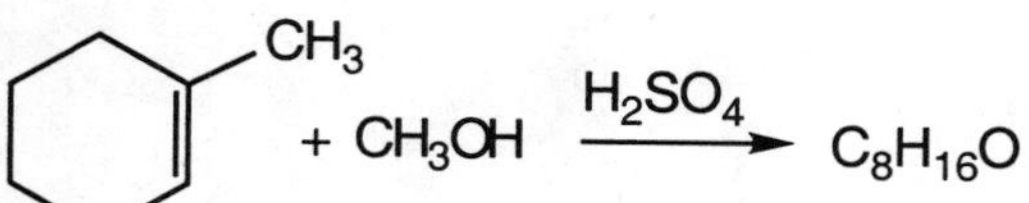

The reaction mechanism is the same as in Problem 6.25. Only the starting compound and the product are different.

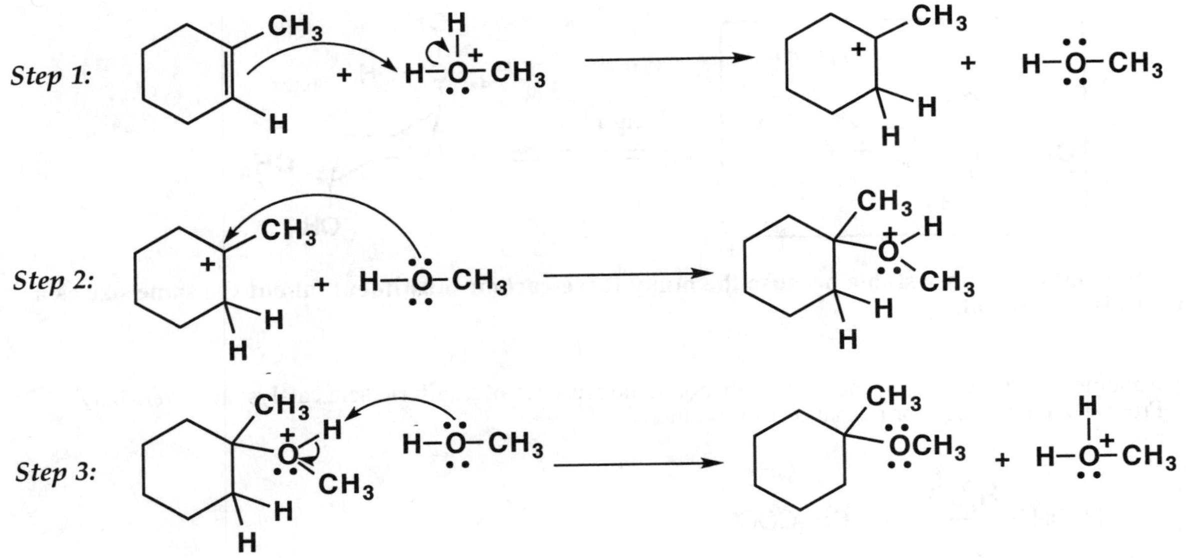

Step 1:

Step 2:

Step 3:

Oxidation-Reduction

Problem 6.27 Use balanced half-reactions to show which transformations involve oxidation, which involve reduction, and which involve neither oxidation nor reduction.

(a) $CH_3\overset{\overset{\displaystyle OH}{|}}{C}HCH_3 \longrightarrow CH_3\overset{\overset{\displaystyle O}{||}}{C}CH_3$ $CH_3\overset{\overset{\displaystyle OH}{|}}{C}HCH_3 \longrightarrow CH_3\overset{\overset{\displaystyle O}{||}}{C}CH_3 + 2H^+ + 2e^-$

(b) $CH_3\overset{\overset{\displaystyle OH}{|}}{C}HCH_3 \longrightarrow CH_3CH{=}CH_2$ $CH_3\overset{\overset{\displaystyle OH}{|}}{C}HCH_3 \longrightarrow CH_3CH{=}CH_2 + H_2O$

(c) $CH_3CH{=}CH_2 \longrightarrow CH_3CH_2CH_3$ $CH_3CH{=}CH_2 + 2H^+ + 2e^- \longrightarrow CH_3CH_2CH_3$

Part (a) is an oxidation because the electrons appear on the right-hand side. Part (c) is a reduction because the electrons appear on the left-hand side. Part (b) is neither a reduction nor an oxidation because no electrons are involved on either side, only H_2O.

Problem 6.28 Write a balanced equation for the combustion of 2-methylpropene in air to give carbon dioxide and water. The oxidizing agent is O_2, which makes up approximately 20% of air.

$$CH_3\overset{\overset{\displaystyle CH_3}{|}}{C}{=}CH_2 + 6\ O_2 \longrightarrow 4\ CO_2 + 4\ H_2O$$

Problem 6.29 Draw the product formed by treatment of each alkene with aqueous $OsO_4/ROOH$.
(a) 1-Methylcyclopentene

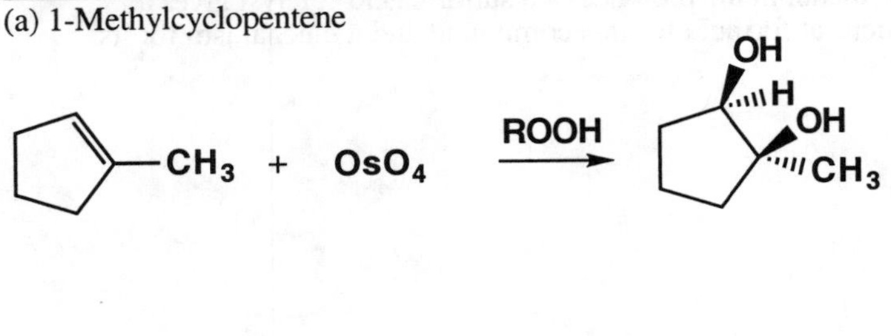

(b) Vinylcyclohexane

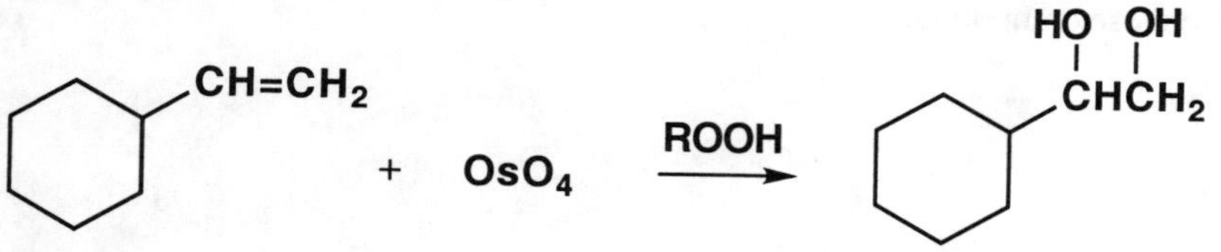

(c) *cis*-2-Pentene

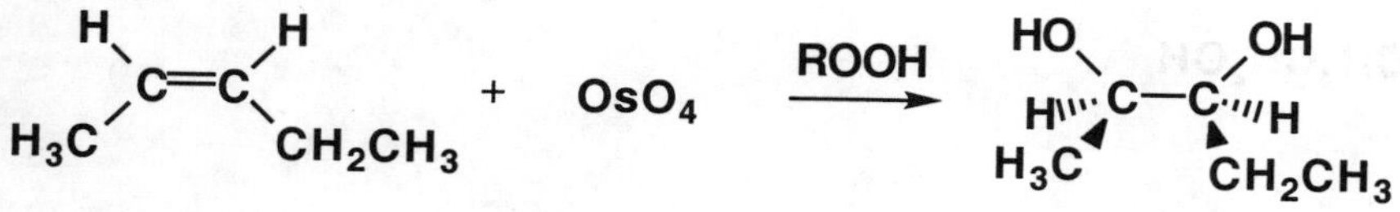

Problem 6.30 What alkene, when treated with OsO4/ROOH, gives each glycol?

(a)

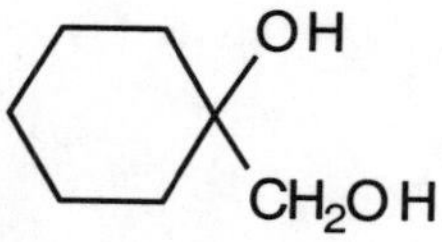

(b)

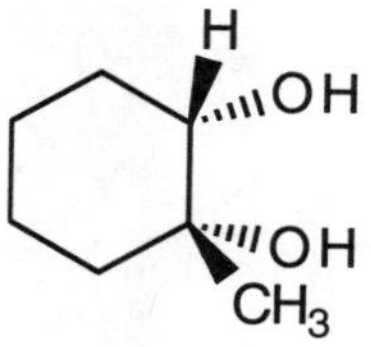

(c) 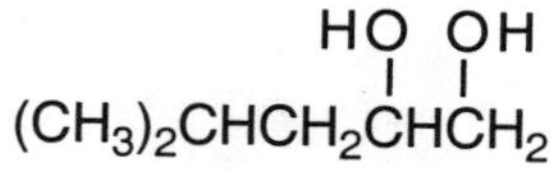 $(CH_3)_2CHCH_2CHCH_2$ with HO OH

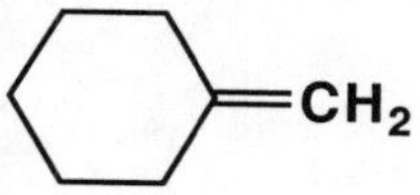

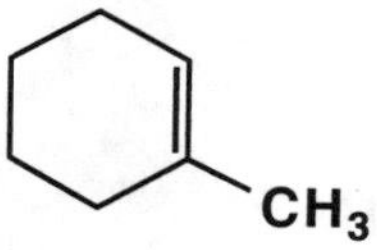

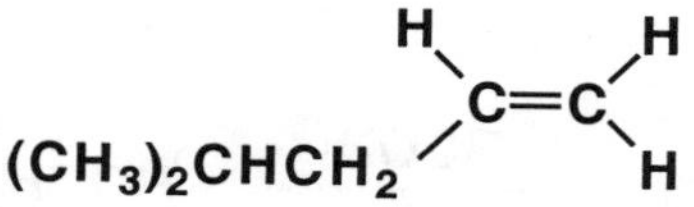

Problem 6.31 Draw the product formed by treatment of each alkene with H_2/Ni.

(a)

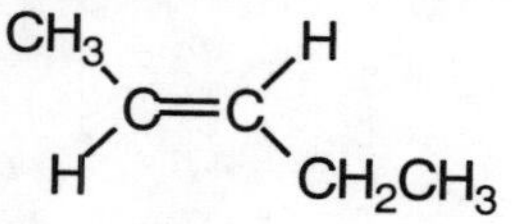

(b)

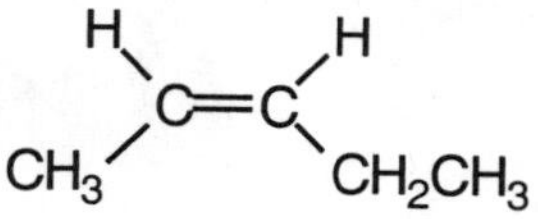

(c)

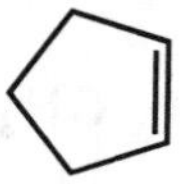

(d)

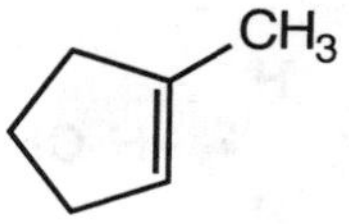

$CH_3CH_2CH_2CH_2CH_3$ $CH_3CH_2CH_2CH_2CH_3$

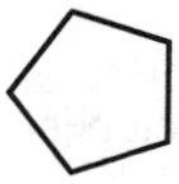

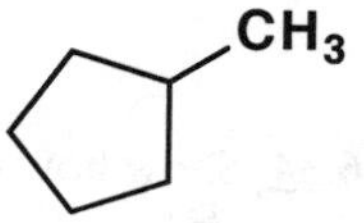

Problem 6.32 Hydrocarbon A, C_5H_8, reacts with 2 mol of Br_2 to give 1,2,3,4-tetrabromo-2-methylbutane. What is the structure of hydrocarbon A?

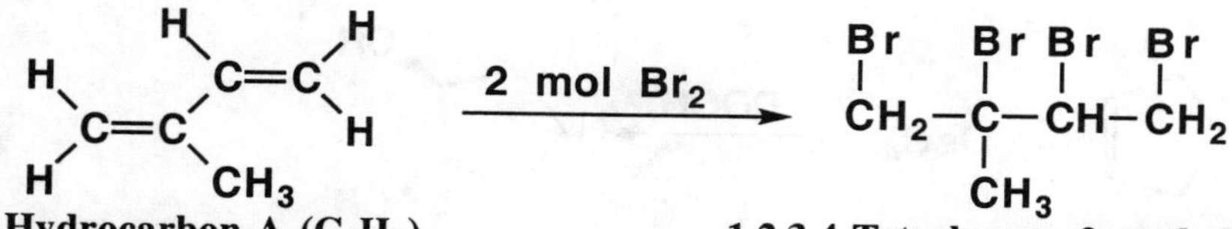

2 mol Br₂ →

Hydrocarbon A (C_5H_8)
2-Methyl-1,3-butadiene

1,2,3,4-Tetrabromo-2-methylbutane

Synthesis

Problem 6.33 Show how to convert ethylene to these compounds.

(a) Ethane

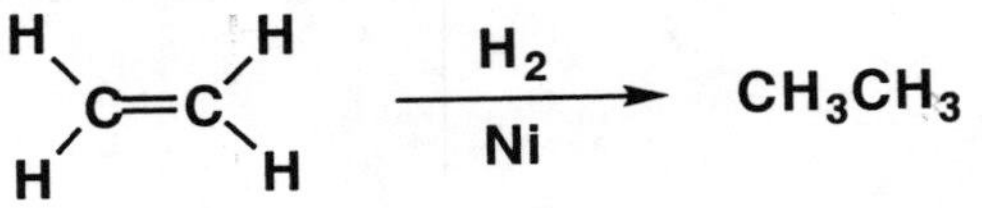
H
H
C=C
H
H
H2
Ni
CH3CH3

(b) Ethanol

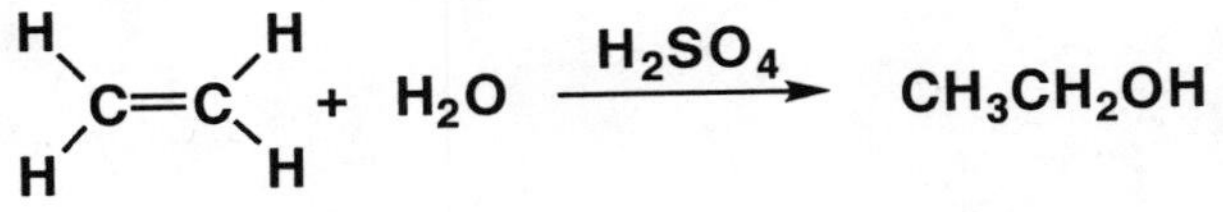
H
H
C=C
H
H
+ H2O
H2SO4
CH3CH2OH

(c) Bromoethane

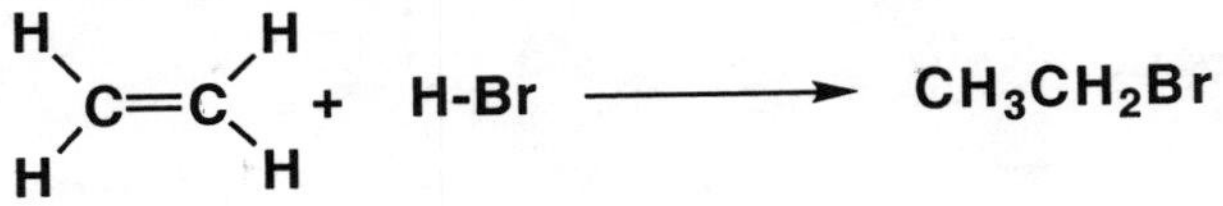
H
H
C=C
H
H
+ H-Br
CH3CH2Br

(d) 1,2-Dibromoethane

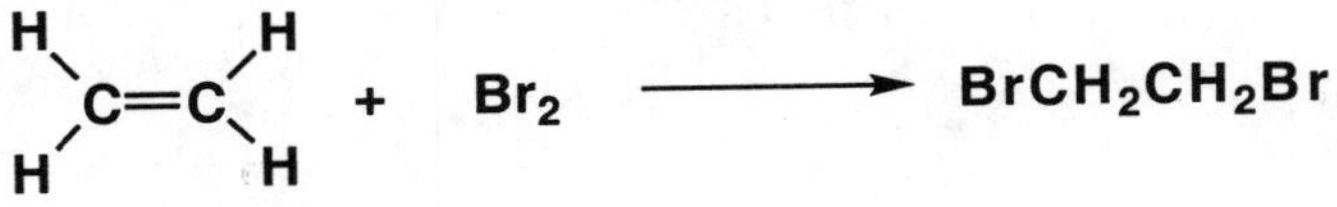
H
H
C=C
H
H
+ Br2
BrCH2CH2Br

(e) 1,2-Ethanediol

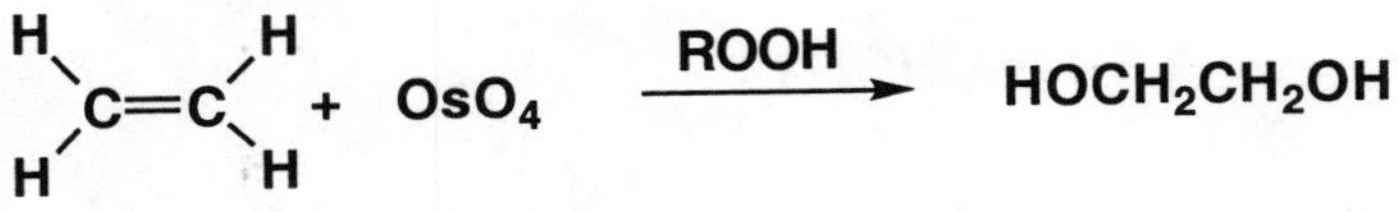
H
H
C=C
H
H
+ OsO4
ROOH
HOCH2CH2OH

(f) Chloroethane

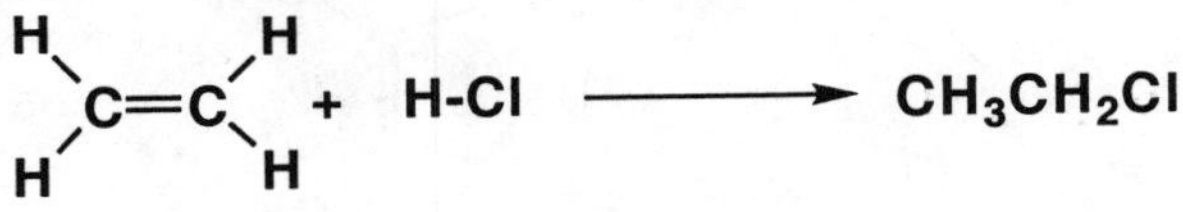
H
H
C=C
H
H
+ H-Cl
CH3CH2Cl

Problem 6.34 Show how to convert cyclopentene into these compounds.

(a) 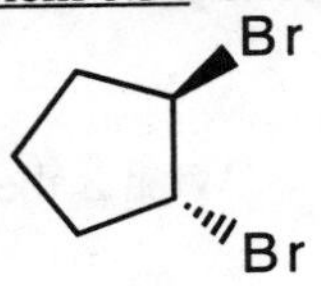
Br
Br

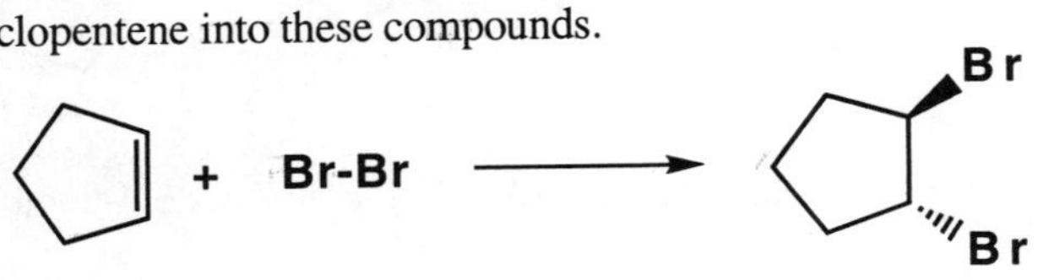
+ Br-Br
Br
Br

(b) 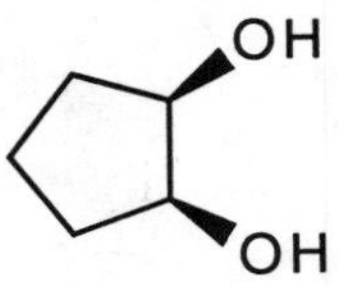
OH
OH

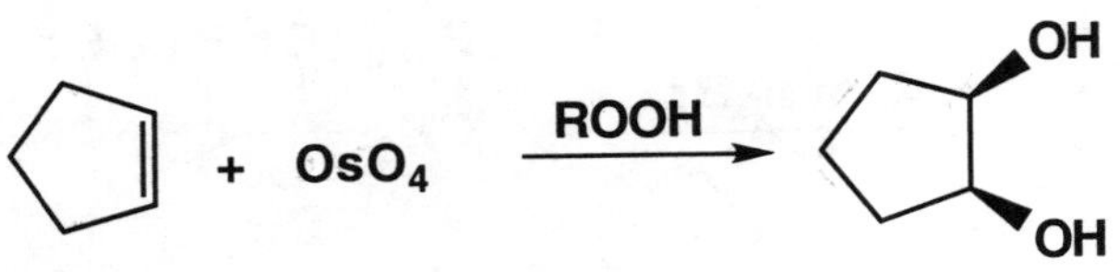
+ OsO4
ROOH
OH
OH

(c) 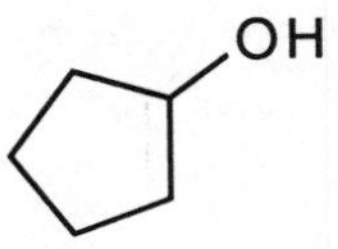
OH

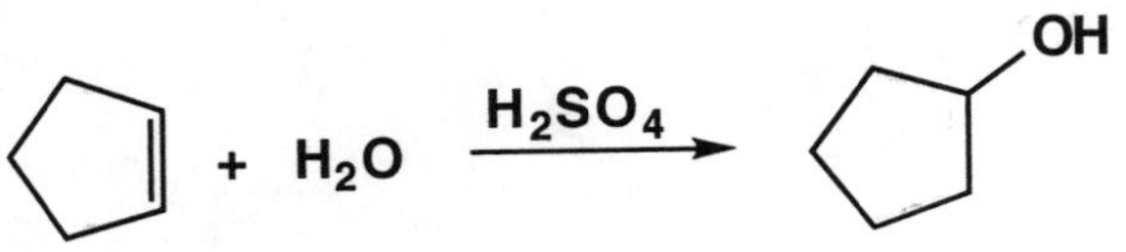
+ H2O
H2SO4
OH

(d)

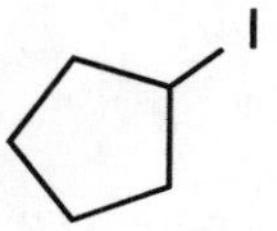

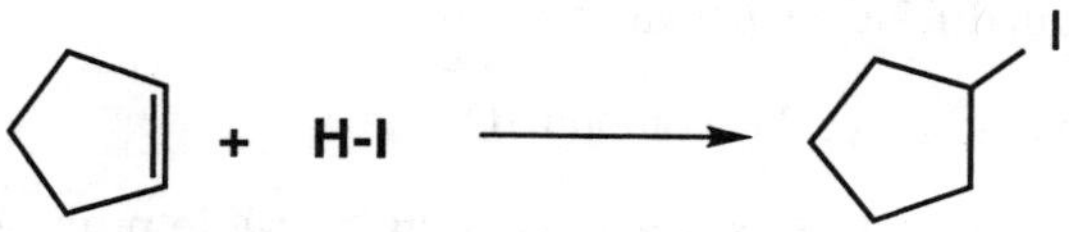

(e)

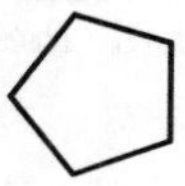

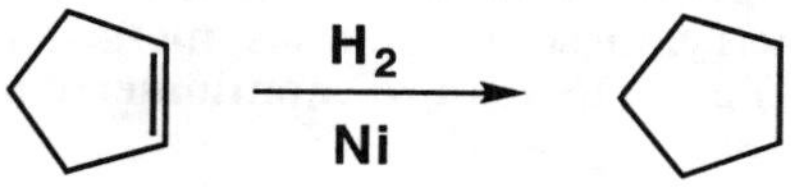

Reactions that Produce Chiral Compounds

Problem 6.35 Show that acid-catalyzed hydration of 1-butene gives a racemic mixture of (R)-2-butanol and (S)-2-butanol and that the product is, therefore, optically inactive.

The reaction mechanism involves an initial protonation to form a planar carbocation intermediate.

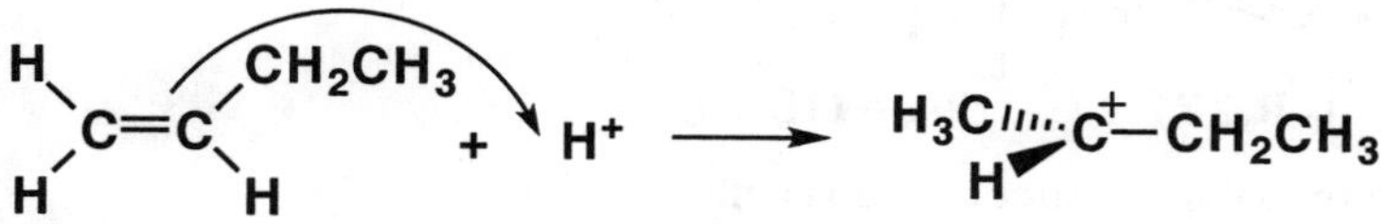

Water could approach the carbocation from either side with equal probability, so a racemic mixture will be formed. Thus, the product will be optically inactive.

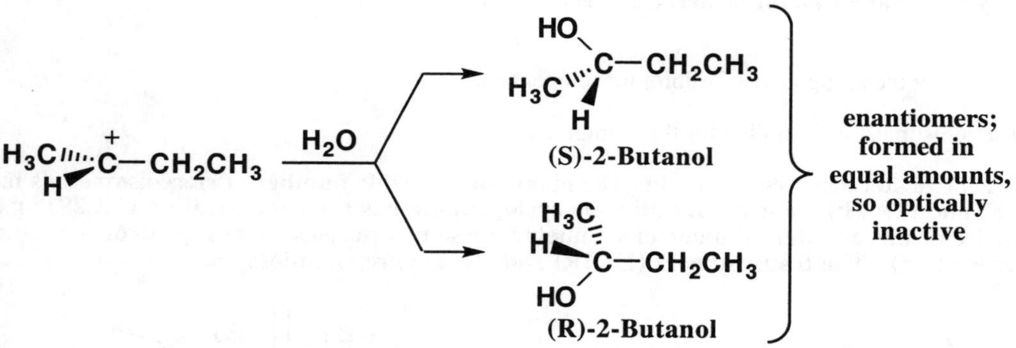

(S)-2-Butanol

(R)-2-Butanol

enantiomers;
formed in
equal amounts,
so optically
inactive

Problem 6.36 Consider the compound 1,2-cyclopentanediol.

(a) How many stereoisomers are possible for this compound?

The molecule has two stereocenters and, thus, the maximum possible number of stereoisomers is four: (1R,2S), (1S,2R), (1S,2S) and (1R,2R). cis-1,2-Cyclopentanediol has the configuration (1R,2S) or (1S,2R). Because the molecule has an internal plane of symmetry, these two stereoisomers represent the same compound (the cis isomer is meso). The trans isomers (1R,2R) and (1S,2S) are enantiomers.

(1S,2R) or (1R,2S)
(meso)

(1R,2R)   (1S,2S)

(NOT a product of the reaction)

(b) Which of the possible stereoisomers are formed when cyclopentene is treated with OsO4/ROOH?

Because osmylation results in syn addition of two hydroxyl groups across a double bond, only the achiral *cis*-1,2-dihydroxycyclopentane (the meso stereoisomer) will be obtained.

Problem 6.37 Consider the compound 1,2-dibromocyclopentane.

(a) How many stereoisomers are possible for this compound?

The molecule has two stereocenters and, thus, the maximum possible number of stereoisomers is four: (1R,2S), (1S,2R), (1S,2S) and (1R,2R). The *cis*-1,2-dibromocyclopentane has the configuration (1R,2S) or (1S,2R). Because the molecule has an internal plane of symmetry, these two stereoisomers represent the same molecule (the cis isomer is meso). The trans isomers (1R,2R) and (1S,2S) are enantiomers.

(1R,2R)   (1S,2S)

(1S,2R) or (1R,2S)
(meso)
(NOT a product of the reaction)

(b) Which of the possible stereoisomers are formed when cyclopentene is treated with bromine, Br2?

Because reaction with Br2 results in anti addition of bromine across a double bond, only the enantiomeric (1R,2R) and (1S,2S) stereoisomers (*trans*-1,2-dibromocyclopentane) will be formed. These will be formed in equal amounts because intial reaction of the alkene with Br2 can occur on either side of the double bond.

CHAPTER 7
Solutions to Problems

Problem 7.1 Write the IUPAC name for each compound.

(a) $(CH_3)_2C$=$CHCH_2Cl$

(b) [structure: cyclohexane with CH_3 and Br at position 1]

1-Chloro-3-methyl-2-butene

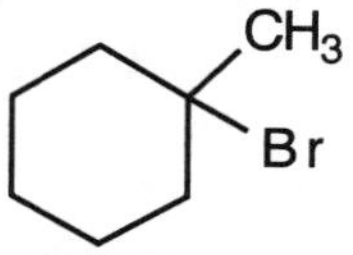

1-Bromo-1-methylcyclohexane

(c) 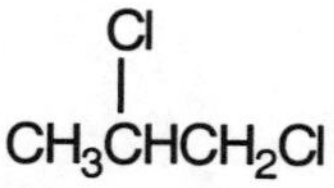 CH_3CHCH_2Cl with Cl on middle carbon

1,2-Dichloropropane

(d) 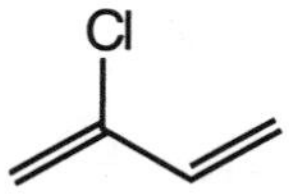 [structure: 2-chloro-1,3-butadiene]

2-Chloro-1,3-butadiene

Problem 7.2 Complete these nucleophilic substitution reactions.

(a)  [cyclopentane]—Br + $CH_3CH_2S^-$ Na^+ $\longrightarrow$ [cyclopentane]—SCH_2CH_3 + Na^+Br^-

(b)  [cyclopentane]—Br + $CH_3CH_2O^-$ Na^+ $\longrightarrow$ [cyclopentane]—OCH_2CH_3 + Na^+Br^-

Problem 7.3 Write the expected product for each nucleophilic substitution reaction and predict the mechanism by which it is formed.

(a) 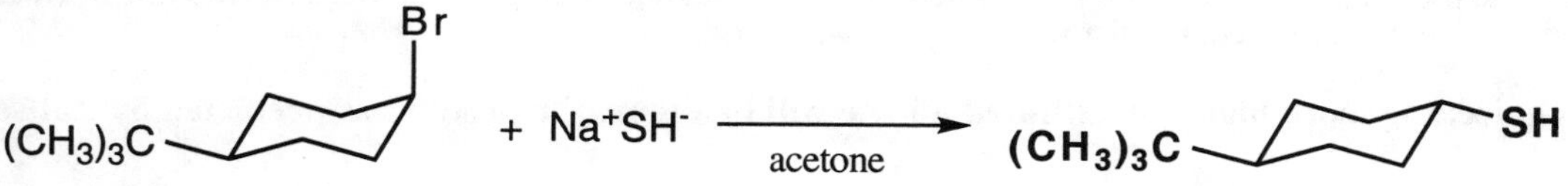 [structure: $(CH_3)_3C$—cyclohexane—Br] + Na^+SH^- $\xrightarrow{\text{acetone}}$ $(CH_3)_3C$—[cyclohexane]—SH

The SH⁻ is a good nucleophile and, because the reaction involves a secondary alkyl halide with a good leaving group, the reaction mechanism is S$_N$2. The mechanism is as follows:

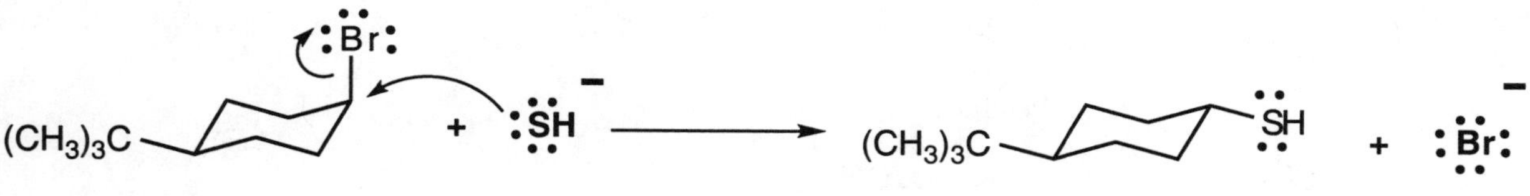

[mechanism structure] $(CH_3)_3C$—[cyclohexane with :Br:] + :$\overset{-}{SH}$ $\longrightarrow$ $(CH_3)_3C$—[cyclohexane]—$\overset{..}{SH}$ + :$\overset{-}{Br}$:

(b) 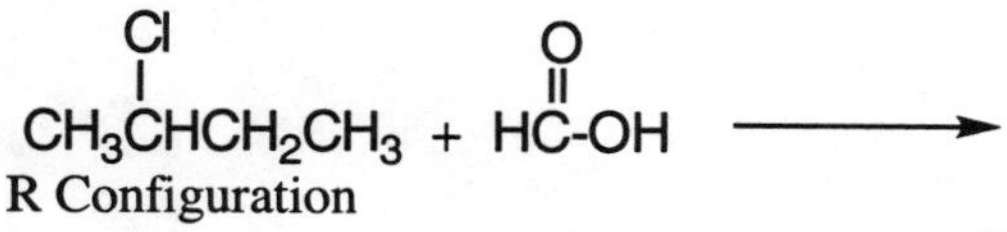 $CH_3CHCH_2CH_3$ with Cl + $HC\overset{O}{-}OH$ $\longrightarrow$
     R Configuration

[product structures]

$\overset{O}{\underset{OCH}{\parallel}}$ CH_3CH_2—$\overset{}{\underset{H}{C}}$'''$CH_3$ + CH_3CH_2—$\overset{CH_3}{\underset{H}{C}}$''' with HCO / $\overset{O}{\parallel}$

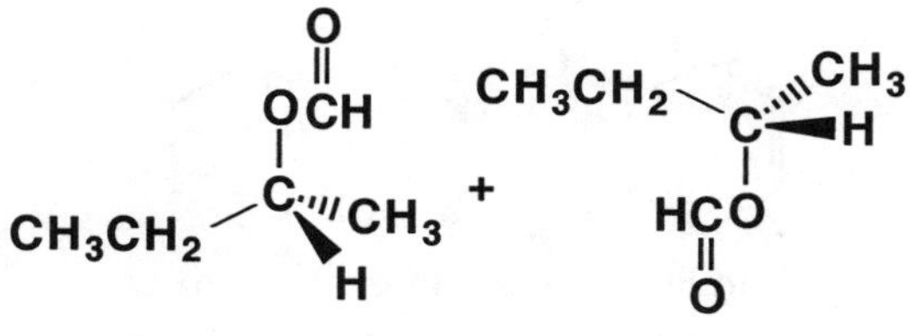

R enantiomer   **S enantiomer**

The alkyl halide is secondary and chloride is a good leaving group. Formic acid is an excellent ionizing solvent and a poor nucleophile. Therefore, substitution takes place by an S_N1 mechanism and leads to racemization.

Step 1:

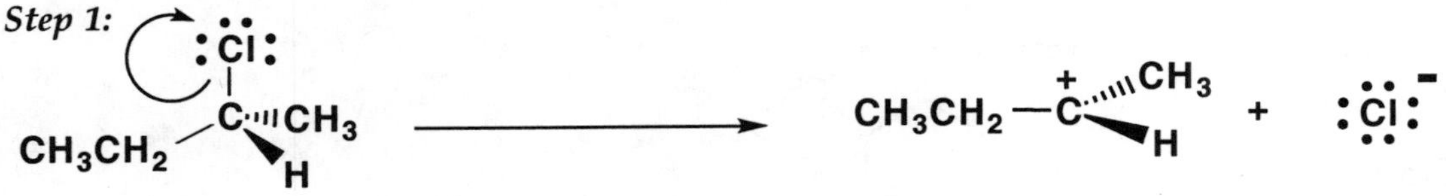

Step 2:

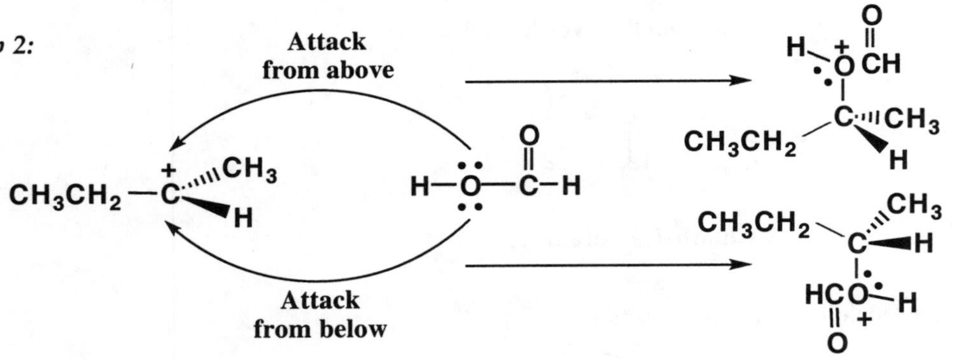

Step 3:

Deprotonation yields the two enantiomers.

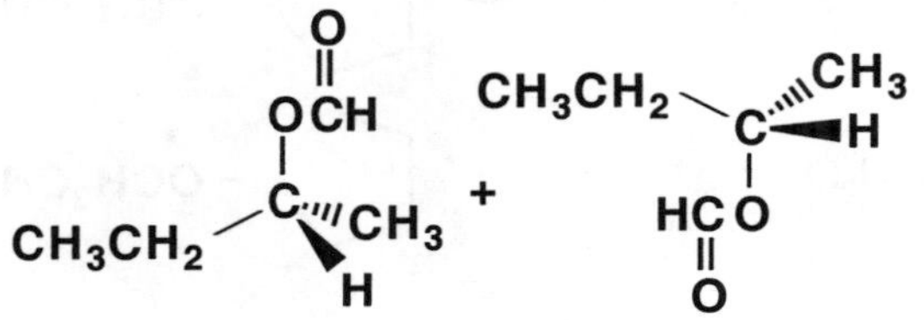

Problem 7.4 Predict the β-elimination product formed when each chloroalkane is treated with sodium ethoxide in ethanol. If two or more products might be formed, predict which is the major product.

When there is a choice, the more highly substituted alkene will be the major product as predicted by Zaitsev's rule.

(a)

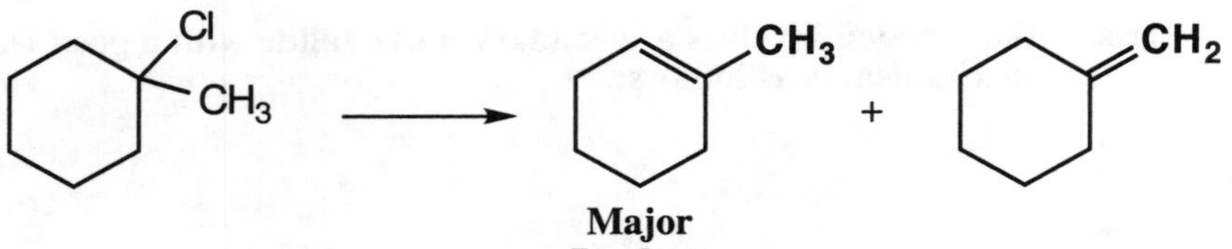

Major Product

(b)

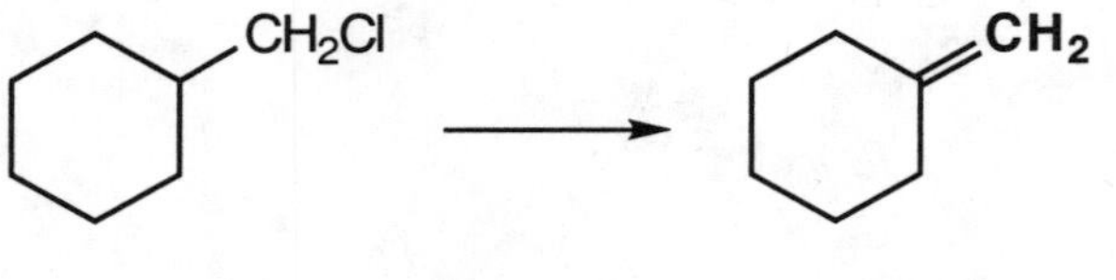

(c)

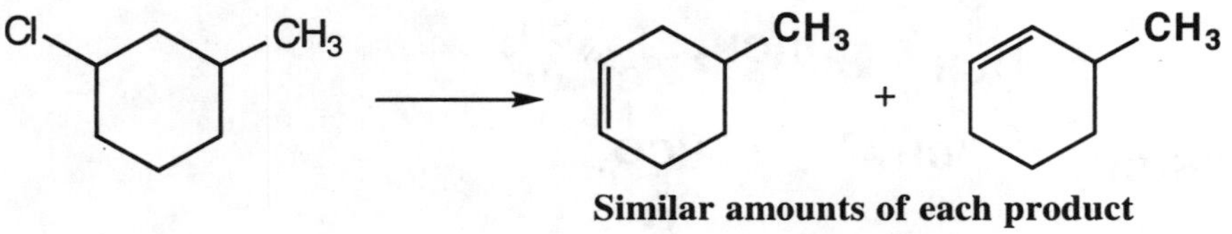

Similar amounts of each product

Problem 7.5 Predict whether each elimination reaction proceeds predominantly by an E1 or E2 mechanism. Write a structural formula for the major organic product.

(a) $CH_3CH_2CHCH_2CH_3$ + $CH_3O^-Na^+$ $\xrightarrow[CH_3OH]{}$ $CH_3CH=CHCH_2CH_3$

(E2)

(b)

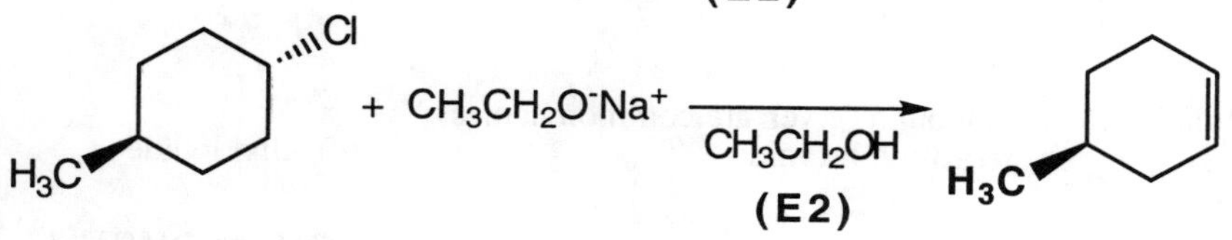

$+ \ CH_3CH_2O^-Na^+ \xrightarrow[CH_3CH_2OH]{}$

(E2)

In each reaction the alkyl halide is secondary and there is a strong base present. Thus, even though both reactions take place in solvents that are moderate to good in their ability to stabilize carbocations, both will occur by an E2 process.

Nomenclature

Problem 7.6 Write the IUPAC name for each compound.

(a) $CH_2{=}CF_2$

1,1-Difluoroethene

(b) ⬡—Br

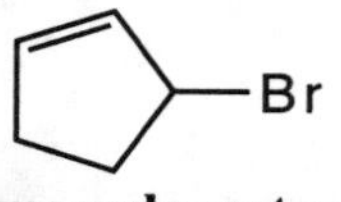

3-Bromocyclopentene

(c) $(CH_3)_2CHCH_2CH_2CHCH_3$ (with Cl)

2-Chloro-5-methylhexane

(d) $Cl(CH_2)_6Cl$

1,6-Dichlorohexane

(e) CF_2Cl_2

Dichlorodifluoromethane

(f) $CH_3CH_2CCH_2CH_3$ (with Br and CH_2CH_3)

3-Bromo-3-ethylpentane

Problem 7.7 Write the IUPAC name for each compound. Be certain to include a designation of configuration in your answer.

(a)

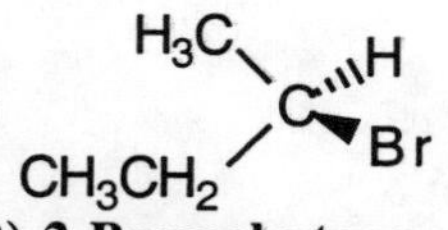

(R)-2-Bromobutane

(b)

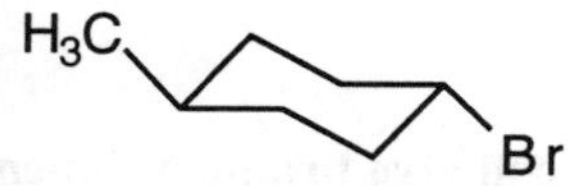

trans-1-Bromo-4-methyl-cyclohexane

(c)

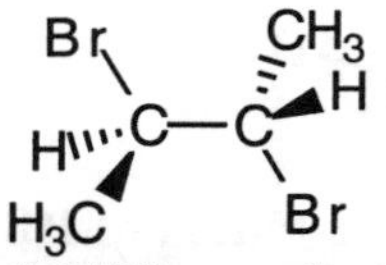

meso-2,3-Dibromobutane

(d)

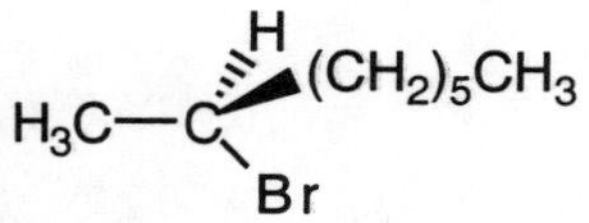

(S)-2-Bromooctane

(e)

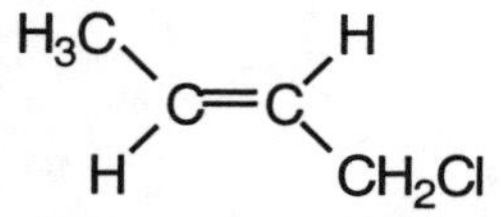

(E)-1-Chloro-2-butene

(f)

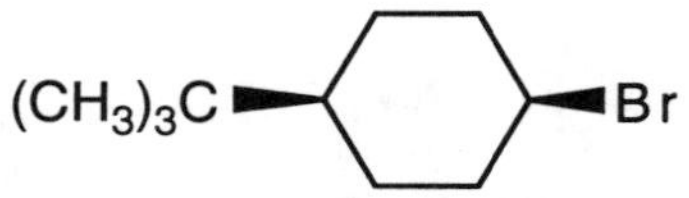

cis-1-Bromo-4-tert-butylcyclohexane

Problem 7.8 Draw a structural formula for each compound (given are IUPAC names).

(a) 3-Bromopropene

(b) (R)-2-Chlorobutane

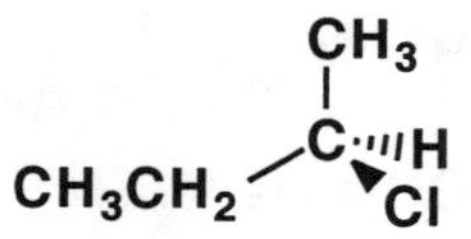

(c) meso-2,3-Dibromohexane

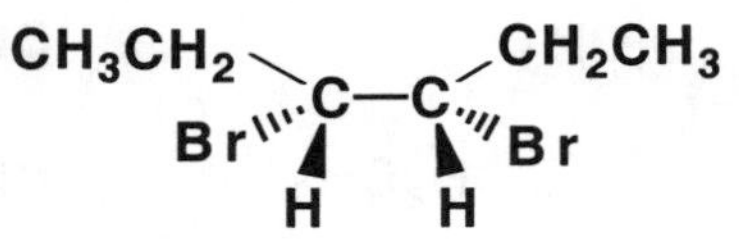

$CH_2{=}CHCH_2Br$

(d) *trans*-1-Bromo-3-isopropylcyclohexane   (e) 1,2-Dichloroethane   (f) Bromocyclobutane

$$CH_3CH_2 \quad (structures)$$

Cl   Cl
| |
H_2C—CH_2

Problem 7.9 Draw a structural formula for each compound (given are common names).

(a) Isopropyl chloride

Cl
|
CH_3CHCH_3

(b) *sec*-Butyl bromide

Br
|
$CH_3CH_2CHCH_3$

(c) Allyl iodide

CH_2═$CHCH_2I$

(d) Methylene chloride

CH_2Cl_2

(e) Chloroform

$CHCl_3$

(f) *tert*-Butyl chloride

$(CH_3)_3CCl$

(g) Isobutyl chloride

$(CH_3)_2CHCH_2Cl$

Problem 7.10 Which compounds are secondary (2°) alkyl halides?

(a) Isobutyl chloride

$(CH_3)_2CHCH_2Cl$

(b) 2-Iodooctane

I
|
$CH_3(CH_2)_5CHCH_3$

(c) *trans*-1-Chloro-4-methylcyclohexane

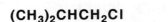

As can be seen in these structures, (b) and (c) are secondary alkyl halides, while (a) is a primary alkyl halide.

Synthesis of Alkyl Halides

Problem 7.11 What alkene or alkenes and reaction conditions give each alkyl halide in good yield?

(a)

Electrophilic addition of HBr to cyclopentene will give bromocyclopentane.

$$\text{(cyclopentene)} \quad + \quad HBr \quad \longrightarrow \quad \text{(bromocyclopentane)}$$

(b)

$$CH_3 \\ | \\ CH_3CCH_2CH_2CH_3 \\ | \\ Br$$

Electrophilic addition of HBr to either 2-methyl-1-pentene or 2-methyl-2-pentene will give 2-bromo-2-methylpentane in accord with Markovnikov's rule.

$$CH_2{=}C \overset{CH_3}{\underset{CH_2CH_2CH_3}{}} \quad \text{or} \quad \overset{CH_3}{\underset{CH_3}{}}C{=}C\overset{H}{\underset{CH_2CH_3}{}} \quad + \quad HBr \quad \longrightarrow \quad \overset{CH_3}{\underset{Br}{CH_3CCH_2CH_2CH_3}}$$

(c)

Electrophilic addition of HCl to either of the two alkenes shown here will give 1-chloro-1-methylcyclohexane in accord with Markovnikov's rule.

Problem 7.12 Show reagents and conditions to bring about these conversions.

(a)

(b) $CH_3CH_2CH{=}CH_2$ + HI $\longrightarrow$ $CH_3CH_2CHCH_3$ (with Cl substituent)

(c) $CH_3CH{=}CHCH_3$ + HCl $\longrightarrow$ $CH_3CHCH_2CH_3$ (with Cl substituent)

(d)   + HBr $\longrightarrow$

Nucleophilic Aliphatic Substitution
Problem 7.13 Write structural formulas for these common organic solvents.

(a) Methylene chloride

CH_2Cl_2

(b) Acetone

$CH_3\overset{O}{\overset{\|}{C}}CH_3$

(c) Ethanol

CH_3CH_2OH

(d) Diethyl Ether

$CH_3CH_2OCH_2CH_3$

(e) Dimethyl sulfoxide

$CH_3\overset{O}{\overset{\|}{S}}CH_3$

Problem 7.14 Arrange these protic solvents in order of increasing polarity.
(a) H_2O (b) CH_3CH_2OH (c) CH_3OH

Alkyl groups decrease the polarity of a solvent. In order of increasing polarity, they are:

$$CH_3CH_2OH < CH_3OH < H_2O$$

Problem 7.15 Arrange these aprotic solvents in order of increasing polarity.
(a) Acetone   (b) Pentane   (c) Diethyl ether

The carbonyl group of acetone is a polar functional group, so acetone is the most polar of the three. The oxygen atom of diethyl ether adds polarity to this solvent compared to the hydrocarbon pentane. In order of increasing polarity, they are:

Pentane < Diethyl ether < Acetone

Problem 7.16 From each pair, select the stronger nucleophile.

(a) H_2O or $\boxed{OH^-}$ (b) $CH_3CO_2^-$ or $\boxed{OH^-}$ (c) CH_3SH or $\boxed{CH_3S^-}$

The stronger nucleophile in each pair is circled.

Problem 7.17 Which statements are true for S_N2 reactions of alkyl halides?
(a) Both the alkyl halide and the nucleophile are involved in the transition state. **True**
(b) Reaction proceeds with inversion of configuration at the substitution center. **True**
(c) Reaction proceeds with retention of optical activity. **True**
(d) The order of reactivity is 3° > 2° > 1° > methyl. **False**
(e) The nucleophile must have an unshared pair of electrons and bear a negative charge. **False**
(f) The greater the nucleophilicity of the nucleophile, the greater the rate of the reaction. **True**

Statement (d) is false because 3° alkyl halides are the least reactive and methyl halides are the most reactive.
Statement (e) is false because a nucleophile does not have to possess a negative charge.

Problem 7.18 Complete these S_N2 reactions.

(a) Na^+I^- + $CH_3CH_2CH_2Cl$ $\xrightarrow{\text{acetone}}$ $CH_3CH_2CH_2I$ + Na^+Cl^-

(b) NH_3 + (cyclohexyl)—Br $\xrightarrow{\text{ethanol}}$ (cyclohexyl)—NH_3^+ Br^-

(c) $CH_3CH_2O^-$ Na^+ + $CH_2{=}CHCH_2Cl$ $\xrightarrow{\text{ethanol}}$ $CH_2{=}CHCH_2OCH_2CH_3$ + Na^+Cl^-

Problem 7.19 Complete these S_N2 reactions.

(a) (cyclohexyl-Cl) + $CH_3C(=O)O^-$ Na^+ $\xrightarrow{\text{ethanol}}$ (cyclohexyl ester) + Na^+Cl^-

(b) $CH_3CHICH_2CH_3$ + $CH_3CH_2S^-$ Na^+ $\xrightarrow{\text{acetone}}$ $CH_3CH(SCH_2CH_3)CH_2CH_3$ + Na^+I^-

(c) $CH_3CH(CH_3)CH_2CH_2Br$ + Na^+I^- $\xrightarrow{\text{acetone}}$ $CH_3CH(CH_3)CH_2CH_2I$ + Na^+Br^-

(d) $(CH_3)_3N$ + CH_3I $\xrightarrow{\text{acetone}}$ $(CH_3)_4N^+I^-$

(e) (phenyl)—CH_2Br + CH_3O^- Na^+ $\xrightarrow{\text{methanol}}$ (phenyl)—CH_2OCH_3 + Na^+Br^-

(f) CH_3(cyclohexyl)—Cl + CH_3S^- Na^+ $\xrightarrow{\text{ethanol}}$ CH_3(cyclohexyl)—SCH_3 + Na^+Cl^-
Notice the inversion of configuration due to the S_N2 mechanism

(g)   [structure: piperidine] $NH + CH_3(CH_2)_6CH_2Cl \xrightarrow{\text{ethanol}}$ [structure: product] $\overset{+}{N}H(CH_2)_7CH_3$ Cl^-

(h)   [cyclopentane]$-CH_2Cl + NH_3 \xrightarrow{\text{ethanol}}$ [cyclopentane]$-CH_2\overset{+}{N}H_3$ Cl^-

Problem 7.20 You were told that each reaction in Problem 7.19 proceeds by an S_N2 mechanism. Suppose you were not told the mechanism. Describe how you could conclude from the structure of the alkyl halide, the nucleophile, and the solvent that each reaction is in fact an S_N2 reaction.

Keep in mind that CH_3^+ and primary carbocations are so unstable that they are never observed in solution. Thus, methyl and primary alkyl halides never undergo S_N1 reactions. Because of extreme crowding around their reaction centers, tertiary alkyl halides never undergo S_N2 reactions.

(a) A secondary halide, a moderate nucleophile and a moderately ionizing solvent all favor S_N2.
(b) The alkyl iodide is secondary, so both S_N1 and S_N2 are possible, but the solvent is only weakly ionizing, which favors S_N2. Ethyl sulfide is a good nucleophile but a weak base, which further favors an S_N2 pathway.
(c) A primary alkyl halide, a good nucleophile that is a weak base, and a weakly ionizing solvent all strongly favor S_N2.
(d) Trimethylamine is a moderate nucleophile. But a methyl halide in acetone, which is a weakly ionizing solvent, work together to favor S_N2.
(e) A primary halide with a strong nucleophile favors S_N2.
(f) Although the halide is secondary and ethanol is moderately ionizing, S_N2 is favored because methyl thiolate is a good nucleophile.
(g) Piperidine is a moderate nucleophile and ethanol is a moderately ionizing solvent. S_N2 is particularly favored in this case because the halide is primary.
(h) Ammonia is a moderate nucleophile and ethanol is a moderately ionizing solvent. S_N2 is favored in this case because the halide is primary.

Problem 7.21 In these reactions, an alkyl halide is treated with a compound that has two nucleophilic sites. Select the more nucleophilic site in each part and show the product of each S_N2 reaction.

(a) $HOCH_2CH_2NH_2 + CH_3I \xrightarrow{\text{ethanol}}$ $HOCH_2CH_2\overset{+}{N}H_2$ $\overset{CH_3}{\underset{}{|}}$ I^-

(b)   [morpholine structure with O and N-H] $+ CH_3I \xrightarrow{\text{ethanol}}$ [morpholine product with $\overset{+}{N}$ bonded to CH_3 and H] I^-

(c) $HOCH_2CH_2SH + CH_3I \xrightarrow{\text{ethanol}}$ $HOCH_2CH_2SCH_3 + HI$

Problem 7.22 Which statements are true for S_N1 reactions of alkyl halides?
(a) Both the alkyl halide and the nucleophile are involved in the transition state of the rate limiting step. **False**
(b) Reaction at a stereocenter proceeds with retention of configuration. **False**
(c) Reaction at a stereocenter proceeds with loss of optical activity. **True**
(d) The order of reactivity is $3° > 2° > 1° >$ methyl. **True**
(e) The greater the steric crowding around the reactive center, the slower the rate of reaction. **False**
(f) Rate of reaction is greater with good nucleophiles compared with poor nucleophiles. **False**

Statements (a), (e), and (f) are false because only the alkyl halide is involved with the rate-limiting step. Statement (b) is false because optical activity is lost due to the achiral trigonal planar carbocation intermediate that is formed.

<u>Problem 7.23</u> Draw a structural formula for the product of each S_N1 reaction.

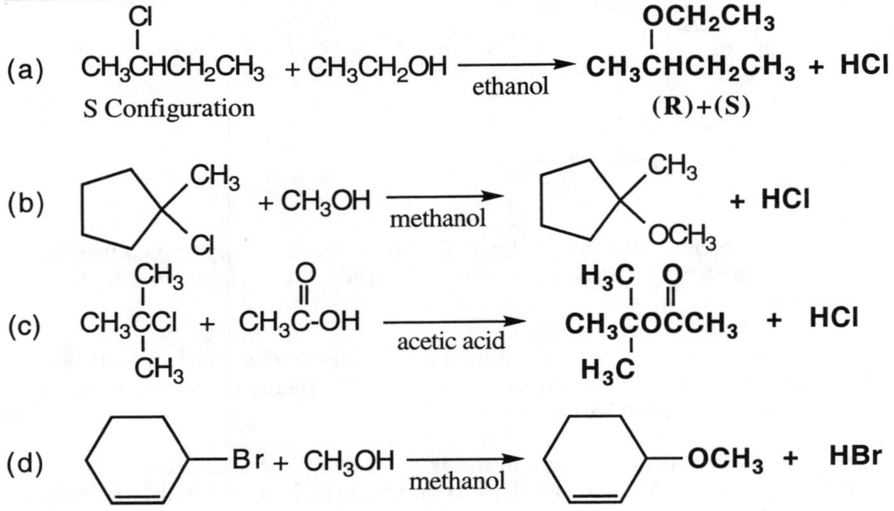

(a)

$$\underset{\text{S Configuration}}{\overset{\overset{\displaystyle Cl}{|}}{CH_3CHCH_2CH_3}} + CH_3CH_2OH \xrightarrow{\text{ethanol}} \underset{\text{(R)+(S)}}{\overset{\overset{\displaystyle OCH_2CH_3}{|}}{CH_3CHCH_2CH_3}} + HCl$$

(b)

(c)

$$\underset{\overset{\displaystyle |}{CH_3}}{\overset{\overset{\displaystyle CH_3}{|}}{CH_3CCl}} + CH_3\overset{\overset{\displaystyle O}{||}}{C}\text{-OH} \xrightarrow{\text{acetic acid}} \underset{\overset{\displaystyle |}{H_3C}}{\overset{\overset{\displaystyle H_3C \; O}{| \; \; ||}}{CH_3COCCH_3}} + HCl$$

(d)

<u>Problem 7.24</u> You were told that each reaction in Problem 7.23 proceeds by an S_N1 mechanism. Suppose that you were not told the mechanism. Describe how you could conclude from the structure of the alkyl halide, the nucleophile, and the solvent that each reaction is in fact an S_N1 reaction.

(a) Chloride is a good leaving group and the resulting secondary carbocation is a relatively stable carbocation intermediate. The key is the nucleophile/solvent: ethanol is a moderately ionizing solvent and a poor nucleophile.
(b) Methanol is a good ionizing solvent and a poor nucleophile. Chloride is a good leaving group and the resulting carbocation is tertiary. Remember that tertiary alkyl halides are unable to undergo S_N2 reactions.
(c) Acetic acid is a strongly ionizing solvent and a poor nucleophile. Chloride is a good leaving group and the resulting carbocation is tertiary.
(d) Methanol is a good ionizing solvent and a poor nucleophile. Bromide is a good leaving group, and the resulting carbocation is secondary.

<u>Problem 7.25</u> Select the member of each pair that undergoes nucleophilic substitution in aqueous ethanol more rapidly.

Aqueous ethanol is an ionizing solvent and there are no strong nucleophiles present in the solution, so the following reactions will proceed by an S_N1 mechanism. Thus, the alkyl halide of each pair that can form the more stable carbocation (i.e., the carbocation with more alkyl groups attached) will react more rapidly. The molecule which will react faster in each pair is circled.

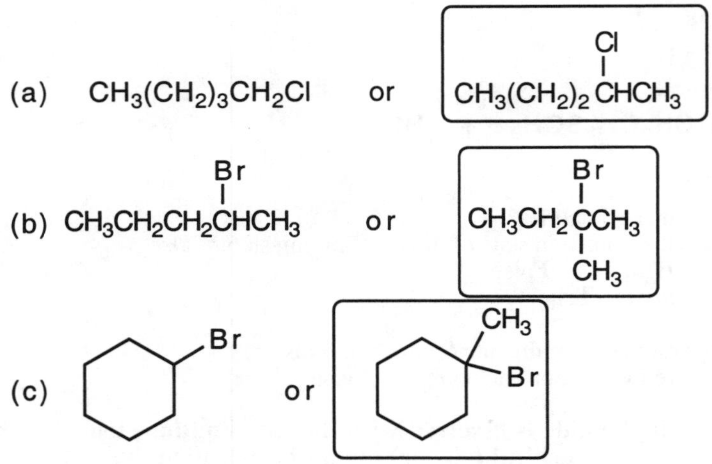

(a) $CH_3(CH_2)_3CH_2Cl$ or $CH_3(CH_2)_2\overset{\overset{\displaystyle Cl}{|}}{CH}CH_3$

(b) $CH_3CH_2CH_2\overset{\overset{\displaystyle Br}{|}}{CH}CH_3$ or $CH_3CH_2\underset{\overset{\displaystyle |}{CH_3}}{\overset{\overset{\displaystyle Br}{|}}{C}}CH_3$

(c)

<u>Problem 7.26</u> Propose a mechanism for the formation of the products (but not their relative percentages) in this reaction.

$$CH_3CCl(CH_3)(CH_3) \xrightarrow[\substack{25\ °C}]{\substack{20\%\ H_2O \\ 80\%\ CH_3CH_2OH}} CH_3COCH_2CH_3(CH_3)(CH_3) + CH_3COH(CH_3)(CH_3) + CH_3C{=}CH_2(CH_3) + HCl$$

$$85\% \qquad\qquad 15\%$$

All of the reaction products shown can be produced from the same intermediate, a carbocation that arises from ionization of the carbon-halogen bond. Notice that the reaction is run in an ionizing solvent, there are no good nucleophiles present, and the tertiary alkyl halide starting material gives a very stable (for a carbocation!) tertiary carbocation intermediate. All of these factors work in concert to favor S_N1 and E1 reaction mechanisms.

<u>Problem 7.27</u> The rate of reaction in Problem 7.26 increases by 140 times when carried out in 80% water: 20% ethanol compared with 40% water: 60% ethanol. Account for this rate difference.

The reaction mechanism involves formation of a carbocation intermediate and water is a more ionizing solvent than ethanol. In other words, the carbocation forms more easily in water. Thus, the higher the percentage of water in the solvent, the faster the rate-limiting formation of the carbocation and the faster the rate of the reaction.

<u>Problem 7.28</u> Show how you might synthesize each compound from an alkyl halide and a nucleophile:

(a) cyclohexyl—NH_2

Treat bromocyclohexane with two moles of ammonia. The first mole of ammonia is for displacement of bromide. The second mole of ammonia is to neutralize the HBr formed in the substitution reaction.

cyclohexyl—Br + 2NH_3 $\longrightarrow$ cyclohexyl—NH_2 + $NH_4{}^+Br^-$

(b)

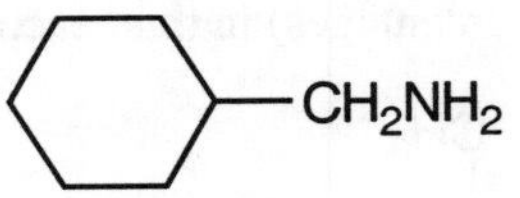

Treat chloromethylcyclohexane with two moles of ammonia. The first mole of ammonia is for displacement of chloride. The second mole of ammonia is to neutralize the HCl formed in the substitution reaction.

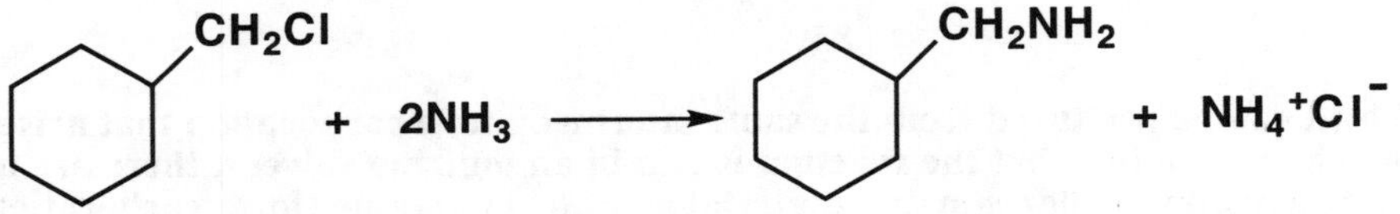

(c)

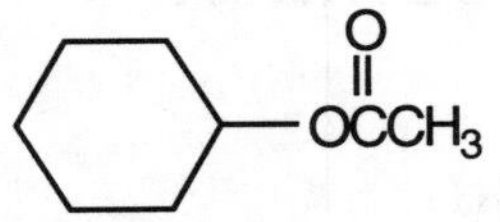

Treatment of a halocyclohexane with the sodium salt of acetic acid.

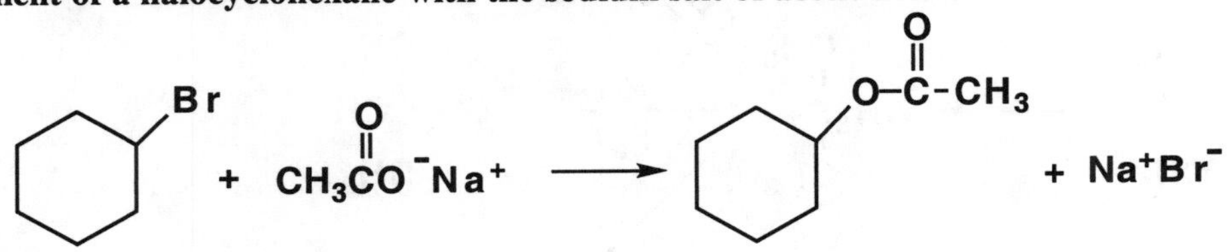

(d) $CH_3(CH_2)_3CH_2SH$

Treatment of 1-halopentane with sodium hydrosulfide.

$$CH_3(CH_2)_3CH_2Br + HS^- Na^+ \longrightarrow CH_3(CH_2)_3CH_2SH + Na^+Br^-$$

(e)

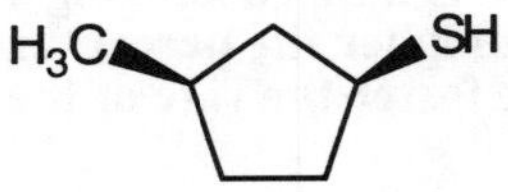

Treatment of the appropriate halocyclopentane with sodium hydrosulfide. Note how the stereocenter is inverted upon reaction with the nucleophile.

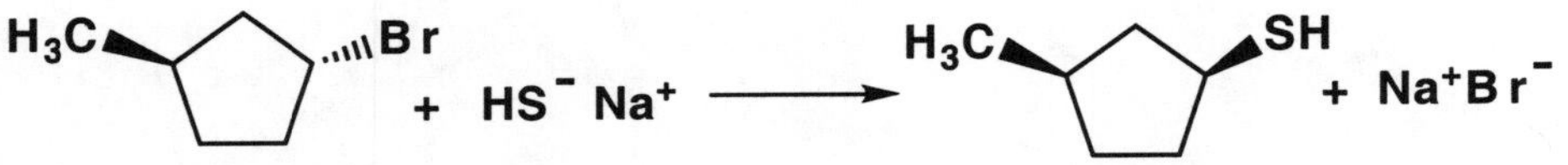

(f) $CH_3CH_2OCH_2CH_3$

Treatment of a haloethane with sodium or potassium ethoxide in ethanol.

$$CH_3CH_2O^- Na^+ + CH_3CH_2I \xrightarrow{CH_3CH_2OH} CH_3CH_2OCH_2CH_3 + Na^+I^-$$

β-Eliminations

Problem 7.29 Draw structural formulas for the alkene(s) formed by treatment of each alkyl halide with sodium ethoxide in ethanol. Assume that elimination is by an E2 mechanism. Where two alkenes are possible, use Zaitsev's rule to predict which alkene is the major product.

(a)

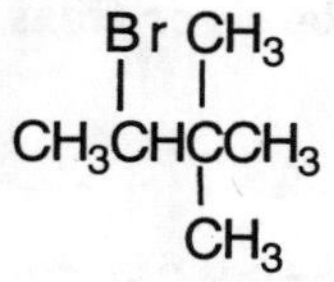

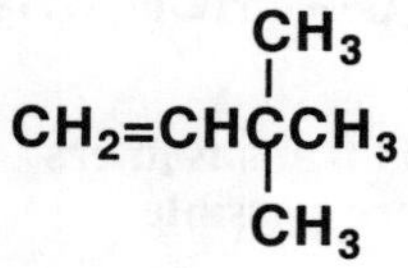

$$\underset{\underset{\underset{CH_3}{|}}{|}}{CH_3CHCCH_3}$$ Br CH$_3$

$$\underset{\underset{CH_3}{|}}{CH_2=CHCCH_3}$$ CH$_3$

(b)

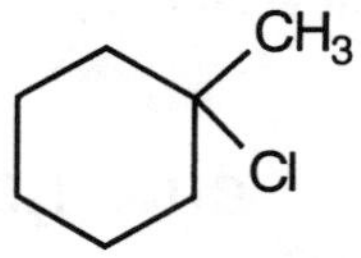

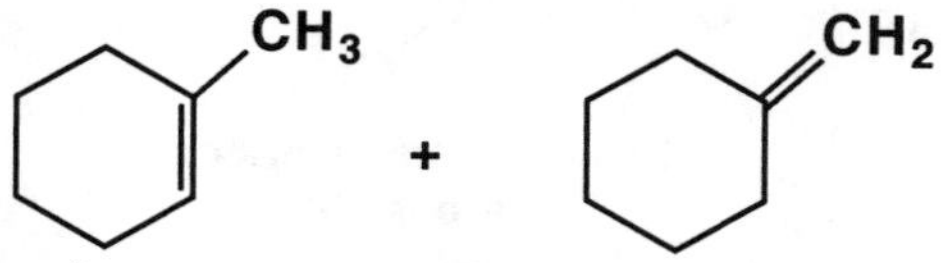

major   +   minor

(c)

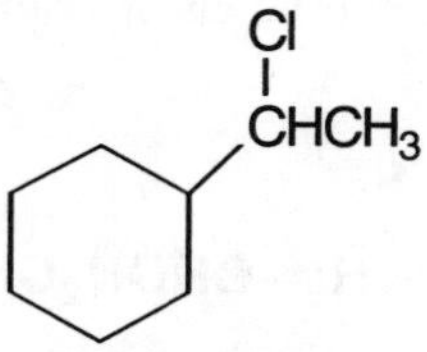

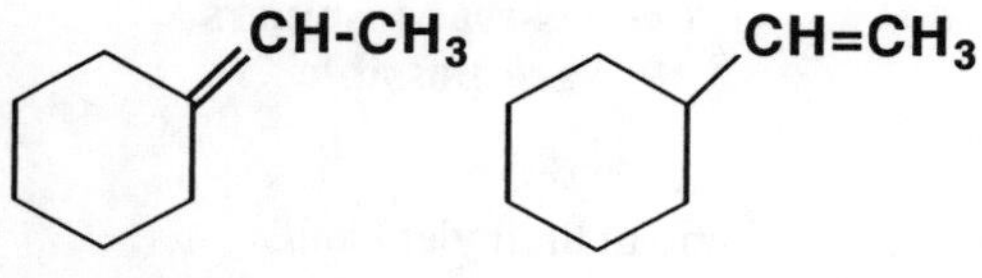

major   minor

(d) 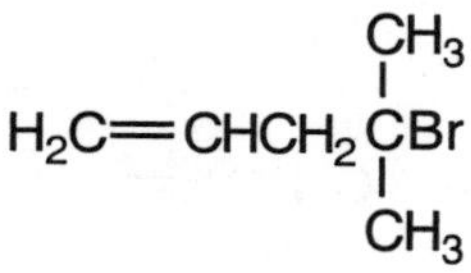 $\underset{\underset{CH_3}{|}}{H_2C=CHCH_2\overset{\overset{CH_3}{|}}{C}Br}$

$$H_2C=CH-CH=C\underset{CH_3}{\overset{CH_3}{<}}$$ $$H_2C=CH-CH_2-C\underset{CH_3}{\overset{CH_2}{<}}$$

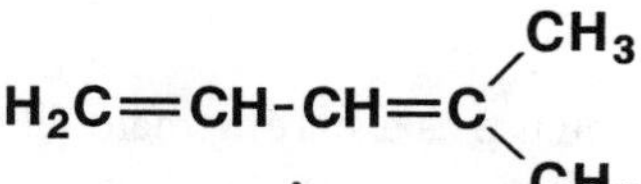

major 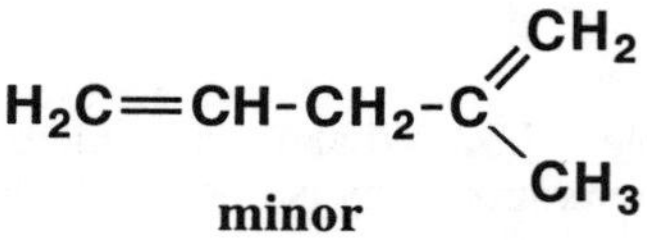   minor

Problem 7.30 Which alkyl halides undergo dehydrohalogenation to give alkenes that do not show cis-trans isomerism?
(a) 2-Chloropentane

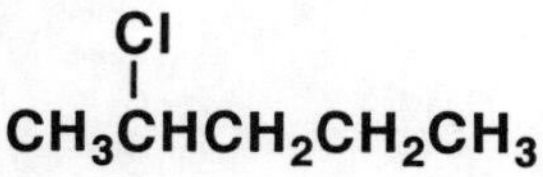

$$\underset{\underset{Cl}{|}}{CH_3CHCH_2CH_2CH_3} \longrightarrow CH_3CH=CHCH_2CH_3$$

major Zaitsev product
cis-trans isomers are possible

(b) 2-Chlorobutane

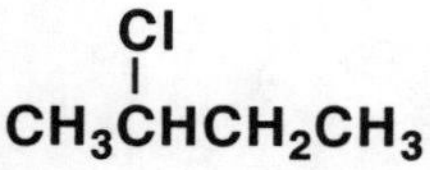

$$\underset{\underset{Cl}{|}}{CH_3CHCH_2CH_3} \longrightarrow CH_3CH=CHCH_3$$

major Zaitsev product
cis-trans isomers are possible

(c) Chlorocyclohexane

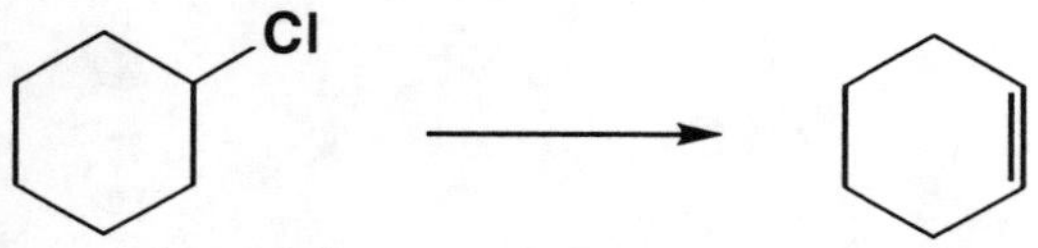

no cis-trans isomers are possible

(d) Isobutyl chloride

$$(CH_3)_2CHCH_2Cl \longrightarrow (CH_3)_2C=CH_2$$

no cis-trans isomers are possible

Problem 7.31 How many isomers, including cis-trans isomers, are possible for the major product of dehydrohalogenation of each haloalkane?

(a) 3-Chloro-3-methylhexane

There are two major alkenes of approximately equal stability formed, and each of those can be cis or trans. Thus, there are 2 x 2 = 4 isomers possible.

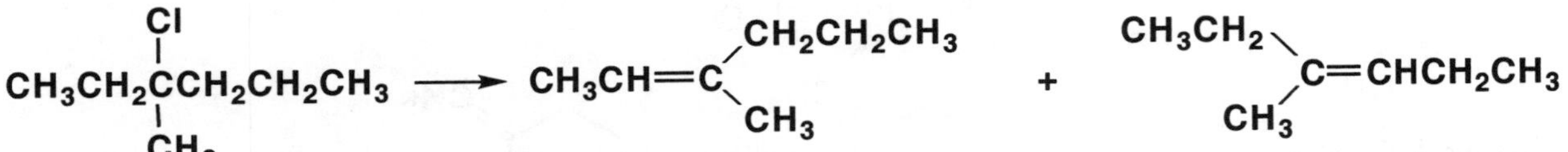

$$CH_3CH_2\underset{\underset{CH_3}{|}}{\overset{\overset{Cl}{|}}{C}}CH_2CH_2CH_3 \longrightarrow CH_3CH=\underset{\underset{CH_3}{}}{\overset{\overset{CH_2CH_2CH_3}{}}{C}} \quad + \quad \underset{\underset{CH_3}{}}{\overset{\overset{CH_3CH_2}{}}{C}}=CHCH_2CH_3$$

two cis-trans isomers   **two cis-trans isomers**
are possible   **are possible**

(b) 3-Bromohexane

There are two major alkenes of approximately equal stability formed, and each of those can be cis or trans. Thus, there are 2 x 2 = 4 isomers possible.

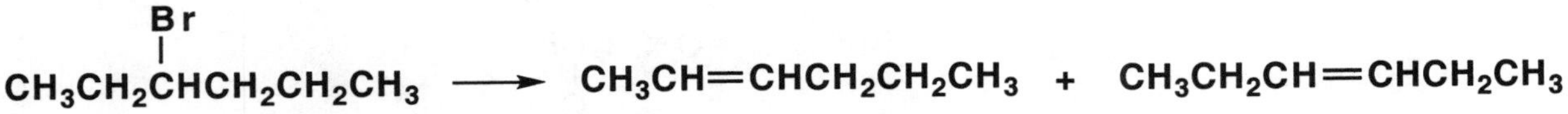

$$CH_3CH_2\overset{\overset{Br}{|}}{C}HCH_2CH_2CH_3 \longrightarrow CH_3CH=CHCH_2CH_2CH_3 \quad + \quad CH_3CH_2CH=CHCH_2CH_3$$

two cis-trans isomers   **two cis-trans isomers**
are possible   **are possible**

Problem 7.32 What alkyl halide might you use as a starting material to produce each alkene in high yield and uncontaminated by isomeric alkenes?

(a) ⬡=CH₂   (b) $CH_3\overset{\overset{CH_3}{|}}{C}HCH_2CH=CH_2$

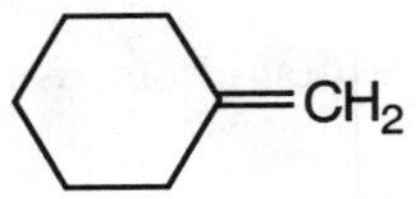

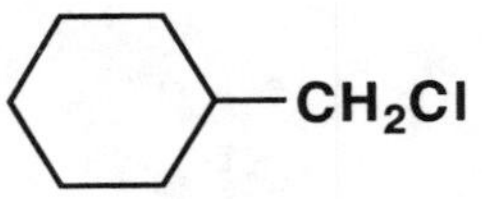

⬡—CH₂Cl $CH_3\overset{\overset{CH_3}{|}}{C}HCH_2CH_2CH_2Cl$

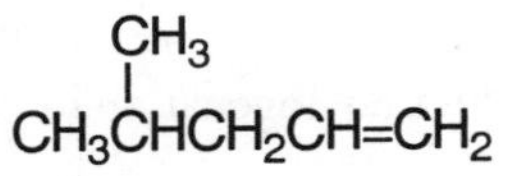

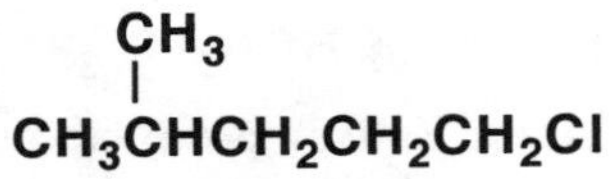

CHAPTER 8
Solutions to Problems

Problem 8.1 Write the IUPAC name for each alcohol.

(a)
$$CH_2OH$$
$$C^{....}CH_2CH_3$$
$$H_3C \quad H$$
(S)-2-Methyl-1-butanol

(b)
$$HO \quad CH_3$$
1-Methylcyclopentanol

(c) $(CH_3)_3CCH_2OH$
2,2-Dimethyl-1-propanol

Problem 8.2 Classify each alcohol as primary, secondary, or tertiary.

(a)
$$CH_3$$
$$CH_3CCH_2OH$$
$$CH_3$$
Primary

(b) ▷—OH
Secondary

(c) $CH_2=CHCH_2OH$
Primary

(d)
$$CH_3$$
$$OH$$
Tertiary

Problem 8.3 Write the IUPAC name for each unsaturated alcohol:

(a) $CH_2=CHCH_2CH_2OH$
3-Buten-1-ol

(b) —OH
2-Cyclopenten-1-ol

Problem 8.4 Write IUPAC and common names for each ether.

(a)
$$CH_3$$
$$CH_3CHCH_2OCH_2CH_3$$
1-Ethoxy-2-methylpropane
Ethyl isobutyl ether

(b) —OCH$_3$
Methoxycyclopentane
Cyclopentyl methyl ether

Problem 8.5 Write a name for each compound.

(a)
$$CH_3$$
$$CH_3CHCH_2CH_2SH$$
3-Methyl-1-butanethiol

(b) $CH_3SCH_2CH_3$
Ethyl methyl sulfide

(c)
$$H \quad\quad H$$
$$C=C \quad SH$$
$$H_3C \quad CH_2CHCH_3$$
(Z)-4-Hexene-2-thiol
cis-**4-Hexene-2-thiol**

Problem 8.6 Arrange these compounds in order of increasing boiling point:

$$CH_3OCH_2CH_2OCH_3 \quad\quad HOCH_2CH_2OH \quad\quad CH_3OCH_2CH_2OH$$

In order of increasing boiling point they are:

$CH_3OCH_2CH_2OCH_3$	$CH_3OCH_2CH_2OH$	$HOCH_2CH_2OH$
1,2-Dimethoxyethane	2-Methoxyethanol	1,2-Ethanediol
(Methyl cellosolve)		(Ethylene glycol)
bp 84°C	bp 125°C	bp 198°C

Hydrogen bonding, or lack of it, is the key. Although 1,2-dimethoxyethane is a polar molecule, there is little intermolecular association between molecules in the pure liquid because centers of positive and negative charge on adjacent molecules cannot approach each other close enough to develop appreciable dipole-dipole interactions. However, the fact that its boiling point is higher than that of hexane (bp 69°C), a nonpolar hydrocarbon of comparable molecular weight, indicates that there is some dipole-dipole interaction. Both 2-methoxyethanol and 1,2-ethanediol can associate by hydrogen bonding. Because 1,2-ethanediol has more sites for hydrogen bonding, it has a higher boiling point than 2-methoxyethanol.

Problem 8.7 Predict the position of equilibrium for this acid-base reaction. (Hint: review Section 2.4).

$$CH_3CH_2O^- \ Na^+ \ + \ CH_3\overset{\overset{\displaystyle O}{\|}}{C}OH \ \rightleftharpoons \ CH_3CH_2OH \ + \ CH_3\overset{\overset{\displaystyle O}{\|}}{C}O^- \ Na^+$$

Acetic acid is a much stronger acid than ethanol. Thus, the equilibrium lies very far to the right.

$$CH_3CH_2O^- \ Na^+ \ + \ CH_3\overset{\overset{\displaystyle O}{\|}}{C}OH \ \longrightarrow \ CH_3CH_2OH \ + \ CH_3\overset{\overset{\displaystyle O}{\|}}{C}O^- \ Na^+$$

| | pK_a 4.76 | | pK_a 15.9 | |
| (stronger base) | (stronger acid) | | (weaker acid) | (weaker base) |

Problem 8.8 Draw structural formulas for the alkenes formed by acid-catalyzed dehydration of each alcohol. For each, predict which is the major product.

(a)
$$\underset{\underset{\displaystyle OH}{|}}{\overset{\overset{\displaystyle CH_3}{|}}{CH_3CCH_2CH_3}} \quad \xrightarrow[\text{dehydration}]{\overset{\displaystyle H_2SO_4}{\text{acid-catalyzed}}} \quad \overset{\overset{\displaystyle CH_3}{|}}{CH_3C}=CHCH_3 \ + \ CH_2=\overset{\overset{\displaystyle CH_3}{|}}{C}CH_2CH_3$$

major product

(b)

$$\xrightarrow[\text{dehydration}]{\overset{\displaystyle H_2SO_4}{\text{acid-catalyzed}}}$$

major product

Problem 8.9 Draw the product of treatment of each alcohol in Example 8.9 with chromic acid.

(a) $CH_3(CH_2)_4\overset{\overset{\displaystyle O}{\|}}{C}OH$ (b) $CH_3(CH_2)_3\overset{\overset{\displaystyle O}{\|}}{C}CH_3$ (c)

Problem 8.10 Draw the structural formula of the epoxide formed by treating 1,2-dimethylcyclopentene with a peroxycarboxylic acid.

$$\xrightarrow[\text{CH}_2\text{Cl}_2]{\text{RCO}_3\text{H}}$$

Problem 8.11 Show how to convert cyclohexene to *cis*-1,2-cyclohexanediol.

$$\text{cyclohexene} \quad + \quad OsO_4 \quad \xrightarrow[\text{or ROOH}]{H_2O_2} \quad \text{cis-1,2-cyclohexanediol (OH, OH)}$$

Structure and Nomenclature

Problem 8.12 Which are secondary alcohols?

(a) [cyclohexane with CH₃ and OH]

(b) $(CH_3)_3COH$

(c) [chain with OH]

(d) [cyclopentane with OH]

The secondary alcohols are (c) and (d). Molecules (a) and (b) are tertiary alcohols.

Problem 8.13 Name these compounds.

(a) $CH_3CH_2CH_2CH_2CH_2OH$ (b) $HOCH_2CH_2CH_2OH$ (c) $CH_2{=}CHCH_2CH_2OH$
      1-Pentanol   **1,3-Propanediol**   **3-Buten-1-ol**

(d) $HOCH_2CH_2\overset{\underset{\displaystyle |}{CH_3}}{C}HCH_3$ (e) [cyclohexane with OH, OH]   (f) $CH_3CH_2CH_2CH_2SH$

3-Methyl-1-butanol   ***cis*-1,2-Cyclohexanediol**   **1-Butanethiol**
(Isopentyl alcohol)

Problem 8.14 Draw a structural formula for each alcohol:

(a) Isopropyl alcohol   (b) Propylene glycol   (c) (R)-5-Methyl-2-hexanol

$$CH_3\overset{\overset{\displaystyle OH}{\displaystyle |}}{C}HCH_3$$

$$CH_3\overset{\overset{\displaystyle OH}{\displaystyle |}}{C}HCH_2OH$$

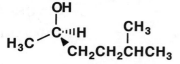

(d) 2-Methyl-2-propyl-1,3-propanediol   (e) 2,2-Dimethyl-1-propanol   (f) 2-Mercaptoethanol

$$HOCH_2\overset{\overset{\displaystyle CH_3}{\displaystyle |}}{\underset{\underset{\displaystyle CH_2CH_2CH_3}{\displaystyle |}}{C}}CH_2OH$$

$$CH_3\overset{\overset{\displaystyle CH_3}{\displaystyle |}}{\underset{\underset{\displaystyle CH_3}{\displaystyle |}}{C}}CH_2OH$$

$$HSCH_2CH_2OH$$

(g) 1,4-Butanediol   (h) (Z)-5-Methyl-2-hexen-1-ol

$$HOCH_2CH_2CH_2CH_2OH$$

[structure: $HOCH_2$ and $CH_2\overset{\overset{\displaystyle CH_3}{\displaystyle |}}{C}HCH_3$ on a C=C double bond with H, H]

(i) *cis*-3-Penten-1-ol

HOCH$_2$CH$_2$ C=C CH$_3$ / H / H

(j) *trans*-1,4-Cyclohexanediol

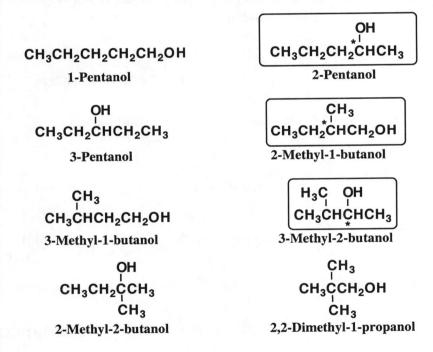

Problem 8.15 Write names for these ethers.

(a) Cyclopentoxypentane
(Dicyclopentyl ether)

(b) [CH$_3$(CH$_2$)$_4$]$_2$O

Pentoxypentane
(Dipentyl ether)

(c) CH$_3$CH$_2$OCH$_2$CH$_2$OH

2-Ethoxyethanol

Problem 8.16 Name and draw structural formulas for the eight isomeric alcohols of molecular formula C$_5$H$_{12}$O. Which are chiral?

The chiral alcohols are circled and the stereocenters are marked with asterisks.

CH$_3$CH$_2$CH$_2$CH$_2$CH$_2$OH
1-Pentanol

OH
CH$_3$CH$_2$CH$_2$C*HCH$_3$
2-Pentanol

OH
CH$_3$CH$_2$CHCH$_2$CH$_3$
3-Pentanol

CH$_3$
CH$_3$CH$_2$C*HCH$_2$OH
2-Methyl-1-butanol

CH$_3$
CH$_3$CHCH$_2$CH$_2$OH
3-Methyl-1-butanol

H$_3$C OH
CH$_3$CHC*HCH$_3$
3-Methyl-2-butanol

OH
CH$_3$CH$_2$CCH$_3$
CH$_3$
2-Methyl-2-butanol

CH$_3$
CH$_3$CCH$_2$OH
CH$_3$
2,2-Dimethyl-1-propanol

Physical Properties
Problem 8.17 Arrange these compounds in order of increasing boiling point. (Values in °C are -42, 78, 117, and 198.)
(a) CH$_3$CH$_2$CH$_2$CH$_2$OH (b) CH$_3$CH$_2$OH (c) HOCH$_2$CH$_2$OH (d) CH$_3$CH$_2$CH$_3$

In order of increasing boiling point they are:

CH$_3$CH$_2$CH$_3$ CH$_3$CH$_2$OH   CH$_3$CH$_2$CH$_2$CH$_2$OH   HOCH$_2$CH$_2$OH

Propane   Ethanol   1-Butanol   Ethylene glycol
bp -42°C   bp 78°C   bp 117°C   bp 198°C

The keys for this problem are hydrogen bonding and size. Propane cannot form any hydrogen bonds, so it has the lowest boiling point. Ethanol and 1-butanol can each form hydrogen bonds through their single -OH group. However, 1-butanol is larger, so it has a higher boiling point than ethanol. Ethylene glycol has two -OH groups per molecule with which to form hydrogen bonds, so it has the highest boiling point of this set.

Problem 8.18 Arrange these compounds in order of increasing boiling point. (Values in °C are -42, -24, 78, and 118)
(a) CH_3CH_2OH　(b) CH_3OCH_3　(c) $CH_3CH_2CH_3$　(d) CH_3CO_2H

In order of increasing boiling point they are:

$CH_3CH_2CH_3$　　　**CH_3OCH_3**　　　**CH_3CH_2OH**　　　**CH_3CO_2H**
Propane　　　　Dimethyl ether　　　　Ethanol　　　　Acetic acid
bp -42°C　　　　bp -24°C　　　　bp 78°C　　　　bp 118°C

The keys for this problem are hydrogen bonding, polarity, and size. We know from the last problem that propane and ethanol have boiling points of -42°C and 78°C, respectively. Dimethyl ether is polar, but cannot form hydrogen bonds. Therefore, it makes sense that dimethyl ether has a boiling point (-24°C) that is higher than propane, but lower than ethanol. Acetic acid can participate in hydrogen bonding through both its carbonyl oxygen and OH group and has a higher molecular weight than ethanol, so it makes sense that acetic acid has the highest boiling point of this set.

Problem 8.19 Propanoic acid and methyl acetate are constitutional isomers and both are liquids at room temperature. One of these compounds has a boiling point of 141°C, the other has a boiling point of 57°C. Which compound has which boiling point?

$$\overset{\displaystyle O}{\overset{\displaystyle \|}{CH_3CH_2C}}-OH$$　　　　$$\overset{\displaystyle O}{\overset{\displaystyle \|}{CH_3C}}-OCH_3$$
Propanoic acid　　　　　　　　　Methyl acetate

Propanoic acid has the higher boiling point. Because the carboxyl group of propanoic acid can function as both a hydrogen bond donor (through the O-H group) and a hydrogen bond acceptor (through the C=O and C-O groups) there is a high degree of intermolecular association between molecules of propanoic acid in the liquid state. Methyl acetate, in the pure liquid state, cannot associate by hydrogen bonding.

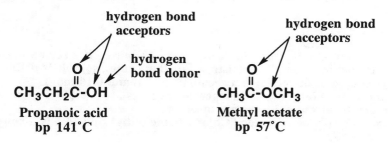

hydrogen bond acceptors　　　　　**hydrogen bond acceptors**
hydrogen bond donor

CH_3CH_2C-OH　　　　　**CH_3C-OCH$_3$**
Propanoic acid　　　　　**Methyl acetate**
bp 141°C　　　　　**bp 57°C**

Problem 8.20 Compounds that contain an -NH group associate by hydrogen bonding.
(a) Do you expect this association to be stronger or weaker than that beween compounds containing an -OH group?

Weaker. The O-H bond is more polar, because the difference in electronegativity between N and H (3.0-2.1) is less than the difference between O and H (3.5-2.1). Thus, the degree of intermolecular interaction between compounds containing an -NH group is less than that between compounds containing an -OH group.

(b) Based on your answer to part (a), which would you predict to have the higher boiling point, 1-butanol or 1-butanamine?

$CH_3CH_2CH_2CH_2OH$　　　　　$CH_3CH_2CH_2CH_2NH_2$
1-Butanol　　　　　　　　1-Butanamine
bp 117°C　　　　　　　　bp 78°C

The stronger the hydrogen bonds, the higher the boiling point since hydrogen bonds in the liquid state must be broken upon boiling. Therefore, 1-butanol, with the stronger hydrogen bonds, has the higher boiling point.

<u>Problem 8.21</u> Following are structural formulas for 1-butanol and 1-butanethiol. One of these compounds has a boiling point of 98.5° C, the other has a boiling point of 117°C. Which compound has which boiling point?

CH₃CH₂CH₂CH₂OH   CH₃CH₂CH₂CH₂SH
1-Butanol   1-Butanethiol
bp 117°C   **bp 98.5°C**

The S-H bond of thiols is much less polar than the O-H bond of alcohols. Therefore, thiols do not form hydrogen bonds that are as strong as the hydrogen bonds formed by alcohols. Hydrogen bonds hold molecules of a liquid together, so a thiol, such as 1-butanethiol, has a lower boiling point than a corresponding alcohol, such as 1-butanol.

<u>Problem 8.22</u> From each pair of compounds, select the one that is more soluble in water.

(a)   CH₂Cl₂   or   [CH₃OH]

Methanol, CH₃OH, is soluble in all proportions in water. Dichloromethane, CH₂Cl₂, is insoluble. The highly polar -OH group of methanol is capable of participating both as a hydrogen bond donor and hydrogen bond acceptor with water and, therefore, interacts strongly with water by intermolecular association. No such interaction is possible with dichloromethane.

(b) $\begin{matrix} O \\ \| \\ CH_3CCH_3 \end{matrix}$ or $\begin{matrix} CH_2 \\ \| \\ CH_3CCH_3 \end{matrix}$

Propanone (acetone), CH₃COCH₃, is soluble in water in all proportions. 2-Methylpropene (isobutylene) is insoluble in water. Acetone has a large dipole moment and can function as a hydrogen bond acceptor from water.

(c)   CH₃CH₂Cl   or   [NaCl]

NaCl is the more soluble. Chloroethane is insoluble in water. Following is a review of some of the general water solubility rules developed in General Chemistry. For these rules, <u>soluble</u> is defined as dissolving greater than 0.10 mol/L. <u>Slightly soluble</u> is dissolving between 0.01 mol/L and 0.10 mol/L.
1. Sodium, potassium, and ammonium salts of halogens or nitrates are soluble.
2. Silver, lead, and mercury(I) salts of halogens are insoluble.
Thus, applying Rule 1, NaCl is soluble in water. Chloroethane (ethyl chloride) is a nonpolar organic compound and insoluble in water.

(d)   CH₃CH₂CH₂SH   or   [CH₃CH₂CH₂OH]

Sulfur is less electronegative than oxygen, so an S-H bond is less polarized than an O-H bond. Hydrogen bonding is, therefore, weaker with thiols than alcohols, so the alcohol is more able to interact with water molecules through hydrogen bonding. The alcohol is more soluble in water.

(e) $\begin{matrix} OH \\ | \\ CH_3CH_2CHCH_2CH_3 \end{matrix}$ or $\begin{matrix} O \\ \| \\ CH_3CH_2CCH_2CH_3 \end{matrix}$

The alcohol group has both a hydrogen bond donor and acceptor (the oxygen and hydrogen atoms of the -OH group, respectively), while the ketone has only a hydrogen bond acceptor (the oxygen atom). Thus, the alcohol is better able to interact with water through hydrogen bonding, and is more soluble in water.

Problem 8.23 Arrange the compounds in each set in order of decreasing solubility in water.
(a) Ethanol; butane; diethyl ether

$$CH_3CH_2OH \qquad CH_3CH_2OCH_2CH_3 \qquad CH_3CH_2CH_2CH_3$$

| soluble in all proportions | 8 g/100 mL water | insoluble in water |

In general, the more strongly a molecule can take part in hydrogen bonding with water, the greater the molecule will interact with the water molecules and dissolve. Only the ethanol can act both as a donor and acceptor of hydrogen bonds with water. Diethyl ether can act as an acceptor of hydrogen bonds. Butane can act as neither a donor nor an acceptor of hydrogen bonds.

(b) 1-Hexanol; 1,2-hexanediol; hexane

$$\underset{\underset{OH\ \ OH}{|\ \ \ |}}{CH_2CHCH_2CH_2CH_2CH_3} \qquad CH_3CH_2CH_2CH_2CH_2CH_2OH \qquad CH_3CH_2CH_2CH_2CH_2CH_3$$

Molecules of 1,2-hexanediol can take part in more hydrogen bonding with water than the 1-hexanol, because the diol has two -OH groups. Molecules of hexane have no polar bonds and, thus, cannot take part in any dipole-dipole interactions with water molecules.

Problem 8.24 Each compound given in this problem is a common organic solvent. From each pair of compounds, select the solvent with the greater solubility in water.

The more soluble compound in each pair is circled. Solubility in water increases with increasing hydrogen bonding ability and decreases with increasing surface area of hydrophobic groups, such as alkyl groups.

(a) CH_2Cl_2 or $\boxed{CH_3CH_2OH}$ (b) $CH_3CH_2OCH_2CH_3$ or $\boxed{CH_3CH_2OH}$

(c) $\boxed{\underset{CH_3CCH_3}{\overset{\overset{O}{\parallel}}{}}}$ or $CH_3CH_2OCH_2CH_3$ (d) $\boxed{CH_3CH_2OCH_2CH_3}$ or $CH_3(CH_2)_3CH_3$

Synthesis of Alcohols
We have encountered three reactions for the synthesis of alcohols, including glycols.
    (1) Acid-catalyzed hydration of alkenes (Section 6.3B).
    (2) Oxidation of alkenes to glycols by OsO_4 (Section 6.4B).
    (3) Acid-catalyzed ring opening of epoxides to give glycols (Section 8.7).

Problem 8.25 Give the structural formula of an alkene or alkenes from which each alcohol or glycol can be prepared.

(a) 2-Butanol

(b) 1-Methylcyclohexanol

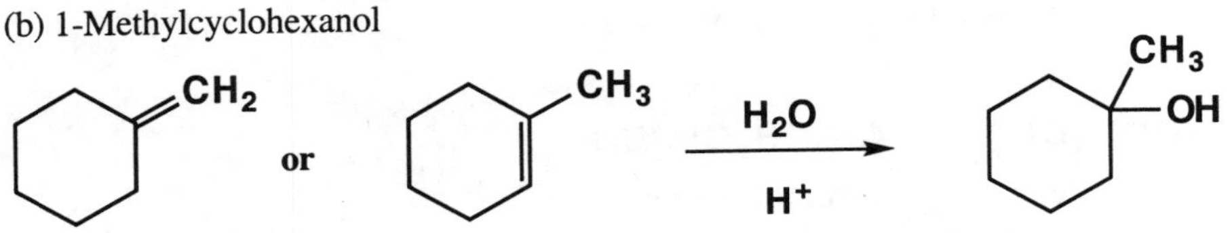

1-Methylcyclohexanol

(c) 3-Hexanol

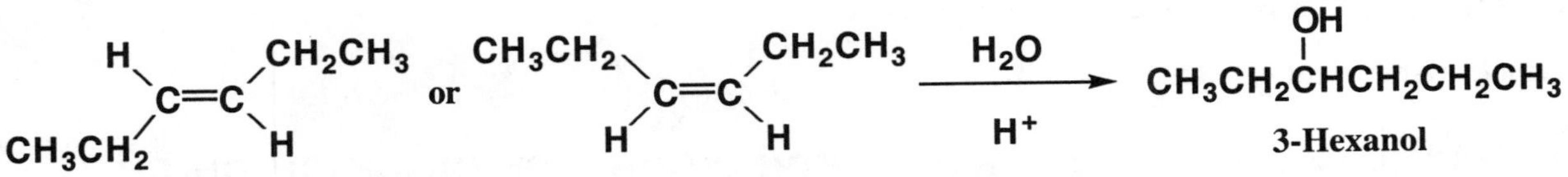

3-Hexanol

(d) 2-Methyl-2-pentanol

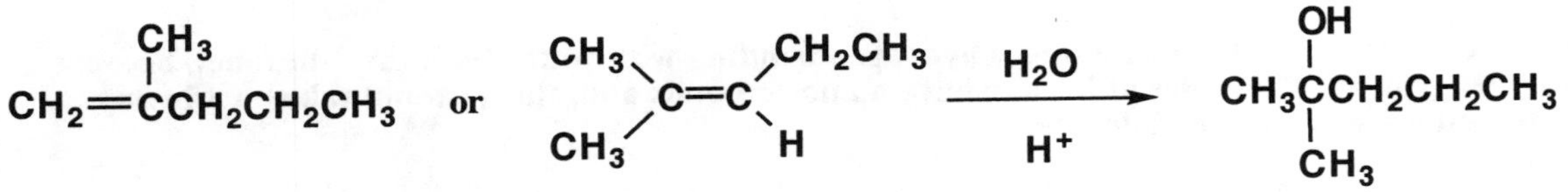

2-Methyl-2-pentanol

(e) Cyclopentanol

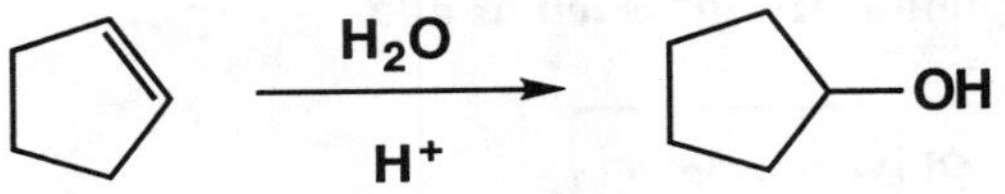

Cyclopentanol

(f) 1,2-Propanediol

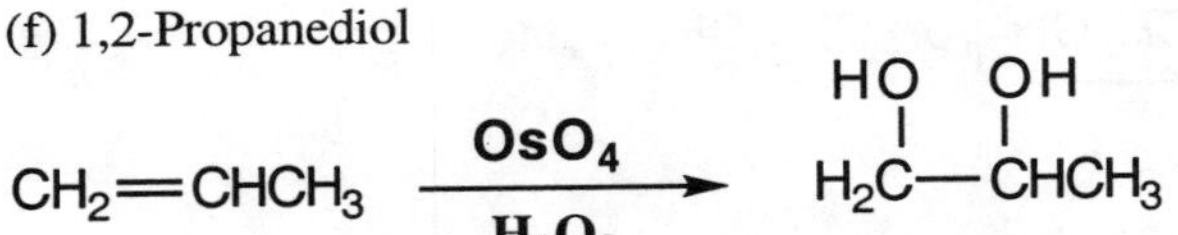

1,2-Propanediol

Problem 8.26 Addition of bromine to cyclopentene and acid-catalyzed hydrolysis of cyclopentene oxide are both stereoselective; each gives a trans product. Compare the mechanisms of these two reactions and show how each accounts for the formation of the trans product.

Notice how, in both mechanisms, the nucleophile (Br⁻ or H_2O) must attack a three-membered ring. One face of this three member ring is blocked by Br or OH, so the nucleophile must attack from the opposite side as shown. The result in both cases is that a trans product is formed.

Reaction of an alkene with bromine:

Step 1:

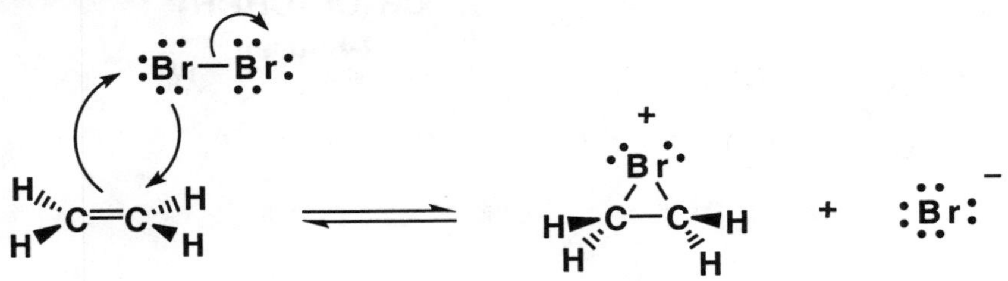

Step 2:

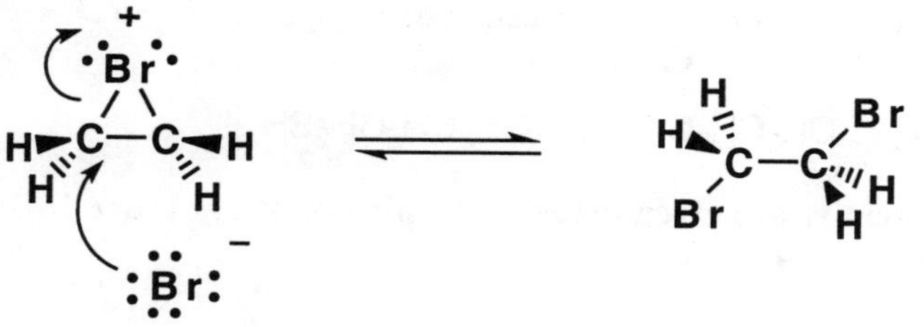

Acid-catalyzed ring opening of epoxides:

Step 1:

Step 2:

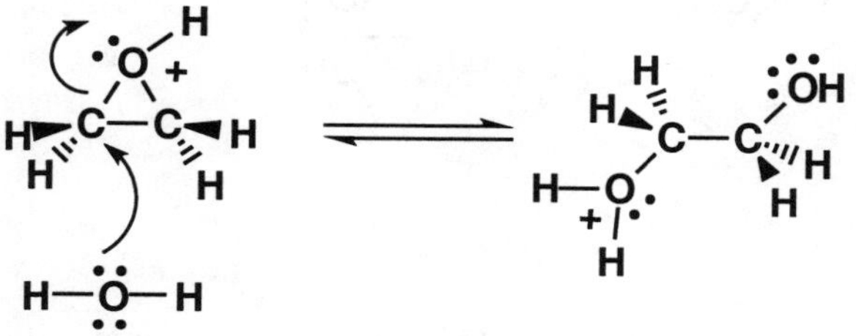

Step 3:

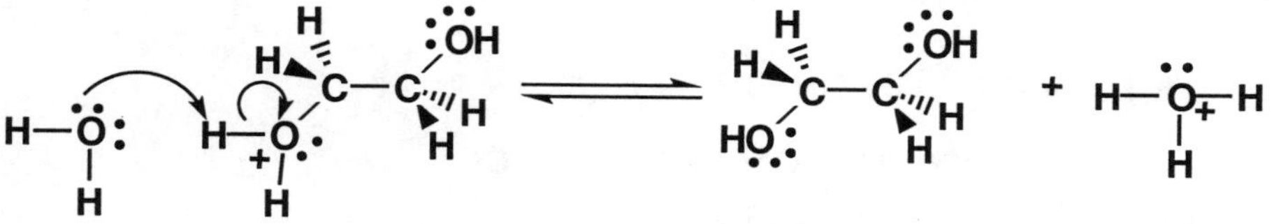

Acidity of Alcohols and Thiols

Problem 8.27 From each pair, select the stronger acid. For each stronger acid, write a structural formula for its conjugate base.

(a) H_2O or H_2CO_3 (b) CH_3OH or CH_3CO_2H (c) CH_3CH_2OH or CH_3CH_2SH

Under each acid is given its pK_a. The stronger acid has the smaller value of pK_a.

	weaker acid	stronger acid	conjugate base of stronger acid
(a)	H_2O pK_a 15.7	H_2CO_3 pK_a 6.36	HCO_3^-
(b)	CH_3OH pK_a 15.5	CH_3CO_2H pK_a 4.76	$CH_3CO_2^-$
(c)	CH_3CH_2OH pK_a 15.9	CH_3CH_2SH pK_a 8.5	$CH_3CH_2S^-$

Problem 8.28 Arrange these compounds in order of increasing acidity (weakest to strongest).

$$CH_3CH_2CH_2OH \qquad CH_3CH_2\overset{\overset{O}{\|}}{C}\text{-}OH \qquad CH_3CH_2CH_2SH$$

The compounds listed in order of increasing acidity are given below. The pK$_a$ of alcohols are 16-18, the pK$_a$ of thiols are ~8.5 and the pK$_a$ of carboxylic acids are ~4.0-5.0.

$$\textbf{CH}_3\textbf{CH}_2\textbf{CH}_2\textbf{OH} \qquad \textbf{CH}_3\textbf{CH}_2\textbf{CH}_2\textbf{SH} \qquad \textbf{CH}_3\textbf{CH}_2\overset{\overset{O}{\|}}{\textbf{C}}\textbf{OH}$$

Problem 8.29 From each pair, select the stronger base. For each stronger base, write the structural formula of its conjugate acid.

The stronger base in each pair is circled. Its conjugate acid is shown below it in bold type.

(a) boxed[OH⁻] or CH₃O⁻   (b) CH₃CH₂S⁻ or boxed[CH₃CH₂O⁻]

 H₂O   **CH₃CH₂OH**

(c) CH₃CH₂O⁻ or boxed[NH₂⁻]

 NH₃

Problem 8.30 Label the stronger acid, stronger base, weaker acid, and weaker base in each equilibrium. Also predict the position of each equilibrium. For pK$_a$ values, see Table 2.1.

(a) $CH_3CH_2O^- \; + \; HCl \; \rightleftharpoons \; CH_3CH_2OH \; + \; Cl^-$

$CH_3CH_2O^- \; + \; HCl \; \longrightarrow \; CH_3CH_2OH \; + \; Cl^-$

stronger base	pK$_a$ -7 stronger acid	pK$_a$ 15.9 weaker acid	weaker base

(b) $CH_3\overset{\overset{O}{\|}}{C}\text{-}OH \; + \; CH_3CH_2O^- \; \rightleftharpoons \; CH_3\overset{\overset{O}{\|}}{C}O^- \; + \; CH_3CH_2OH$

$CH_3\overset{\overset{O}{\|}}{C}\text{-}OH \; + \; CH_3CH_2O^- \; \longrightarrow \; CH_3\overset{\overset{O}{\|}}{C}\text{-}O^- \; + \; CH_3CH_2OH$

pK$_a$ 4.76 stronger acid	stronger base	weaker base	pK$_a$ 15.9 weaker acid

Problem 8.31 Predict the position of equilibrium for each acid-base reaction, that is, does it lie considerably to the left, considerably to the right, or are concentrations evenly balanced?

(a) $CH_3CH_2OH \; + \; Na^+OH^- \; \rightleftharpoons \; CH_3CH_2O^- \, Na^+ \; + \; H_2O$

$CH_3CH_2OH \; + \; Na^+OH^- \; \rightleftharpoons \; CH_3CH_2O^- \, Na^+ \; + \; H_2O$

pK$_a$ 15.9 weaker acid	weaker base	stronger base	pK$_a$ 15.7 stronger acid

The difference in acidity between water and ethanol is relatively small. So, even though the equilibrium lies toward the left, there will still be significant amounts of hydroxide present. This is indicated in the equation by the use of two arrows of different lengths.

(b) CH_3CH_2SH + Na^+OH^- $\rightleftharpoons$ $CH_3CH_2S^-$ Na^+ + H_2O

$$CH_3CH_2SH + Na^+OH^- \longrightarrow CH_3CH_2S^- Na^+ + H_2O$$

pK_a 8.5			pK_a 15.7
stronger acid	stronger base	weaker base	weaker acid

The acid ionization constant (K_a) for water is $>10^7$ times higher than for ethanethiol (i.e. it is a much weaker acid). Thus, the equilibrium lies almost completely to the right.

(c) CH_3CH_2OH + $CH_3CH_2S^-$ Na^+ $\rightleftharpoons$ $CH_3CH_2O^-$ Na^+ + CH_3CH_2SH

$$CH_3CH_2OH + CH_3CH_2S^-Na^+ \longleftarrow CH_3CH_2O^-Na^+ + CH_3CH_2SH$$

pK_a 15.9			pK_a 8.5
weaker acid	weaker base	stronger base	stronger acid

The K_a for ethanol is $>10^7$ times higher than for ethanethiol. Thus, this equilibrium lies almost completely to the left.

(d) $CH_3CH_2S^-$ Na^+ + $CH_3\overset{O}{\overset{\|}{C}}OH$ $\rightleftharpoons$ CH_3CH_2SH + $CH_3\overset{O}{\overset{\|}{C}}O^-$

$$CH_3CH_2S^- Na^+ + CH_3\overset{O}{\overset{\|}{C}}OH \longrightarrow CH_3CH_2SH + CH_3\overset{O}{\overset{\|}{C}}O^-$$

	pK_a 4.76	pK_a 8.5	
stronger base	stronger acid	weaker acid	weaker base

The K_a for ethanethiol is about 10^4 times higher than the K_a for acetic acid. Thus, the equilibrium lies far to the right.

Reactions of Alcohols

Problem 8.32 Show how to distinguish between cyclohexanol and cyclohexene by a simple chemical test. Hint: Treat each with Br_2 in CCl_4 and watch what happens.

Cyclohexene discharges the color of bromine in CCl_4. Cyclohexanol does not react with bromine, so the color of the solution would be unchanged.

(colorless)   (red)   (colorless)

Alternatively, only cyclohexanol reacts with active metals, such as sodium and potassium, to liberate hydrogen gas, which then bubbles from the test solution.

Problem 8.33 Write equations for the reaction of 1-butanol, a primary alcohol, with these reagents.

(a) Na metal

$$2 \ CH_3CH_2CH_2CH_2OH \ + \ 2 \ Na \longrightarrow 2 \ CH_3CH_2CH_2CH_2O^- \ Na^+ \ + \ H_2$$

(b) HBr, heat

$$CH_3CH_2CH_2CH_2OH \ + \ HBr \xrightarrow[heat]{} CH_3CH_2CH_2CH_2Br \ + \ H_2O$$

(c) $K_2Cr_2O_7$, H_2SO_4, heat

$$CH_3CH_2CH_2CH_2OH \ + \ K_2Cr_2O_7 \xrightarrow[heat]{H_2SO_4} \overset{\displaystyle O}{\overset{\|}{CH_3CH_2CH_2C}}\text{-OH} \ + \ Cr^{3+}$$

(d) $SOCl_2$

$$CH_3CH_2CH_2CH_2OH \ + \ SOCl_2 \longrightarrow CH_3CH_2CH_2CH_2Cl \ + \ SO_2 \ + \ HCl$$

(e) Pyridinium chlorochromate (PCC)

$$CH_3CH_2CH_2CH_2OH \ + \ PCC \longrightarrow \overset{\displaystyle O}{\overset{\|}{CH_3CH_2CH_2C}}\text{-H} \ + \ Cr^{3+}$$

Problem 8.34 Write equations for the reaction of 2-butanol, a secondary alcohol, with these reagents.

(a) Na metal

$$2 \ CH_3CH_2\overset{\displaystyle OH}{\overset{|}{C}}HCH_3 \ + \ 2 \ Na \longrightarrow 2 \ CH_3CH_2\overset{\displaystyle O^-Na^+}{\overset{|}{C}}HCH_3 \ + \ H_2$$

(b) H_2SO_4, heat

$$CH_3CH_2\overset{\displaystyle OH}{\overset{|}{C}}HCH_3 \xrightarrow[heat]{H_2SO_4} CH_3CH{=}CHCH_3 \ + \ H_2O$$

(c) HBr, heat

$$CH_3CH_2\overset{\displaystyle OH}{\overset{|}{C}}HCH_3 \ + \ HBr \xrightarrow[heat]{} CH_3CH_2\overset{\displaystyle Br}{\overset{|}{C}}HCH_3 \ + \ H_2O$$

(d) $K_2Cr_2O_7$, H_2SO_4, heat

$$CH_3CH_2\overset{\displaystyle OH}{\overset{|}{C}}HCH_3 \ + \ K_2Cr_2O_7 \xrightarrow[heat]{H_2SO_4} \overset{\displaystyle O}{\overset{\|}{CH_3CH_2C}}CH_3 \ + \ Cr^{3+}$$

(e) $SOCl_2$

$$CH_3CH_2\overset{\displaystyle OH}{\overset{|}{C}}HCH_3 \ + \ SOCl_2 \xrightarrow{pyridine} CH_3CH_2\overset{\displaystyle Cl}{\overset{|}{C}}HCH_3 \ + \ SO_2 \ + \ HCl$$

(f) Pyridinium chlorochromate (PCC)

$$CH_3CH_2\overset{\displaystyle OH}{\overset{|}{C}}HCH_3 \ + \ PCC \longrightarrow \overset{\displaystyle O}{\overset{\|}{CH_3CH_2C}}CH_3 \ + \ Cr^{3+}$$

Problem 8.35 When (R)-2-butanol is left standing in aqueous acid, it slowly loses its optical activity. When the organic material is recovered from the aqueous solution, only 2-butanol is found. Account for the observed loss of optical activity.

Protonation of alcohols converts the hydroxyl group, which is a very poor leaving group, into a good leaving group (water). Thus, secondary alcohols such as (R)-2-butanol can undergo acid-catalyzed SN1 (and E1) reactions. Loss of water leads to a planar carbocation. Because the molecule is dissolved in water, water is the predominant nucleophile. The water will approach the carbocation from either side, leading to loss of optical activity over time, without generation of a new organic product.

Step 1:

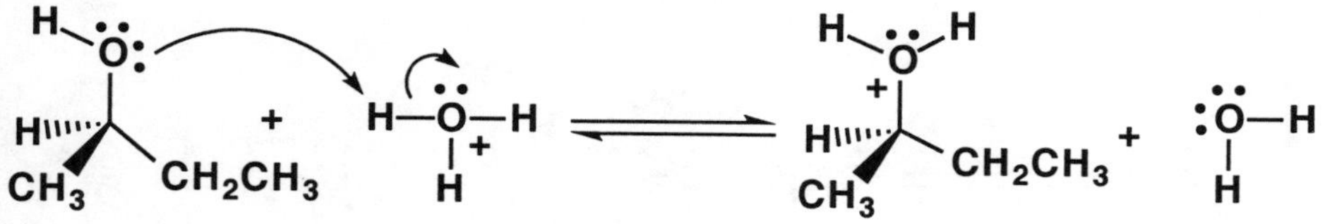

Step 2:

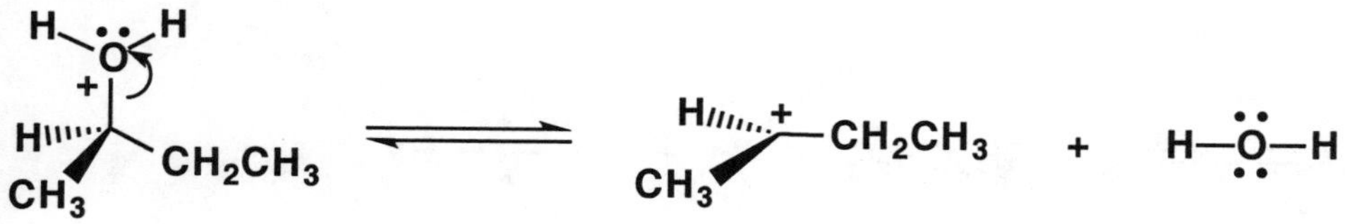

Step 3:

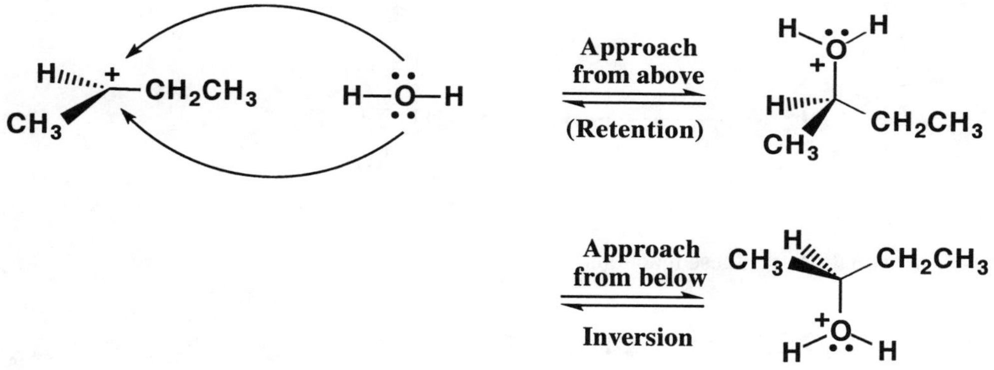

Approach from above
(Retention)

Approach from below

Inversion

Step 4:

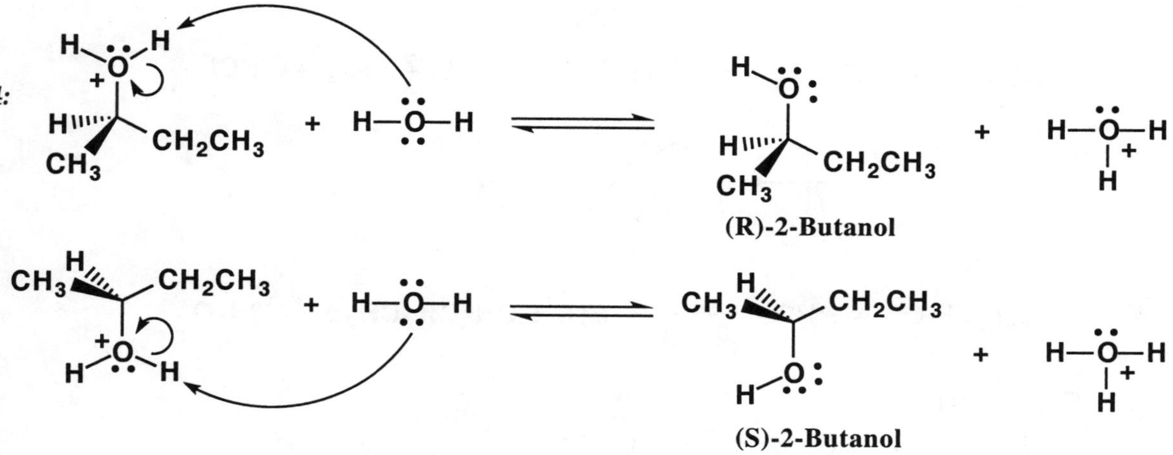

(R)-2-Butanol

(S)-2-Butanol

Problem 8.36 What is the most likely mechanism of this reaction? Draw a structural formula for any intermediate formed during the reaction.

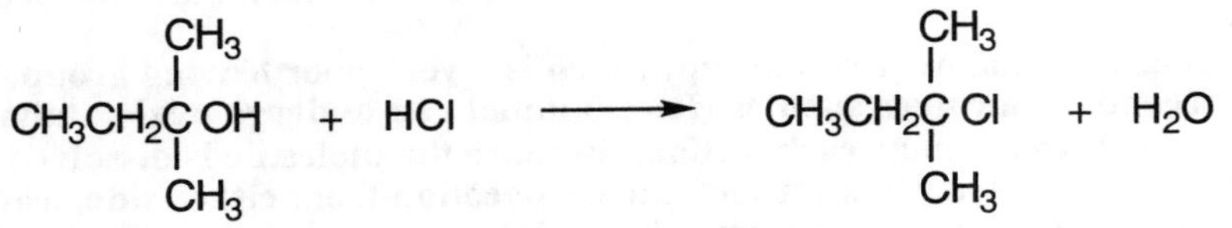

$$
\underset{\underset{CH_3}{|}}{\overset{\overset{CH_3}{|}}{CH_3CH_2C}}OH \ + \ HCl \ \longrightarrow \ \underset{\underset{CH_3}{|}}{\overset{\overset{CH_3}{|}}{CH_3CH_2C}}Cl \ + \ H_2O
$$

This mechanism, like the one in Problem 8.35, proceeds by an S$_N$1 pathway. It differs in that a tertiary carbocation is generated, and this carbocation reacts with chloride ion rather than water.

Step 1:

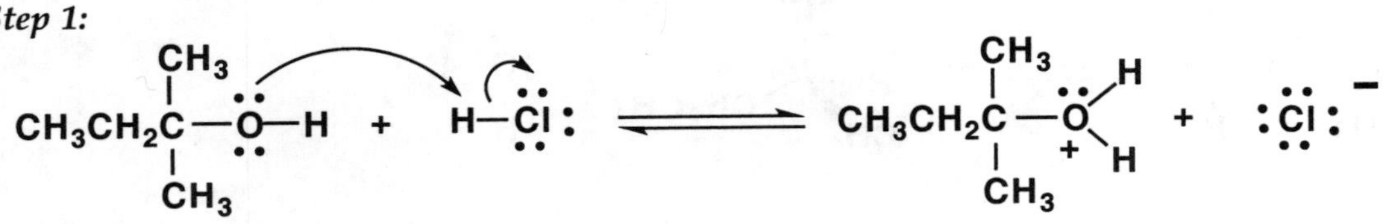

Step 2:

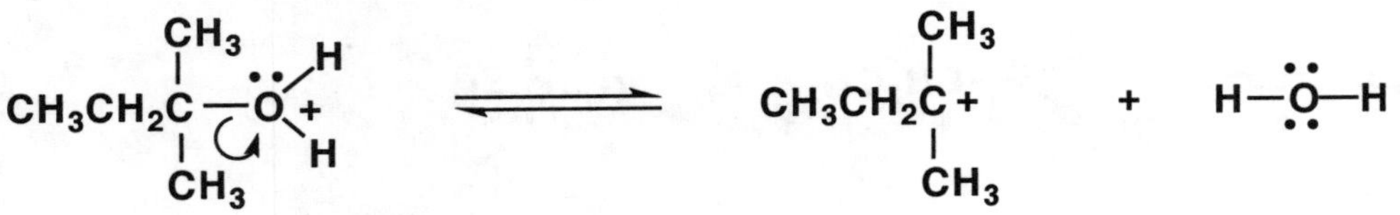

Step 3:

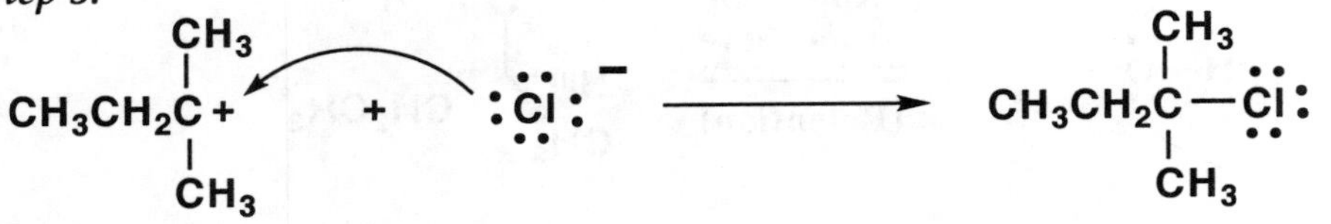

Problem 8.37 Complete the equations for these reactions.

(a) $CH_3CH_2CH_2OH \ + \ H_2CrO_4 \ \longrightarrow \ CH_3CH_2\overset{\overset{\displaystyle O}{\|}}{C}OH \ + \ Cr^{3+}$

(b) $\underset{\underset{CH_3}{|}}{CH_3CHCH_2CH_2OH} \ + \ SOCl_2 \ \longrightarrow \ \underset{\underset{CH_3}{|}}{CH_3CHCH_2CH_2Cl} \ + \ SO_2 \ + \ HCl$

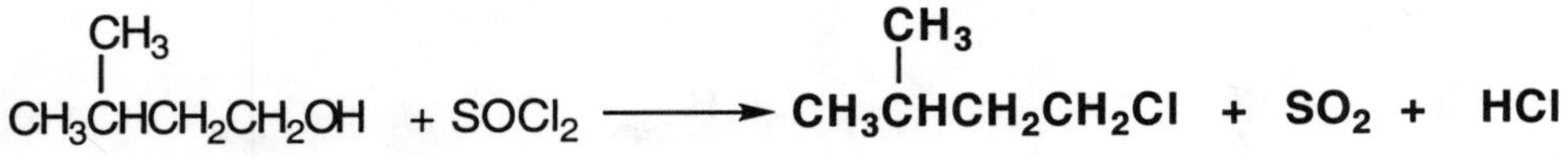

(c)   + HCl $\longrightarrow$ + H_2O

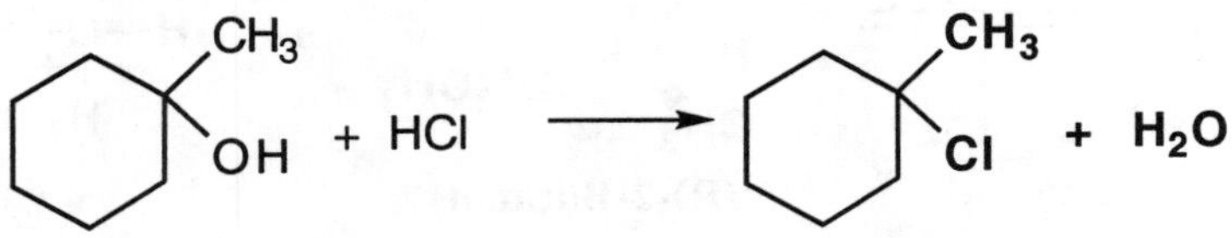

(d) $HOCH_2CH_2CH_2CH_2OH + 2\,HBr \ \xrightarrow{\text{heat}} \ BrCH_2CH_2CH_2CH_2Br \ + \ 2H_2O$

(e) $+ \ H_2CrO_4 \ \longrightarrow$ $+ \ Cr^{3+}$

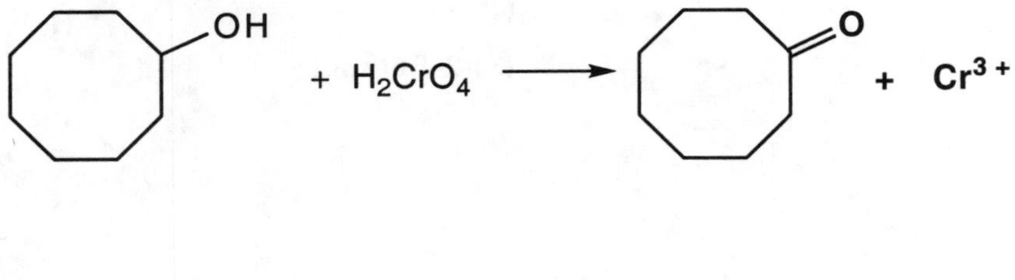

(f)

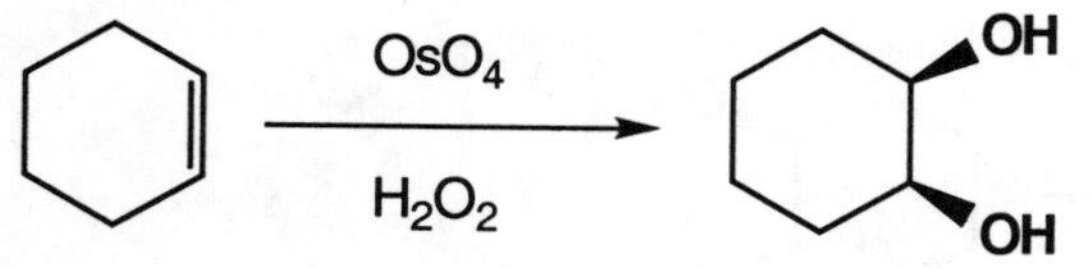

Problem 8.38 In the commercial synthesis of methyl *tert*-butyl ether (MTBE), an antiknock, octane-improving gasoline additive, 2-methylpropene and methanol are passed over an acid catalyst to give the ether. Propose a mechanism for this reaction.

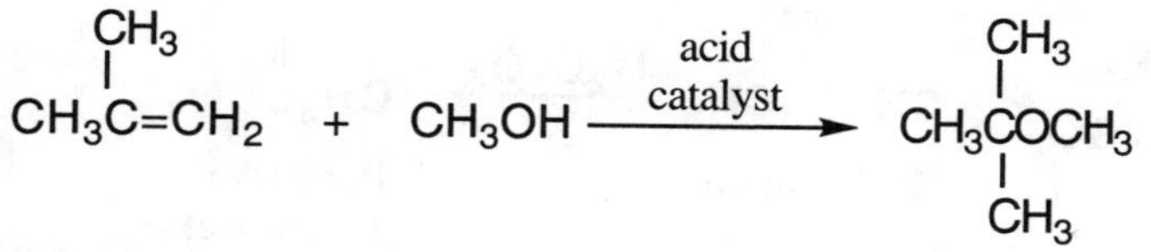

2-Methylpropene   Methanol   2-Methoxy-2-methylpropane
(Isobutylene)   (Methyl *tert*-butyl ether)

The mechanism for this reaction is analogous to acid-catalyzed hydration of an alkene (Section 6.3). The first step involves protonation of the alkene to produce a carbocation intermediate. Methanol reacts as a nucleophile with the carbocation intermediate, followed by loss of a proton, to give the final product.

Step 1:

Step 2:

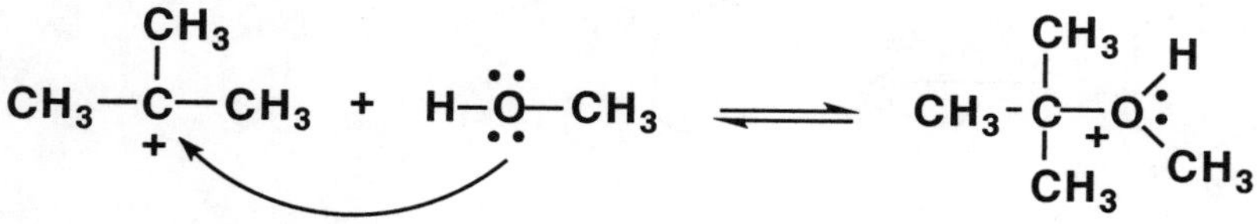

Step 3:

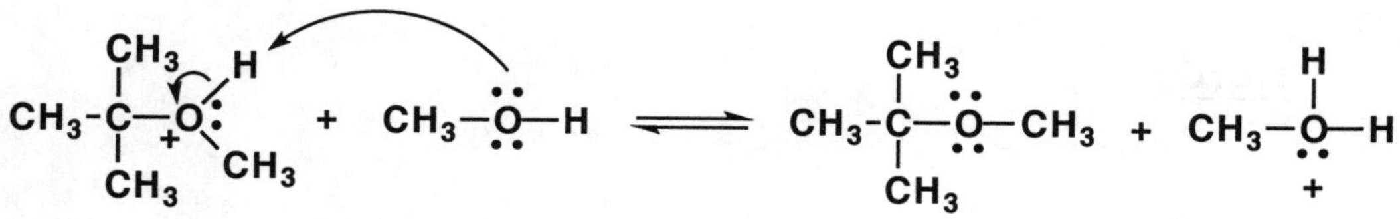

Syntheses

Problem 8.39 Show how to convert:
(a) 1-Propanol to 2-propanol in two steps.

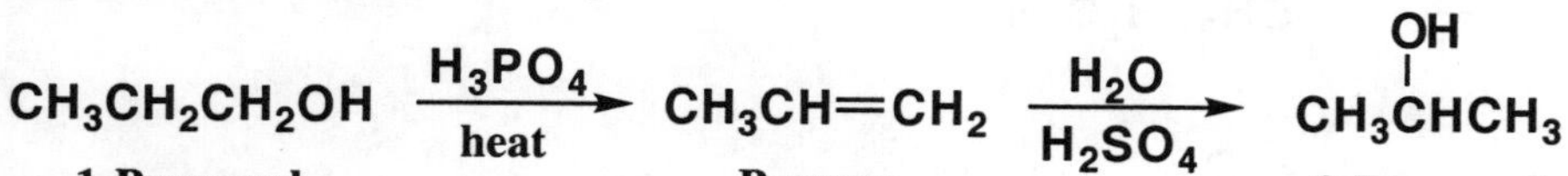

$$CH_3CH_2CH_2OH \xrightarrow[\text{heat}]{H_3PO_4} CH_3CH=CH_2 \xrightarrow[H_2SO_4]{H_2O} CH_3CHCH_3$$

1-Propanol   Propene   2-Propanol

Note how the hydration of propene will give 2-propanol as the major (Markovnikov) product.

(b) Cyclohexene to cyclohexanone in two steps.

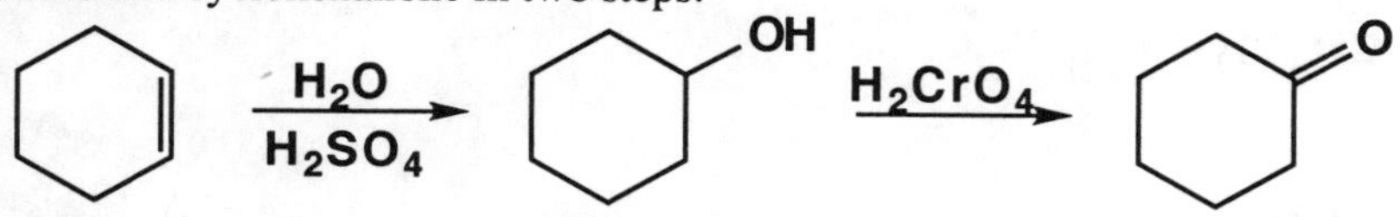

Cyclohexene   Cyclohexanol   Cyclohexanone

(c) Cyclohexanol to *cis*-1,2-cyclohexanediol in two steps.

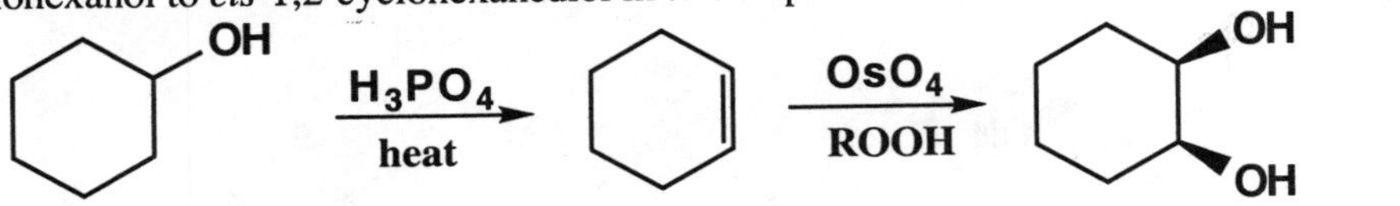

Cyclohexanol **Cyclohexene** ***cis*-1,2-Cyclohexanediol**

(d) Propene to propanone (acetone) in two steps.

$$CH_3CH{=}CH_2 \xrightarrow[H_2SO_4]{H_2O} CH_3\underset{}{C}H(OH)CH_3 \xrightarrow{H_2CrO_4} CH_3\overset{O}{\overset{\|}{C}}CH_3$$

Propene **2-Propanol** **Propanone (Acetone)**

Problem 8.40 Show how to convert cyclohexanol to these compounds:

(a) Cyclohexene

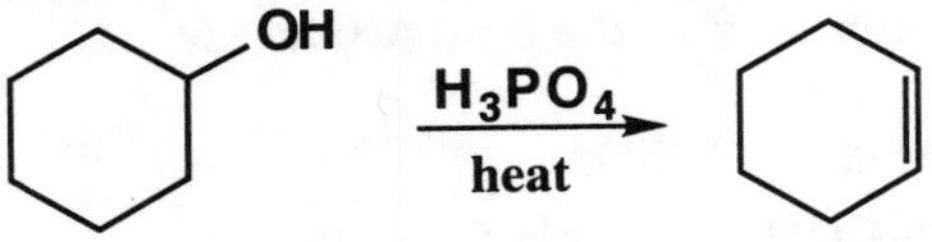

(b) Cyclohexane

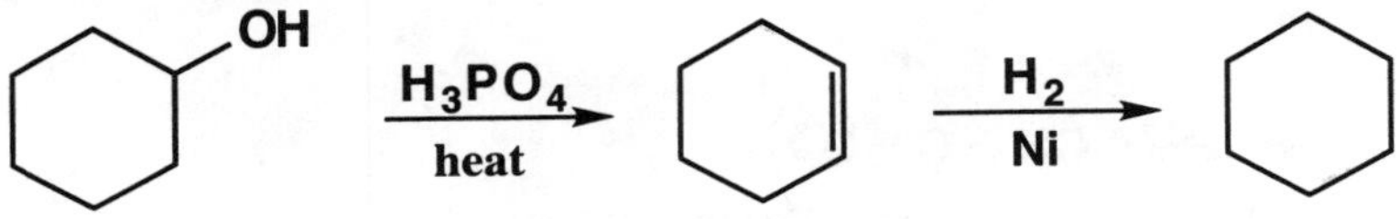

(c) Cyclohexanone

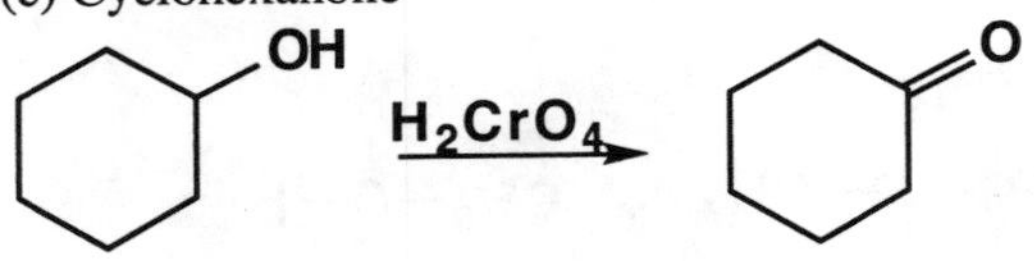

Problem 8.41 Show reagents and experimental conditions to synthesize these compounds from 1-propanol. Any derivative of 1-propanol prepared in an earlier part of this problem may then be used for a later synthesis.

(a) Propanal

$$CH_3CH_2CH_2OH + PCC \longrightarrow CH_3CH_2\overset{O}{\overset{\|}{C}}\text{-}H$$

(b) Propanoic acid

$$CH_3CH_2CH_2OH \xrightarrow{H_2CrO_4} CH_3CH_2\overset{O}{\overset{\|}{C}}\cdot OH$$

(c) Propene

$$CH_3CH_2CH_2OH \xrightarrow[heat]{H_3PO_4} CH_3CH{=}CH_2$$

(d) 2-Propanol

$$CH_3CH{=}CH_2 \xrightarrow[H_2O]{H_2SO_4} CH_3\overset{OH}{\overset{|}{C}}HCH_3$$

From (c)

(e) 2-Bromopropane

OH
|
CH₃CHCH₃   or   H₂C=CHCH₃   →HBr→   CH₃CHCH₃
From (d)   From (c) |
                                              Br

(f) 1-Chloropropane

CH₃CH₂CH₂OH   →SOCl₂→   CH₃CH₂CH₂Cl

(g) Propanone

OH   O
| ‖
CH₃CHCH₃   →H₂CrO₄→   CH₃CCH₃
From (d)

(h) 1,2-Propanediol

                          HO OH
                          | |
CH₃CH=CH₂   →OsO₄ / ROOH→   CH₃CHCH₂
From (c)

Problem 8.42 Show how to prepare each compound from 2-methyl-1-propanol (isobutyl alcohol). For any preparation involving more than one step, show each intermediate compound formed.

        CH₃   CH₃   CH₃
        | | |
(a) CH₃C=CH₂   CH₃CHCH₂OH   →H₃PO₄ / heat→   CH₃C=CH₂

        CH₃   CH₃   CH₃   CH₃
        | | | |
(b) CH₃CCH₃   CH₃CHCH₂OH →H₃PO₄/heat→ CH₃C=CH₂ →H₂O / H₂SO₄→ CH₃CCH₃
        | |
        OH   OH

        CH₃   CH₃   CH₃   CH₃
        | | | |
(c) CH₃C-CH₂   CH₃CHCH₂OH →H₃PO₄/heat→ CH₃C=CH₂ →OsO₄ / H₂O₂→ CH₃C-CH₃
        | | | |
        HO OH   HO OH

        CH₃   CH₃   CH₃
        | | |
(d) CH₃CHCO₂H   CH₃CHCH₂OH   →H₂CrO₄→   CH₃CHCO₂H

Problem 8.43 Show how to prepare each compound from 2-methylcyclohexanol. For any preparation involving more than one step, show each intermediate compound formed.

(a) <image: cyclohexene ring with CH₃ on double-bond carbon>   <image: cyclohexane ring with CH₃ and OH on adjacent carbons>   →H₃PO₄ / heat→   <image: cyclohexene ring with CH₃>

(b)

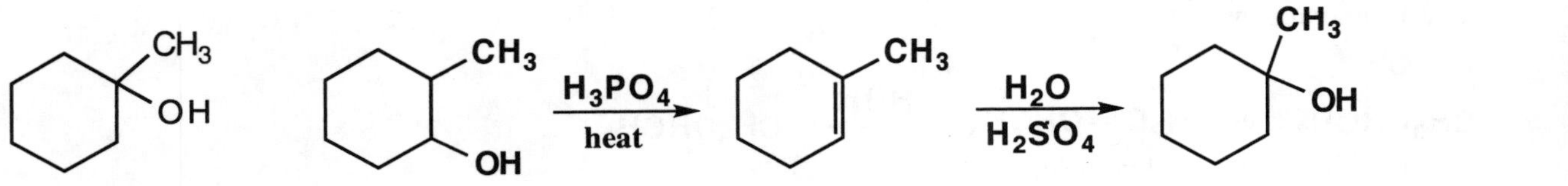

(c)

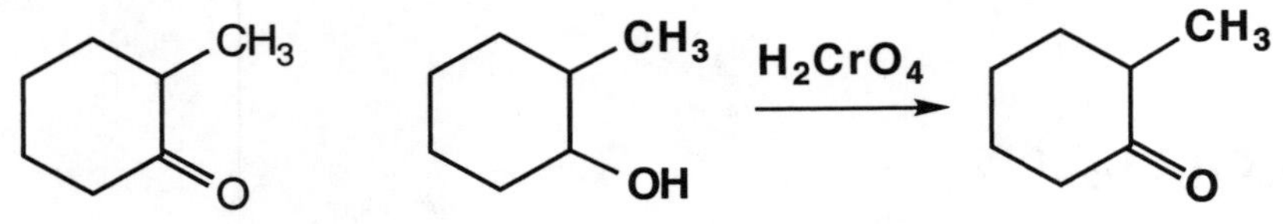

(d)

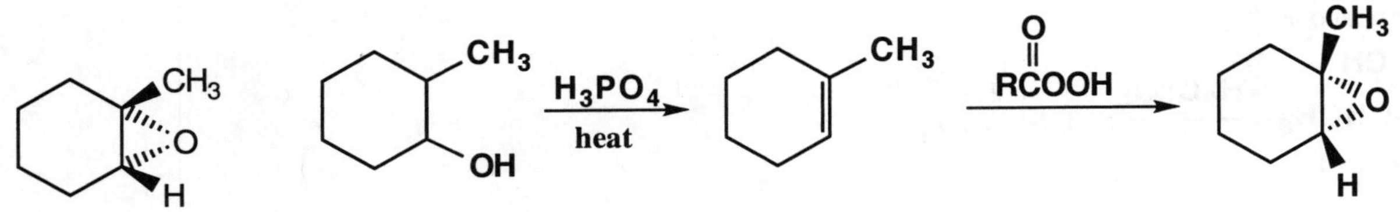

(e)

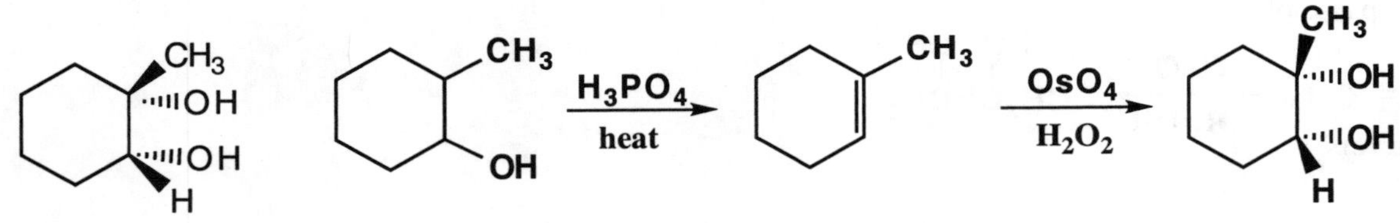

(f)

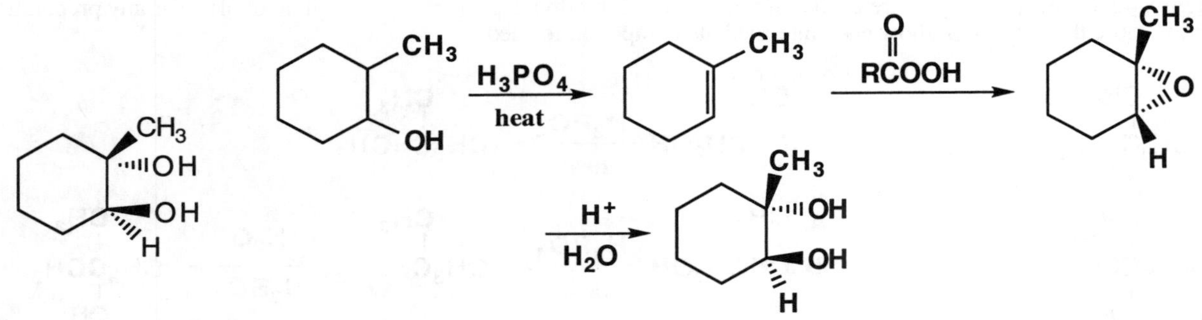

Notice that in parts (e) and (f), it is the stereochemistry of the products that determines which reagents must be used. In the case of (e), the *cis*-diol product is desired, so OsO₄ can be used. In the case of (f), the *trans*-diol product is desired, and this can be produced by reaction of the epoxide with H⁺ and H₂O.

Problem 8.44 Show how to convert the compound on the left to compounds (a), (b), and (c).

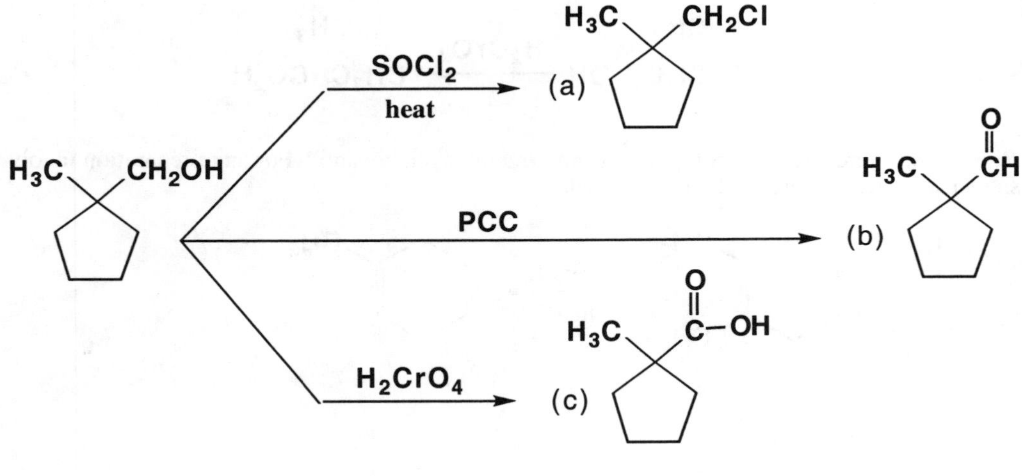

<u>Problem 8.45</u> Disparlure, a sex attractant of the gypsy moth (*Porthetria dispar*), has been synthesized in the laboratory from the following (Z)-alkene.

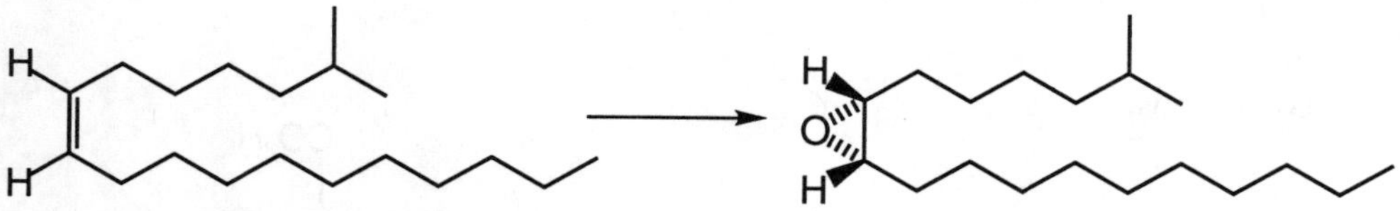

(Z)-2-Methyl-7-octadecene   Disparlure

(a) How might the (Z)-alkene be converted to disparlure?

Epoxides can be generated from alkenes using peroxycarboxylic acids. The configuration about the carbon-carbon double bond is maintained in the reaction.

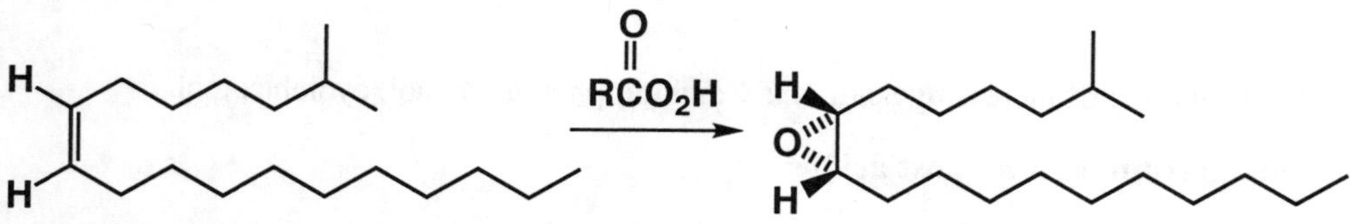

(Z)-2-Methyl-7-octadecene   Disparlure

(b) How many stereoisomers are possible for disparlure? How many are formed in the sequence you chose?

There are two stereocenters in disparlure, marked with asterisks in the structure below. Thus, there are $2^2 = 4$ possible stereoisomers: (7R,8R), (7S,8S), (7R,8S) and (7S,8R). Because the configuration about the double bond of the alkene is maintained in the reaction, only two stereoisomers can be formed. The stereoisomer shown below is (7R,8S). Because the peroxycarboxylic acid can approach the alkene from either side, an equal amount of the (7S,8R) enantiomer will be formed.

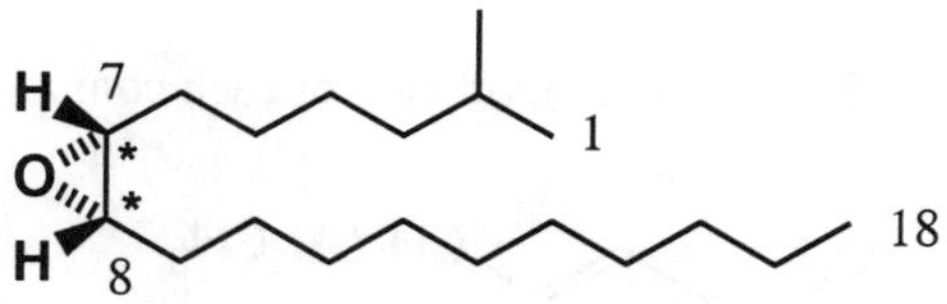

Disparlure

<u>Problem 8.46</u> The chemical name for bombykol, the sex pheromone secreted by the female silkworm moth to attract male silkworm moths, is *trans*-10-*cis*-12-hexadecadien-1-ol. (It has one hydroxyl group and two carbon-carbon double bonds in a 16-carbon chain.)

(a) Draw a structural formula for bombykol, showing the correct configuration about each carbon-carbon double bond.

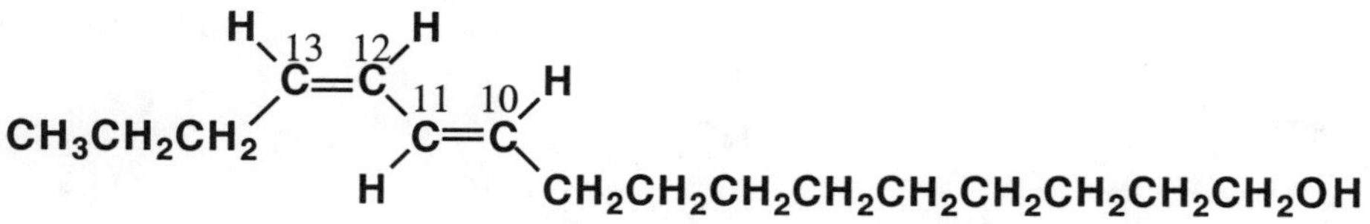

Bombykol

(b) How many cis-trans isomers are possible for this structural formula? All possible cis-trans isomers have been synthesized in the laboratory, but only the one named bombykol is produced by the female silkworm moth, and only it attracts male silkworm moths.

Bombykol has two double bonds, each of which can be either cis or trans. Thus, four stereoisomers of 10,12-hexadecadien-1-ol are possible.

CHAPTER 9
Solutions to Problems

Problem 9.1 Write names for these compounds.

(a)

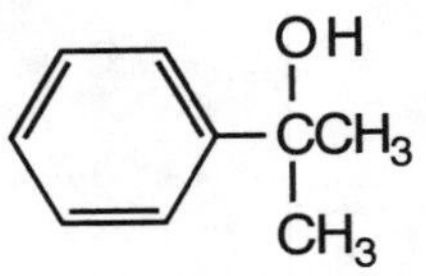

OH
|
–CCH₃
|
CH₃

2-Phenyl-2-propanol

(b)

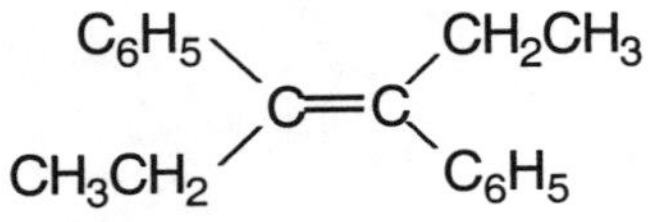

C_6H_5 \ / CH_2CH_3
  C=C
CH_3CH_2 / \ C_6H_5

(E)-3,4-Diphenyl-3-hexene

(c)

CO_2H

CH_3

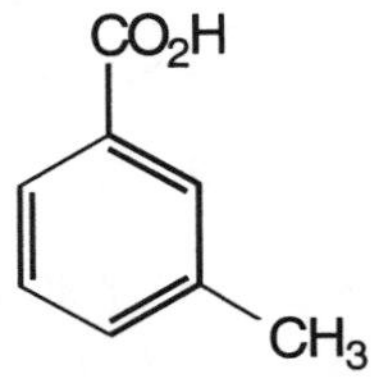

3-Methylbenzoic acid
***m*-Methylbenzoic acid**

Problem 9.2 Arrange these compounds in order of increasing acidity: 2,4-dichlorophenol, phenol, cyclohexanol.

The following compounds are ranked from least to most acidic:

Cyclohexanol   <   Phenol   <   2,4-Dichlorophenol

A good way to predict relative acidities between related compounds is to keep track of the anionic conjugate bases produced upon deprotonation. In general, the more stable the conjugate base anion, the stronger the acid. Anions become increasingly stabilized as the negative charge is more delocalized around the molecule. Thus, phenol is more acidic than an aliphatic alcohol like cyclohexanol. Resonance involving the aromatic ring of phenol leads to increased charge delocalization and stabilization of the phenoxide anion compared to the cyclohexylalkoxide anion. The electronegative chlorine atoms of 2,4-dichlorophenol withdraw electron density from the aromatic ring and help to stabilize the 2,4-dichlorophenoxide anion even more, compared to phenoxide.

Problem 9.3 Predict the products resulting from vigorous oxidation of each compound by $K_2Cr_2O_7$ in aqueous H_2SO_4.

(a)

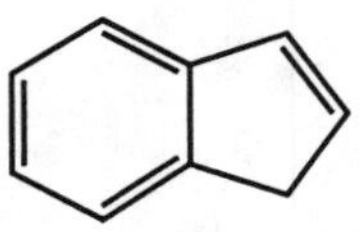

(b)

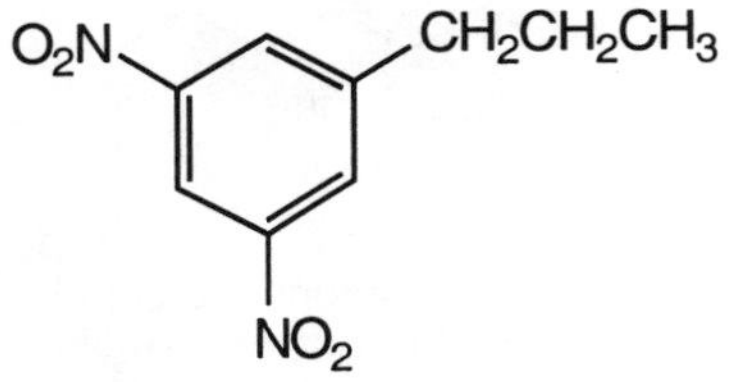

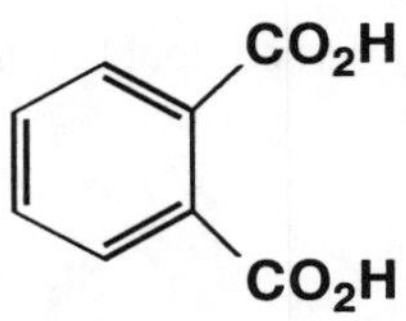

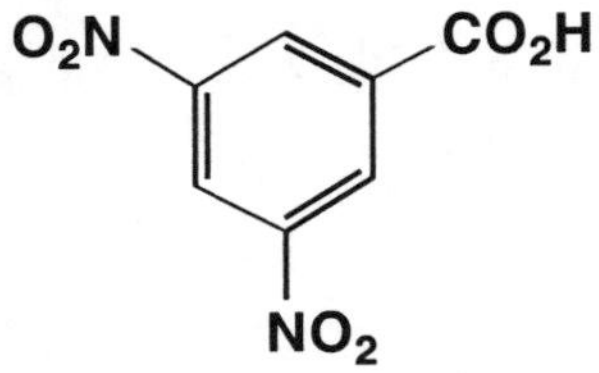

O_2N \   / $CH_2CH_2CH_3$

NO_2

CO_2H

CO_2H

O_2N \   / CO_2H

NO_2

<u>Problem 9.4</u> Write a stepwise mechanism for the sulfonation of benzene.

The electrophile HSO₃⁺ is produced in step 1 from the reaction of sulfuric acid with a proton. In step 2, reaction of benzene with the electrophile yields a resonance-stabilized cation. In step 3, this intermediate loses a proton to complete the reaction.

Step 1:

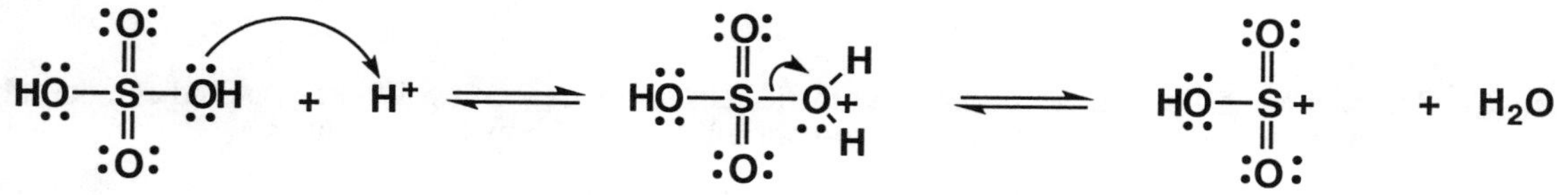

Step 2:

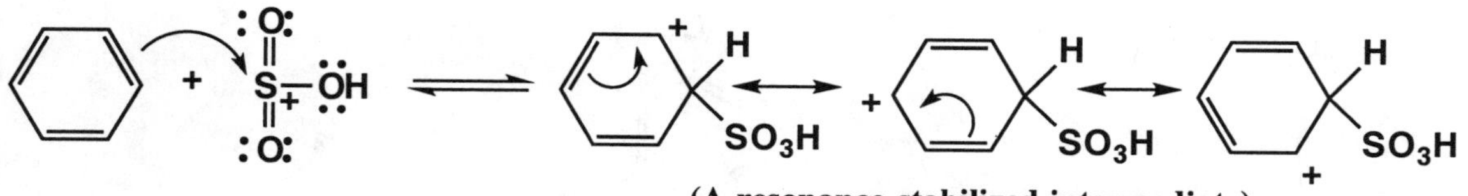

(A resonance-stabilized intermediate)

Step 3:

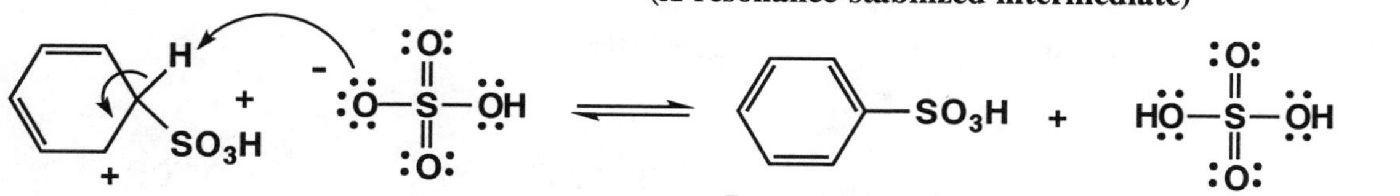

Benzenesulfonic acid

<u>Problem 9.5</u> Write a structural formula for the product formed from Friedel-Crafts alkylation or acylation of benzene with:

(a) 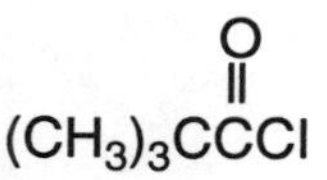 $(CH_3)_3CCCl$ (with O double bonded to C)

(b) $(CH_3)_2CHCl$

(c)

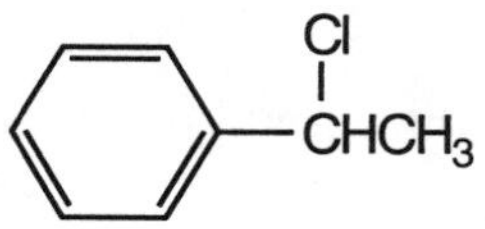

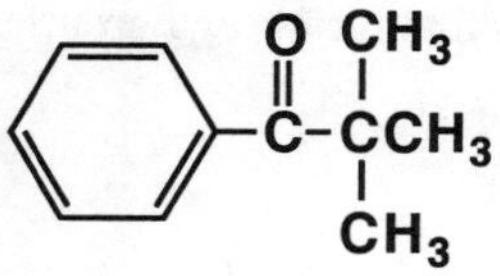

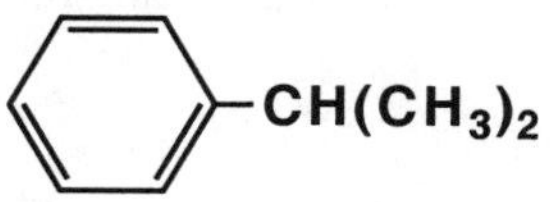

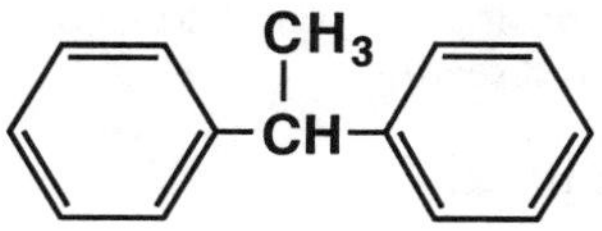

<u>Problem 9.6</u> Write a mechanism for the formation of *tert*-butylbenzene from benzene and *tert*-butyl alcohol in the presence of phosphoric acid.

Step 1:

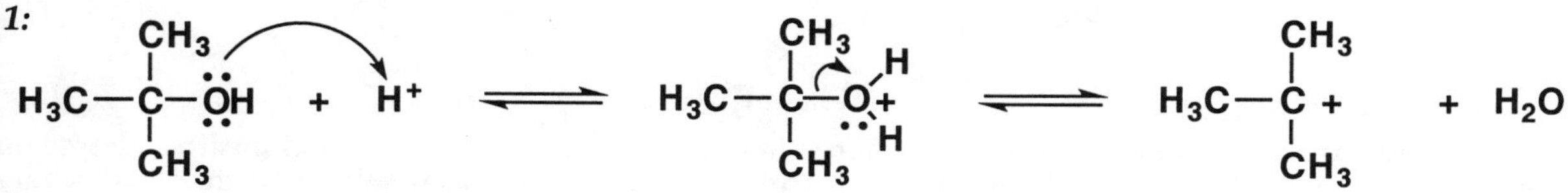

Step 2:

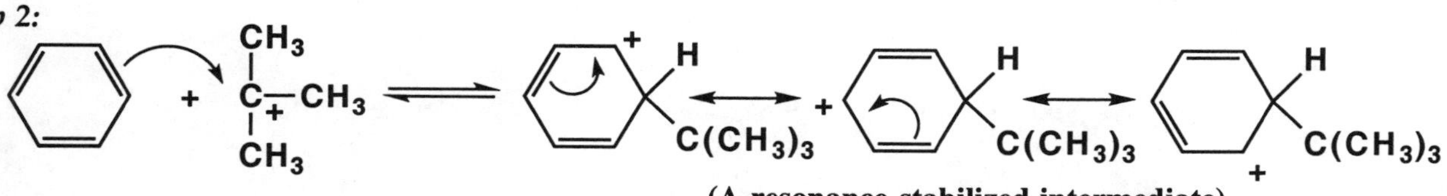

(A resonance-stabilized intermediate)

Step 3:

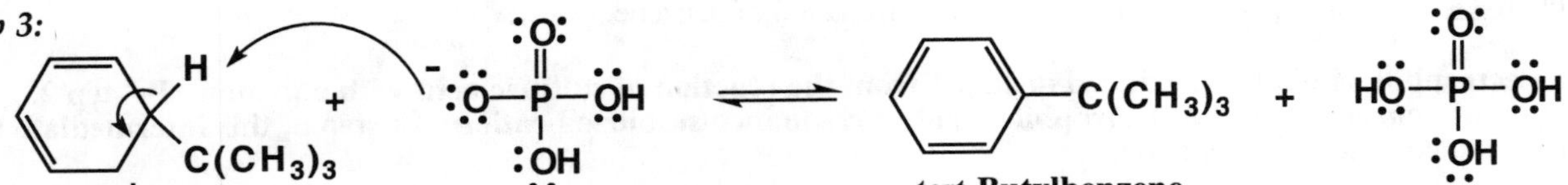

tert-Butylbenzene

Problem 9.7 Complete the following electrophilic aromatic substitution reactions. Where you predict meta substitution, show only the meta product. Where you predict ortho-para substitution, show both products.

(a)

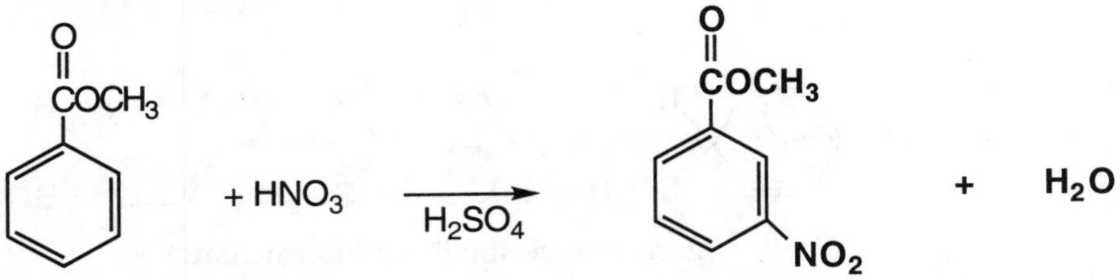

$+ HNO_3$ → H_2SO_4 → $+ H_2O$

The carboxymethyl group is meta directing and deactivating.

(b)

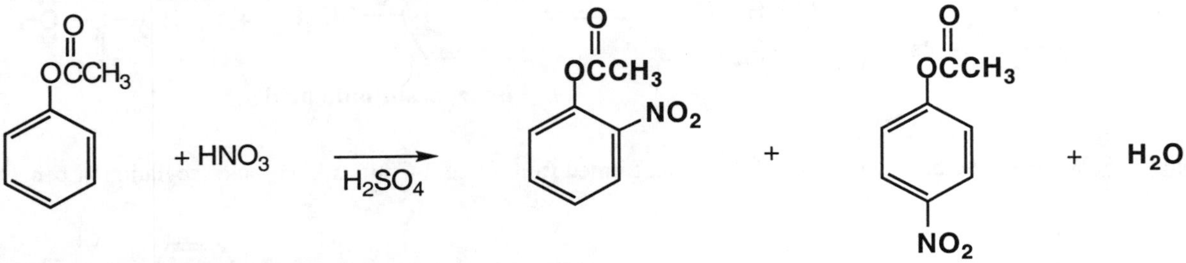

$+ HNO_3$ → H_2SO_4 → $+$ $+ H_2O$

The acetoxy group is ortho-para directing and activating.

Problem 9.8 Because the electronegativity of oxygen is greater than that of carbon, the carbon of a carbonyl group bears a partial positive charge, and its oxygen bears a partial negative charge. Using this information, show that a carbonyl group is meta- directing.

Para attack:

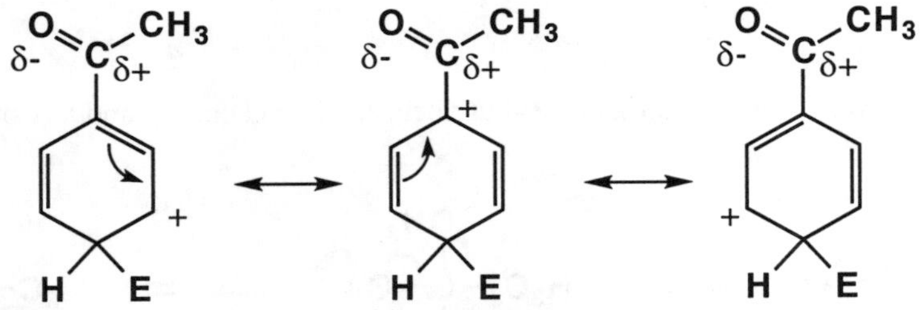

The middle resonance form places the positive charge in the ring adjacent to the partial positive charge on the carbonyl carbon atom. This is a destabilizing interaction that raises the overall energy of the intermediate, so para attack is disfavored.

Ortho attack:

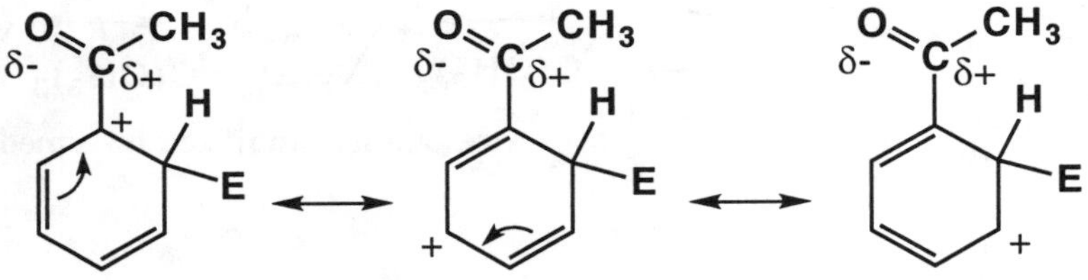

The first resonance form places the positive charge in the ring adjacent to the partial positive charge on the carbonyl carbon atom. This is a destabilizing interaction that raises the overall energy of the intermediate, so ortho attack is disfavored.

Meta attack never places the positive charge on the ring adjacent to the carbonyl carbon atom, so meta products predominate in electrophilic aromatic substitution reactions of benzene rings with attached carbonyl groups.

Nomenclature and Structural Formulas
Problem 9.9 Name these compounds.

(a)

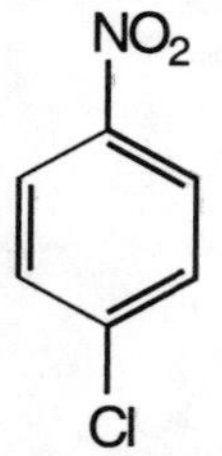

4-Chloronitrobenzene
(***p*-Chloronitrobenzene**)

(b)

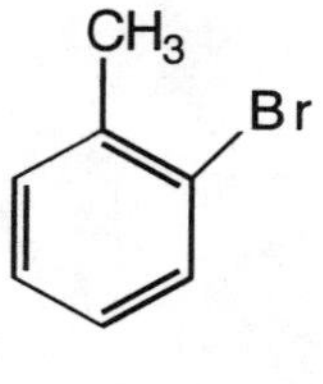

2-Bromotoluene
(***o*-Bromotoluene**)

(c) $C_6H_5CH_2CH_2CH_2OH$

3-Phenyl-1-propanol

(d) 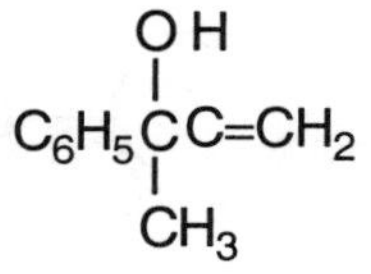 $C_6H_5\overset{OH}{\underset{CH_3}{C}}C=CH_2$

2-Phenyl-3-buten-2-ol

(e)

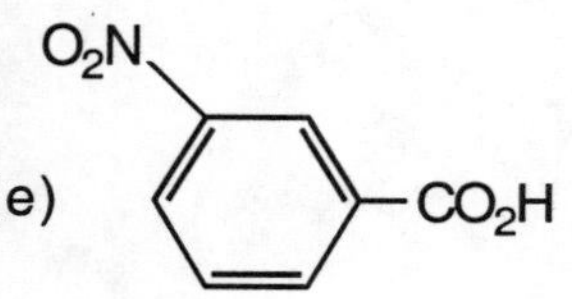

3-Nitrobenzoic acid
(***m*-Nitrobenzoic acid**)

(f)

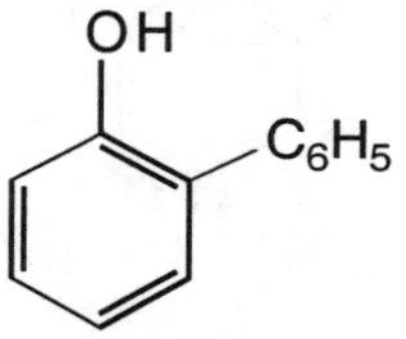

2-Phenylphenol
(***o*-Phenylphenol**)

(g)

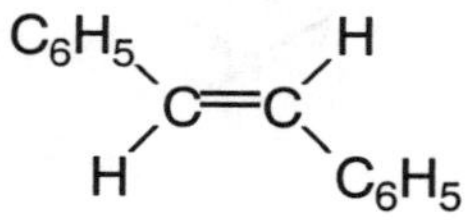

(E)-1,2-Diphenylethylene
(***trans*-1,2-Diphenylethylene**)

(h)

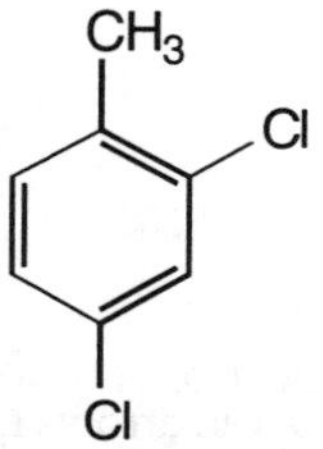

2,4-Dichlorotoluene

Problem 9.10 Draw structural formulas for these compounds.
(a) 1-Bromo-2-chloro-4-ethylbenzene   (b) 4-Iodo-1,2-dimethylbenzene 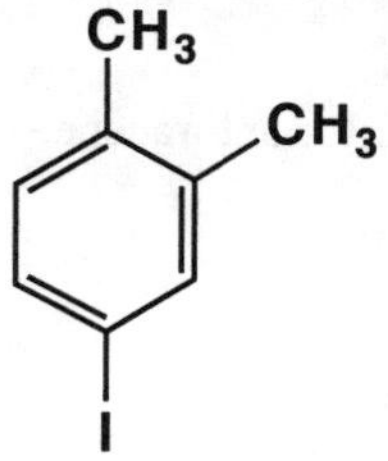   (c) 2,4,6-Trinitrotoluene

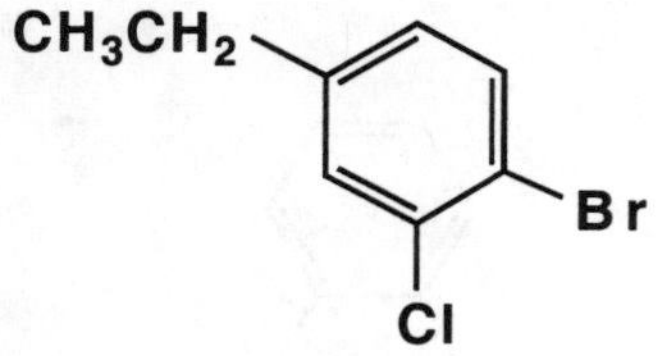

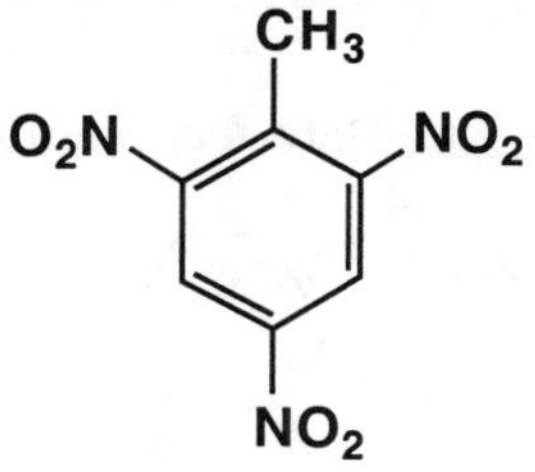

(d) 4-Phenyl-2-pentanol 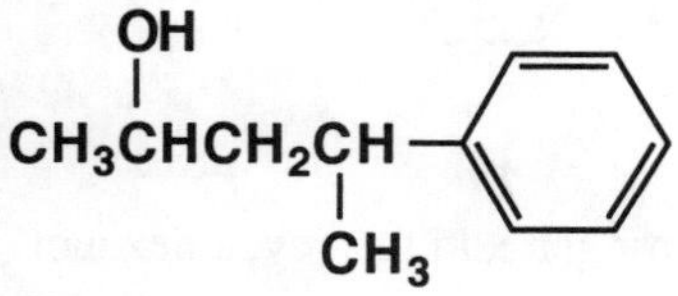   (e) *p*-Cresol    (f) 2,4-Dichlorophenol 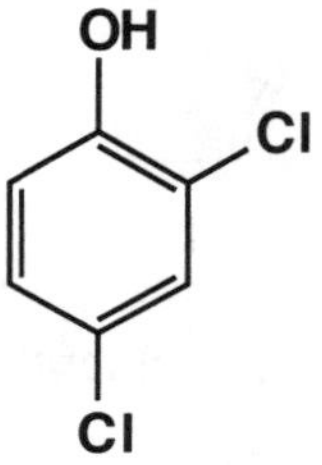   (g) 1-Phenylcyclopropanol

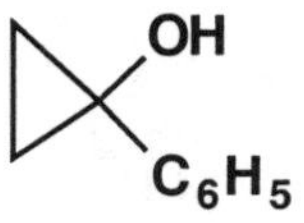

(h) Styrene (phenylethylene)

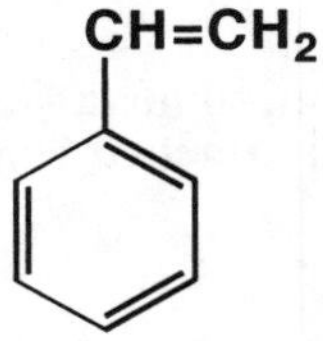

(i) *m*-Bromophenol

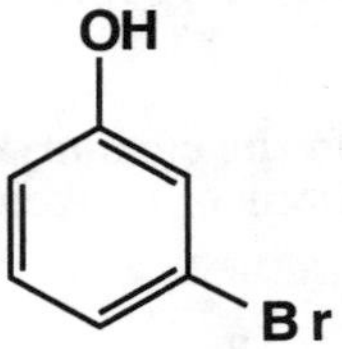

(j) 2,4-Dibromoaniline

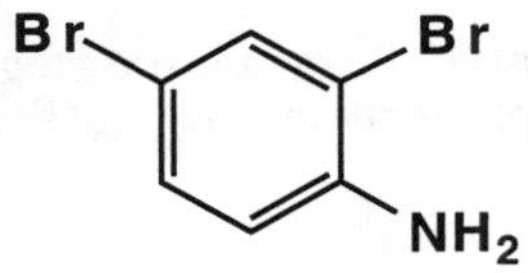

(k) Isobutylbenzene

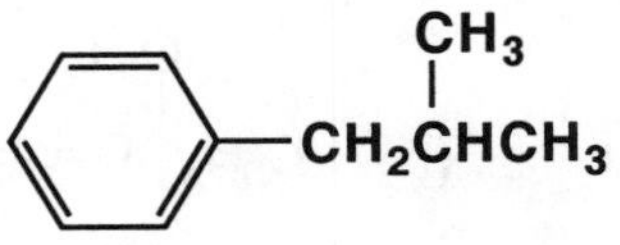

(l) *m*-Xylene

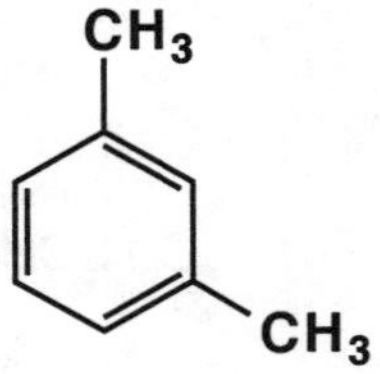

Problem 9.11 Show that pyridine can be represented as a hybrid of two equivalent contributing structures.

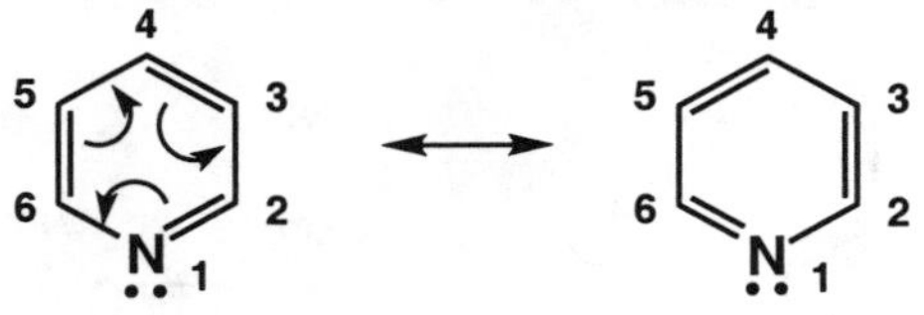

Problem 9.12 Show that naphthalene can be represented as a hybrid of three contributing structures, and show by the use of curved arrows how one contributing structure is converted to the next.

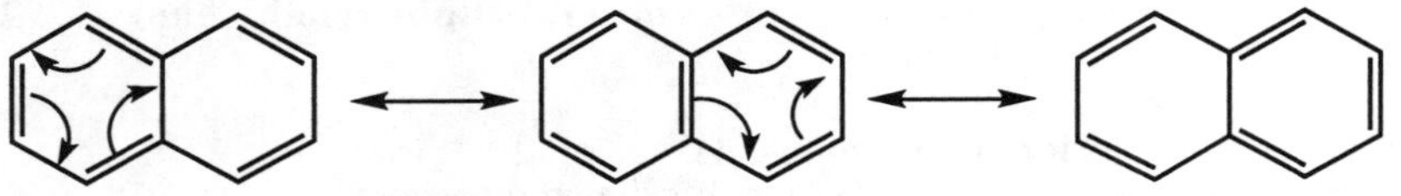

Problem 9.13 Draw four contributing structures for anthracene.

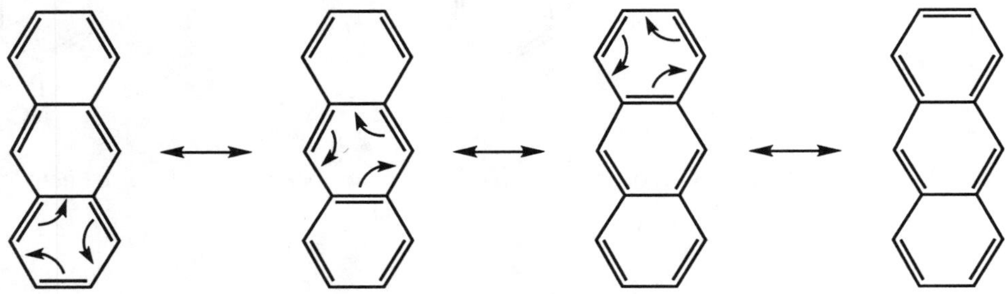

Acidity of Phenols

Problem 9.14 Use the resonance theory to account for the fact that phenol (pK$_a$ 9.95) is a stronger acid than cyclohexanol (pK$_a$ approximately 18).

A good way to predict relative acidities between related compounds is to keep track of the anionic conjugate bases produced upon deprotonation. In general, the more stable the conjugate base anion, the stronger the acid. Anions become increasingly stabilized as the negative charge is more delocalized around the molecule. Thus, phenol is significantly more acidic than an aliphatic alcohol like cyclohexanol. Resonance involving the

aromatic ring of phenol leads to increased charge delocalization and stabilization of the phenoxide anion compared to the cyclohexoxide anion. Four contributing structures for the phenoxide anion are shown below, illustrating how the negative charge is delocalized into the aromatic ring.

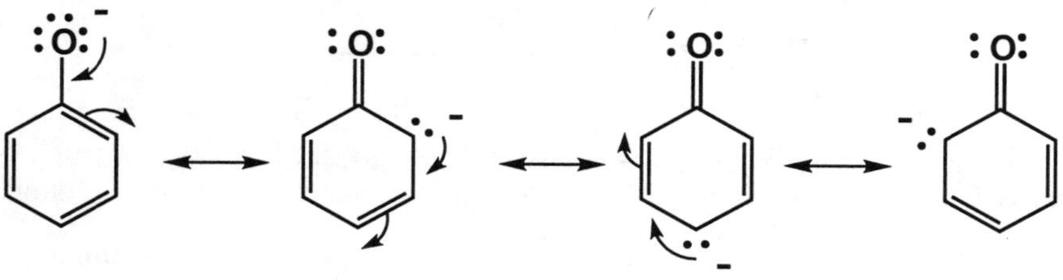

Problem 9.15 Arrange the compounds in each set in order of increasing acidity (from least acidic to most acidic).

(a) 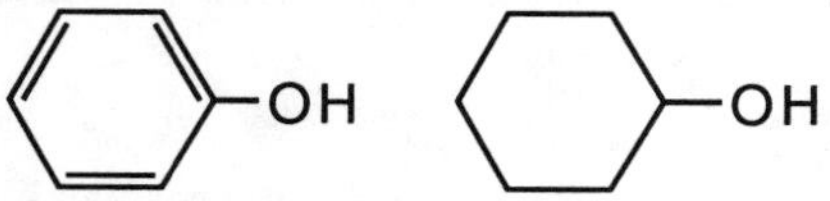 CH_3CO_2H

To arrange these in order of increasing acidity, refer to Table 2.1. For those compounds not listed in Table 2.1, estimate pKa using values for compounds that are given in the table.

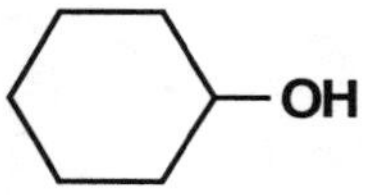	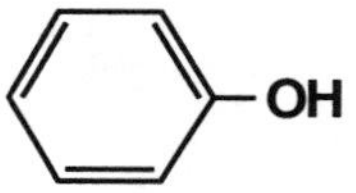	$\mathbf{CH_3CO_2H}$
pK_a ~ 18	**pK_a 9.95**	**pK_a 4.76**

(b) 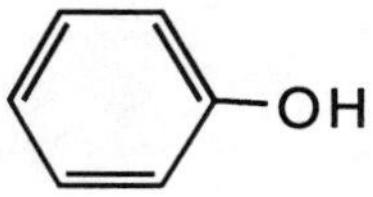 $NaHCO_3$ H_2O

$\mathbf{H_2O}$	$\mathbf{HCO_3^-}$	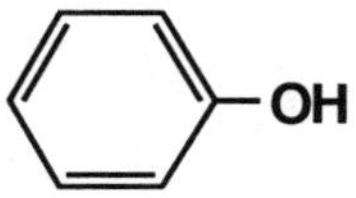
pK_a 15.7	**pK_a 10.33**	**pK_a 9.95**

(c) 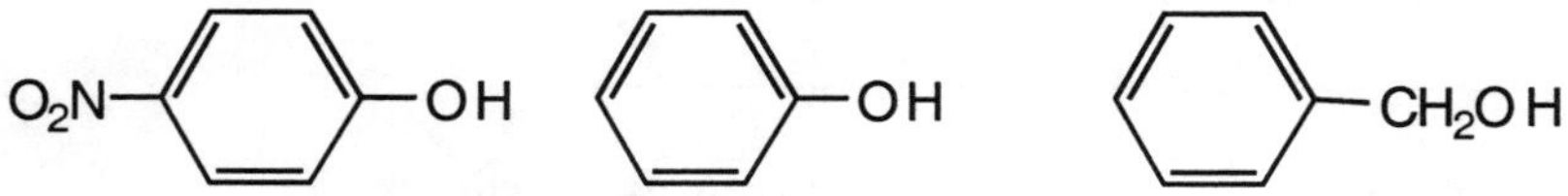

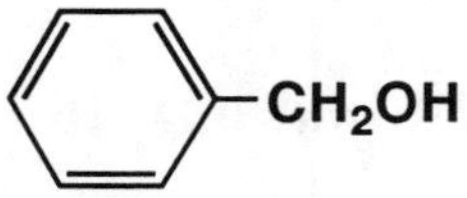	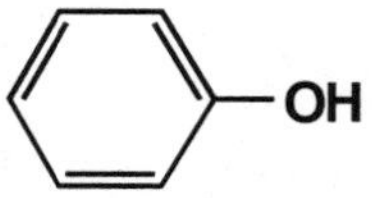	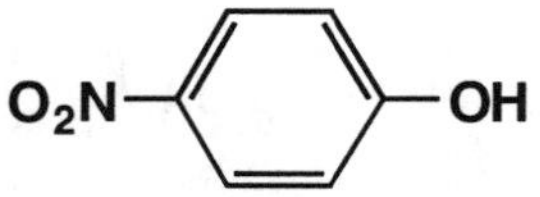
pKa ~ 18	**pKa 9.95**	**pKa 7.15**

Problem 9.16 From each pair, select the stronger base.

The stronger base in each pair is circled. To estimate which is the stronger base, first determine which conjugate acid is the weaker acid. The weaker the acid, the stronger its conjugate base.

(a) 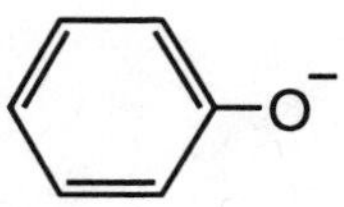 or OH^- (b) 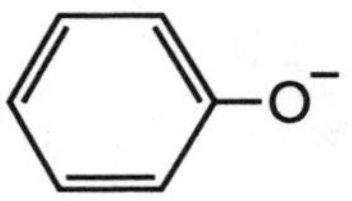 or 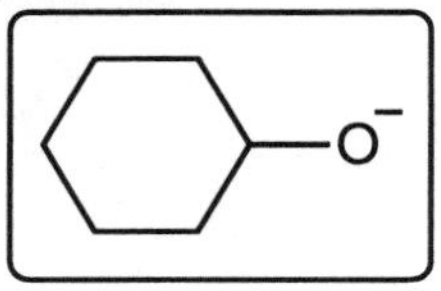

(c) [benzene ring]—O⁻   or   [HCO₃⁻ in box]   (d) [benzene ring in box]—O⁻   or   CH₃CO₂⁻

Problem 9.17 Account for the fact that water-insoluble carboxylic acids (pK$_a$ 4-5) dissolve in 10% sodium bicarbonate with the evolution of a gas, but water-insoluble phenols (pK$_a$ 9.5-10.5) do not dissolve in this solution.

Carbonic acid (H$_2$CO$_3$), the conjugate acid of bicarbonate, has a pK$_a$ of 6.36. Bicarbonate is therefore basic enough to deprotonate carboxylic acids and, in the process, generate the water-soluble, negatively charged conjugate base of the carboxylic acid (a carboxylate ion) and carbonic acid. As we saw in Chapter 2, carbonic acid can dissociate to give a molecule of water and a molecule of carbon dioxide (CO$_2$), which is a gas and, therefore, forms bubbles as it leaves solution. Because the pK$_a$'s of phenols are higher than the pK$_a$ of carbonic acid, the bicarbonate ion is not a strong enough base to deprotonate a phenol to generate its water-soluble conjugate base (a phenolate ion).

Problem 9.18 Describe a procedure to separate a mixture of 1-hexanol and 2-methylphenol (o-cresol) and recover each in pure form. Each is insoluble in water but soluble in diethyl ether.

CH₃(CH₂)₄CH₂OH   [benzene ring with CH₃ and OH]

1-Hexanol   o-Cresol

Following is a flow chart for an experimental method for separating these two compounds. Separation is based on the facts that both are insoluble in water, but soluble in diethyl ether and that o-cresol reacts with 10% NaOH to form a water-soluble phenoxide salt, while 1-hexanol does not.

CH₃(CH₂)₄CH₂OH + [benzene ring with CH₃ and OH]

dissolve in diethyl ether

mix with 0.1M NaOH

**ether layer containing   aqueous layer containing
1-hexanol   sodium salt of o-cresol**

distill ether   acidify with 0.1M HCl

CH₃(CH₂)₄CH₂OH   [benzene ring with CH₃ and OH]

1-Hexanol   o-Cresol

Electrophilic Aromatic Substitution: Monosubstitution

Problem 9.19 Draw a structural formula for the compound formed by treatment of benzene with each combination of reagents.

(a) CH_3CH_2Cl / $AlCl_3$ (b) $CH_2=CH_2$ / H_2SO_4 (c) CH_3CH_2OH / H_2SO_4

All of these reagents will react with benzene to give the same product, ethyl benzene.

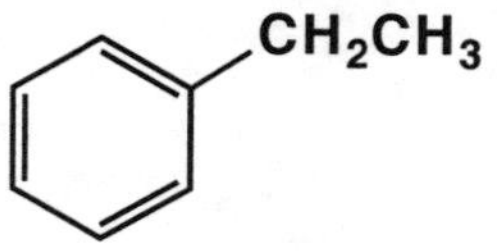

Ethyl benzene

Problem 9.20 Show three different combinations of reagents you might use to convert benzene to isopropylbenzene.

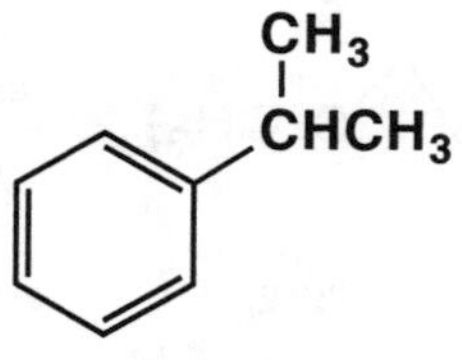

Isopropylbenzene

(1) $CH_3\overset{Cl}{\underset{|}{C}HCH_3}$ / $AlCl_3$ (2) $CH_3CH=CH_2$ / H_2SO_4 (3) $CH_3\overset{OH}{\underset{|}{C}HCH_3}$ / H_2SO_4

Note that you could use H_3PO_4 in place of H_2SO_4 for (2) and (3).

Problem 9.21 How many monochlorination products are possible when naphthalene is treated with Cl_2 / $AlCl_3$?

There are two monochlorination products, and these are shown below:

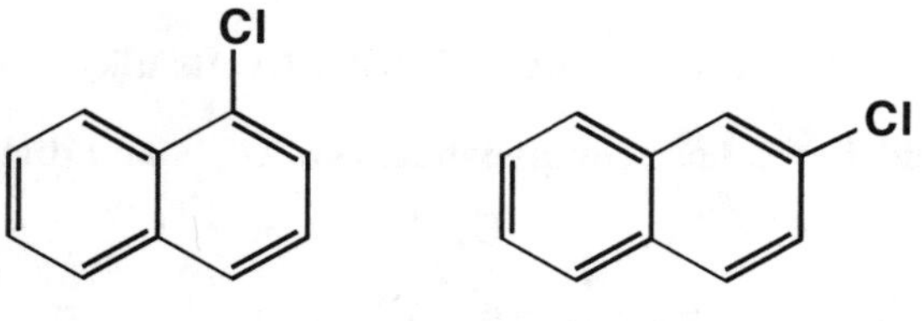

Problem 9.22 Write a stepwise mechanism of this reaction. Use curved arrows to show the flow of electrons in each step.

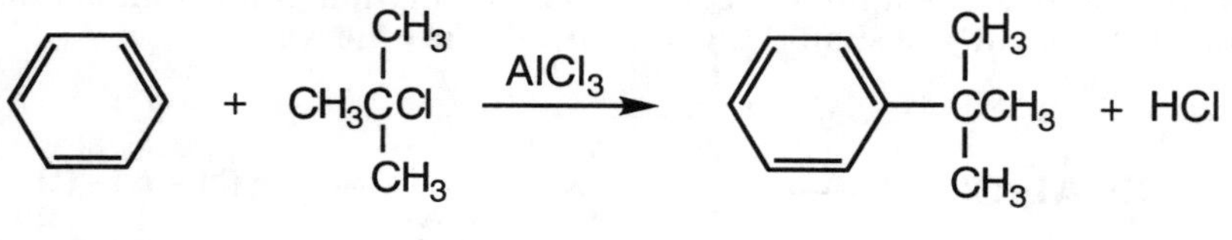

The reaction of benzene with 2-chloro-2-methylpropane in the presence of aluminum chloride involves initial formation of a complex between 2-chloro-2-methylpropane and aluminum chloride, followed by formation of an alkyl carbocation. This cationic species is the electrophile that undergoes further reaction with benzene.

Step 1: **Formation of a complex between 2-chloro-2-methylpropane (a Lewis base) and aluminum chloride (a Lewis acid).**

Step 2: **Formation of *tert*-butyl cation.**

Step 3: **Electrophilic attack on the aromatic ring.**

Resonance-stabilized carbocation intermediate

Step 4: **Proton transfer to regenerate the aromatic ring.**

Problem 9.23 Write a stepwise mechanism for the preparation of diphenylmethane by treating benzene with dichloromethane in the presence of an aluminum chloride catalyst.

Formation of diphenylmethane involves two successive Friedel-Crafts alkylations.

Step 1: **Formation of a Lewis acid-Lewis base complex between AlCl₃ and dichloromethane.**

Step 2: **Nucleophilic attack of benzene upon the complex and formation of a resonance stabilized carbocation (only one resonance form is shown; you should try to draw all the others).**

Step 3: Deprotonation of the cation to give benzyl chloride, HCl and AlCl$_3$.

Step 4: Formation of a Lewis acid-Lewis base complex between benzyl chloride and AlCl$_3$.

Step 5: Dissociation of the complex to give the resonance stabilized benzyl cation (only one contributor is shown) and AlCl$_4^-$.

Resonance-stabilized
benzyl cation

Step 6: Reaction of benzene with the benzyl cation to form another resonance stabilized carbocation (again, only one resonance contributor is shown).

Resonance-stabilized
carbocation intermediate

Step 7: Deprotonation of the carbocation intermediate to give diphenylmethane, HCl and AlCl$_3$,

Electrophilic Aromatic Substitution - Disubstitution

Problem 9.24 When treated with Cl$_2$ / AlCl$_3$, 1,2-dimethylbenzene (*o*-xylene) gives a mixture of two products. Draw structural formulas for these products.

For this problem, assume that only one chlorine atom is added to the ring. Because of symmetry considerations, there are two different products that can be produced when *o*-xylene undergoes chlorination.

Problem 9.25 How many monosubstitution products are possible when 1,4-dimethylbenzene (*p*-xylene) is treated with Cl_2 / $AlCl_3$? When *m*-xylene is treated with Cl_2 / $AlCl_3$?

There is only one monosubstitution product formed from *p*-xylene:

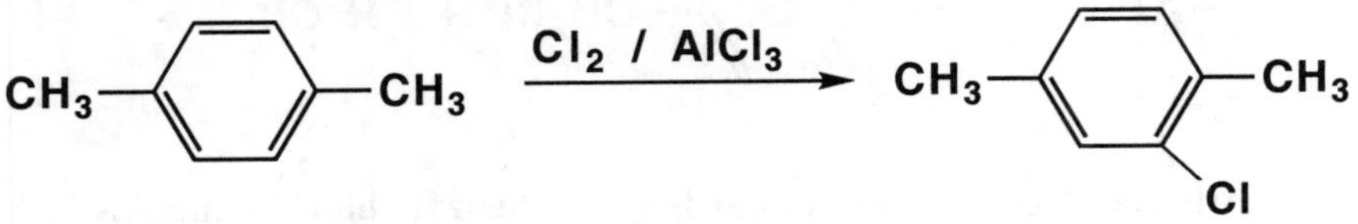

There are two products formed from *m*-xylene:

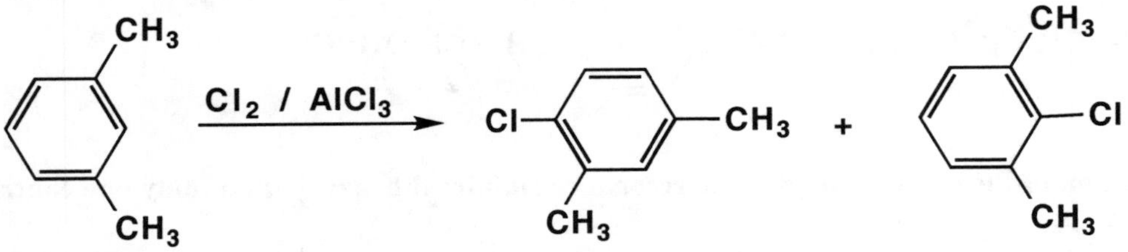

Note that the last product, the one with the chlorine atom between the methyl groups, will be formed in the smaller amount because of the combined steric effects of the two methyl groups.

Problem 9.26 Draw the structural formula for the major product formed on treatment of each compound with Cl_2 / $AlCl_3$.
(a) Toluene

The methyl group of toluene is weakly activating and ortho-para directing.

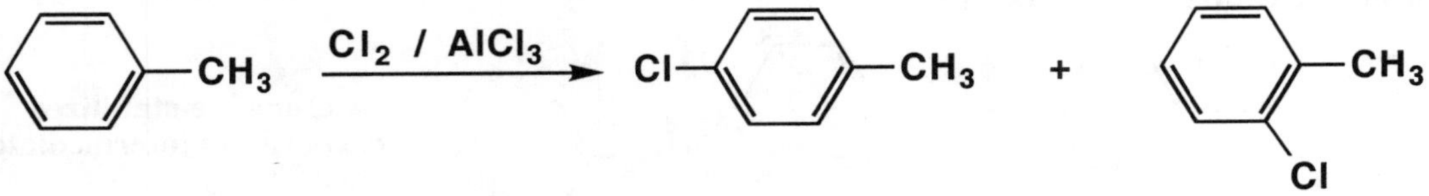

(b) Nitrobenzene

The nitro group of nitrobenzene is strongly deactivating and meta directing.

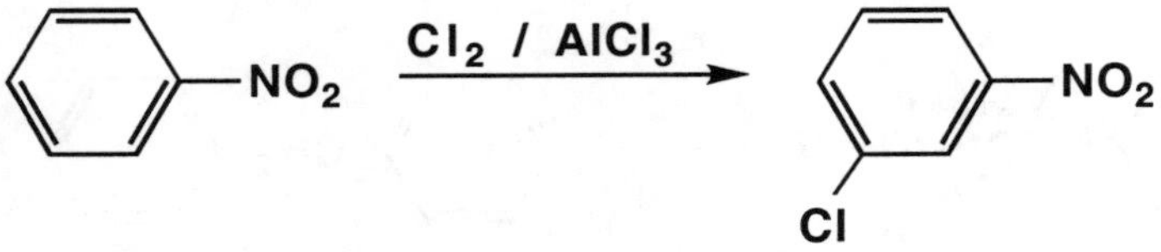

(c) Benzoic acid

The carboxyl group of benzoic acid is moderately deactivating and meta directing.

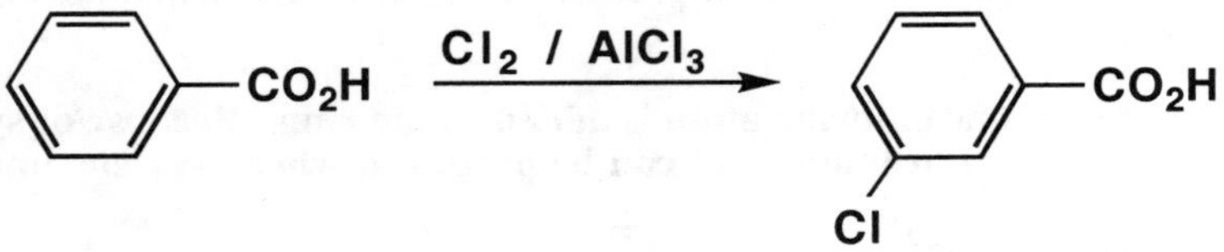

(d) Chlorobenzene

The chlorine atom of chlorobenzene is weakly deactivating and ortho-para directing.

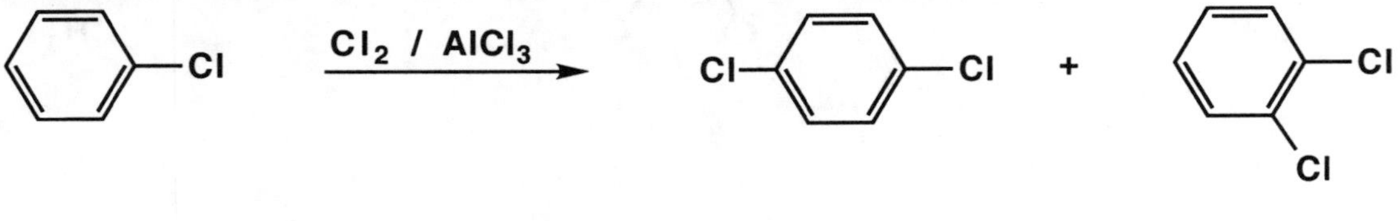

(e) *tert*-Butylbenzene

The *tert*-butyl group of *tert*-butylbenzene is weakly activating and ortho-para directing.

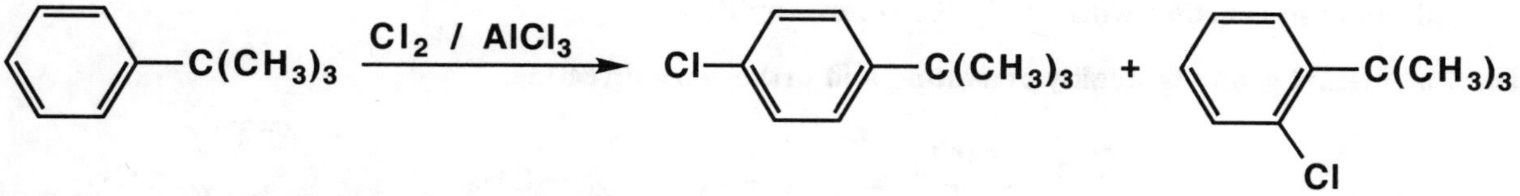

The ortho product will be formed in very small amounts because of the steric influence of the *tert*-butyl group.

(f)

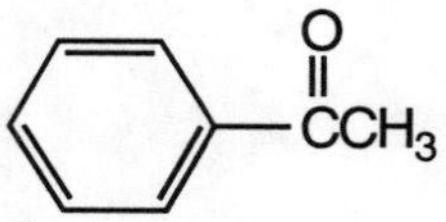

The carbonyl group of a ketone is moderately deactivating and meta directing.

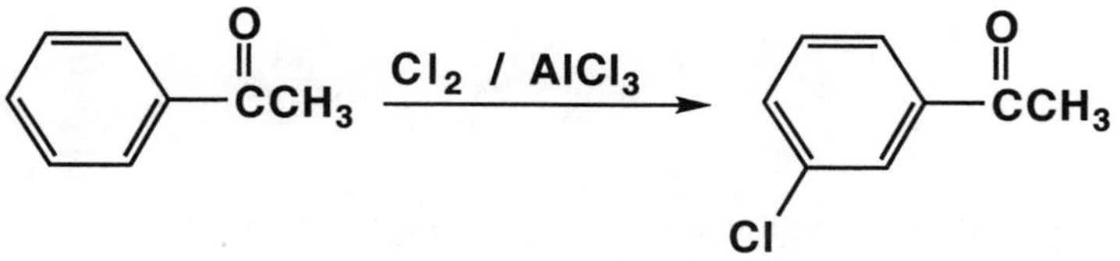

(g)

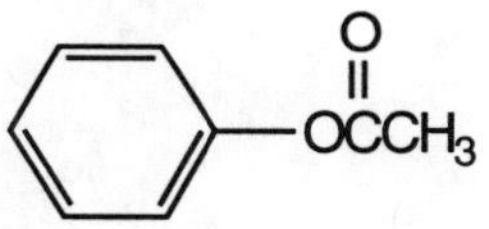

An ester with the oxygen atom attached to the ring is moderately activating and ortho-para directing.

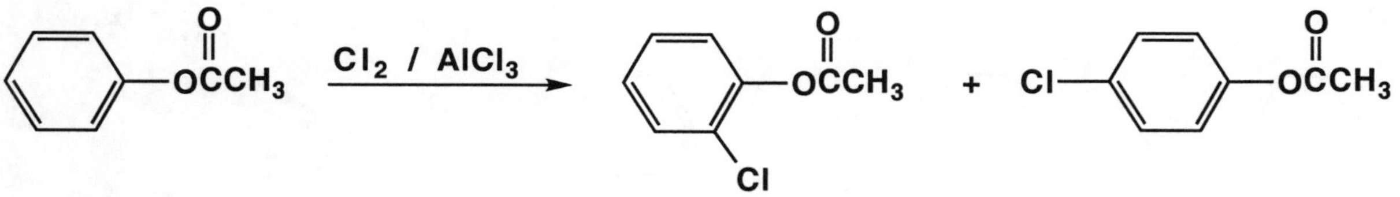

(h)

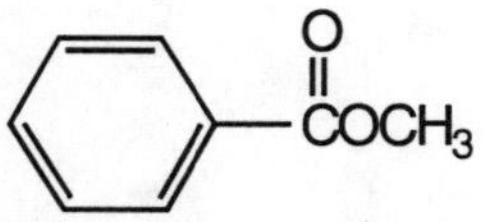

A carbonyl group is moderately deactivating and meta directing.

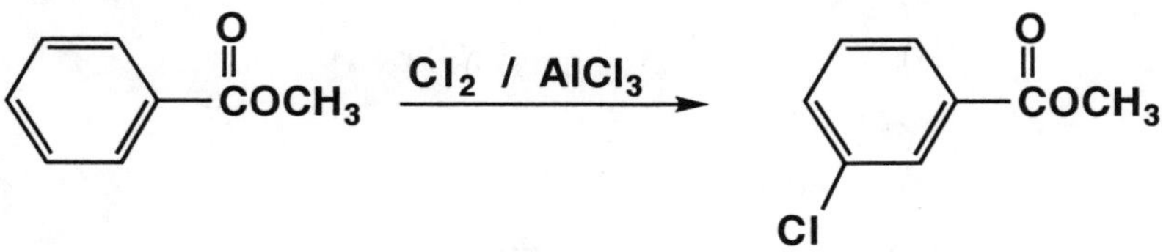

<u>Problem 9.27</u> Which compound undergoes electrophilic aromatic substitution more rapidly when treated with Cl_2 / $AlCl_3$, chlorobenzene or toluene? Explain and draw structural formulas for the major product(s) from each reaction.

The electron-withdrawing effect of the chlorine atom makes chlorobenzene less reactive than toluene. Thus, toluene will undergo reaction with Cl_2 / $AlCl_3$ more rapidly.

Toluene: the methyl group is weakly activating and ortho-para directing.

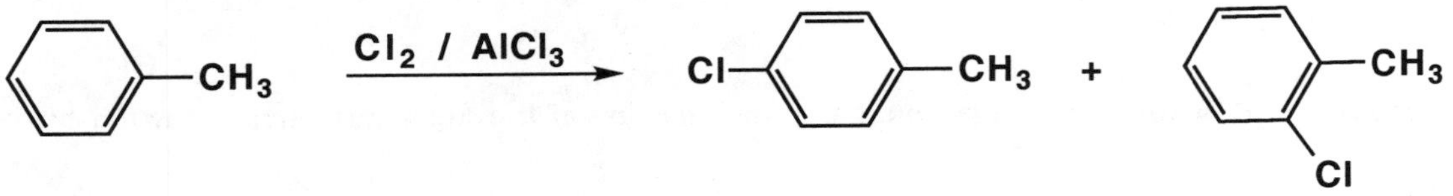

Chlorobenzene: the chlorine atom is weakly deactivating and ortho-para directing.

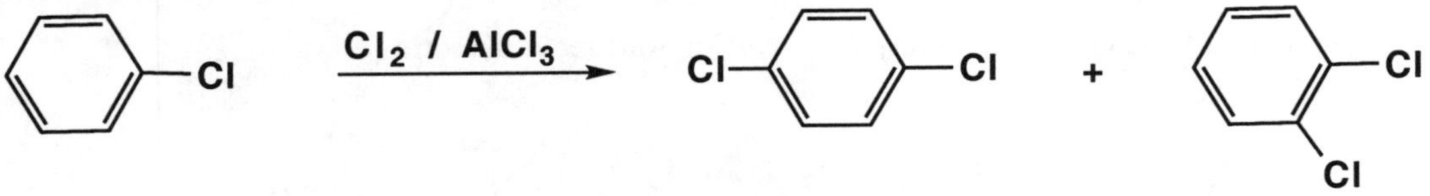

<u>Problem 9.28</u> Arrange the compounds in each set in order of decreasing reactivity (fastest to slowest) toward electrophilic aromatic substitution.

(a)

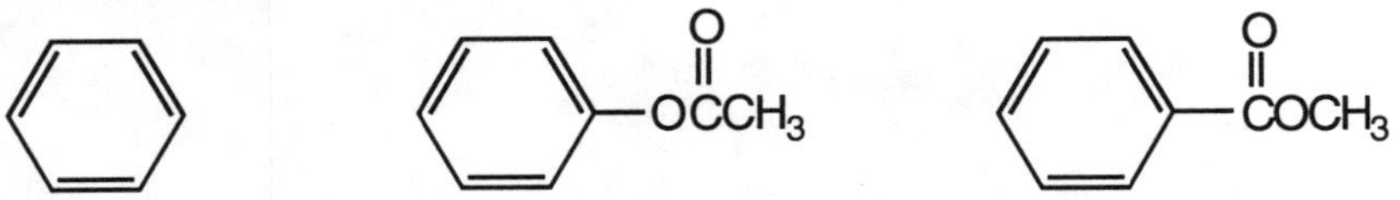

(A)   (B)   (C)

B>A>C

(b)

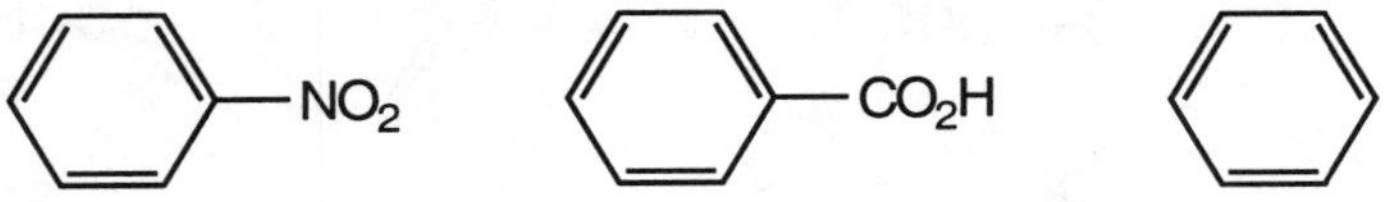

(A)   (B)   (C)

C>B>A

(c)

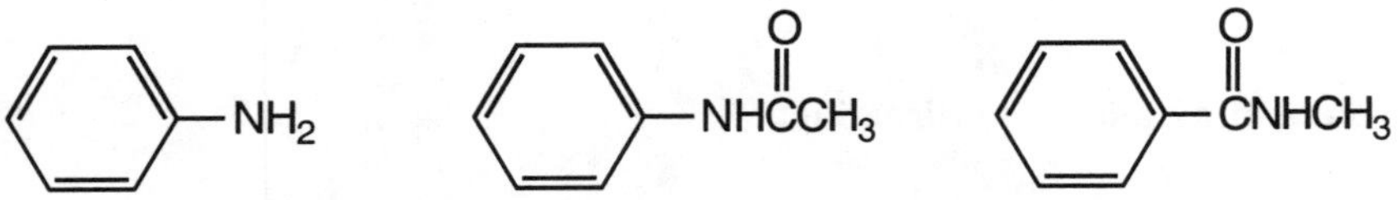

(A)   (B)   (C)

A>B>C

(d)

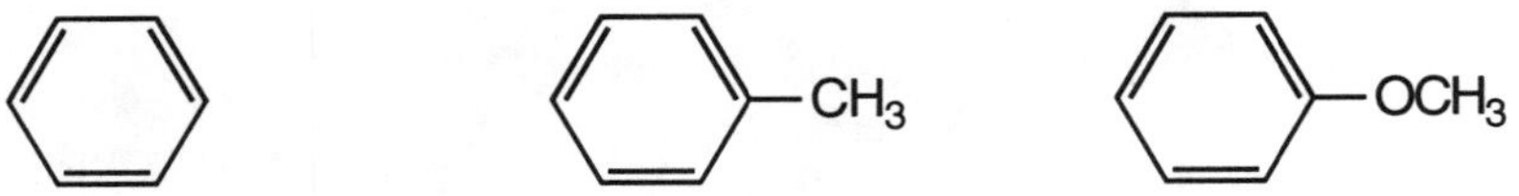

(A)   (B)   (C)

C>B>A

<u>Problem 9.29</u> Account for the observation that the trifluoromethyl group is meta directing as shown in the following example.

The trifluoromethyl group is highly electron withdrawing because the very electronegative flourine atoms withdraw electron density from the attached carbon atom, making it electron deficient. To compensate, this carbon atom withdraws electron density from the pi system of the aromatic ring, making it a poorer nucleophile (deactivating the ring toward electrophilic aromatic substitution). Following are contributing structures for meta and para attack of the electrophile. For meta attack, three contributing structures can be drawn which all make approximately equal contributions to the hybrid. Three contributing structures can also be drawn for ortho-para attack, one of which places a positive charge on carbon bearing the trifluoromethyl group; this structure makes only a negligible contribution to the hybrid because of adjacent positive charges. Thus, for meta attack, the positive charge on the aryl cation intermediate can be delocalized almost equally over three atoms of the ring, giving this cation's formation a lower energy of activation. For ortho-para attack, the positive charge on the aryl cation intermediate is delocalized over only two carbons of the ring, giving this cation's formation a higher energy of activation.

<u>meta attack:</u>

<u>ortho-para attack:</u>

adjacent positive
charges

<u>Problem 9.30</u> Show how to convert toluene to these carboxylic acids.
(a) 4-Chlorobenzoic acid

Methyl is ortho-para directing. Therefore, toluene can be chlorinated and then oxidized with $K_2Cr_2O_7$ in aqueous H_2SO_4 to convert the methyl group into the carboxyl group.

4-Chlorobenzoic acid

(b) 3-Chlorobenzoic acid

The reaction sequence is very similar to the last one, except now the order of the reactions is reversed because the carboxylic acid group is a meta director.

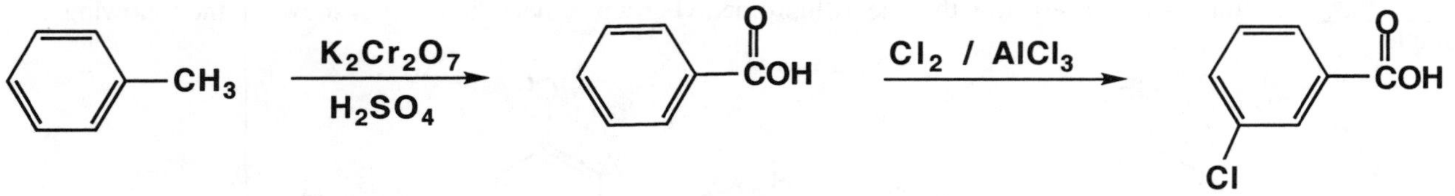

3-Chlorobenzoic acid

<u>Problem 9.31</u> Show reagents and conditions to bring about these conversions.

(a)

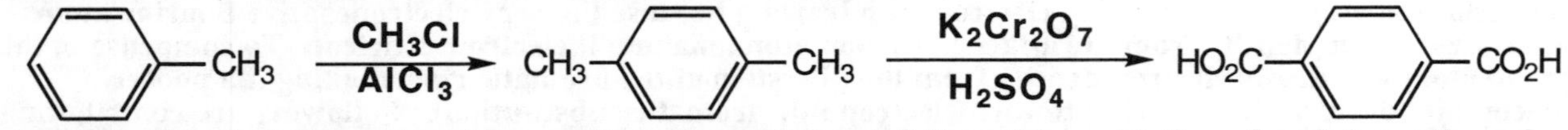

A Friedel-Crafts alkylation using chloromethane is carried out, followed by oxidation of each benzylic carbon to a carboxyl group.

(b)

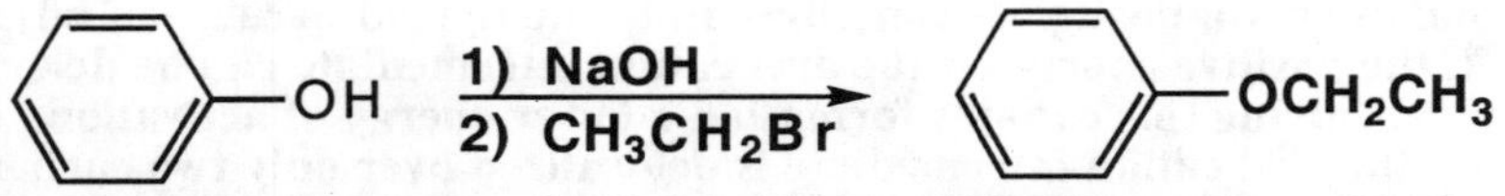

Phenol is first converted into the highly nucleophilic phenoxide anion by treatment with base, then the phenoxide is alkylated by ethyl bromide.

(c)

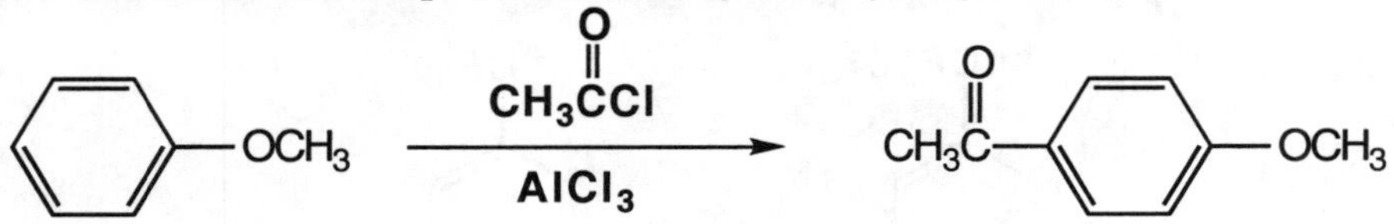

(d)

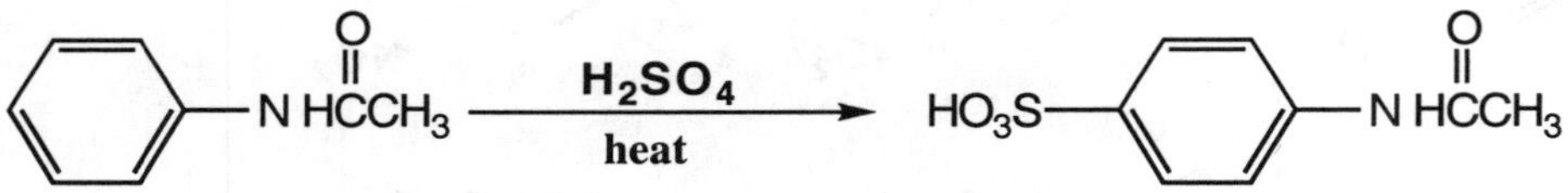

<u>Problem 9.32</u> Propose a synthesis of triphenylmethane from benzene as the only source of aromatic rings. Use any other necessary reagents.

Reaction of three moles of benzene with one mole of trichloromethane (chloroform) in the presence of aluminum chloride will give triphenylmethane.

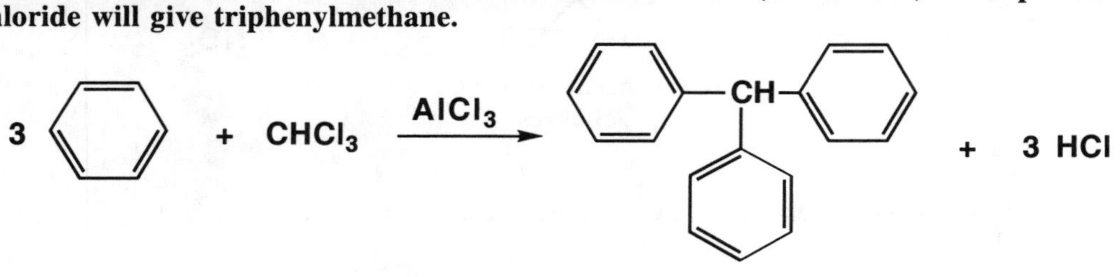

<u>Problem 9.33</u> Reaction of phenol with acetone in the presence of an acid catalyst gives bisphenol A, a compound used in the production of polycarbonate and epoxy resins (Sections 15.4C and 15.4E). Propose a mechanism for the formation of bisphenol A.

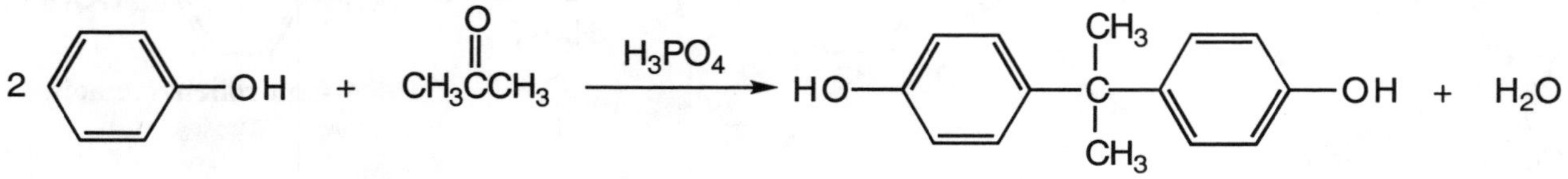

Bisphenol A

Step 1: The reaction begins with protonation of acetone to form its conjugate acid, which may be written as a hybrid of two contributing structures.

$$CH_3-\overset{\overset{\displaystyle :O:}{\|}}{C}-CH_3 \;+\; H^+ \longrightarrow CH_3-\overset{\overset{\displaystyle :\overset{+}{O}{\diagup}^H}{\|}}{C}-CH_3 \longleftrightarrow CH_3-\overset{\overset{\displaystyle :\ddot{O}{\diagup}^H}{|}}{\underset{+}{C}}-CH_3$$

Step 2: The conjugate acid of acetone is an electrophile and reacts with phenol at the para position to give a resonance-stabilized carbocation (only one contributor is shown; you should try to draw all the others).

Step 3: Deprotonation of the carbocation gives the intermediate 2-(4-hydroxyphenyl)-2-propanol.

Step 4: The tertiary alcohol is protonated.

Step 5: The protonated alcohol loses a molecule of water to give a very stable carbocation intermediate (only one contributor is shown; four others are possible).

Step 6: Attack of phenol upon the carbocation gives a new resonance-stabilized carbocation intermediate.

Step 7: Deprotonation of the carbocation intermediate yields the product bisphenol A.

Bisphenol A

Problem 9.34 2,6-Di-*tert*-butyl-4-methylphenol, more commonly known as butylated hydroxytoluene or BHT, is used as an antioxidant in foods to "retard spoilage". BHT is synthesized industrially from 4-methylphenol (*p*-cresol) by reaction with 2-methylpropene in the presence of phosphoric acid. Propose a mechanism for this reaction.

4-Methylphenol
(p-cresol)

2,6-Di-tert-butyl-4-methylphenol
"butylated hydroxytoluene"
(BHT)

Step 1: **Protonation of 2-methylpropene by phosphoric acid gives the electrophilic *tert*-butyl cation.**

Step 2: **The *tert*-butyl cation reacts with the aromatic ring ortho to the strongly activating -OH group, yielding a resonance-stabilized carbocation (only one resonance form is shown; you should try to draw the others).**

Step 3: **Deprotonation of the carbocation intermediate gives 2-*tert*-butyl-4-methylphenol, which can react further.**

Step 4: **The intermediate product reacts with *tert*-butyl cation to give a resonance stabilized carbocation intermediate (three other contributors can be drawn).**

Step 5: Deprotonation yields BHT.

Problem 9.35 The first widely used herbicide for control of weeds was 2,4-dichlorophenoxyacetic acid (2,4-D). Show how this compound might be synthesized from 2,4-dichlorophenol and 2-chloroacetic acid, ClCH$_2$CO$_2$H.

2,4-Dichlorophenol   2,4-Dichlorophenoxyacetic acid
                                                        (2,4-D)

Note that you need to use the salt of 2-chloroacetic acid, as the conjugate base of 2,4-dichlorophenol (pK$_a$ of 2,4-dichlorophenol = 7.85) will be strong enough to deprotonate 2-chloroacetic acid (pK$_a$ 2.87). The carboxylic acid is obtained by adding an acid, such as HCl, at the end of the S$_N$2 reaction.

Syntheses

Problem 9.36 Using styrene, $C_6H_5CH=CH_2$, as the only aromatic starting material, show how to synthesize these compounds. In addition to styrene, use any other necessary organic or inorganic chemicals. Any compound synthesized in one part of this problem may be used to make any other compound in the problem.

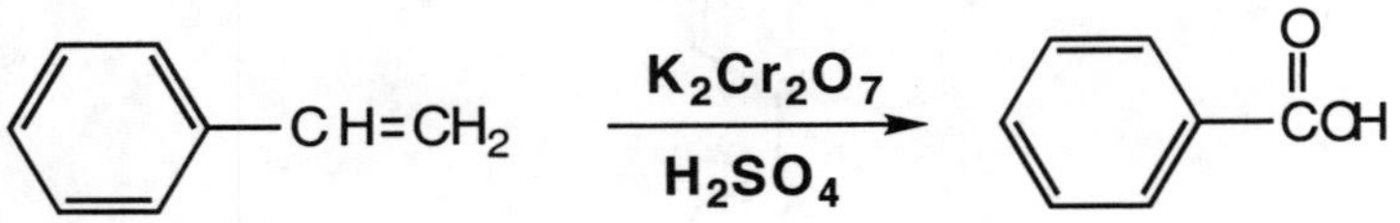

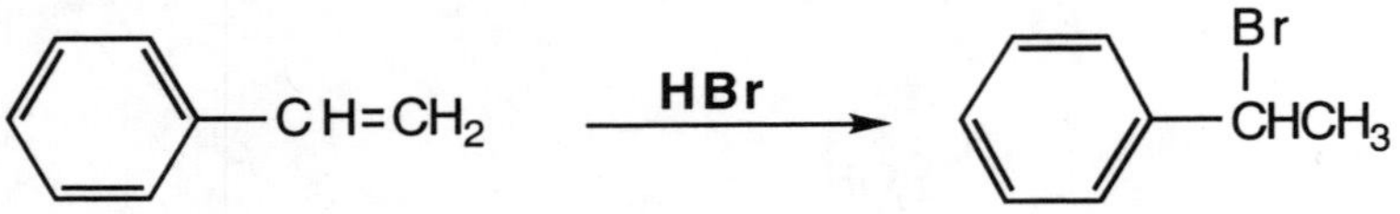

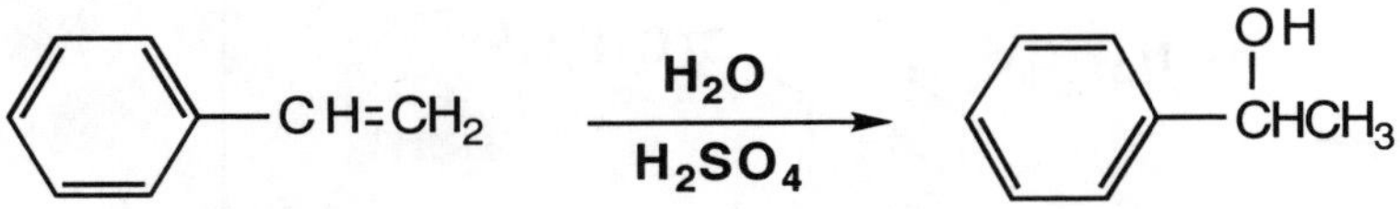

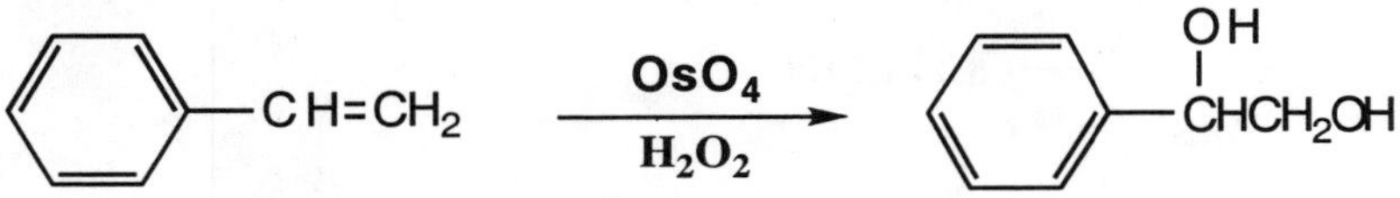

Problem 9.37 Starting with benzene, toluene, or phenol as the only sources of aromatic rings, show how to synthesize these compounds. Assume in all syntheses that mixtures of ortho-para products can be separated to give the desired isomer in pure form.

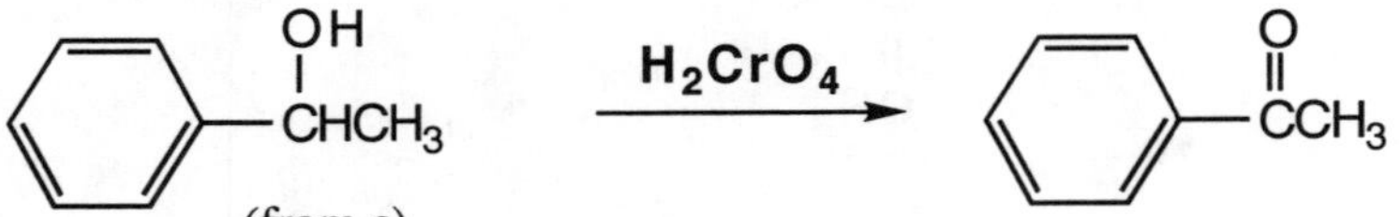

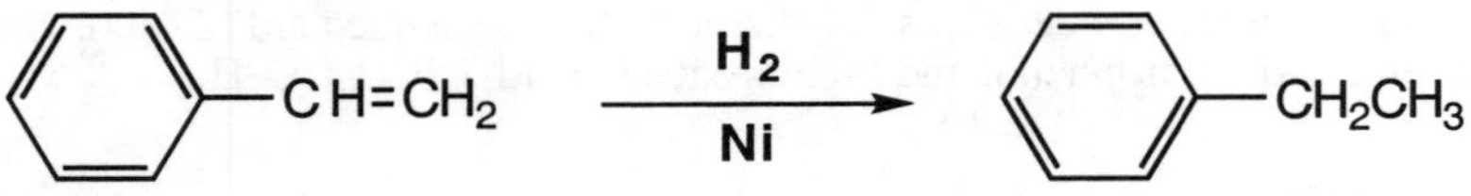

(a) *m*-Bromonitrobenzene

Nitro is meta directing; bromine is ortho-para directing. Therefore, to have the two substituents meta to each other, carry out nitration first, followed by bromination.

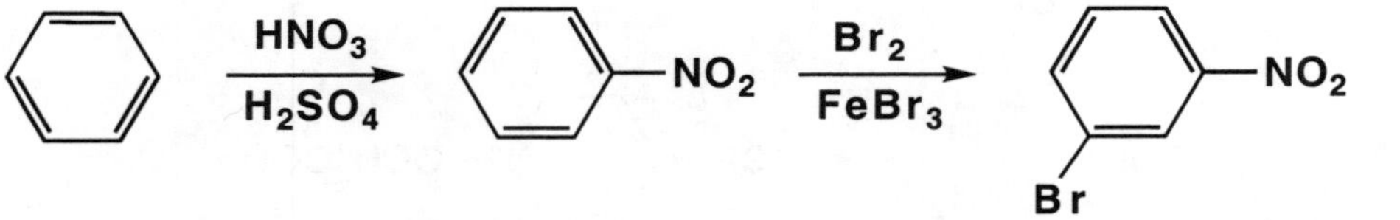

Nitrobenzene

1-Bromo-3-nitrobenzene
***m*-Bromonitrobenzene**

(b) 1-Bromo-4-nitrobenzene

Reverse the order of steps from part (a). Nitro is meta directing; bromine is ortho-para directing. Therefore, to have the two substituents para to each other, carry out bromination first, followed by nitration.

Bromobenzene

1-Bromo-4-nitrobenzene
***p*-Bromonitrobenzene**

(c) 2,4,6-Trinitrotoluene (TNT)

The methyl group is ortho-para directing. Therefore, nitrate toluene three successive times. This works well because the nitro group is strongly meta-directing. Thus, the first nitro group will direct the second one meta.

2,4,6-Trinitrotoluene

(d) *m*-Bromobenzoic acid

The carboxyl group and bromine atom are meta to each other, an orientation best accomplished by bromination of benzoic acid (the carboxyl group is meta-directing). Oxidation of toluene with $K_2Cr_2O_7$ in aqueous H_2SO_4 gives benzoic acid. Treatment of benzoic acid with bromine in the presence of aluminum chloride or ferric chloride gives the desired product.

Toluene   Benzoic acid   Br   3-Bromobenzoic acid

(e) *p*-Bromobenzoic acid

Start with toluene. Methyl is weakly activating and directs bromination to the ortho-para positions. Separate the desired para isomer. Then oxidize the methyl group to a carboxyl group using $K_2Cr_2O_7$ in aqueous H_2SO_4 to give 4-bromobenzoic acid.

Toluene   4-Bromotoluene   4-Bromobenzoic acid

(f) *p*-Dichlorobenzene

Treatment of benzene with chlorine in the presence of aluminum chloride gives chlorobenzene. The chlorine atom is ortho-para directing. Treatment with chlorine in the presence of aluminum chloride a second time gives 1,4-dichlorobenzene.

Benzene   Chlorobenzene   *p*-Dichlorobenzene

(g) *m*-Nitrobenzenesulfonic acid

Both the sulfonic acid group and the nitro group are meta directors. Therefore, the two electrophilic aromatic substitution reactions may be carried out in either order. The sequence shown is nitration followed by sulfonation.

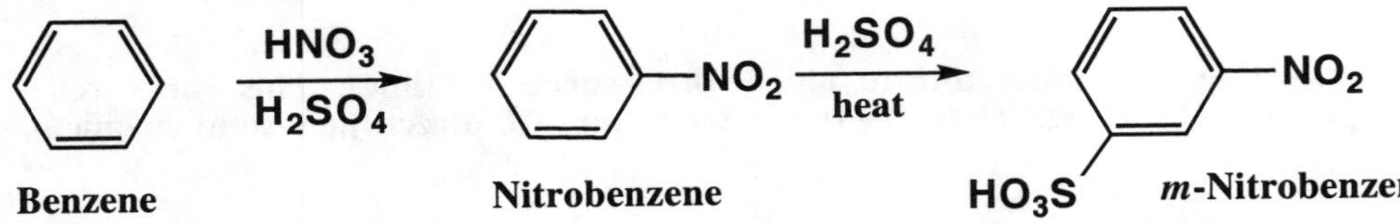

Benzene → HNO₃ / H₂SO₄ → Nitrobenzene → H₂SO₄ / heat → m-Nitrobenzenesulfonic acid

(h) 1-Chloro-3-nitrobenzene

The nitro group is meta directing, while the chlorine atom is ortho-para directing. Therefore, in order to obtain the desired meta product, the nitration must be carried out prior to the chlorination.

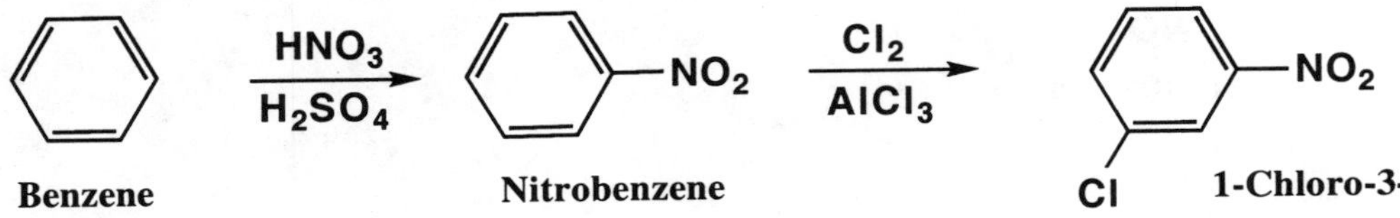

Benzene → HNO₃ / H₂SO₄ → Nitrobenzene → Cl₂ / AlCl₃ → 1-Chloro-3-nitrobenzene

<u>Problem 9.38</u> Starting with benzene or toluene as the only sources of aromatic rings, show how to synthesize these aromatic ketones. Assume in all syntheses that mixtures of ortho-para products can be separated to give the desired isomer in pure form.

(a)

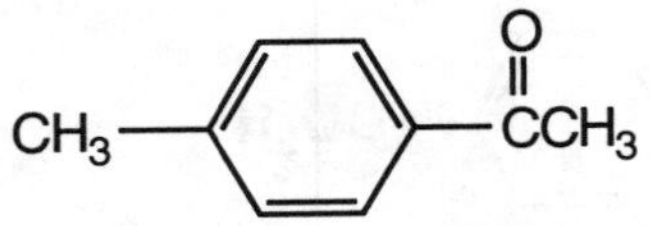

The methyl group of toluene is ortho-para directing, so a Friedel-Crafts acylation of toluene will give the desired product in good yield.

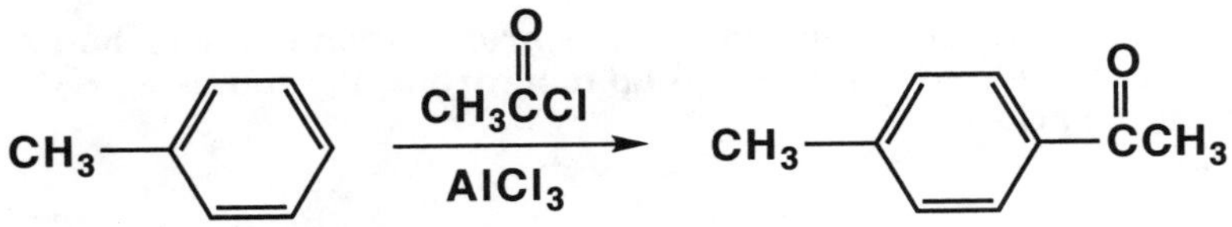

(b)

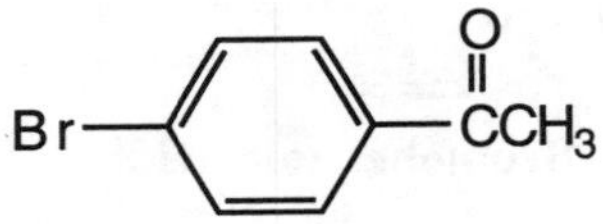

The bromine atom of bromobenzene is ortho-para directing, so a Friedel-Crafts acylation following bromination of benzene will give the desired product in good yield.

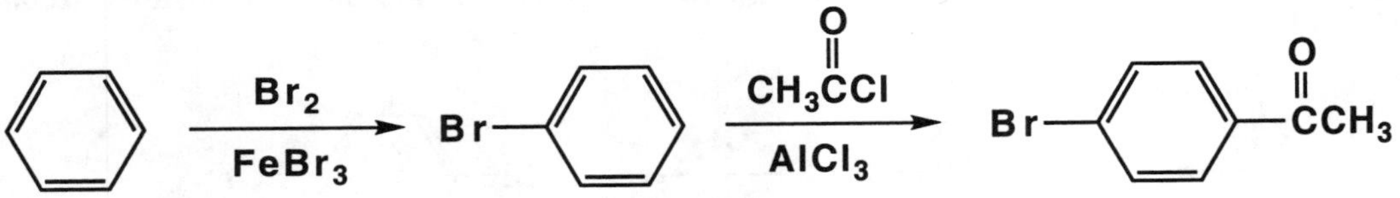

(c)

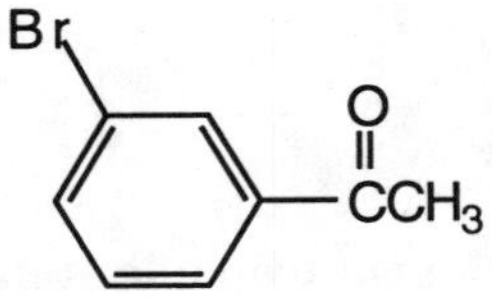

The ketone group is meta directing, so that bromination following a Friedel-Crafts acylation will give the desired product in good yield.

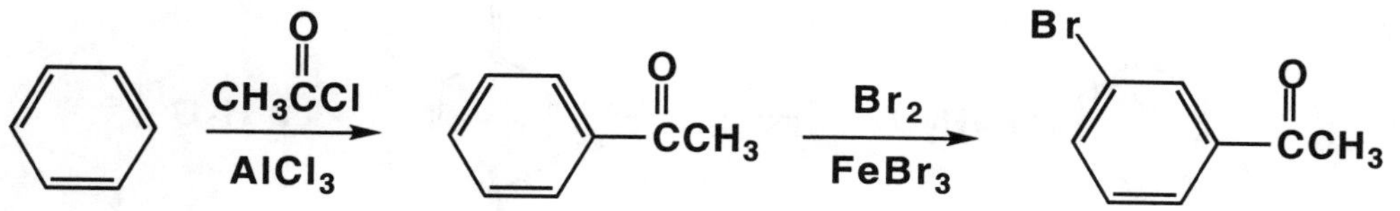

Problem 9.39 The following ketone, isolated from the roots of several members of the iris family, has an odor like that of violets and is used as a fragrance in perfumes. Describe the synthesis of this ketone from benzene.

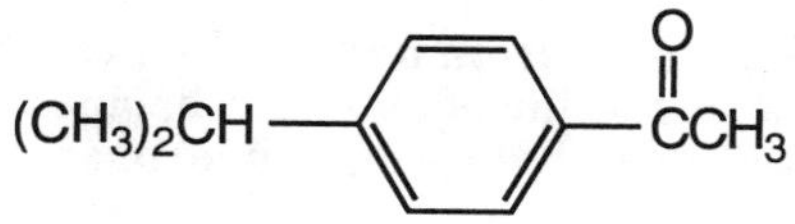

4-Isopropylacetophenone

The isopropyl group is weakly activating and ortho-para directing; the carbonyl group of the acetyl group is deactivating and meta directing. Therefore, start with benzene and convert it to isopropylbenzene. Then, carry out a Friedel-Crafts acylation using acetyl chloride in the presence of aluminum chloride. Friedel-Crafts alkylation of benzene can be accomplished using a 2-halopropane, 2-propanol, or propene, each in the presence of an appropriate catalyst.

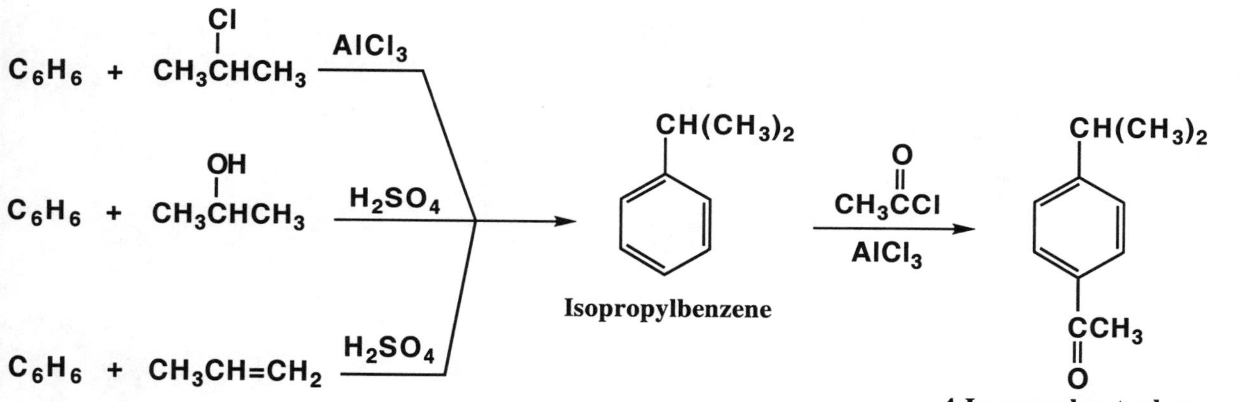

Isopropylbenzene

4-Isopropylacetophenone

Problem 9.40 The bombardier beetle generates *p*-quinone, an irritating chemical, by the enzyme-catalyzed oxidation of hydroquinone using hydrogen peroxide as the oxidizing agent. Heat generated in this oxidation produces superheated steam which is ejected, along with *p*-quinone, with explosive force.

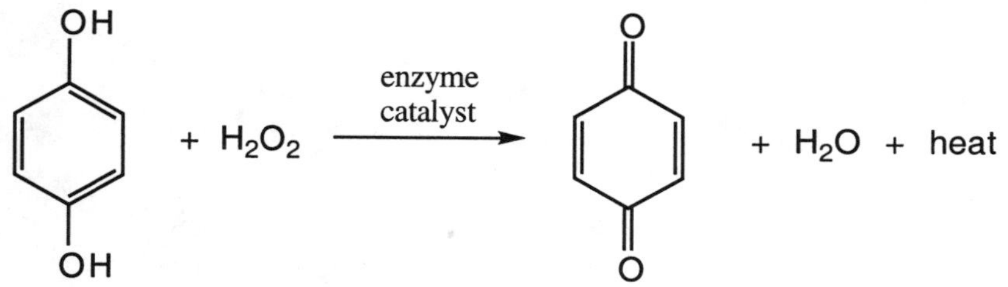

(a) Balance the above equation.

(b) Show by a balanced half-reaction that conversion of hydroquinone to *p*-quinone is a two-electron oxidation (Hint: review section 6.4A on the use of half-reactions.)

Two hydrogens must be lost to generate quinone from hydroquinone. Therefore, two protons and two electrons are required on the right hand side of the equation to balance it, making this a two electron oxidation reaction. There is a corresponding two electron reduction of hydrogen peroxide to give the two molecules of water.

$$H_2O_2 + 2 H+ \ 2 e^- \longrightarrow 2 H_2O$$

CHAPTER 10
Solutions to Problems

Problem 10.1 Identify all stereocenters in coniine, nicotine, and cocaine.

The stereocenters are circled in the following structures:

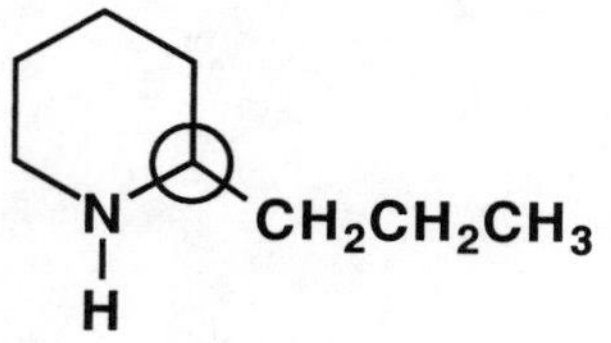

Coniine 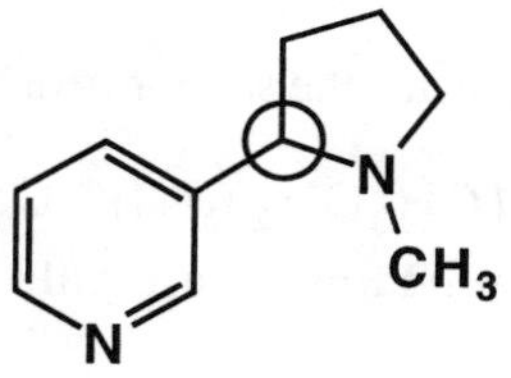   **Nicotine** 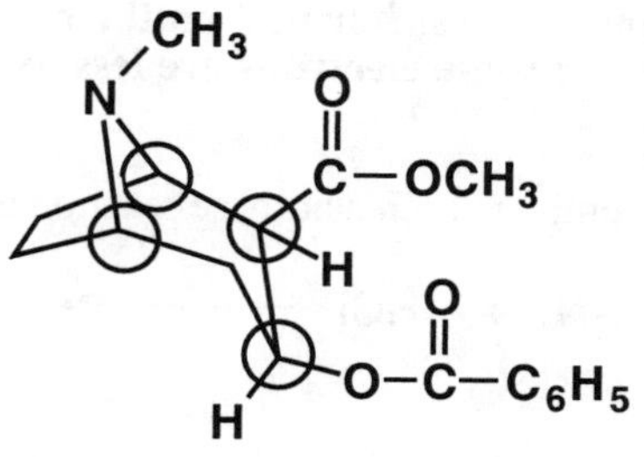   **Cocaine**

Problem 10.2 Write a structural formula for each amine.
(a) 2-Methyl-1-propanamine   (b) Cyclohexanamine   (c) (R)-2-Butanamine

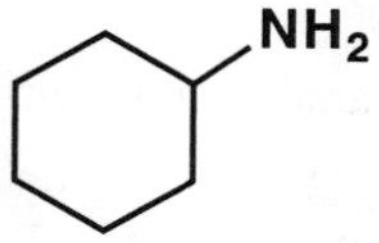

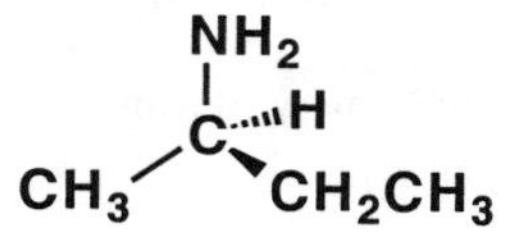

Problem 10.3 Write a structural formula for each amine.
(a) Isobutylamine   (b) Triphenylamine   (c) Diisopropylamine

Problem 10.4 Predict the position of equilibrium for this acid-base reaction.

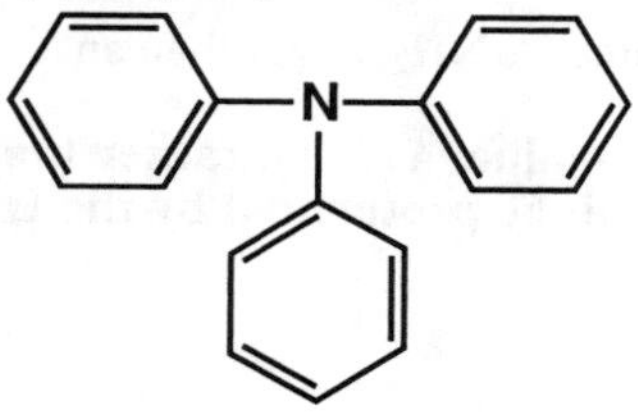

$$CH_3NH_3^+ \quad + \quad H_2O \quad \rightleftharpoons \quad CH_3NH_2 \quad + \quad H_3O^+$$

pK$_a$ 10.64   **Weaker**   **Stronger**   pK$_a$ -1.74
Weaker acid   **base**   **base**   **Stronger acid**

Equilibrium favors formation of the weaker acid, so the equilibrium lies to the left.

Problem 10.5 Select the stronger acid from each pair of ions.

(a) 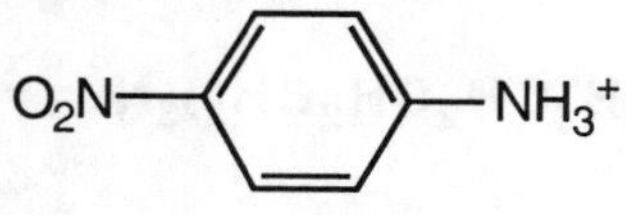 O_2N—⟨benzene ring⟩—NH_3^+ or CH_3—⟨benzene ring⟩—NH_3^+

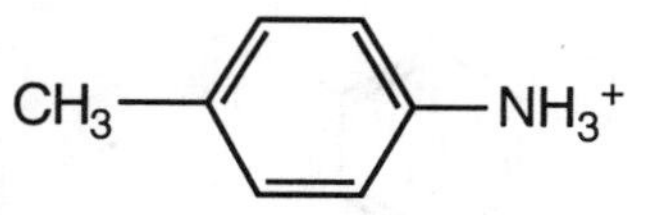

              (A)   (B)

4-Nitroaniline (pK$_b$ 13.0) is a weaker base than 4-methylaniline (pK$_b$ 8.92). The decreased basicity of 4-nitroaniline is due to the electron-withdrawing effect of the *para* nitro group. Because 4-nitroaniline is the weaker base, its conjugate acid (A) is the stronger acid.

(b)   [pyridinium cation structure] $\overset{+}{N}H$ or   [cyclohexyl structure] $-NH_3^+$

                    (C)   (D)

Aromatic heterocycles, such as pyridine (compound C), are much weaker bases (pK_b 8.75), so the conjugate acids are stronger than those of aliphatic amines, such as compound D, cyclohexanamine (pK_b 3.34). The lone pair of electrons in the sp^2 orbital on the nitrogen atom of pyridine (C) has more s character than the sp^3 nitrogen of (D), so these electrons are less available for bonding to a proton.

Problem 10.6 Complete each acid-base reaction and name the salt formed.

(a) $(CH_3CH_2)_3N$ + HCl $\longrightarrow$ $(CH_3CH_2)_3NH^+ \, Cl^-$

                                                    Triethylammonium chloride

(b) [piperidine ring] NH + CH_3CO_2H $\longrightarrow$ [piperidinium ring] $\overset{H}{\underset{H}{N^+}}$ $CH_3CO_2^-$

                                                    Piperidinium acetate

Problem 10.7 In what way(s) might the results of the separation and purification procedure outlined in Example 10.7 be different if

(a) Aqueous NaOH is used in place of aqueous $NaHCO_3$?

If NaOH is used in place of aqueous $NaHCO_3$, the phenol will be deprotonated along with the carboxylic acid, so they will be isolated together in fraction A.

(b) The starting mixture contains an aromatic amine, $ArNH_2$, rather than an aliphatic amine, RNH_2?

If the starting mixture contains an aromatic amine, $ArNH_2$, rather than an aliphatic amine, RNH_2, the results will be the same. The aromatic amine will still be protonated by the HCl wash and deprotonated by the NaOH treatment.

Structure and Nomenclature
Problem 10.8 Draw a structural formula for each amine.
(a) (R)-2-Butanamine   (b) 1-Octanamine   c) 2,2-Dimethyl-1-propanamine

[structure (a): C with NH₂, H, CH₃, CH₂CH₃]

$CH_3(CH_2)_6CH_2NH_2$

[structure (c):]
$$CH_3\underset{\underset{CH_3}{|}}{\overset{\overset{CH_3}{|}}{C}}CH_2NH_2$$

(d) 1,5-Pentanediamine   e) 2-Bromoaniline   (f) Tributylamine

$H_2NCH_2CH_2CH_2CH_2CH_2NH_2$ [benzene ring with NH_2 and Br] $(CH_3CH_2CH_2CH_2)_3N$

(g) N,N-Dimethylaniline

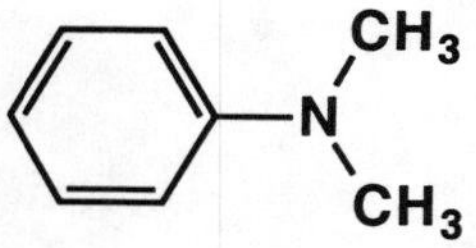

(h) Benzylamine

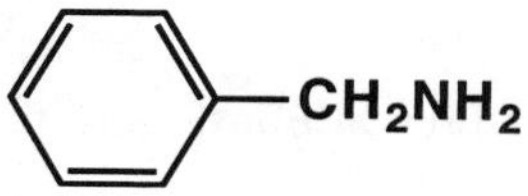

(i) *tert*-Butylamine

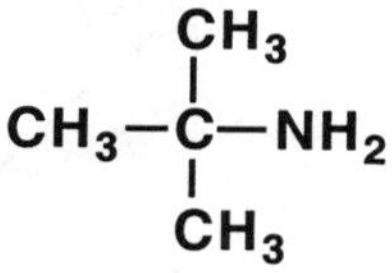

(j) N-Ethylcyclohexanamine

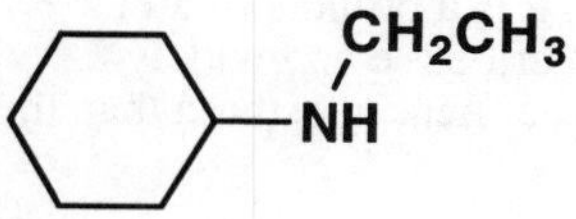

(k) Diphenylamine

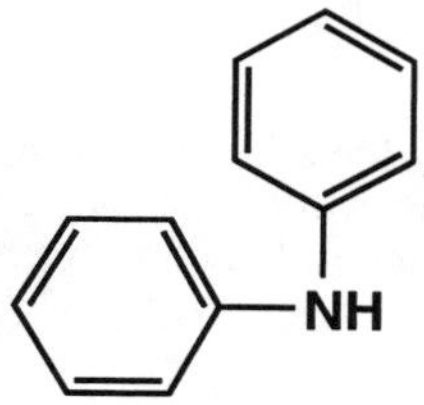

(l) Isobutylamine

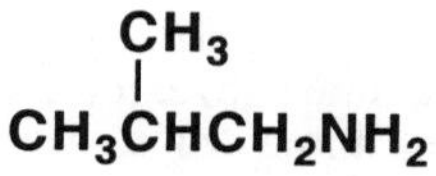

Problem 10.9 Draw a structural formula for each amine.
(a) 4-Aminobutanoic acid

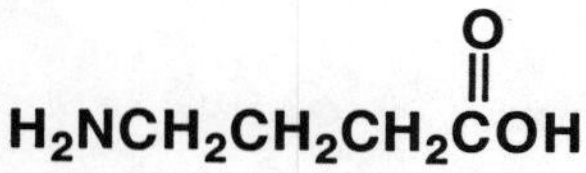

(b) 2-Aminoethanol
(Ethanolamine)

(c) 2-Aminobenzoic acid

$H_2NCH_2CH_2CH_2COH$

$H_2NCH_2CH_2OH$

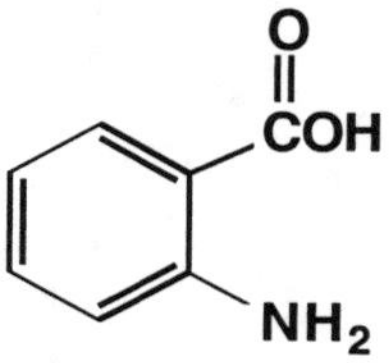

(S)-2-Aminopropanoic acid
(Alanine)

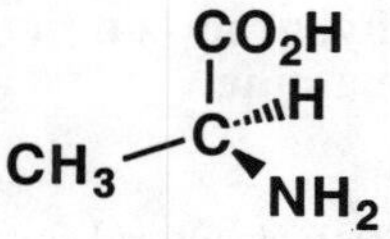

(e) 4-Aminobutanal

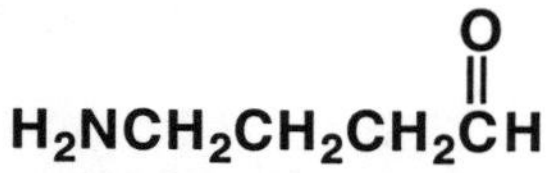

(f) 4-Amino-2-butanone

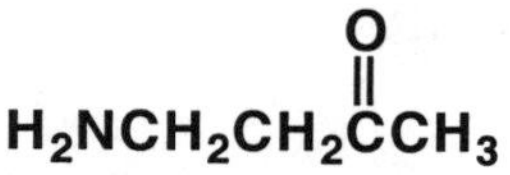

$H_2NCH_2CH_2CH_2CH$

$H_2NCH_2CH_2CCH_3$

Problem 10.10 Classify each amine as primary, secondary, or tertiary; as aliphatic or aromatic.

(a)

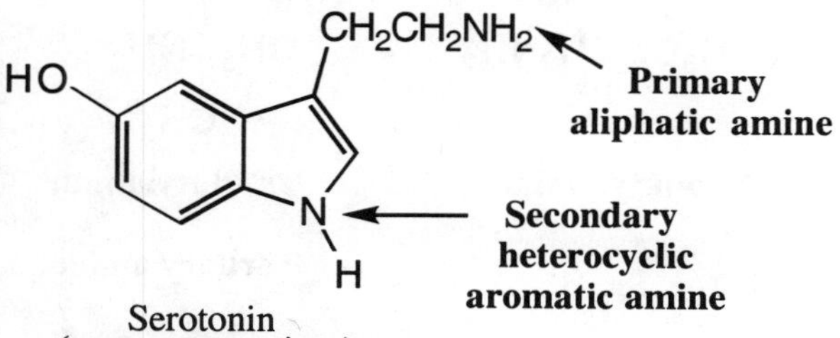

CH₂CH₂NH₂ → **Primary aliphatic amine**

→ **Secondary heterocyclic aromatic amine**

Serotonin
(a neurotransmitter)

(b) **Primary aromatic amine**

H_2N — COCH₂CH₃

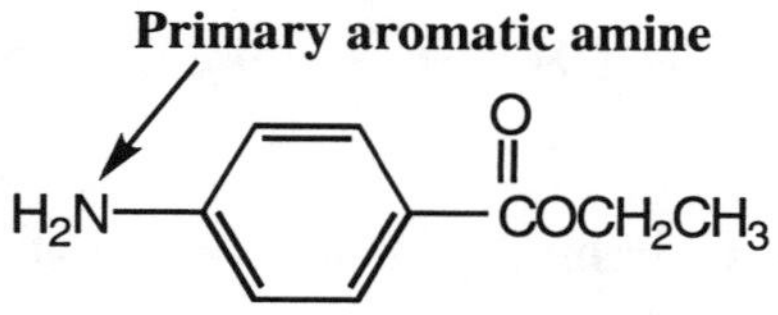

Benzocaine
(a topical anesthetic)

$$CH_3$$
$$NHCH(CH_2)_3N(C_2H_5)_2$$

Tertiary aliphatic amine

(c)

Cl

Secondary aromatic amine

N

Chloroquine
(a drug for the
treatment of malaria)

Heterocyclic aromatic amine

Problem 10.11 Epinephrine is a hormone secreted by the adrenal medulla. Among its actions, it is a bronchodilator. Albuterol, sold under several trade names, including Proventil and Salbumol, is one of the most effective and widely prescribed antiasthma drugs. The R enantiomer of albuterol is 68 times more effective in the treatment of asthma than the S enantiomer.

Secondary aliphatic amine   **Secondary aliphatic amine**

H   OH   H   OH
C   C
CH_2NHCH_3   CH_2NHC(CH_3)_3

HO   HO

(R)-Epinephrine   (R)-Albuterol
(Adrenaline)

OH   CH_2OH

(a) Classify each amino group as primary, secondary, or tertiary.
(b) List the similarities and differences between the structural formulas of these compounds.

The parts of the molecules that are identical are indicated in bold on the structures. As far as differences are concerned, epinephrine posseses a second hydroxyl group on the aromatic ring and a methyl group on the amine, while (R)-albuterol has a hydroxymethyl on the ring and a *tert*-butyl group on the amine.

Problem 10.12 There are eight constitutional isomers of molecular formula $C_4H_{11}N$. Name and draw structural formulas for each. Classify each amine as primary, secondary, or tertiary.

Primary amines:

$$CH_3$$ $$CH_3$$ $$CH_3$$

$CH_3CH_2CH_2CH_2NH_2$ $CH_3CH_2CHNH_2$ $CH_3CHCH_2NH_2$ CH_3CNH_2
$$CH_3$$

Butylamine   ***sec*-Butylamine**   **Isobutylamine**   ***tert*-Butylamine**

Secondary amines: **Tertiary amine:**

$$CH_3$$

$CH_3CH_2CH_2NCH_3$ $CH_3CH_2NCH_2CH_3$ CH_3CHNCH_3 $CH_3NCH_2CH_3$
H   H   H CH_3

Methylpropylamine   **Diethylamine**   **Isopropylmethylamine** **Ethyldimethylamine**

Problem 10.13 Draw a structural formula for each compound of the given molecular formula.
(a) A 2° arylamine, C_7H_9N (b) A 3° arylamine, $C_8H_{11}N$ 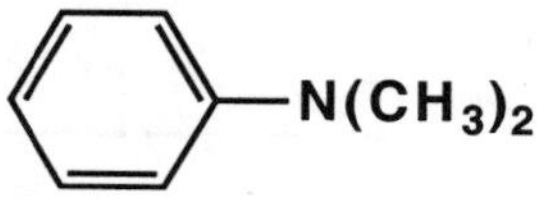   (c) A 1° aliphatic amine, C_7H_9N

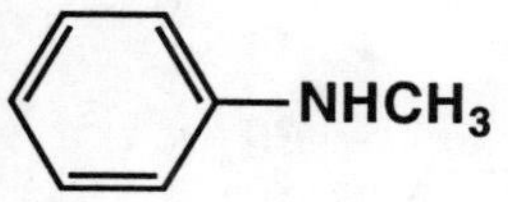

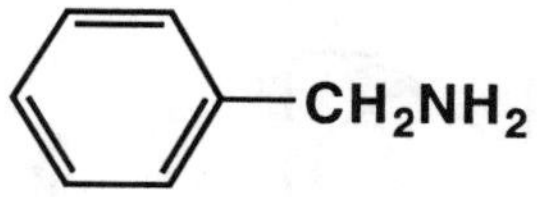

(d) A chiral 1° amine, $C_4H_{11}N$ 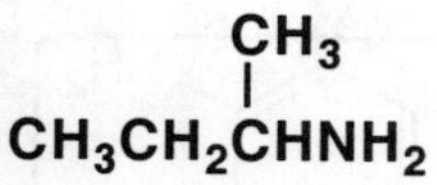   (e) A 3° heterocyclic amine, $C_5H_{11}N$ 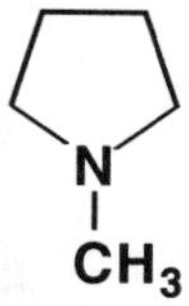   (f) A trisubstituted 1° arylamine, $C_9H_{13}N$

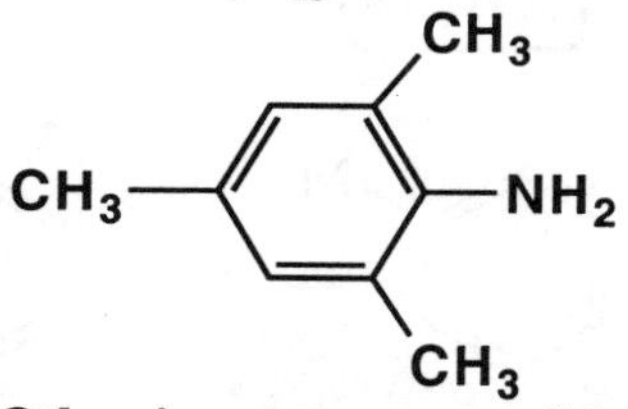

(Other isomers are possible)

(g) A chiral quaternary ammonium salt, $C_9H_{22}NCl$

There are two answers to this question. For the structure on the left, the stereocenter is the nitrogen atom. For the structure on the right, the stereocenter is the carbon atom marked with an asterisk.

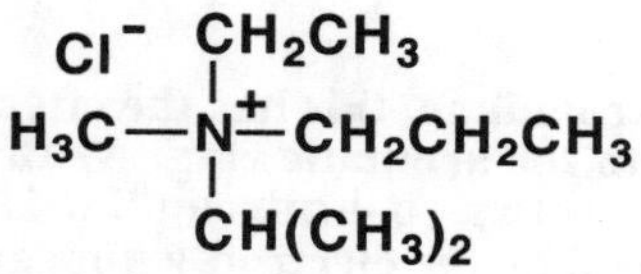

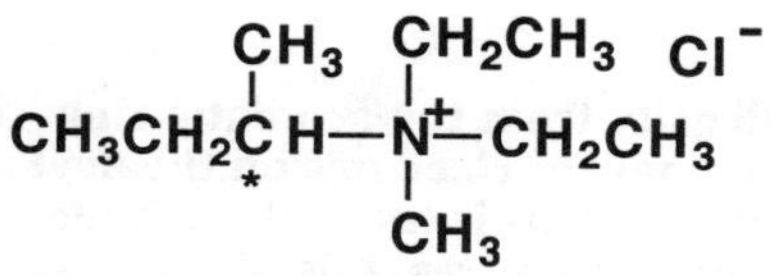

Physical Properties
Problem 10.14 Propylamine, ethylmethylamine, and trimethylamine are constitutional isomers of molecular formula C_3H_9N. Account for the fact that trimethylamine has the lowest boiling point of the three.

$$CH_3CH_2CH_2NH_2 \qquad CH_3CH_2NHCH_3 \qquad (CH_3)_3N$$
$$\text{bp } 48°C \qquad\qquad \text{bp } 37°C \qquad\qquad \text{bp } 3°C$$

Trimethylamine does not have any hydrogen atoms attached to the nitrogen atom, so this compound cannot form intermolecular hydrogen bonds. The other two compounds do have hydrogens attached to nitrogen, so they can hydrogen bond. The ability to hydrogen bond increases attraction between molecules and, thus, raises boiling points.

Problem 10.15 Account for the fact that 1-butanamine has a lower boiling point than 1-butanol.

$$CH_3CH_2CH_2CH_2OH \qquad\qquad CH_3CH_2CH_2CH_2NH_2$$
$$\text{bp } 117°C \qquad\qquad\qquad \text{bp } 78°C$$

An N-H----N hydrogen bond is not as strong as an O-H----O hydrogen bond, because the difference in electronegativity between nitrogen and hydrogen is not as great as the difference in electronegativity between oxygen and hydrogen. Stronger hydrogen bonds between the molecules of 1-butanol lead to increased attraction between the molecules and a higher boiling point.

Basicity of Amines
Problem 10.16 Account for the fact that amines are more basic than alcohols.

Nitrogen is less electronegative than oxygen, so a lone pair of electrons on nitrogen is more available to interact with a proton than a lone pair of electrons on oxygen. Thus, amines are more basic than alcohols.

Problem 10.17 From each pair of compounds, select the stronger base.

The more basic amine is circled.

(a)   [piperidine] or [pyridine]

(b)   [cyclohexyl–N(CH₃)₂] or [phenyl–N(CH₃)₂]

(c)   [3-methylaniline] or [benzylamine]

(d)   [4-nitroaniline] or [4-methylaniline]

Problem 10.18 Account for the fact that substitution of a nitro group makes an aromatic amine a weaker base but makes a phenol a stronger acid. For example, 4-nitroaniline is a weaker base than aniline, but 4-nitrophenol is a stronger acid than phenol.

The nitro group withdraws electron density from the aromatic ring. For 4-nitroaniline, this has the effect of increasing the delocalization of the lone pair of electrons on the aryl amine into the aromatic ring. With greater delocalization, the lone pair of electrons is less able to interact with a proton, so 4-nitroaniline is less basic than aniline. In terms of resonance structures, this idea can be understood as the electron withdrawing effect of the nitro group acting to increase the relative contribution of the resonance form on the right.

In the case of the relative acidities of the phenols, the key interactions involve the deprotonated phenoxide anions. For the 4-nitrophenoxide anion, the electron withdrawing effect of the nitro group helps withdraw more of the negative charge on the oxygen atom into the aromatic ring. Delocalization of a charge leads to stabilization, so this increased delocalization of the negative charge increases the stabilization of the 4-nitrophenoxide anion compared to the phenoxide anion. By stabilizing the 4-nitrophenoxide anion, the equilibrium is moved more toward deprotonation of the 4-nitrophenol; in other words, 4-nitrophenol is more acidic. In terms of resonance structures, this idea can be understood as the electron withdrawing effect of the nitro group acting to increase the relative contribution of the resonance form on the right.

Problem 10.19 Select the stronger base in this pair of compounds.

[phenyl–CH₂N(CH₃)₂]   or   [phenyl–CH₂N⁺(CH₃)₃ OH⁻]

The stronger base is the molecule on the right, as indicated by the circle around it, because the basic site is the hydroxide anion. The quaternary ammonium ion is simply the counterion. For the molecule on the left, the basic site is the tertiary amine. Hydroxide is a stronger base ($pK_b = -1.7$) than a tertiary amine ($pK_b = \sim3 - 4$).

Problem 10.20 Following are two structural formulas for alanine (2-aminopropanoic acid), one of the building blocks of proteins (Chapter 18). Is alanine better represented by structural formula (A) or structural formula (B)? Explain.

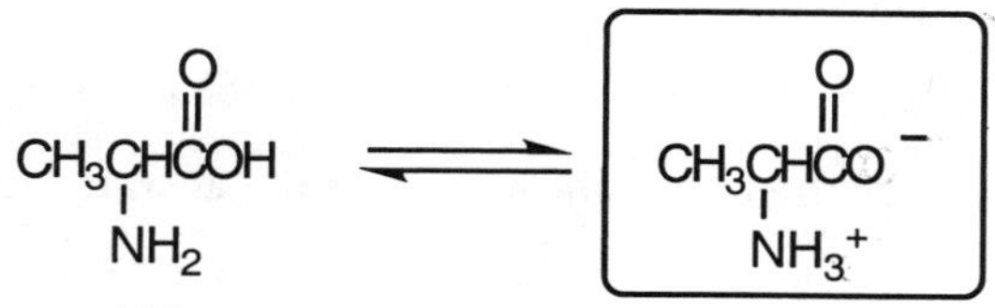

(A) (B)

Carboxylic acids are acidic enough ($pK_a = \sim3-5$) and primary amines are basic enough ($pK_b = \sim3 - 4$) that equilibrium favors the deprotonation of a carboxylic acid by a primary amine to form an ammonium cation and carboxylate anion. The structure on the right, structure B, is the better representation of alanine.

Problem 10.21 Complete the following acid-base reactions and predict the position of equilibrium for each. Justify your prediction by citing values of pK_a for the stronger and weaker acid in each equilibrium. For values of acid ionization constants, consult Table 2.1 (pK_a values of some inorganic and organic acids), Table 8.4 (pK_a values of alcohols), Section 9.4B (acidity of phenols) and Table 10.2 (base strengths of amines). Where no ionization constants are given, make the best estimate from the information given in the reference tables and sections.

In all cases, the equilibrium favors formation of the weaker acid (higher pK_a) and weaker base (higher pK_b). Recall that $pK_a + pK_b = 14$ for any conjugate acid-base pair.

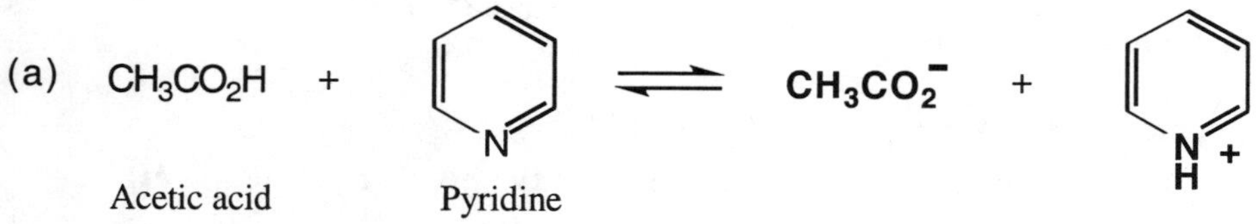

(a)

Acetic acid   Pyridine
pK_a 4.76 pK_b 8.75 pK_b 9.24 pK_a 5.25

Equilibrium lies to the right, because the acetate anion and pyridinium ion are the weaker base and acid, respectively.

(b)

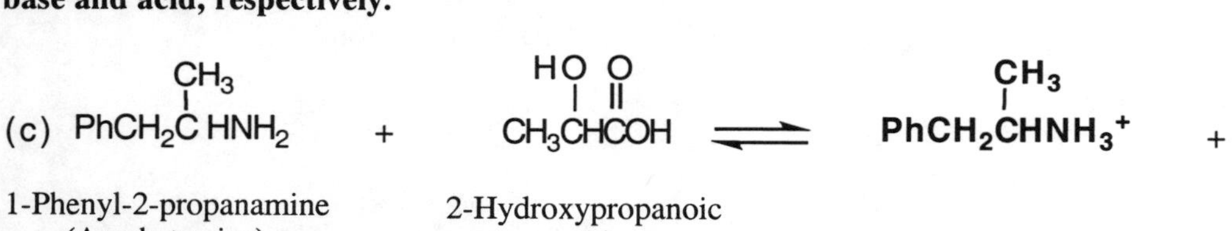

Phenol   Triethylamine
pK_a 9.95 pK_b 3.25 pK_b 4.05 pK_a 10.75

Equilibrium lies to the right, because the phenoxide anion and the triethylammonium species are the weaker base and acid, respectively.

(c)

1-Phenyl-2-propanamine (Amphetamine)   2-Hydroxypropanoic acid (Lactic acid)

pK_b ~3 pK_a 3.08 pK_a ~11 pK_b 9.92

Equilibrium lies to the right, since the ammonium ion and the carboxylate ion are the weaker acid and base, respectively.

(d) PhCH$_2$CHNHCH$_3$ + CH$_3$CO$_2$H $\rightleftharpoons$ PhCH$_2$CHNH$_2$CH$_3$ + CH$_3$CO$_2^-$

(with CH$_3$ substituents on the CH carbons, and the + on the product ammonium group)

Methamphetamine   Acetic acid

pK$_b$ ~3   pK$_a$ 4.76   pK$_a$ ~11   pK$_b$ 9.24

Equilibrium again lies to the right, since the ammonium ion and the acetate ion are the weaker acid and base, respectively.

Problem 10.22 The pK$_a$ of morpholinium ion is 8.33.

Morpholinium ion + H$_2$O $\rightleftharpoons$ Morpholine + H$_3$O$^+$ pK$_a$= 8.33

(a) Calculate the ratio of morpholine to morpholinium ion in aqueous solution at pH 7.0.

$$K_a = \frac{[\text{Morpholine}][\text{H}^+]}{[\text{Morpholinium Ion}]} = 10^{-8.33}$$ At pH 7.0 $[\text{H}^+] = 10^{-7}$

$$\frac{[\text{Morpholine}]}{[\text{Morpholinium Ion}]} = \frac{10^{-8.33}}{[\text{H}^+]} = \frac{10^{-8.33}}{10^{-7.0}} = 10^{-1.33} = 0.047$$

(b) At what pH are the concentrations of morpholine and morpholinium ion equal?

The concentrations of morpholine and morpholinium ion will be equal when the pK$_a$ is equal to the pH, namely, at pH 8.33.

Problem 10.23 The pK$_b$ of amphetamine (Example 10.2) is approximately 3.2. Calculate the ratio of amphetamine to its conjugate acid at pH 7.4, the pH of blood plasma.

We know that [OH$^-$][H$^+$] = 10^{-14}   At pH 7.4, $[\text{OH}^-] = \frac{10^{-14}}{[\text{H}^+]} = \frac{10^{-14}}{10^{-7.4}} = 10^{-6.6}$

$$K_b = \frac{[\text{Conjugate Acid}][\text{OH}^-]}{[\text{Amphetamine}]} = 10^{-3.2}$$

$$\frac{[\text{Conjugate Acid}]}{[\text{Amphetamine}]} = \frac{10^{-3.2}}{[\text{OH}^-]} = \frac{10^{-3.2}}{10^{-6.6}} = 10^{3.4}$$

Therefore, the ratio of [Amphetamine] : [Conjugate Acid] at pH 7.4 is 1 : 10$^{3.4}$ (1: 2,500).

Problem 10.24 Calculate the ratio of amphetamine to its conjugate acid at pH 1.0, such as might be present in stomach acid.

$$\textbf{We know that } [OH^-][H^+] = 10^{-14} \qquad \textbf{At pH 1.0,} \quad [OH^-] = \frac{10^{-14}}{[H^+]} = \frac{10^{-14}}{10^{-1}} = 10^{-13}$$

$$\frac{[\textbf{Conjugate Acid}]}{[\textbf{Amphetamine}]} = \frac{10^{-3.2}}{[OH^-]} = \frac{10^{-3.2}}{10^{-13}} = 10^{9.8}$$

Therefore, the ratio of [Amphetamine] : [Conjugate Acid] at pH 1.0 is 1 : $10^{9.8}$.

Problem 10.25 Following is the structural formula of pyridoxamine, one form of vitamin B_6.

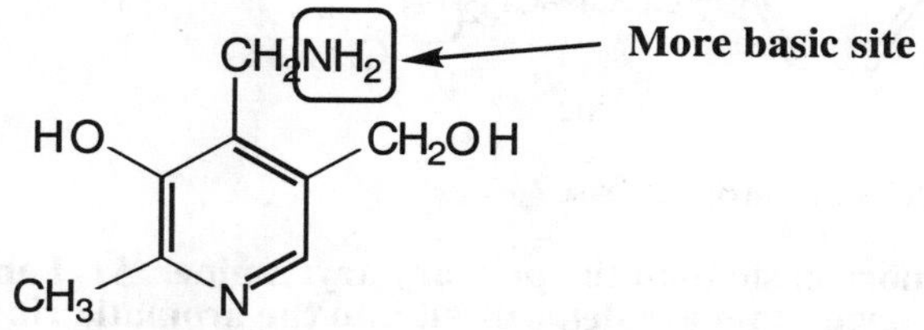

Pyridoxamine
(Vitamin B_6)

(a) Which nitrogen atom of pyridoxamine is the stronger base?

The primary amine indicated on the structure is more basic than the pyridine nitrogen atom, because the primary amine nitrogen atom is sp^3 hybridized, while the pyridine nitrogen is sp^2 hybridized. The sp^2 hybridized nitrogen atom has a greater percentage of s character, so the electrons are less available for interactions with protons.

(b) Draw the structural formula of the hydrochloride salt formed when pyridoxamine is treated with 1 mol of HCl.

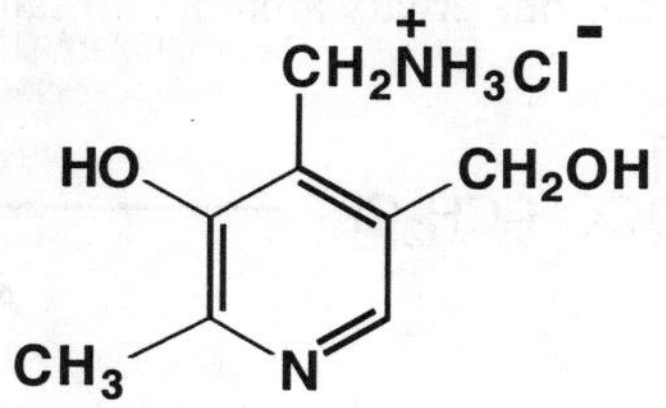

Problem 10.26 Epibatidine, a colorless oil isolated from the skin of the Ecuadorian poison frog *Epipedobates tricolor*, has several times the analgesic potency of morphine. It is the first chlorine-containing, nonopoid (nonmorphine-like in structure) analgesic ever isolated from a natural source.

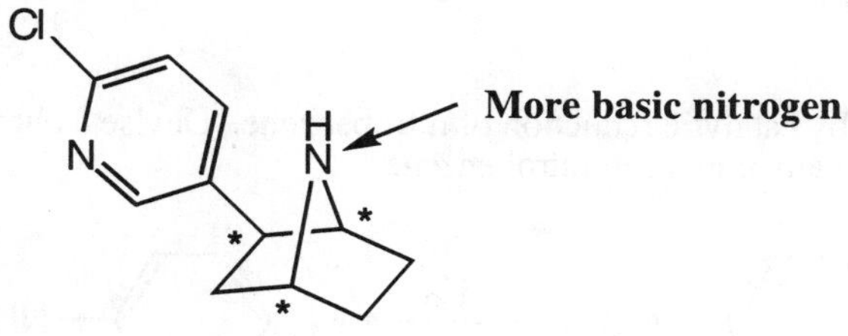

Epibatidine

(a) Which of the two nitrogen atoms of epibatidine is the more basic?

The secondary aliphatic amine indicated in the structure is more basic than the pyridyl nitrogen, because the secondary amine is sp³ hybridized, while the pyridine nitrogen is sp² hybridized. The lone pair electrons of the secondary aliphatic amine are more available for bonding to a proton.

(b) Mark all stereocenters in this molecule.

The stereocenters are indicated with asterisks in the structure.

Problem 10.27 Procaine was one of the first local anesthetics for infiltration and regional anesthesia. Its hydrochloride salt is marketed as Novocaine.

Procaine

(a) Which nitrogen atom of procaine is the stronger base?

The tertiary aliphatic amine is more basic than the primary aryl amine. Aryl amines are less basic because the lone pair of electrons on the nitrogen atom are delocalized into the aromatic ring, making them less available for interaction with a proton.

(b) Draw the formula of the salt formed by treating procaine with 1 mol of HCl.

Problem 10.28 Treatment of trimethylamine with 2-chloroethyl acetate gives the neurotransmitter acetylcholine as its chloride salt. Propose a structural formula for this quaternary ammonium salt and a mechanism for its formation.

$$(CH_3)_3N \;+\; CH_3\overset{O}{\overset{\|}{C}}OCH_2CH_2Cl \longrightarrow C_7H_{16}ClNO_2$$

Acetylcholine chloride

Acetylcholine chloride, a quaternary amine, is formed by S_N2 reaction between trimethylamine and 2-chloroethylacetate.

Problem 10.29 Aniline is prepared by catalytic reduction of nitrobenzene. Devise a chemical procedure based on the basicity of aniline to separate it from any unreacted nitrobenzene.

These compounds can first be dissolved in an organic solvent, such as diethyl ether, and then separated by extraction with an acidic (pH of ~4 or below) aqueous solution. Only the aniline will be protonated, thereby making it soluble in the aqueous layer as the anilinium ion. The hydrophobic nitrobenzene has no basic sites, so it will not be protonated and will remain in the ether layer. The aniline can be recovered by making the aqueous solution basic, thereby deprotonating aniline, and extracting with diethyl ether.

Problem 10.30 Suppose that you have a mixture of these three compounds. Devise a chemical procedure based on their relative acidity or basicity to separate and isolate each in pure form.

H_3C—⟨benzene ring⟩—NO_2 H_3C—⟨benzene ring⟩—NH_2 H_3C—⟨benzene ring⟩—OH

4-Nitrotoluene   4-Methylaniline   4-Methylphenol
(*p*-Nitrotoluene)   (*p*-Toluidine)   (*p*-Cresol)

These molecules can be separated by extraction into different aqueous solutions. First, the mixture is dissolved in an organic solvent such as ether in which all three compounds are soluble. Then, the ether solution is extracted with dilute aqueous HCl. Under these conditions, 4-methylaniline (a weak base) is converted to its protonated form and dissolves in the aqueous solution. The aqueous solution is separated, treated with dilute NaOH, and the water-insoluble 4-methylaniline separates, and is recovered. The ether solution containing the other two components is then treated with dilute aqueous NaOH. Under these conditions, 4-methylphenol (a weak acid) is converted to its phenoxide ion and dissolves in the aqueous solution. Acidification of this aqueous solution with dilute HCl forms water-insoluble 4-methylphenol that is then isolated. Evaporation of the remaining ether solution gives the 4-nitrotoluene.

CHAPTER 11
Solutions to Problems

Problem 11.1 Write the IUPAC name for each compound. Specify the configuration of (c).

(a)
$$CH_3CCH_2CHO$$
with CH$_3$ above and CH$_3$ below the central C

3,3-Dimethylbutanal

(b)

3-Hydroxycyclohexanone

(c)
$$C_6H_5-C\text{'''}CH_3$$
with CHO above and H below

(R)-2-Phenylpropanal

Problem 11.2 Write structural formulas for all aldehydes of molecular formula $C_6H_{12}O$ and give each its IUPAC name. Which of these aldehydes are chiral?

There are eight aldehydes with this molecular formula. The chiral aldehydes are circled.

$$CH_3CH_2CH_2CH_2CH_2C-H$$
Hexanal

$$CH_3CHCH_2CH_2C-H$$
with CH$_3$
4-Methylpentanal

$$CH_3CH_2CHCH_2C-H$$
with CH$_3$
3-Methylpentanal

$$CH_3CH_2CH_2CHC-H$$
with H$_3$C
2-Methylpentanal

$$CH_3CCH_2C-H$$
with CH$_3$ above and CH$_3$ below
3,3-Dimethylbutanal

$$CH_3CH_2CC-H$$
with H$_3$C above and H$_3$C below
2,2-Dimethylbutanal

$$CH_3CHCHC-H$$
with H$_3$C above and CH$_3$ below
2,3-Dimethylbutanal

$$CH_3CH_2CHC-H$$
with CH$_2$CH$_3$ below
2-Ethylbutanal

Problem 11.3 Write IUPAC names for these compounds, each of which is important in intermediary metabolism. Below each is the name by which it is more commonly known in the biological sciences.

(a) CH_3CHCO_2H with OH above

Lactic acid
(product of anaerobic
glycolysis)

2-Hydroxypropanoic acid

(b) CH_3CCO_2H with O above

Pyruvic acid
(Product of anaerobic
glycolysis)

2-Oxopropanoic acid

(c) $H_2NCH_2CH_2CH_2CO_2H$

γ-Aminobutyric acid
(a neurotransmitter)

4-Aminobutanoic acid

Problem 11.4 Explain how these Grignard reagents react with molecules of their own kind to "self-destruct."

(a) HO—⟨benzene ring⟩—MgBr

(b) $HOCCH_2CH_2CH_2MgBr$ with O above

These Grignard reagents cannot be used because they have acidic functions that would be deprotonated by the basic Grignard reagent portion of the molecule. In (a), the phenol would be deprotonated to give a phenoxide, and in (b), the carboxylic acid would be deprotonated to give the carboxylate.

Problem 11.5 Show how these four compounds can be synthesized from the same Grignard reagent.

(a) [cyclohexene]—CO_2H (b) [cyclohexene]—CH_2OH (c) [cyclohexene]—$\overset{\overset{\displaystyle OH}{|}}{C}HCH_3$ (d) [cyclohexene]—CH_2CH_2OH

All of the products can be obtained using 2-cyclohexenylmagnesium bromide and the other reactants shown.

[cyclohexene]—MgBr $\xrightarrow{\text{1) } CO_2}$ $\xrightarrow[\text{H}_2\text{O}]{\text{2) NH}_4\text{Cl}}$ [cyclohexene]—CO_2H

[cyclohexene]—MgBr $\xrightarrow{\text{1) } H\overset{O}{\overset{||}{C}}H}$ $\xrightarrow[\text{H}_2\text{O}]{\text{2) NH}_4\text{Cl}}$ [cyclohexene]—CH_2OH

[cyclohexene]—MgBr $\xrightarrow{\text{1) } CH_3\overset{O}{\overset{||}{C}}H}$ $\xrightarrow[\text{H}_2\text{O}]{\text{2) NH}_4\text{Cl}}$ [cyclohexene]—$\overset{\overset{\displaystyle OH}{|}}{C}HCH_3$

[cyclohexene]—MgBr $\xrightarrow{\text{1) } H_2C\overset{O}{\overset{\triangle}{-}}CH_2}$ $\xrightarrow[\text{H}_2\text{O}]{\text{2) NH}_4\text{Cl}}$ [cyclohexene]—CH_2CH_2OH

Problem 11.6 Hydrolysis of an acetal forms an aldehyde or ketone and two molecules of alcohol. Following are structural formulas for three acetals. Draw the structural formulas for the products of hydrolysis of each in aqueous acid.

(a) CH_3O—[benzene]—$\overset{\overset{\displaystyle OCH_3}{|}}{C}HOCH_3$ $\xrightarrow{H_3O^+}$ CH_3O—[benzene]—$\overset{O}{\overset{||}{C}}$-H + 2 CH_3OH

(b) $\overset{CH_3}{\underset{CH_3}{>}}\overset{O-CH_2}{\underset{O-CH_2}{<}}$ $\xrightarrow{H_3O^+}$ $CH_3\overset{O}{\overset{||}{C}}CH_3$ + $HOCH_2CH_2OH$

(c) $\underset{CH_3}{\diagdown}$[furan ring]$\underset{OCH_3}{}$ $\xrightarrow{H_3O^+}$ $CH_3\overset{\overset{\displaystyle OH}{|}}{C}HCH_2CH_2\overset{O}{\overset{||}{C}}H$ + CH_3OH

Problem 11.7 Acid-catalyzed hydrolysis of an imine gives an amine and an aldehyde or ketone. When one equivalent of acid is used, the amine is converted to its ammonium salt. Write structural formulas for the products of hydrolysis of each imine using one equivalent of HCl.

(a) CH_3O—[benzene]—$CH=NCH_2CH_3$ + H_2O $\xrightarrow{HCl}$ CH_3O—[benzene]—$\overset{O}{\overset{||}{C}}H$ + $H_3\overset{+}{N}CH_2CH_3$

Hydrolysis of (a) gives an aldehyde and a primary ammonium salt.

(b)

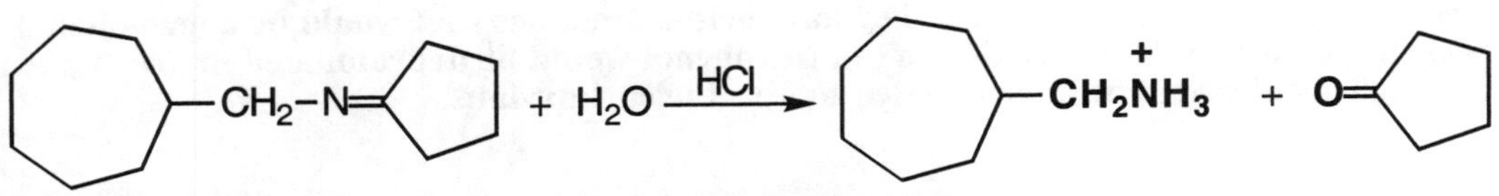

Hydrolysis of (b) gives a ketone and a primary ammonium salt.

Problem 11.8 Draw the structural formula for the keto form of each enol.

(a)

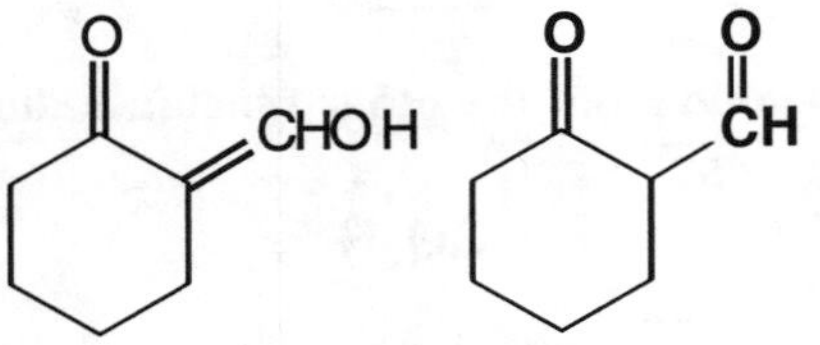

(b)

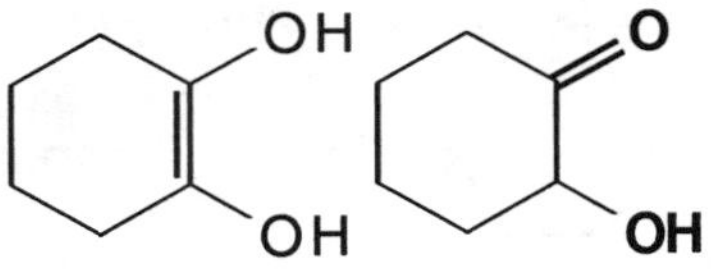

(c)

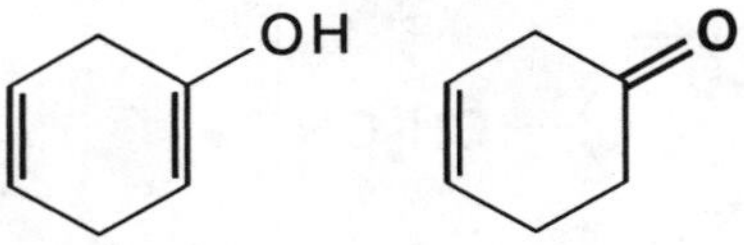

Problem 11.9 Complete these oxidations.
(a) Hexanedial + H_2O_2 $\longrightarrow$

$$HC(CH_2)_4CH + H_2O_2 \longrightarrow HOC(CH_2)_4COH$$

(b) 3-Phenylpropanal + Tollen's reagent $\longrightarrow$

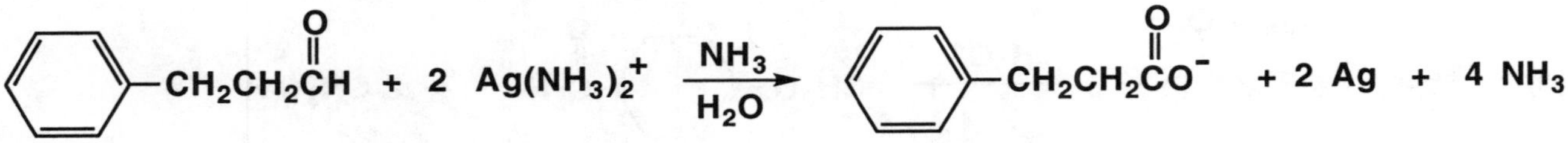

Problem 11.10 What aldehyde or ketone gives these alcohols on reduction with $NaBH_4$?

(a)   (b) $CH_3O-$$-CH_2CH_2OH$ (c) $CH_3CH(CH_2)_3CHCH_3$ with OH OH

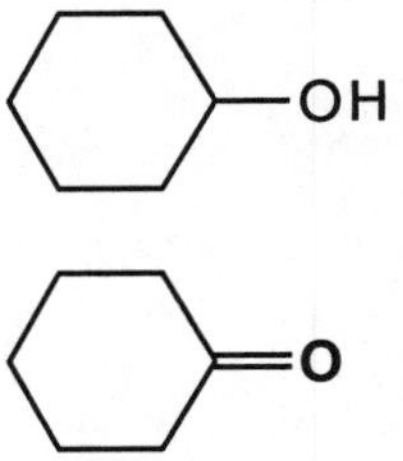

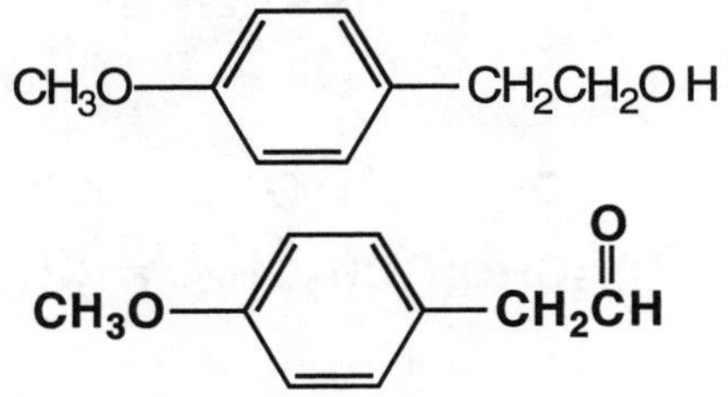

Problem 11.11 Show how you might convert piperidine to these compounds, using any other organic compounds and necessary reagents.

(a)

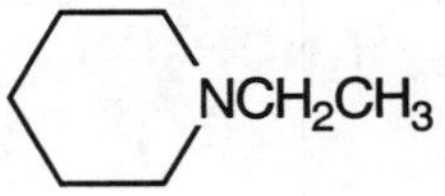

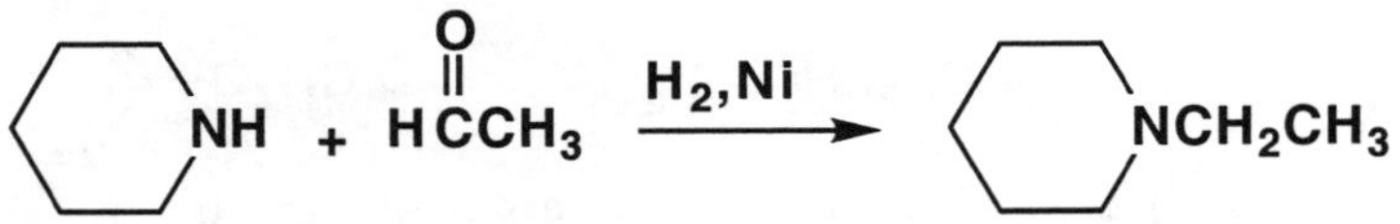

(b)

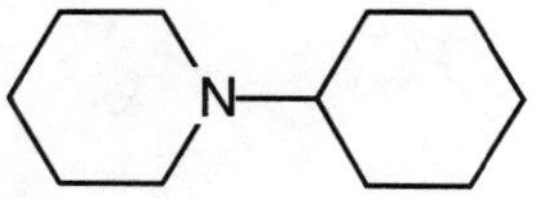

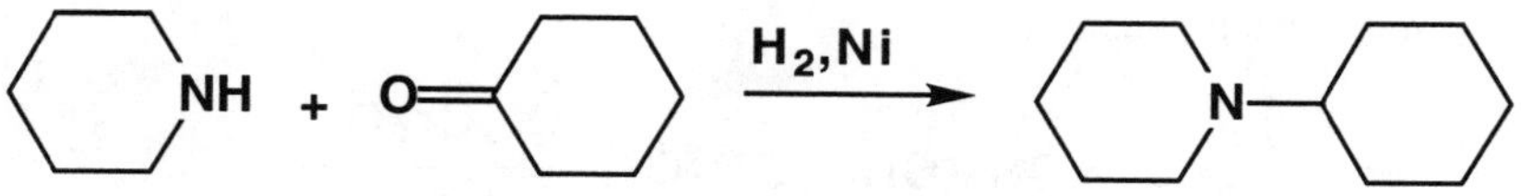

Preparation of Aldehydes and Ketones

The methods covered to this point for preparation of aldehydes and ketones are oxidation of primary and secondary alcohols, (Section 8.4F), and Friedel-Crafts acylation of arenes (Section 9.7C).

Problem 11.12 Complete these reactions.

(a)

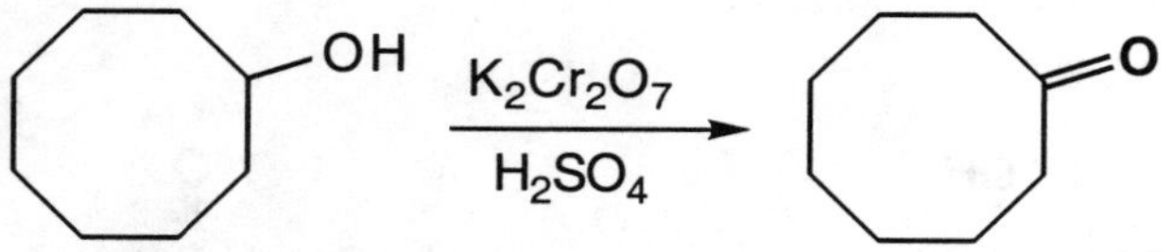

(b)

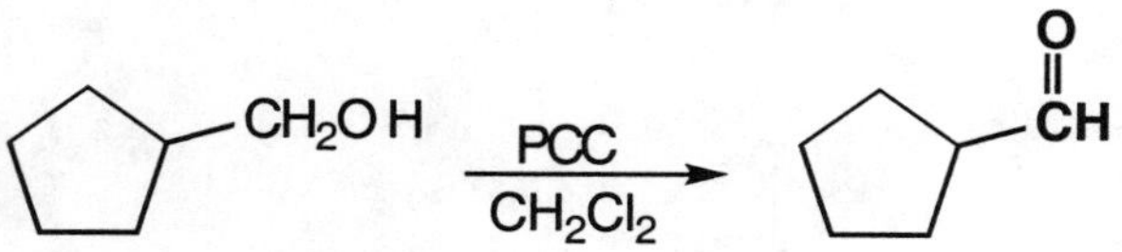

(c)

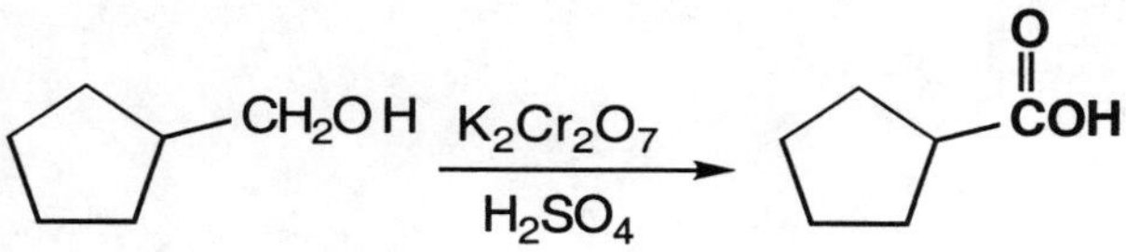

(d)

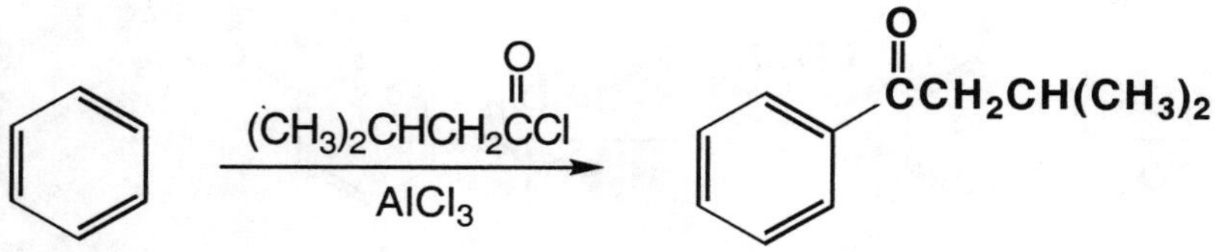

(e)

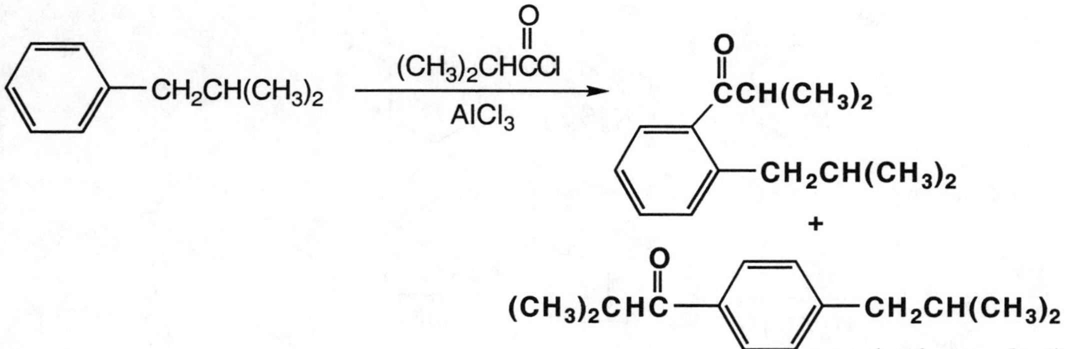

(major product)

Problem 11.13 Show how you would bring about these conversions.
(a) 1-Pentanol to pentanal

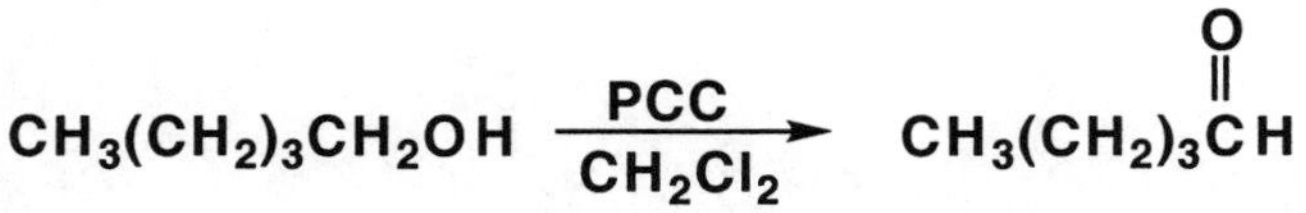

$$CH_3(CH_2)_3CH_2OH \xrightarrow[CH_2Cl_2]{PCC} CH_3(CH_2)_3\overset{O}{\overset{\|}{C}}H$$

(b) 1-Pentanol to pentanoic acid

$$CH_3(CH_2)_3CH_2OH \xrightarrow[H_2SO_4]{K_2Cr_2O_7} CH_3(CH_2)_3\overset{O}{\overset{\|}{C}}OH$$

(c) 2-Pentanol to 2-pentanone

$$CH_3(CH_2)_2\overset{OH}{\overset{|}{C}H}CH_3 \xrightarrow[H_2SO_4]{K_2Cr_2O_7} CH_3(CH_2)_2\overset{O}{\overset{\|}{C}}CH_3$$

(d) 1-Pentene to 2-pentanone

$$CH_3(CH_2)_2CH{=}CH_2 \xrightarrow[H_2SO_4]{H_2O} CH_3(CH_2)_2\overset{OH}{\overset{|}{C}H}CH_3 \xrightarrow[H_2SO_4]{K_2Cr_2O_7} CH_3(CH_2)_2\overset{O}{\overset{\|}{C}}CH_3$$

(e) Benzene to acetophenone

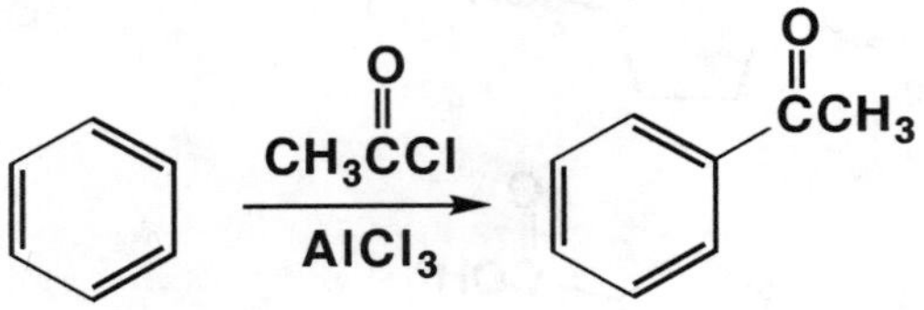

(f) Styrene to acetophenone

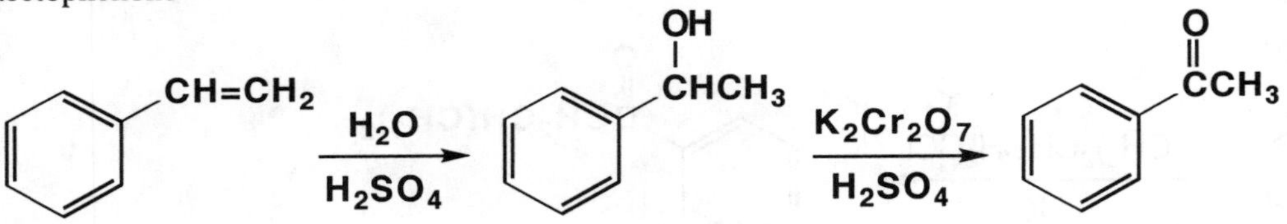

(g) Cyclohexanol to cyclohexanone

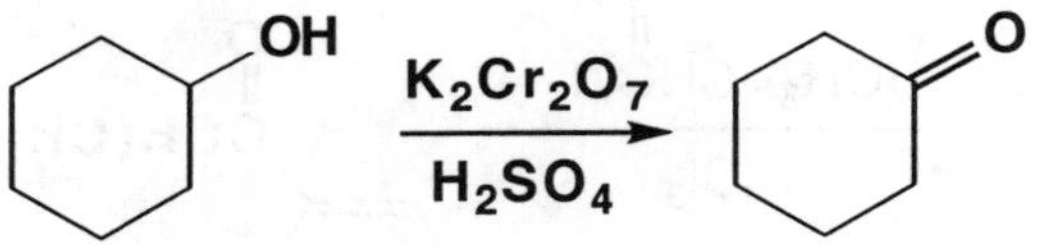

(h) Cyclohexene to cyclohexanone

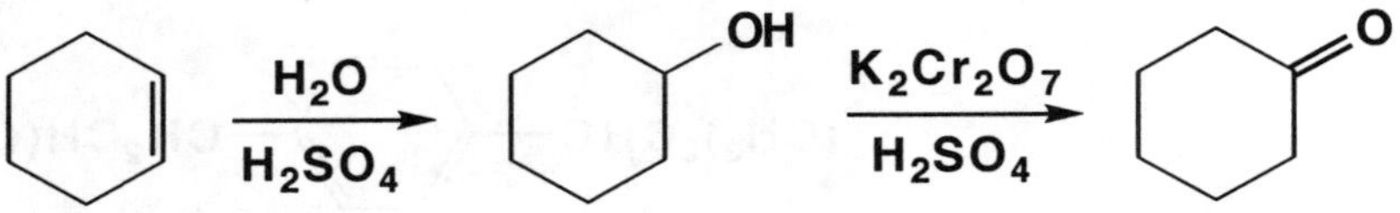

Structure and Nomenclature
<u>Problem 11.14</u> Draw a structural formula for the one ketone of molecular formula C_4H_8O and for the two aldehydes of molecular formula C_4H_8O.

$$CH_3CH_2\overset{O}{\overset{\|}{C}}CH_3 \qquad CH_3CH_2CH_2\overset{O}{\overset{\|}{C}}H \qquad (CH_3)_2CH\overset{O}{\overset{\|}{C}}H$$

Problem 11.15 Draw structural formulas for the four aldehydes of molecular formula $C_5H_{10}O$. Which are chiral?

The chiral aldehyde is circled, and the stereocenter is indicated with an asterisk.

$$CH_3CH_2CH_2CH_2\overset{\overset{\displaystyle O}{\|}}{C}-H$$

$$\boxed{CH_3CH_2\overset{*}{\underset{\underset{\displaystyle CH_3}{|}}{C}}H\overset{\overset{\displaystyle O}{\|}}{C}-H}$$

$$CH_3\underset{\underset{\displaystyle CH_3}{|}}{C}HCH_2\overset{\overset{\displaystyle O}{\|}}{C}-H$$

$$\underset{\underset{\displaystyle H_3C}{|}}{\overset{\overset{\displaystyle H_3C\ \ O}{|\ \ \|}}{CH_3C}}C-H$$

Problem 11.16 Name these compounds.

(a) $(CH_3CH_2CH_2)_2C=O$

**4-Heptanone
(Dipropyl ketone)**

(b)

**(S)-2-Methyl-
cyclopentanone**

(c)

**(Z)-2-Methyl-2-butenal
(*cis*-2-Methyl-2-butenal)**

(d)

(S)-2-Hydroxypropanal

(e) $CH_3O-\!\!\!\!-\!\!\!\!-\overset{\overset{\displaystyle O}{\|}}{C}CH_3$

4-Methoxyacetophenone

(f)

$$HC(CH_2)_4CH$$

Hexanedial

(g) $-CH_2CH_2CH_3$

2-Propyl-1,3-cyclopentanedione

Problem 11.17 Draw structural formulas for these compounds.
(a) 1-Chloro-2-propanone   (b) 3-Hydroxybutanal   (c) 4-Hydroxy-4-methyl-2-pentanone

$$CH_3\overset{\overset{\displaystyle O}{\|}}{C}CH_2Cl$$

$$CH_3\overset{\overset{\displaystyle OH}{|}}{C}HCH_2\overset{\overset{\displaystyle O}{\|}}{C}H$$

$$CH_3\underset{\underset{\displaystyle CH_3}{|}}{\overset{\overset{\displaystyle OH}{|}}{C}}CH_2\overset{\overset{\displaystyle O}{\|}}{C}CH_3$$

(d) 3-Methyl-3-phenylbutanal   (e) 1,3-Cyclohexanedione   (f) 3-Methyl-3-butene-2-one

$$-\underset{\underset{\displaystyle CH_3}{|}}{\overset{\overset{\displaystyle CH_3}{|}}{C}}CH_2\overset{\overset{\displaystyle O}{\|}}{C}H$$

$$CH_2=\underset{\underset{\displaystyle CH_3}{|}}{C}\overset{\overset{\displaystyle O}{\|}}{C}CH_3$$

(g) 5-Oxohexanal

$$CH_3\overset{\displaystyle O}{\overset{\|}{C}}CH_2CH_2CH_2\overset{\displaystyle O}{\overset{\|}{C}}H$$

(h) 2,2-Dimethylcyclohexane-
carbaldehyde

(i) 3-Oxobutanoic acid

$$CH_3\overset{\displaystyle O}{\overset{\|}{C}}CH_2\overset{\displaystyle O}{\overset{\|}{C}}OH$$

Addition of Carbon Nucleophiles

Problem 11.18 Write an equation for the acid-base reaction between phenylmagnesium iodide and a carboxylic acid. Use curved arrows to show the flow of electrons in this reaction. In addition, show that this reaction is an example of a stronger acid and stronger base reacting to form a weaker acid and weaker base.

pK$_a$ 4-5   **Stronger base**   **Weaker base**   pK$_a$ 43
Stronger acid   **Weaker acid**

Problem 11.19 Diethyl ether is prepared on an industrial scale by acid-catalyzed dehydration of ethanol. Explain why diethyl ether used in the preparation of Grignard reagents must be carefully purified to remove all traces of ethanol and water.

$$2\ CH_3CH_2OH \quad \xrightarrow[180°C]{H_2SO_4} \quad CH_3CH_2OCH_2CH_3 \ + \ H_2O$$

All water and alcohol must be removed from diethyl ether prior to adding any Grignard reagent, ethylmagnesium iodide for example, because these impurities would deactivate the Grignard reagent due to the following acid-base reaction.

$$R\overset{\frown}{-O-H} + CH_3CH_2-MgI \longrightarrow R-O^- \ (MgI)^+ \ + \ CH_3CH_3$$

R = CH$_3$CH$_2$ or H

Problem 11.20 Draw structural formulas for the product formed by treatment of each compound with propylmagnesium bromide followed by hydrolysis in aqueous acid.

The products after acid hydrolysis are given in bold.

(a)   CH$_2$O

(b)
$$\overset{\displaystyle H_2C\text{------}CH_2}{\underset{\displaystyle O}{\diagdown\diagup}}$$

(c) $CH_3CH_2\overset{\displaystyle O}{\overset{\|}{C}}CH_2CH_3$

CH$_3$CH$_2$CH$_2$CH$_2$OH

$CH_3CH_2CH_2CH_2\overset{\displaystyle OH}{\overset{\displaystyle |}{C}}H_2$

**$CH_3CH_2\overset{\displaystyle OH}{\overset{\displaystyle |}{C}}CH_2CH_3$
$\ \ \ \ \ \ \ \ \ \ |$
$\ \ \ \ \ \ \ CH_2CH_2CH_3$**

(d)

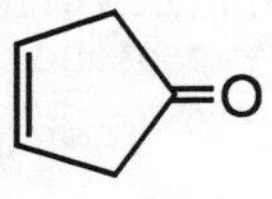

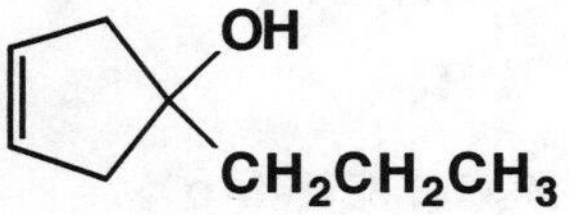

(e) CO_2

(f)

$$CH_3CH_2CH_2\overset{\overset{\displaystyle O}{\|}}{C}OH$$

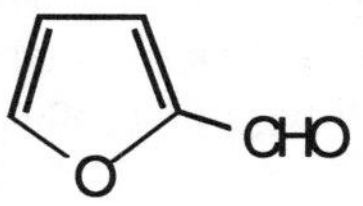

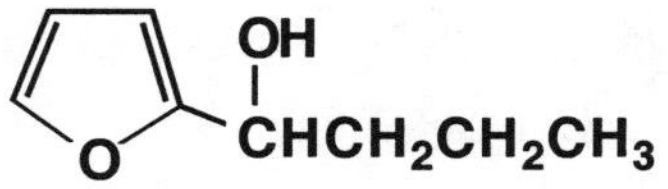

(g)

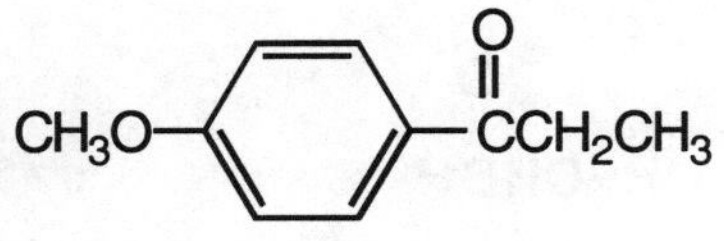

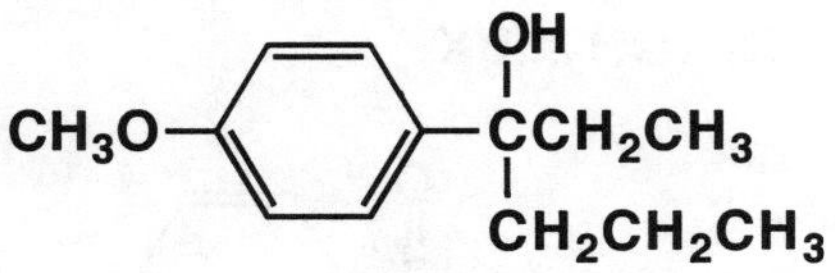

Problem 11.21 Write structural formulas for all combinations of Grignard reagent and aldehyde or ketone that might be used to synthesize each alcohol.

(a) $CH_3CH_2CH_2CH_2\overset{\overset{\displaystyle OH}{|}}{C}HCH_3$

$$CH_3CH_2CH_2CH_2\overset{\overset{\displaystyle O}{\|}}{C}H \ + \ CH_3MgI \ \xrightarrow{\ 2)\ NH_4Cl\ /\ H_2O\ }$$

or

$$CH_3\overset{\overset{\displaystyle O}{\|}}{C}H \ + \ CH_3CH_2CH_2CH_2MgI \ \xrightarrow{\ 2)\ NH_4Cl\ /\ H_2O\ }$$

(b)

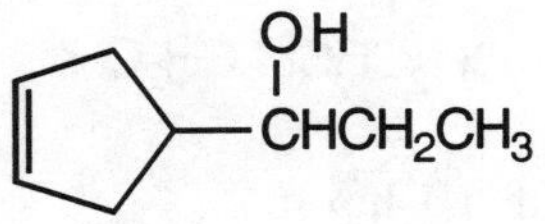

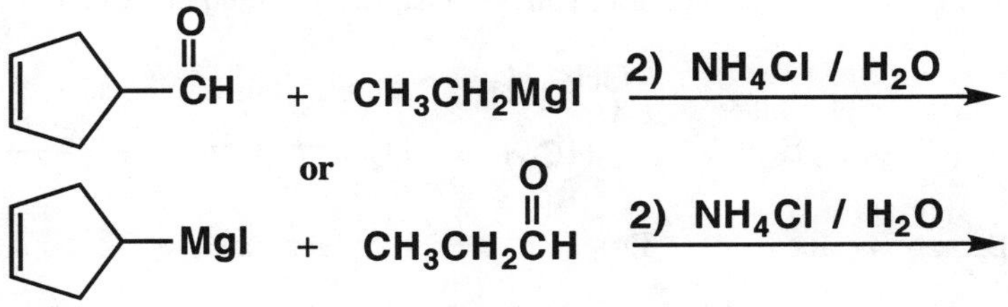

$+ \ CH_3CH_2MgI \ \xrightarrow{\ 2)\ NH_4Cl\ /\ H_2O\ }$

or

$+ \ CH_3CH_2\overset{\overset{\displaystyle O}{\|}}{C}H \ \xrightarrow{\ 2)\ NH_4Cl\ /\ H_2O\ }$

Problem 11.22 Show reagents to bring about this conversion.

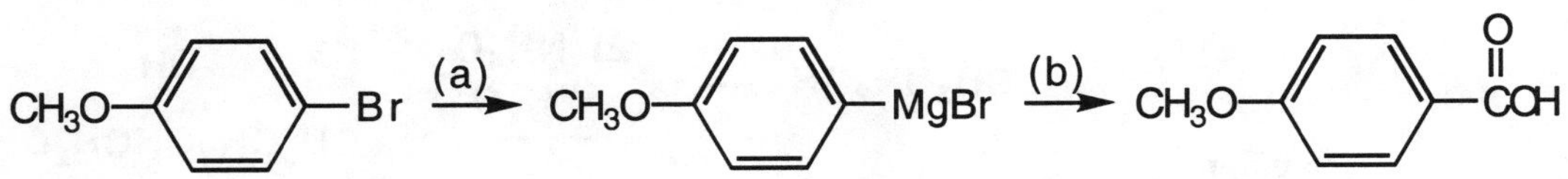

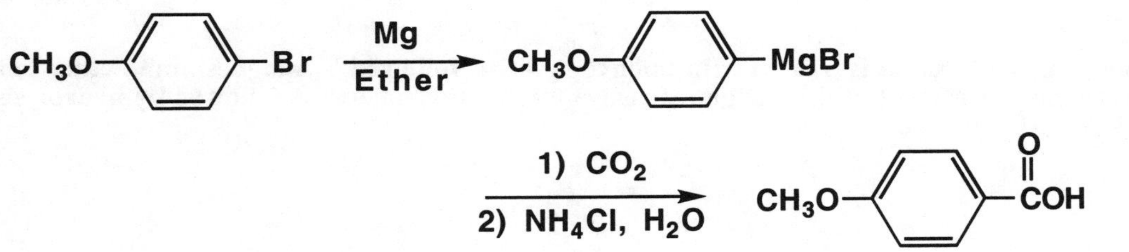

Problem 11.23 Suggest a synthesis for these alcohols starting from an aldehyde or ketone and an appropriate Grignard reagent. In parentheses below each target molecule is shown the number of combinations of Grignard reagent and aldehyde or ketone that might be used.

(a)
$$\underset{\substack{\displaystyle OH \\ | \\ CH_2CH_3}}{CH_3CCH_2CH_2CH_3}$$

(3 combinations)

$$\underset{\substack{O \\ \parallel}}{CH_3CCH_2CH_2CH_3}$$

+ CH_3CH_2MgX

$$\underset{\substack{O \\ \parallel \\ CH_2CH_3}}{CCH_2CH_2CH_3}$$

+ CH_3MgX

$$\underset{\substack{O \\ \parallel \\ CH_3}}{CCH_2CH_3}$$

+ CH_3CH_2CH_2MgX

(b)
$$\underset{\substack{OH \\ |}}{CH_3CH_2CHCH=CHCH_3}$$

(2 combinations)

$$\underset{\substack{O \\ \parallel}}{CH_3CH_2CH}$$
+
CH_3CH=CHMgX

$$\underset{\substack{O \\ \parallel}}{CH_3CH=CHCH}$$
+
CH_3CH_2MgX

(c)
$$CH_3O-\text{(ring)}-\underset{\substack{OH \\ | \\ CH}}{}-\text{(ring)}$$

(2 combinations)

$$CH_3O-\text{(ring)}-\underset{\substack{O \\ \parallel}}{CH}$$
+
(ring)—MgX

$$HC-\text{(ring)}$$ with O
+
CH_3O-(ring)—MgX

Problem 11.24 Show how to synthesize 3-ethyl-1-hexanol using 1-bromopropane, propanal and ethylene oxide as the only sources of carbon atoms. It can be done using each compound only once. (Hint: Carry out one Grignard reaction to form an alcohol, convert the alcohol to an alkyl halide, and then do a second Grignard reaction).

$$CH_3CH_2CH_2Br \;+\; \underset{\substack{CH_2CH_3 \\ | \\ CHO}}{} \;+\; H_2C\!\!-\!\!CH_2 \quad \xrightarrow[\text{steps}]{\text{several}} \quad \underset{\substack{CH_2CH_3 \\ | }}{CH_3CH_2CH_2CHCH_2CH_2OH}$$

1-Bromopropane   Propanal   Ethylene oxide   3-Ethyl-1-hexanol

This synthesis is divided into two stages. In the first stage, 1-bromopropane is treated with magnesium to form a Grignard reagent, then with propanal, followed by hydrolysis in aqueous acid to give 3-hexanol.

$$CH_3CH_2CH_2Br + Mg \xrightarrow{\text{ether}} CH_3CH_2CH_2MgBr \xrightarrow[\substack{O \\ \parallel \\ CH_3CH_2CH}]{\substack{\text{2) } NH_4Cl \\ H_2O}} \underset{\substack{OH \\ |}}{CH_3CH_2CHCH_2CH_2CH_3}$$

3-Hexanol

In the second stage, 3-hexanol is treated with thionyl chloride, followed by magnesium in ether to form a Grignard reagent. Treatment of this Grignard reagent with ethylene oxide, followed by hydrolysis in aqueous acid gives 3-ethyl-1-hexanol.

$$\underset{\textbf{3-Ethyl-1-hexanol}}{\overset{\overset{\displaystyle OH}{|}}{CH_3CH_2CHCH_2CH_2CH_3}} \xrightarrow[\textbf{2) Mg, ether}]{\textbf{1) SOCl}_2} CH_3CH_2\overset{\overset{\displaystyle MgCl}{|}}{CH}CH_2CH_2CH_3 \xrightarrow[\substack{\textbf{2) NH}_4\textbf{Cl} \\ \textbf{H}_2\textbf{O}}]{\textbf{1)} \overset{CH_2-CH_2}{\underset{O}{\diagdown\diagup}}}$$

$$\underset{}{\overset{\overset{\displaystyle CH_2CH_2OH}{|}}{CH_3CH_2CHCH_2CH_2CH_3}}$$

3-Ethyl-1-hexanol

Problem 11.25 1-Phenyl-2-butanol is used in perfumery. Show how to synthesize this alcohol from bromobenzene, 1-butene, and any other needed reagents.

Bromobenzene   1-Butene   1-Phenyl-2-butanol

(a) Bromobenzene is treated with magnesium in diethyl ether to form phenylmagnesium bromide, in preparation for part (c).

(b) Treatment of 1-butene with a peroxycarboxylic acid gives 1,2-epoxybutane.

$$CH_3CH_2CH{=}CH_2 \;+\; RCO_3H \longrightarrow CH_3CH_2\overset{O}{\overset{\diagup\diagdown}{CH-CH_2}} + RCO_2H$$

(c) Treatment of phenylmagnesium bromide with 1,2-epoxybutane followed by hydrolysis in aqueous acid gives 1-phenyl-2-butanol.

Addition of Oxygen Nucleophiles
Problem 11.26 5-Hydroxyhexanal forms a six-membered cyclic hemiacetal, which predominates at equilibrium in aqueous solution.

$$\underset{\substack{|\\OH}}{CH_3\overset{*}{C}HCH_2CH_2CH_2\overset{\overset{\displaystyle O}{\|}}{C}H} \underset{}{\overset{H^+}{\rightleftharpoons}} \quad \text{A cyclic hemiacetal}$$

5-Hydroxyhexanal

(a) Draw a structural formula for this cyclic hemiacetal.

(b) How many stereoisomers are possible for 5-hydroxyhexanal?

5-Hydroxyhexanal has one stereocenter, so there are two stereoisomers possible (a pair of enantiomers).

(c) How many stereoisomers are possible for this cyclic hemiacetal?

Four stereoisomers are possible for the cyclic hemiacetal; two pairs of enantiomers.
Following are planar hexagon formulas for each pair of enantiomers of the cyclic hemiacetal.

(d) Draw alternative chair conformations for each stereoisomer.

Alternative chair conformations are drawn for (A), one of the cis enantiomers, and for (C), one of the trans enantiomers.

Alternative chair conformations of (A)   Alternative chair conformations of (C)

(e) Which alternative chair conformation for each stereoisomer is the more stable?

For (A), the diequatorial chair is the more stable. For (C), the alternative chairs are of approximately equal stability.

Problem 11.27 Draw structural formulas for the hemiacetal and then the acetal formed from each pair of reactants in the presence of an acid catalyst.

Problem 11.28 Draw structural formulas for the products of hydrolysis of each acetal in aqueous acid.

(a)

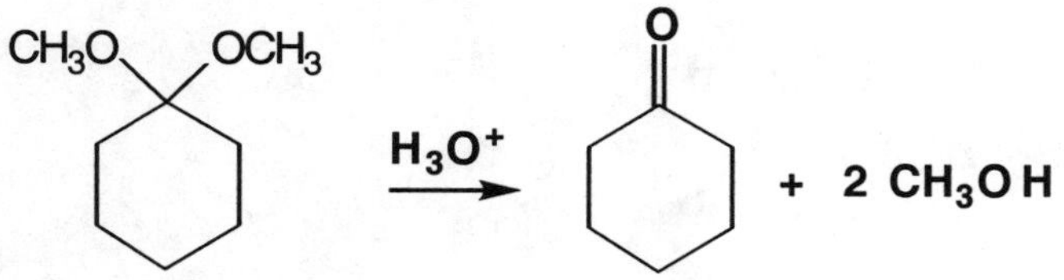

(b)

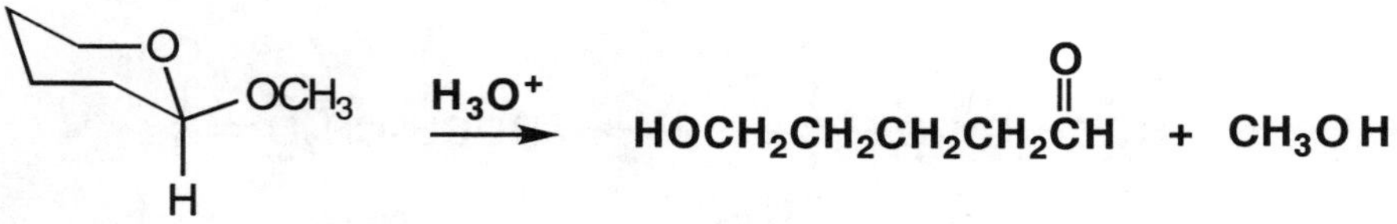

$$HOCH_2CH_2CH_2CH_2CH \overset{O}{\parallel} \quad + \quad CH_3OH$$

(c)

$$\underset{\underset{HO}{|}}{CH_2}\underset{\underset{OH}{|}}{CHCH}\overset{O}{\parallel} \quad + \quad CH_3\overset{O}{\overset{\parallel}{C}}CH_3$$

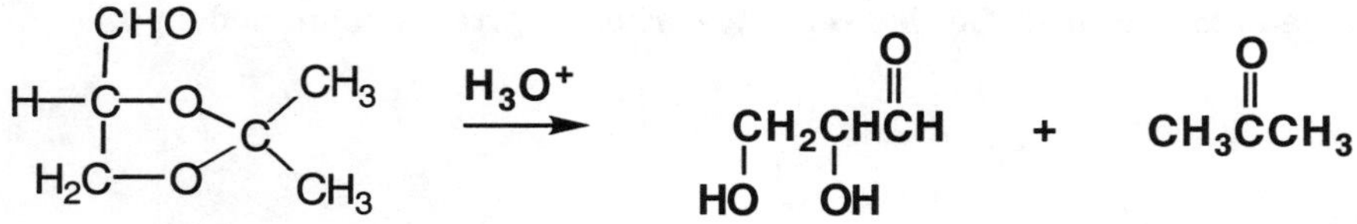

Problem 11.29 Propose a mechanism for formation of the cyclic acetal from treatment of acetone with ethylene glycol in the presence of an acid catalyst. Your mechanism must be consistent with the fact that the oxygen atom of the water molecule is derived from the carbonyl oxygen atom of acetone.

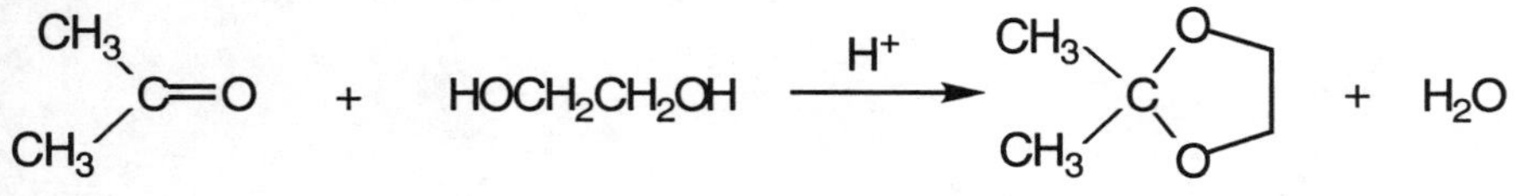

Step 1: Protonation of the carbonyl oxygen atom to form a reactive electrophilic species.

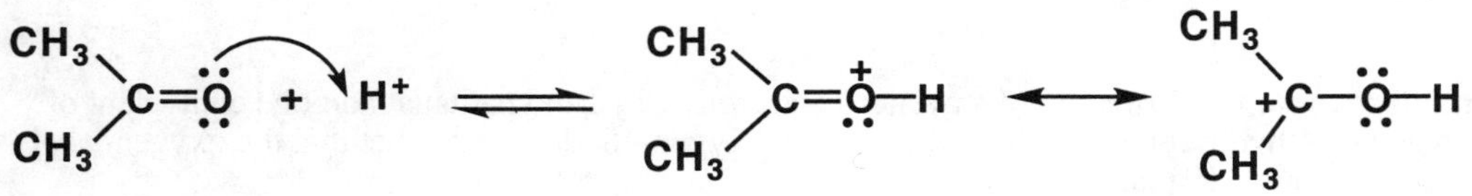

Step 2: One of the nucleophilic oxygen atoms of ethylene glycol attacks the protonated carbonyl species.

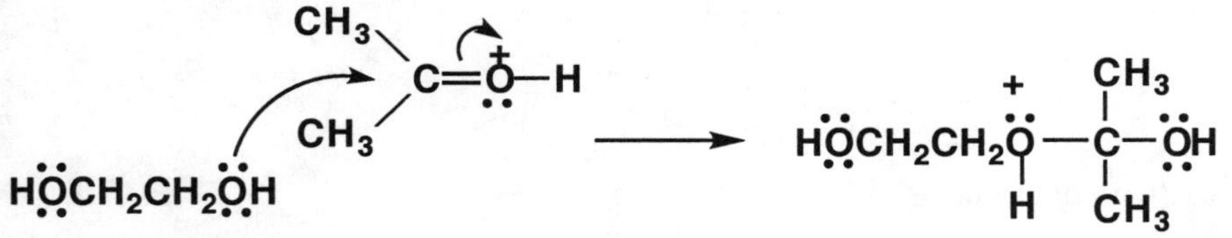

Step 3: A proton is lost to give the hemiacetal intermediate

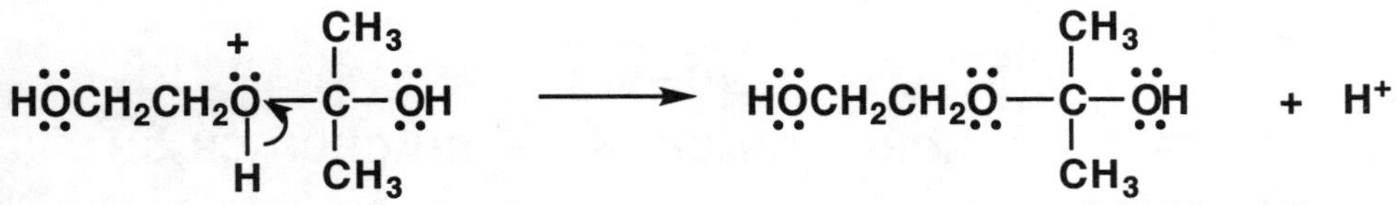

A hemiacetal

Step 4: **The hemiacetal is protonated on the hydroxyl group. Note that protonation on the ether oxygen could also occur, but that would simply be the reverse of the previous step and not productive.**

Step 5: **Water is a good leaving group, so it departs to give another highly electrophilic species.**

Step 6: **The other hydroxyl oxygen atom attacks the electrophilic carbon to give a protonated, cyclic intermediate.**

Step 7: **Loss of a proton gives the final cyclic acetal product.**

<u>Problem 11.30</u> Propose a mechanism for the formation of a cyclic acetal from 4-hydroxypentanal and one equivalent of methanol. If the carbonyl oxygen of 4-hydroxypentanal is enriched with oxygen-18, do you predict that the oxygen label appears in the cyclic acetal or in the water? Explain.

4-Hydroxypentanal   A cyclic acetal

The ^{18}O-enriched oxygen atom of 4-hydroxypentanal is indicated with an asterisk.

Step 1: **As in the mechanism in Problem 11.29, the first step is protonation of the carbonyl oxygen atom to give a resonance-stabilized carbocation.**

Step 2: **The hydroxyl group attacks the carbon atom of the protonated carbonyl group, which forms a protonated cyclic hemiacetal. Loss of a proton gives a hemiacetal intermediate.**

Step 3: Protonation of the hydroxyl group converts it to a good leaving group.

Step 4: Loss of water yields a new resonance-stabilized carbocation. Note that the oxygen 18 label appears in the water.

A resonance-stabilized carbocation

Step 5: Attack of methanol upon the carbon atom of the carbocation yields a protonated acetal, which subsequently loses a proton to give the acetal.

Addition of Nitrogen Nucleophiles

<u>Problem 11.31</u> Show how this secondary amine can be prepared by two successive reductive aminations:

The first step involves a reductive amination of the ketone using ammonia.

The second step uses benzaldehyde along with the primary amine produced in the first step.

Problem 11.32 Show how to convert cyclohexanone into each amine.

(a) 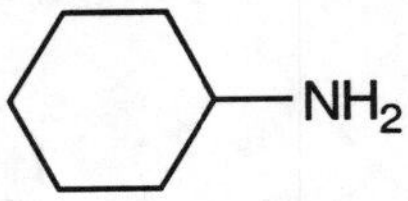 [cyclohexane ring]—NH₂

[cyclohexanone]=O + NH₃ —H₂, Ni→ [cyclohexane]—NH₂

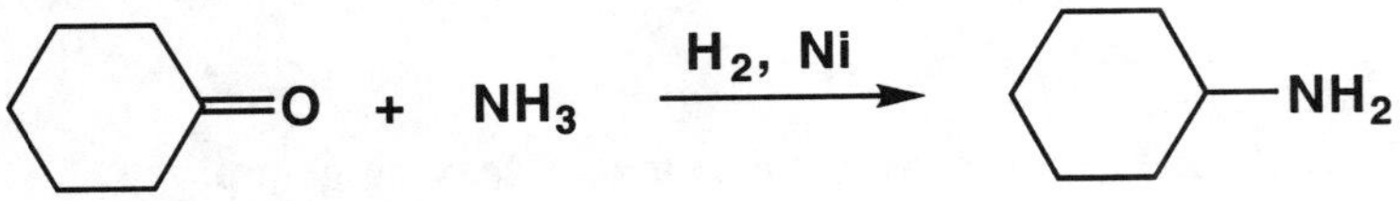

(b) 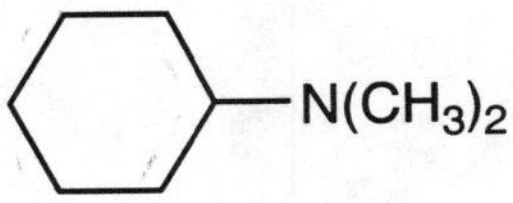 [cyclohexane ring]—N(CH₃)₂

[cyclohexanone]=O + HN(CH₃)₂ —H₂, Ni→ [cyclohexane]—N(CH₃)₂

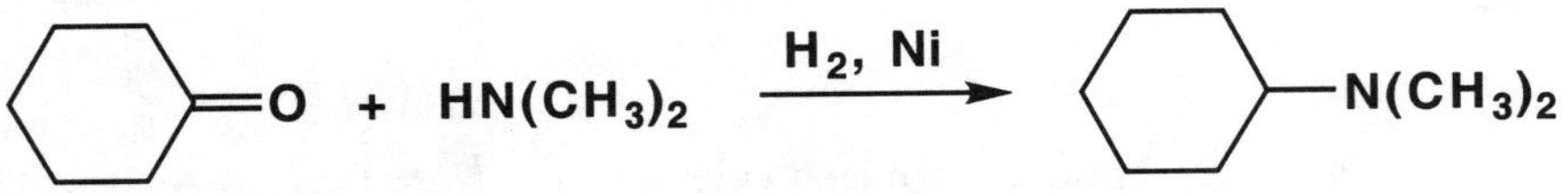

(c) 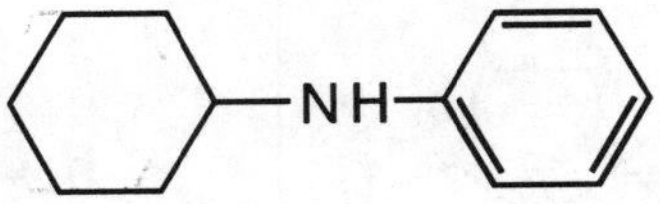 [cyclohexane ring]—NH—[benzene ring]

[cyclohexanone]=O + H₂N—[benzene ring] —H₂, Ni→ [cyclohexane]—NH—[benzene ring]

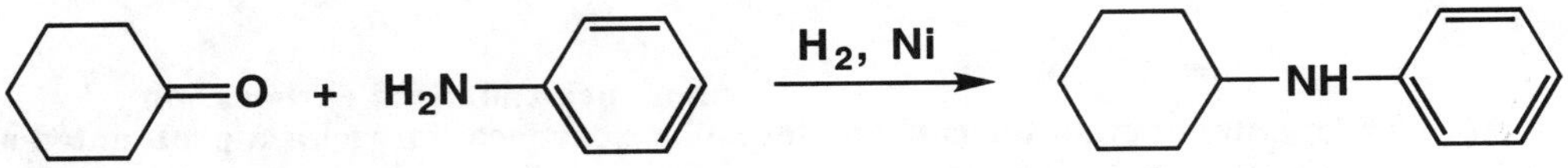

Problem 11.33 Following are structural formulas for amphetamine and methamphetamine. The major central nervous system effects of amphetamine and amphetamine-like drugs are locomotor stimulation, euphoria and excitement, stereotyped behavior, and anorexia. Show how each drug can be synthesized by reductive amination of an appropriate aldehyde or ketone. Assign an R or S configuration to the enantiomer of methamphetamine shown here.

(a) 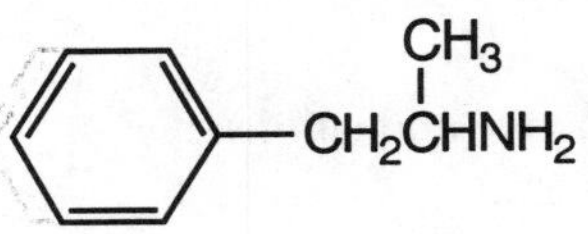 [benzene ring]—CH₂CHNH₂
            |
           CH₃

Amphetamine

(b) 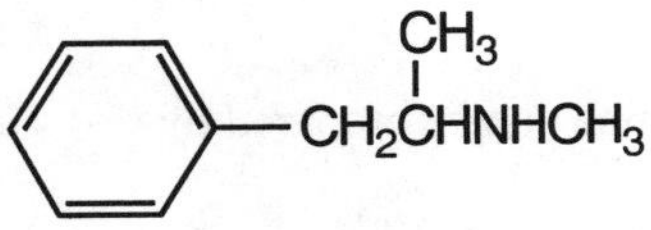 [benzene ring]—CH₂CHNHCH₃
            |
           CH₃

Methamphetamine

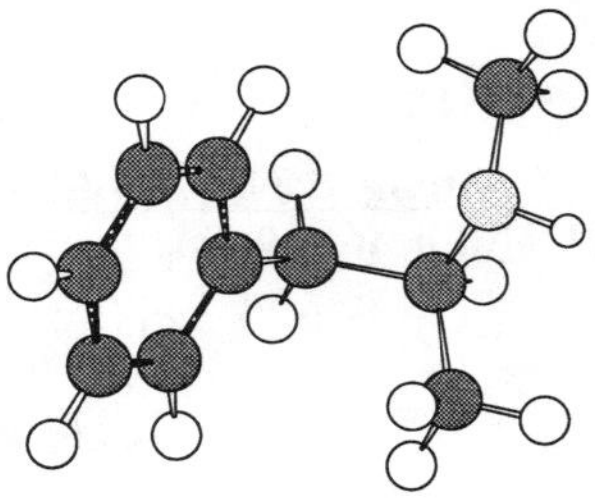

(a) 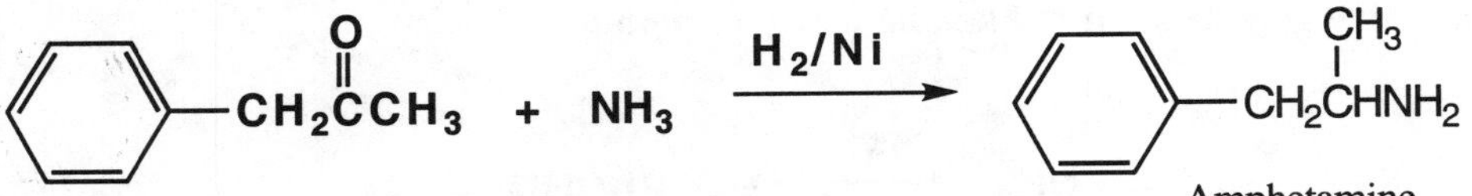 [benzene ring]—CH₂CCH₃ + NH₃ —H₂/Ni→ [benzene ring]—CH₂CHNH₂
                    ‖ |
                    O   CH₃

Amphetamine

(b) 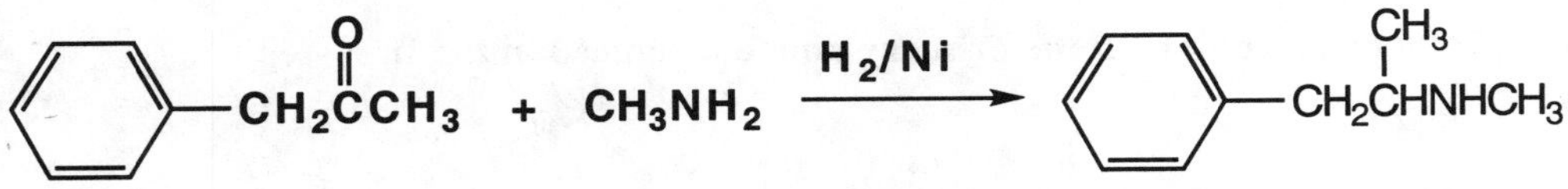 [benzene ring]—CH₂CCH₃ + CH₃NH₂ —H₂/Ni→ [benzene ring]—CH₂CHNHCH₃
                    ‖ |
                    O   CH₃

Methamphetamine

The molecule of methamphetamine shown in the space-filling figure is S.

Problem 11.34 Rimantadine is effective in preventing infections caused by the influenza A virus and in treating established illness. It is thought to exert its antiviral effect by blocking a late stage in the assembly of the virus. Following is the final step in the synthesis of this compound. Describe experimental conditions to bring about this conversion.

Rimantadine
(an antiviral agent)

The final step in the synthesis involves a reductive amination of the ketone using ammonia.

Problem 11.35 Methenamine, a product of the reaction of formaldehyde and ammonia, is a prodrug, a compound that is inactive itself but is converted to an active drug in the body by a biochemical transformation. The strategy behind use of methenamine as a prodrug is that nearly all bacteria are sensitive to formaldehyde at concentrations of 20 mg/mL or higher. Formaldehyde cannot be used directly in medicine, however, because an effective concentration in plasma cannot be achieved with safe doses. Methenamine is stable at pH 7.4 (the pH of blood plasma) but undergoes acid-catalyzed hydrolysis to formaldehyde and ammonium ion under the acidic conditions of the kidneys and the urinary tract. Thus, methenamine can be used as a site-specific drug to treat urinary infections.

Methenamine

(a) Balance the equation for the hydrolysis of methenamine to formaldehyde and ammonium ion.

$$+ \quad 10 \ H_2O \quad \longrightarrow \quad 6 \ CH_2O \quad + \quad 4 \ NH_4^+ \ OH^-$$

(b) Does the pH of an aqueous solution of methenamine increase, remain the same, or decrease as a result of its hydrolysis? Explain.

When methenamine is hydrolyzed, ammonia is released. Ammonia is a base so the pH will increase.

(c) Explain the meaning of the following statement: The functional group in methenamine is the nitrogen analog of an acetal.

With an acetal, a single carbon atom is bonded to two oxygen atoms. In the case of methenamine, each carbon atom is bonded to two nitrogen atoms.

(d) Account for the observation that methenamine is stable in blood plasma but undergoes hydrolysis in the urinary tract.

Blood plasma is buffered to the slightly basic pH of 7.4. Methenamine is relatively stable to hydrolysis at this pH, since it is stable to base. Recall that acetals are also stable to base. On the other hand, both methenamine and acetals are readily hydrolyzed at acidic pH. The urinary tract is more acidic, so the methenamine is hydrolyzed more rapidly there.

Keto-Enol Tautomerism

Problem 11.36 The following molecule belongs to a class of compounds called enediols; each carbon of the double bond carries an -OH group. Draw structural formulas for the α-hydroxyketone and the α-hydroxyaldehyde with which this enediol is in equilibrium.

Following are formulas for the α-hydroxyaldehyde and α-hydroxyketone in equilibrium by way of the enediol intermediate.

Problem 11.37 In dilute aqueous acid, (R)-glyceraldehyde is converted into an equilibrium mixture of (R,S)-glyceraldehyde and dihydroxyacetone. Propose a mechanism for this isomerization.

The key is keto-enol tautomerism. In the presence of acid, the ketone can be protonated to give a resonance-stabilized cation.

Deprotonation of this carbocation at carbon 2 generates an enediol intermediate. The (Z) isomer is shown, but the (E) isomer will be formed as well.

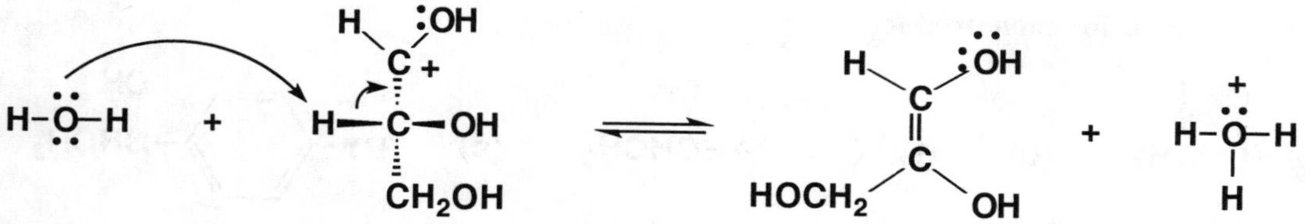

Reprotonation of the enediol at the original carbon can occur from either face of the alkene, thereby leading to a racemic mixture of glyceraldehyde.

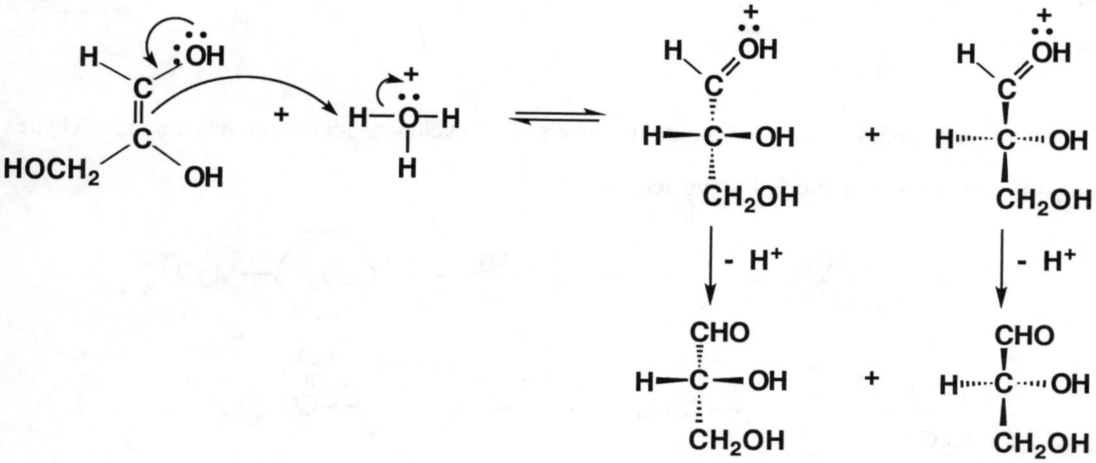

(R)-Glyceraldehyde   (S)-Glyceraldehyde

If instead protonation occurs at carbon one, the achiral molecule dihydroxyacetone is generated.

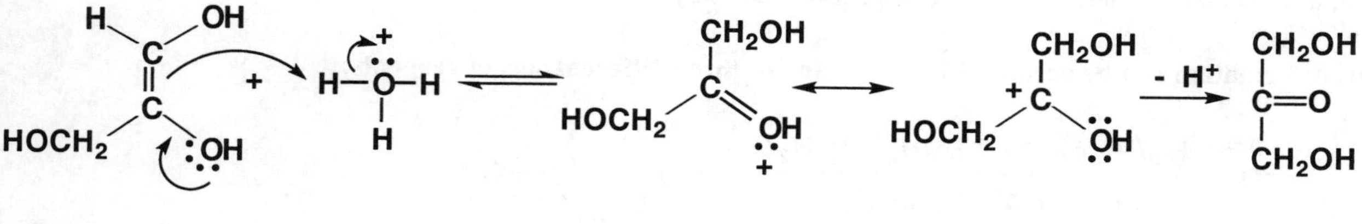

Oxidation/Reduction of Aldehydes and Ketones

Problem 11.38 Draw a structural formula for the product formed by treatment of butanal with each set of reagents.

(a)   LiAlH$_4$ followed by H$_2$O

CH$_3$CH$_2$CH$_2$CH$_2$OH

(b)   NaBH$_4$ in CH$_3$OH/H$_2$O

CH$_3$CH$_2$CH$_2$CH$_2$OH

(c)   H$_2$/Pt

CH$_3$CH$_2$CH$_2$CH$_2$OH

(d) Ag(NH$_3$)$_2$$^+$ in NH$_3$/H$_2$O
    then HCl/H$_2$O

CH$_3$CH$_2$CH$_2$CO$_2$H

(e)   K$_2$Cr$_2$O$_7$/H$_2$SO$_4$

CH$_3$CH$_2$CH$_2$CO$_2$H

(f) C$_6$H$_5$NH$_2$ in the
    presence of H$_2$/Ni

CH$_3$CH$_2$CH$_2$CH$_2$NHPh

Problem 11.39 Draw a structural formula for the product of the reaction of *p*-bromoacetophenone with each set of reagents in Problem 11.38.

Following are structural formulas for each product.

(a)

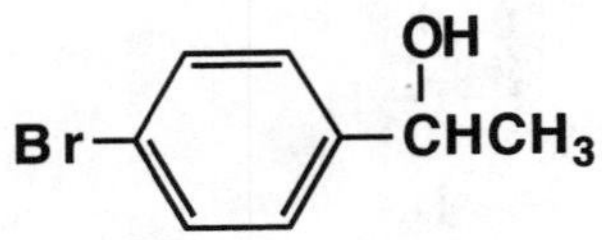

(b)

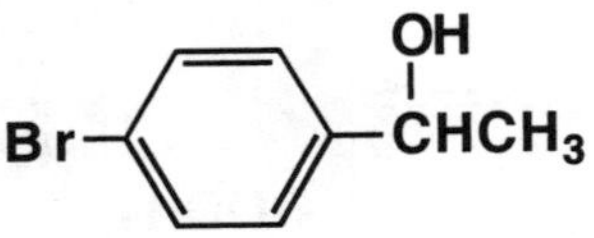

(c)

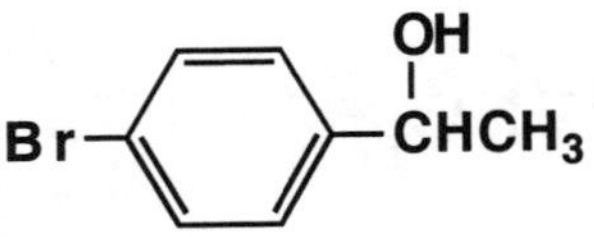

(d) no reaction

(e)

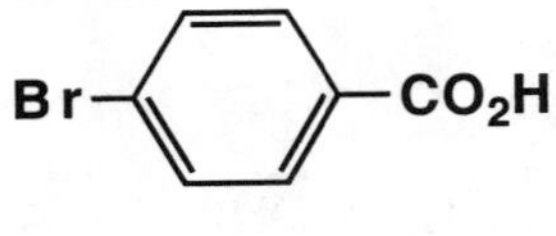

(f)

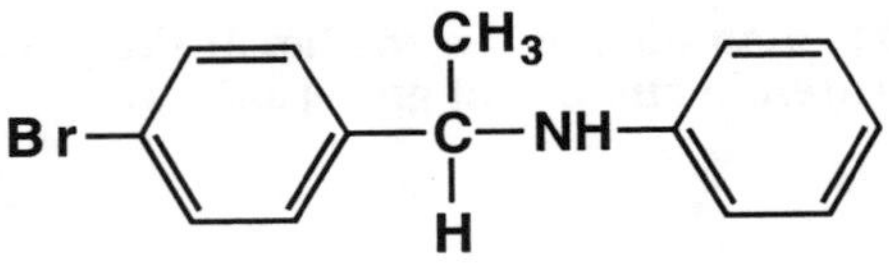

Synthesis
Problem 11.40 Show reagents and conditions to bring about the conversion of cyclohexanol to cyclohexanecarbaldehyde.

The appropriate reagents are written in the following scheme.

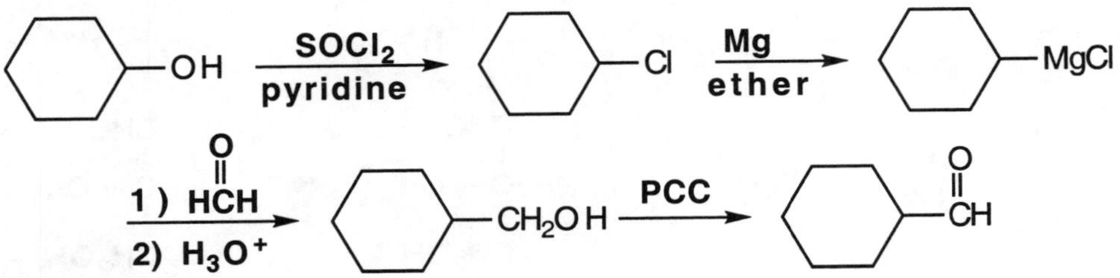

Problem 11.41 Starting with cyclohexanone, show how to prepare these compounds. In addition to the given starting material, use any other organic or inorganic reagents as necessary.
(a) Cyclohexanol

This transformation can be accomplished with any of three different sets of reagents:

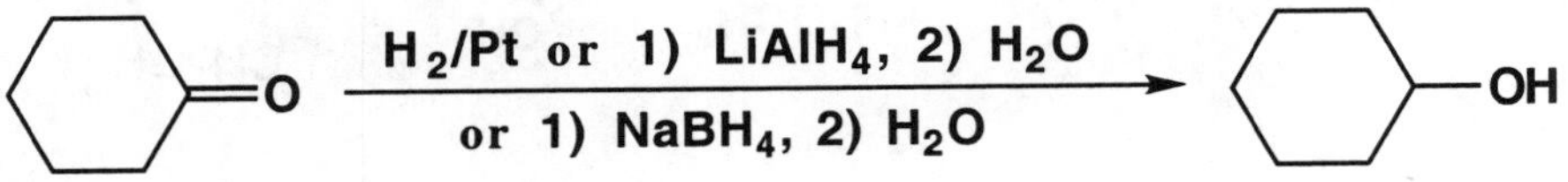

(b) Cyclohexene

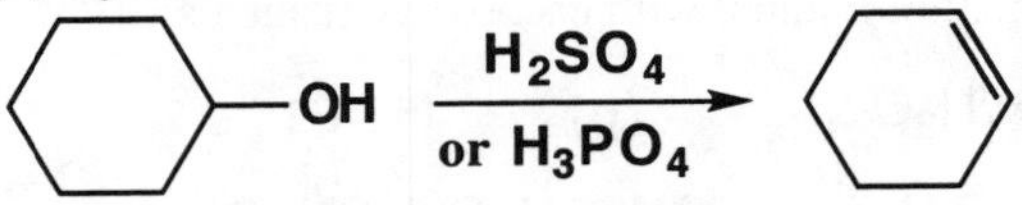

From (a)

(c) *cis*-1,2-Cyclohexanediol

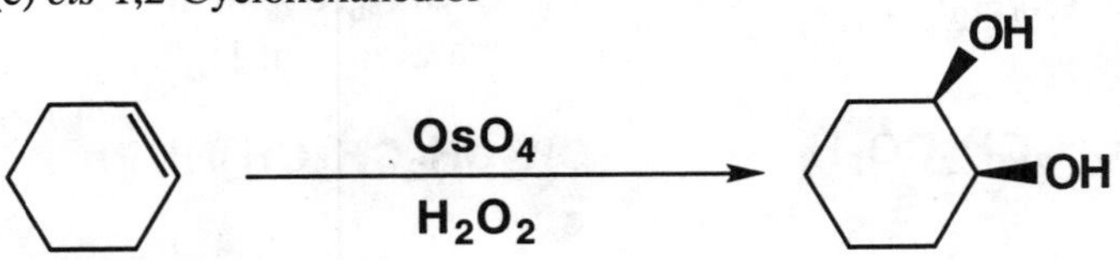

From (b)

(d) 1-Methylcyclohexanol

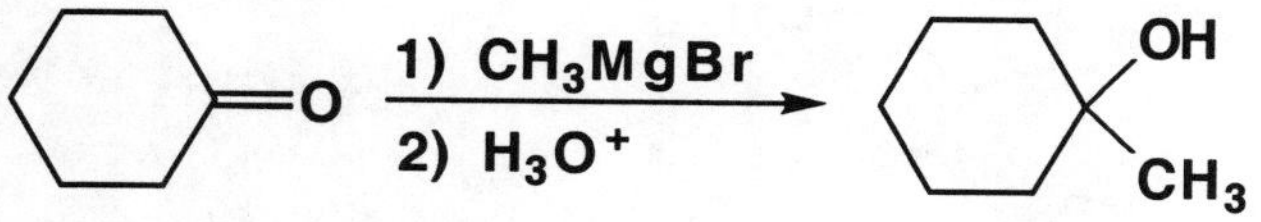

$$\text{1) } CH_3MgBr$$
$$\text{2) } H_3O^+$$

(e) 1-Methylcyclohexene

$$H_2SO_4 \text{ or } H_3PO_4$$

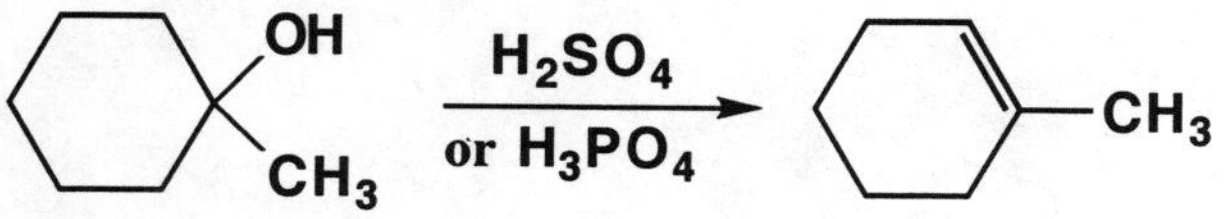

From (d)

(f) 1-Phenylcyclohexanol

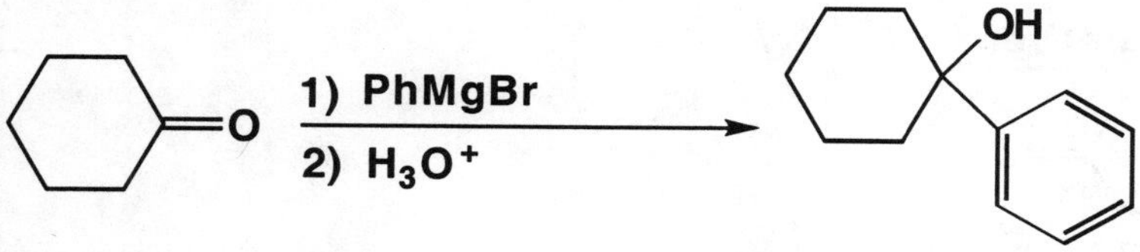

$$\text{1) } PhMgBr$$
$$\text{2) } H_3O^+$$

(g) 1-Phenylcyclohexene

$$H_2SO_4 \text{ or } H_3PO_4$$

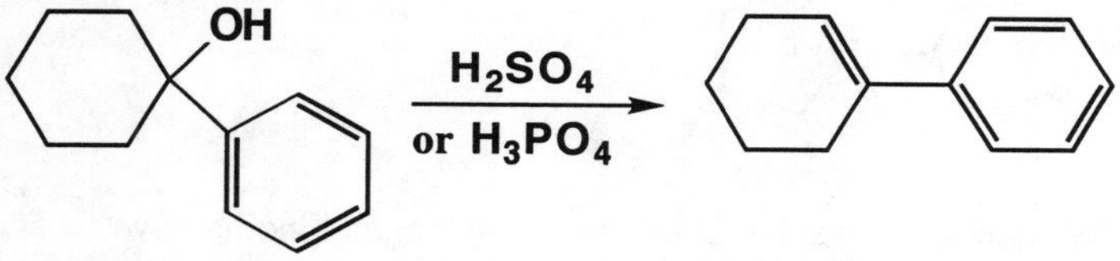

From (f)

(h) Cyclohexene oxide

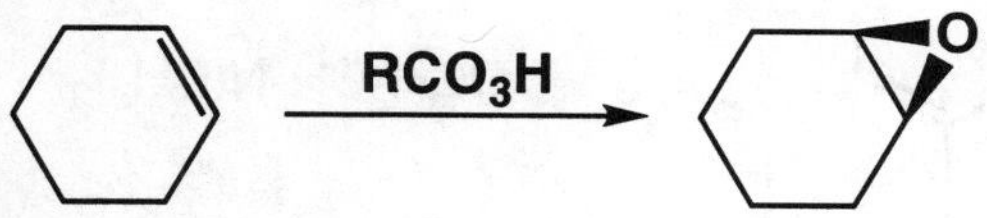

$$RCO_3H$$

From (b)

(i) *trans*-1,2-Cyclohexanediol

Recall that ring-opening of an epoxide in acid gives the desired trans product. Compare this to part (c) of this problem in which the cis product is desired, so that OsO_4 and H_2O_2 are used.

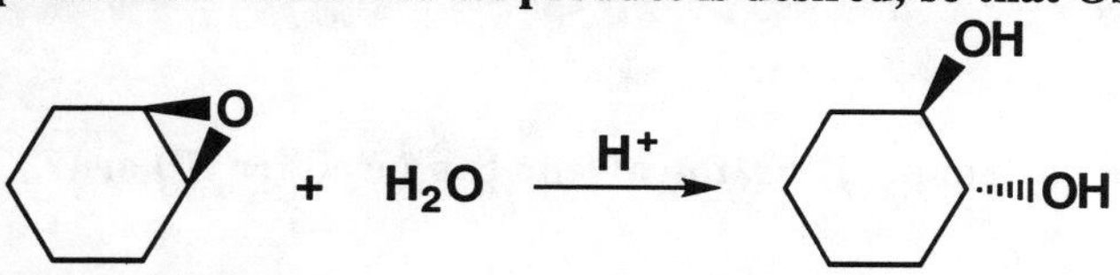

$$+ \ H_2O \xrightarrow{H^+}$$

From (h)

<u>Problem 11.42</u> Show how to bring about these conversions. In addition to the given starting material, use any other organic or inorganic reagents as necessary.

(a) $C_6H_5\overset{O}{\overset{\|}{C}}CH_2CH_3 \xrightarrow[\text{or 1) NaBH}_4\text{, 2) H}_2\text{O}]{\text{H}_2/\text{Pt or 1) LiAlH}_4\text{, 2) H}_2\text{O}} C_6H_5\overset{OH}{\overset{|}{C}}CH_2CH_3 \xrightarrow[\text{or H}_3\text{PO}_4]{\text{H}_2\text{SO}_4} C_6H_5CH\!=\!CHCH_3$

(b) 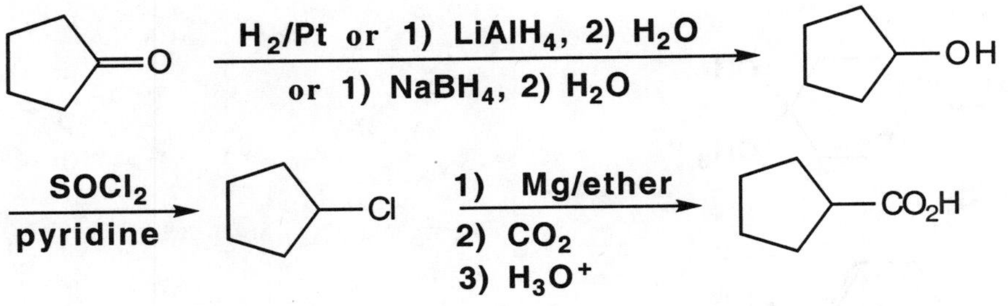
O
H_2/Pt or 1) $LiAlH_4$, 2) H_2O
or 1) $NaBH_4$, 2) H_2O
OH
$SOCl_2$
pyridine
Cl
1) Mg/ether
2) CO_2
3) H_3O^+
CO_2H

(c) 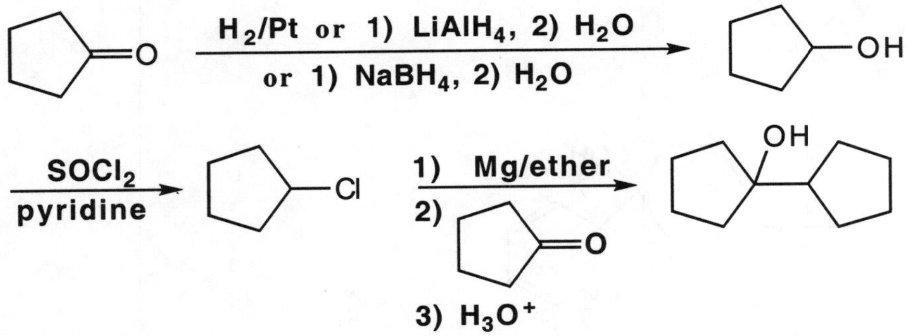
O
H_2/Pt or 1) $LiAlH_4$, 2) H_2O
or 1) $NaBH_4$, 2) H_2O
OH
$SOCl_2$
pyridine
Cl
1) Mg/ether
2)
O
3) H_3O^+
OH

(d) 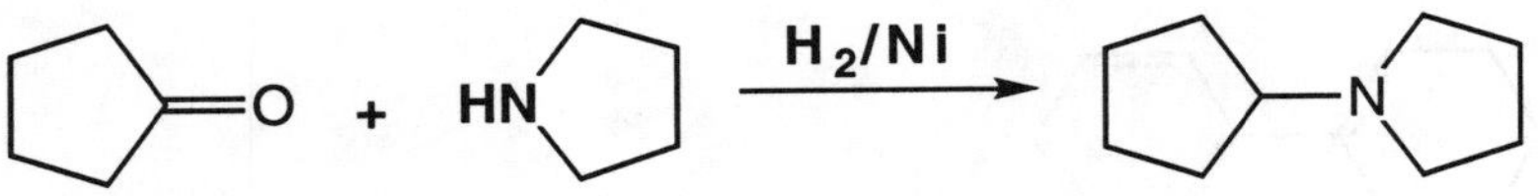
O
+
HN
H_2/Ni
N

Problem 11.43 Many tumors of the breast are estrogen-dependent. Drugs that interfere with estrogen binding have antitumor activity and may even help prevent tumor occurrence. A widely used antiestrogen drug is tamoxifen.

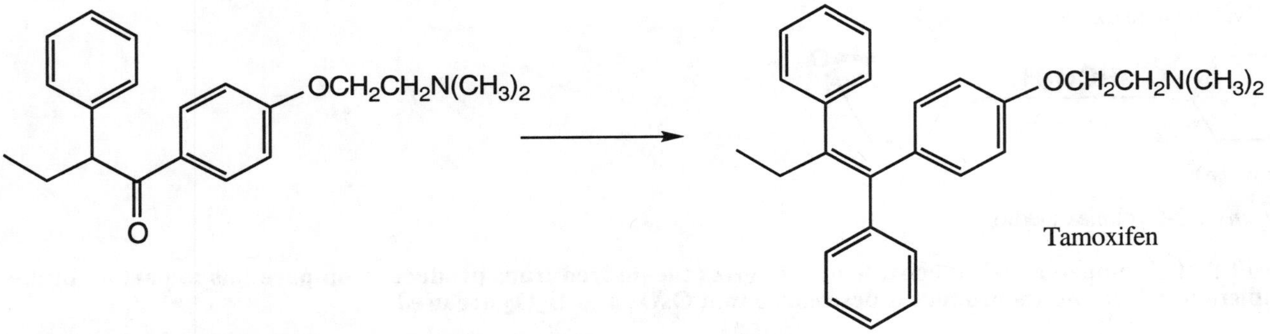
$OCH_2CH_2N(CH_3)_2$
O
$OCH_2CH_2N(CH_3)_2$
Tamoxifen

(a) How many stereoisomers are possible for tamoxifen?

Tamoxifen has a double bond with two different groups at each end. Thus, the alkene can be either (E) or (Z), so two stereoisomers are possible

(b) Specify the configuration of the stereoisomer shown here.

The higher priority groups are attached to the same side of the double bond. Therefore, tamoxifen is the (Z) stereoisomer.

(c) Show how tamoxifen can be synthesized from the given ketone using a Grignard reaction followed by dehydration.

CHAPTER 12
Solutions to Problems

Problem 12.1 Each of these compounds has a well-recognized common name. A derivative of glyceric acid is an intermediate in glycolysis. Maleic acid is an intermediate in the tricarboxylic acid (TCA) cycle. Mevalonic acid is an intermediate in the biosynthesis of steroids (Section 17.4). Write the IUPAC name for each compound. Be certain to show configuration for each.

(a)

CO_2H
H—C—OH
CH_2OH

Glyceric acid

(R)-2,3-Dihydroxy-propanoic acid

(b)

HO_2C CO_2H
       C=C
H   H

Maleic acid

(Z)-2-Butenedioic acid
(*cis*-2-Butenedioic acid)

(c)

HO CH_3
     C
$HOCH_2CH_2$ CH_2CO_2H

Mevalonic acid

(R)-3,5-Dihydroxy-3-methylpentanoic acid

Problem 12.2 Match each compound with its appropriate pK_a value.

CH_3
CH_3CCO_2H
CH_3

CF_3CO_2H

OH
CH_3CHCO_2H

pK_a values = 5.03, 3.08 and 0.22

2,2-Dimethyl-propanoic acid

Trifluoro-acetic acid

2-Hydroxy-propanoic acid
(Lactic acid)

2,2-Dimethylpropanoic acid has a pK_a comparable to that of an unsubstituted aliphatic carboxylic acid. Lactic acid is a stronger acid for two reasons: the inductive effect of the hydroxyl group on an sp^3 carbon atom adjacent to the carboxyl group and the fact that the hydroxyl group can also stabilize the carboxylate anion of the deprotonated acid via an intramolecular hydrogen bond. Trifluoroacetic acid is an even stronger acid because of the combined inductive effects of the three fluorine atoms. In order of increasing acidity, these acids are:

CH_3
CH_3CCO_2H
CH_3

OH
CH_3CHCO_2H

CF_3CO_2H

pK_a **5.03**   **3.08**   **0.22**

increasing acidity

Problem 12.3 Write an equation for the reaction of each acid in Example 12.3 with ammonia and name the salt formed.

(a) $CH_3(CH_2)_2CO_2H$ + NH_3 $\longrightarrow$ $CH_3(CH_2)_2CO_2^-$ NH_4^+
Ammonium butanoate

(b)
OH
CH_3CHCO_2H + NH_3 $\longrightarrow$
OH
$CH_3CHCO_2^-$ NH_4^+
Ammonium 2-hydroxypropanoate
(Ammonium lactate)

Problem 12.4 Complete these Fischer esterification reactions.

(a) $HOCH_2CH_2CH_2\overset{O}{\overset{\|}{C}}OH$ $\xrightarrow{H^+}$ [γ-butyrolactone] $+$ H_2O

(b) $CH_3\underset{\underset{CH_3}{|}}{\overset{O}{\overset{\|}{CH}}}\overset{O}{\overset{\|}{C}}OH$ $+$ HO—[cyclohexyl] $\xrightarrow{H^+}$ $CH_3\underset{\underset{CH_3}{|}}{CH}\overset{O}{\overset{\|}{C}}O$—[cyclohexyl] $+$ H_2O

Problem 12.5 Complete each equation:

(a) [benzene ring with CO_2H and OCH_3] $+$ $SOCl_2$ $\longrightarrow$ [benzene ring with $\overset{O}{\overset{\|}{C}}Cl$ and OCH_3] $+$ SO_2 $+$ HCl

(b) [cyclohexane with OH] $+$ $SOCl_2$ $\longrightarrow$ [cyclohexane with Cl] $+$ SO_2 $+$ HCl

Problem 12.6 Draw the structural formula for the indicated β-ketoacid.

[benzene ring]—$\overset{O}{\overset{\|}{C}}$—$\underset{\underset{CH_3}{|}}{\overset{\overset{CO_2H}{|}}{C}}CH_2CH_3$ $\xrightarrow[\text{Heat}]{-CO_2}$ [benzene ring]—$\overset{O}{\overset{\|}{C}}$—$\underset{\underset{CH_3}{|}}{CH}CH_2CH_3$

Structure and Nomenclature

Problem 12.7 Name and draw structural formulas for the four carboxylic acids of molecular formula $C_5H_{10}O_2$. Which of these carboxylic acids is chiral?

The chiral carboxylic acid is circled, and the stereocenter is indicated with an asterisk.

$CH_3(CH_2)_3\overset{O}{\overset{\|}{C}}OH$ $CH_3\underset{\underset{CH_3}{|}}{CH}CH_2\overset{O}{\overset{\|}{C}}OH$ $\boxed{CH_3CH_2\overset{*}{\underset{\underset{CH_3}{|}}{CH}}\overset{O}{\overset{\|}{C}}OH}$ $CH_3\underset{\underset{H_3C}{|}}{\overset{\overset{H_3C}{|}}{C}}\overset{O}{\overset{\|}{C}}OH$

Pentanoic acid   **3-Methylbutanoic acid**   **2-Methylbutanoic acid**   **2,2-Dimethylpropanoic acid**

Problem 12.8 Write the IUPAC name for each compound.

(a) [cyclohexene ring]—CO_2H (b) $CH_3\underset{\underset{OH}{|}}{CH}CH_2CH_2CO_2H$ (c) [branched diene structure with $\overset{O}{\overset{\|}{}}$ and OH]

1-Cyclohexenecarboxylic acid   **4-Hydroxypentanoic acid**   **(2E) 3,7-Dimethyl-2,6-octadienoic acid**

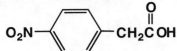

(d)

**1-Methylcyclopentane-
carboxylic acid**

(e) $CH_3(CH_2)_4CO_2^-$ NH_4^+

Ammonium hexanoate

(f)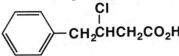

2-Hydroxybutanedioic acid

Problem 12.9 Draw a structural formula for each carboxylic acid.
(a) 4-Nitrophenylacetic acid

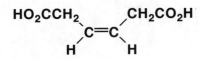

(b) 4-Aminobutanoic acid

$H_2NCH_2CH_2CH_2CO_2H$

(c) 3-Chloro-4-phenylbutanoic acid

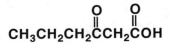

(d) *cis*-3-Hexenedioic acid

HO_2CCH_2 CH_2CO_2H
        C=C
      H   H

(e) 2,3-Dihydroxypropanoic acid

OH OH
 | |
CH_2CHCO_2H

(f) 3-Oxohexanoic acid

     O   O
     || ||
$CH_3CH_2CH_2CCH_2COH$

(g) 2-Oxocyclohexanecarboxylic acid

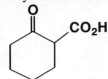

(j) 2,2-Dimethylpropanoic acid

        CH_3
         |
CH_3CCO_2H
         |
        CH_3

Problem 12.10 Megatomoic acid, the sex attractant of the female black carpet beetle, has the structure

$$CH_3(CH_2)_7CH=CHCH=CHCH_2CO_2H$$
Megatomoic acid

(a) What is its UPAC name?

Its IUPAC name is 3,5-Tetradecadienoic acid.

(b) State the number of stereoisomers possible for this compound.

Four stereoisomers are possible; each double bond can have either an E or Z (*trans* or *cis*) configuration.

Problem 12.11 The IUPAC name of ibuprofen is 2-(4-isobutylphenyl)propanoic acid. Draw the structural formula of ibuprofen.

        CH_3 CH_3
         | |
CH_3CHCH_2——⟨   ⟩——$CHCO_2H$

Ibuprofen

Problem 12.12 Draw structural formulas for these salts.

(a) Sodium benzoate

(b) Lithium acetate

(c) Ammonium acetate

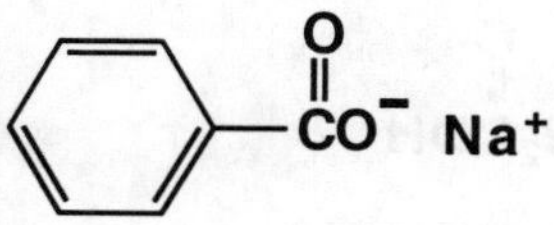

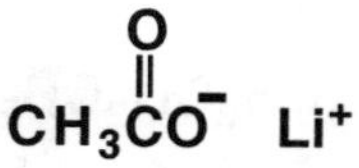

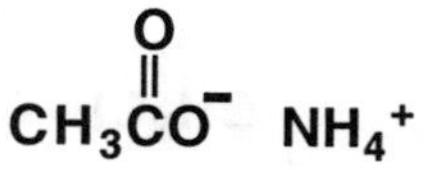

(d) Disodium adipate

(e) Sodium salicylate

(f) Calcium butanoate

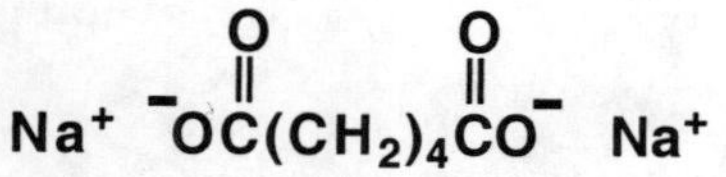

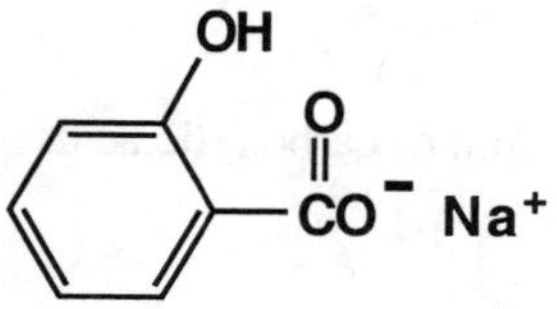

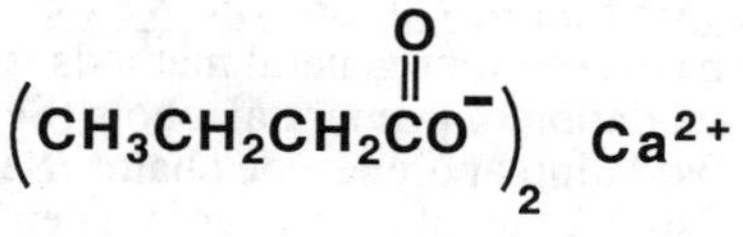

Problem 12.13 The monopotassium salt of oxalic acid is present in certain leafy vegetables, including rhubarb. Both oxalic acid and its salts are poisonous in high concentrations. Draw the structural formula of monopotassium oxalate.

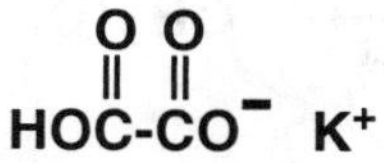

Monopotassium oxalate

Problem 12.14 Potassium sorbate is added as a preservative to certain foods to prevent bacteria and molds from causing food spoilage and to extend the foods' shelf life. The IUPAC name of potassium sorbate is potassium (E,E)-2,4-hexadienoate. Draw a structural formula of potassium sorbate.

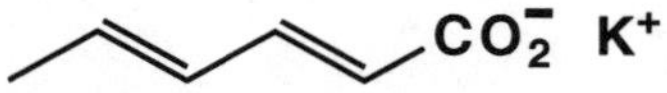

Potassium sorbate

Problem 12.15 Zinc 10-undecenoate, the zinc salt of 10-undecenoic acid, is used to treat certain fungal infections, particularly *tinea pedis* (athlete's foot). Draw a structural formula of this zinc salt.

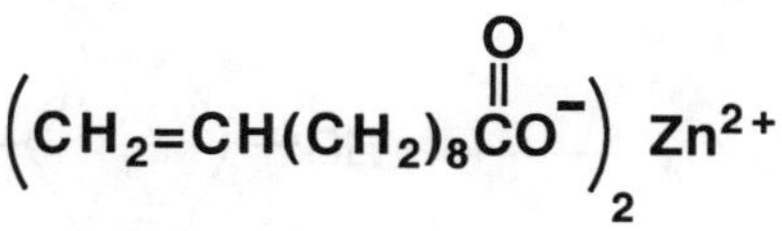

Zinc 10-undecenoate

Physical properties

Problem 12.16 Arrange the compounds in each set in order of increasing boiling point:

(a) 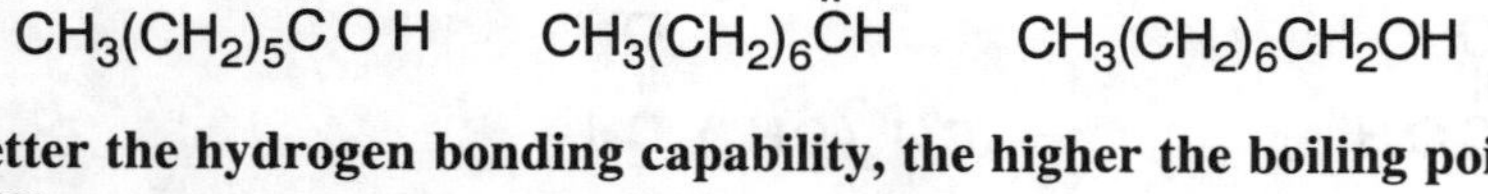 $CH_3(CH_2)_5COH$ $CH_3(CH_2)_6CH$ $CH_3(CH_2)_6CH_2OH$

The better the hydrogen bonding capability, the higher the boiling point. In order of increasing boiling point, they are:

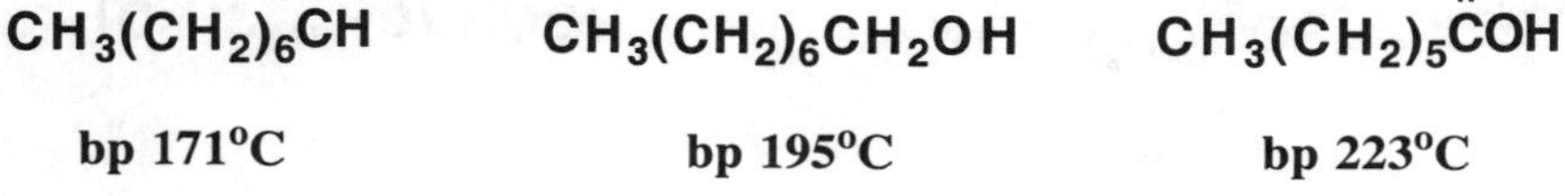

$CH_3(CH_2)_6CH$ $CH_3(CH_2)_6CH_2OH$ $CH_3(CH_2)_5COH$

bp 171°C   **bp 195°C**   **bp 223°C**

(b) $CH_3CH_2\overset{\displaystyle O}{\overset{\displaystyle \|}{C}}OH$ $CH_3CH_2CH_2CH_2OH$ $CH_3CH_2OCH_2CH_3$

$CH_3CH_2OCH_2CH_3$ $CH_3CH_2CH_2CH_2OH$ $CH_3CH_2\overset{\displaystyle O}{\overset{\displaystyle \|}{C}}OH$

bp 35°C   **bp 117°C**   **bp 141°C**

Preparation of Carboxylic Acids

We have seen four general methods for the preparation of carboxylic acids:
(1) Oxidation of primary alcohols (Section 8.4F)
(2) Oxidation of arene side chains (Section 9.5)
(3) Carbonation of a Grignard reagent (Section 11.5B)
(4) Oxidation of aldehydes (Section 11.9A)

<u>Problem 12.17</u> Complete these oxidations.

(a) $CH_3(CH_2)_4CH_2OH$ + $Cr_2O_7{}^{2-}$ $\xrightarrow[\text{heat}]{H^+}$ $CH_3(CH_2)_4CO_2H$ + Cr^{3+}

(b)

+ $Ag(NH_3)_2{}^+$ $\xrightarrow{NH_3,\ H_2O}$ + Ag

Vanillin

(c)

+ H_2CrO_4 $\xrightarrow{\text{heat}}$ + Cr^{3+}

(d) $HO-$⟨benzene⟩$-CH_2OH$ + H_2CrO_4 $\longrightarrow$ $HO-$⟨benzene⟩$-\overset{\displaystyle O}{\overset{\displaystyle \|}{C}}OH$ + Cr^{3+}

<u>Problem 12.18</u> Draw the structural formula of a compound of the given molecular formula that, on oxidation by chromic acid, gives the carboxylic acid or dicarboxylic acid shown.

(a) $C_6H_{14}O$ $\xrightarrow{\text{oxidation}}$ $CH_3(CH_2)_4\overset{\displaystyle O}{\overset{\displaystyle \|}{C}}OH$ $CH_3(CH_2)_4CH_2OH$

(b) $C_6H_{12}O$ $\xrightarrow{\text{oxidation}}$ $CH_3(CH_2)_4\overset{\displaystyle O}{\overset{\displaystyle \|}{C}}OH$ $CH_3(CH_2)_4\overset{\displaystyle O}{\overset{\displaystyle \|}{C}}H$

(c) $C_6H_{14}O_2$ $\xrightarrow{\text{oxidation}}$ $HO\overset{\displaystyle O}{\overset{\displaystyle \|}{C}}(CH_2)_4\overset{\displaystyle O}{\overset{\displaystyle \|}{C}}OH$ $HOCH_2(CH_2)_4CH_2OH$

Problem 12.19 Show reagents and experimental conditions to complete this synthesis.

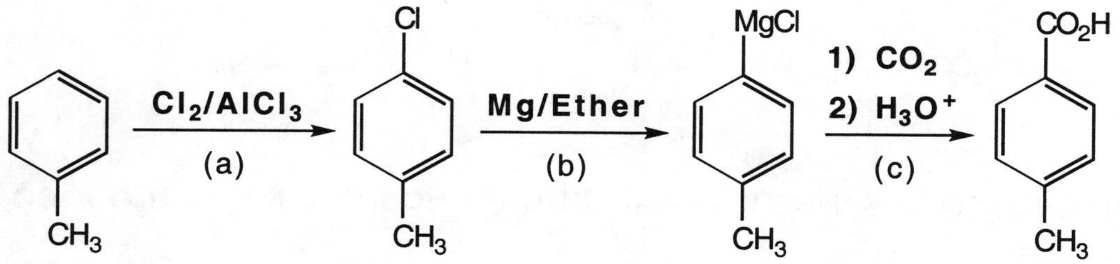

Acidity of Carboxylic Acids
Problem 12.20 Which is the stronger acid in each pair?

(a) Phenol (pK_a 9.95) or benzoic acid (pK_a 4.17)

Recall that pK_a is the negative $\log_{10}$ of K_a. The smaller the pK_a, the stronger the acid, so benzoic acid is the stronger acid.

(b) Lactic acid (K_a 8.4 x 10^{-4}) or ascorbic acid (K_a 7.9 x 10^{-5})

The larger the value of K_a, the stronger the acid, so lactic acid is the stronger acid.

Problem 12.21 Assign the acid in each set its appropriate pK_a.

(a) 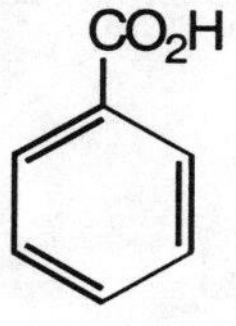   and 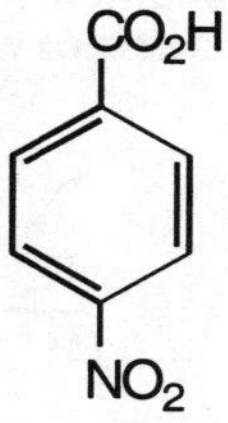   (pK_a 4.19 and 3.14)

4.19   3.14

(b) 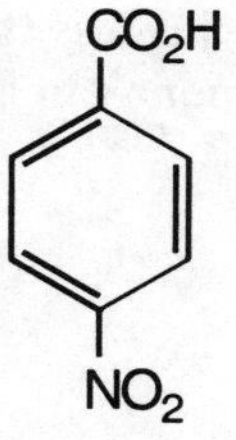   and 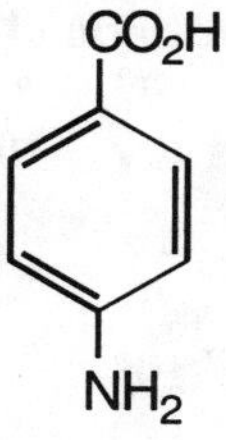   (pK_a 4.92 and 3.14)

3.14   4.92

(c) $CH_3CCH_2CO_2H$ and CH_3CCO_2H (pK_a 3.58 and 2.49)

3.58   2.49

(d) CH_3CHCO_2H and $CH_3CH_2CO_2H$ (pK_a 4.78 and 3.08)

3.08   4.78

<u>Problem 12.22</u> Complete these acid-base reactions:

(a) $C_6H_5\text{-}CH_2CO_2H + NaOH \longrightarrow C_6H_5\text{-}CH_2CO_2^- \ Na^+ + H_2O$

(b) $CH_3CH\text{=}CHCH_2CO_2H + NaHCO_3 \longrightarrow CH_3CH\text{=}CHCH_2CO_2^- \ Na^+ + H_2O + CO_2$

(c) (salicylic acid) $\text{-CO}_2H,\ \text{-OH} + NaHCO_3 \longrightarrow \text{-CO}_2^- \ Na^+,\ \text{-OH} + H_2O + CO_2$

(d)
$$\underset{\text{OH}}{CH_3CHCO_2H} + H_2NCH_2CH_2OH \longrightarrow \underset{\text{OH}}{CH_3CHCO_2^-} + \overset{+}{H_3N}CH_2CH_2OH$$

(e) $CH_3CH\text{=}CHCH_2CO_2^- \ Na^+ + HCl \longrightarrow CH_3CH\text{=}CHCH_2CO_2H + NaCl$

<u>Problem 12.23</u> The normal pH range for blood plasma is 7.35-7.45. Under these conditions, would you expect the carboxyl group of lactic acid (pK_a 4.07) to exist primarily as a carboxyl group or as a carboxylic anion? Explain.

Recall from the definition of K_a that:

$$K_a = \frac{[A^-]\,[H^+]}{[H\text{-}A]} \qquad \text{so dividing both sides by } [H^+] \text{ gives} \qquad \frac{K_a}{[H^+]} = \frac{[A^-]}{[H\text{-}A]}$$

Here, $[H^+]$ is the concentration of H^+, $[H\text{-}A]$ is the concentration of protonated acid (lactic acid in this case), and $[A^-]$ is the concentration of deprotonated acid (lactic acid carboxylate anion in this case). Therefore, if the ratio of $K_a\,/\,[H^+]$ is greater than 1, $[A^-]$ will be the predominant form. If the ratio of $K_a\,/\,[H^+]$ is less than 1, then $[H\text{-}A]$ will be the predominant form. Recall that $pH = -\log_{10} [H^+]$, so a pH of 7.4 corresponds to $[H^+]$ of $10^{-(pH)} = 10^{-(7.4)} = 4.0 \times 10^{-8}$. Similarly, $pK_a = -\log_{10} K_a$, so for lactic acid $K_a = 10^{-(Ka)} = 10^{-(4.07)} = 8.5 \times 10^{-5}$. Using these numbers:

$$\frac{[A^-]}{[H\text{-}A]} = \frac{K_a}{[H^+]} = \frac{8.5 \times 10^{-5}}{4.0 \times 10^{-8}} = 2.1 \times 10^3$$

Therefore, lactic acid will exist primarily as the carboxylate anion in blood plasma.

<u>Problem 12.24</u> The K_a of ascorbic acid (Section 12.6) is 7.94×10^{-5}. Would you expect ascorbic acid dissolved in blood plasma, pH 7.35-7.45, to exist primarily as ascorbic acid or as ascorbate anion? Explain.

Using the same reasoning described in the answer to Problem 12.23:

$$\frac{[A^-]}{[H\text{-}A]} = \frac{K_a}{[H^+]} = \frac{7.9 \times 10^{-5}}{4.0 \times 10^{-8}} = 2.0 \times 10^3$$

Therefore, ascorbic acid will exist primarily as the ascorbate anion in blood plasma.

Problem 12.25 Excess ascorbic acid is excreted in the urine, the pH of which is normally in the range 4.8 - 8.4. What form of ascorbic acid would you expect to be present in urine of pH 8.4, ascorbic acid or ascorbate anion?

At pH 8.4, $[H^+] = 4.0 \times 10^{-9}$. Therefore, using the same reasoning as described in the answer to Problem 12.23:

$$\frac{[A^-]}{[H\text{-}A]} = \frac{K_a}{[H^+]} = \frac{7.9 \times 10^{-5}}{4.0 \times 10^{-9}} = 2.0 \times 10^4$$

Therefore, ascorbic acid will exist primarily as the ascorbate anion in urine of pH 8.4.

Problem 12.26 The pH of human gastric juice is normally in the range 1.0 - 3.0. What form of lactic acid (pK$_a$ 4.07) would you expect to be present in the stomach, lactic acid or its anion?

At pH 3.0, $[H^+] = 1.0 \times 10^{-3}$ and, as described in Problem 12.23, the K_a of lactic acid is 8.5×10^{-5}. Therefore, using the same reasoning as described in the answer to Problem 12.23:

$$\frac{[A^-]}{[H\text{-}A]} = \frac{K_a}{[H^+]} = \frac{8.5 \times 10^{-5}}{1.0 \times 10^{-3}} = 8.5 \times 10^{-2}$$

Therefore, lactic acid will exist primarily in the protonated acid form in gastric juices of pH <3.0.

Reactions of Carboxylic Acids

Problem 12.27 Give the expected organic products formed when phenylacetic acid, PhCH$_2$CO$_2$H, is treated with each reagent.

(a) SOCl$_2$

$$\underset{\substack{||\\ O}}{PhCH_2COH} + SOCl_2 \longrightarrow \underset{\substack{||\\ O}}{PhCH_2CCl} + SO_2 + HCl$$

(b) NaHCO$_3$, H$_2$O

$$\underset{\substack{||\\ O}}{PhCH_2COH} + NaHCO_3 \longrightarrow \underset{\substack{||\\ O}}{PhCH_2CO^-} Na^+ + CO_2 + H_2O$$

(c) NaOH, H$_2$O

$$\underset{\substack{||\\ O}}{PhCH_2COH} + NaOH \longrightarrow \underset{\substack{||\\ O}}{PhCH_2CO^-} Na^+ + H_2O$$

(d) NH$_3$, H$_2$O

$$\underset{\substack{||\\ O}}{PhCH_2COH} + NH_3 \longrightarrow \underset{\substack{||\\ O}}{PhCH_2CO^-} NH_4^+$$

(e) LiAlH$_4$ followed by H$_2$O

$$\underset{\substack{||\\ O}}{PhCH_2COH} \xrightarrow[\text{2) } H_2O]{\text{1) } LiAlH_4} PhCH_2CH_2OH$$

(f) CH$_3$OH + H$_2$SO$_4$ (catalyst)

$$\underset{\substack{||\\ O}}{PhCH_2COH} + CH_3OH \underset{\xrightleftharpoons{}}{\xrightarrow{H_2SO_4}} \underset{\substack{||\\ O}}{PhCH_2COCH_3} + H_2O$$

Problem 12.28 Show how to convert *trans*-3-phenyl-2-propenoic acid (cinnamic acid) to these compounds.

(a)

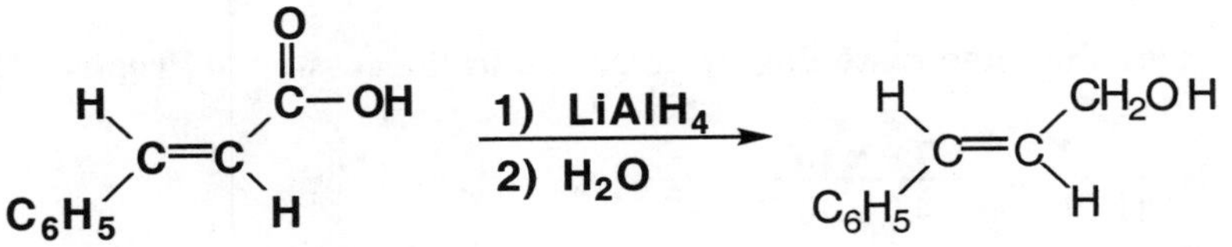

(b)

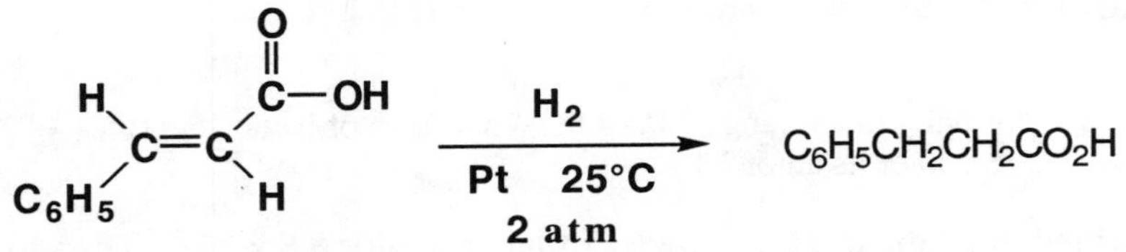

$$C_6H_5CH_2CH_2CO_2H$$

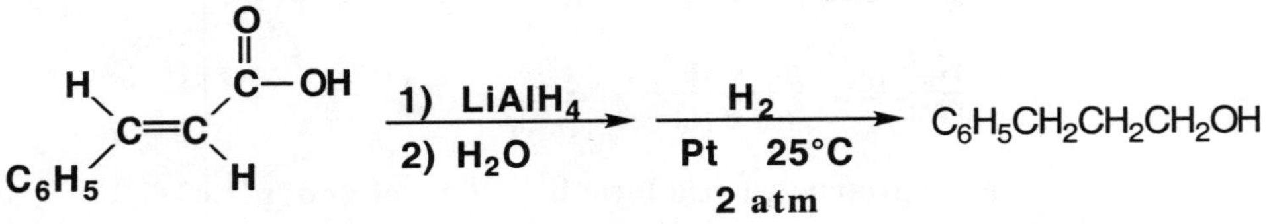

(c)

$$C_6H_5CH_2CH_2CH_2OH$$

Problem 12.29 Show how to convert 3-oxobutanoic acid (acetoacetic acid) to these compounds.

(a) 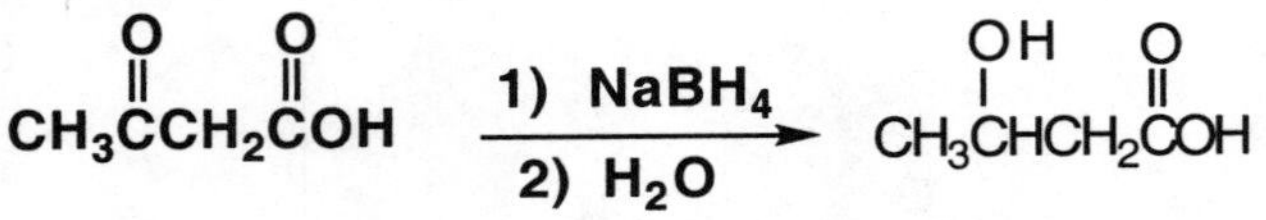 CH_3CCH_2COH $\xrightarrow[\text{2) H}_2\text{O}]{\text{1) NaBH}_4}$ CH_3CHCH_2COH

(b) 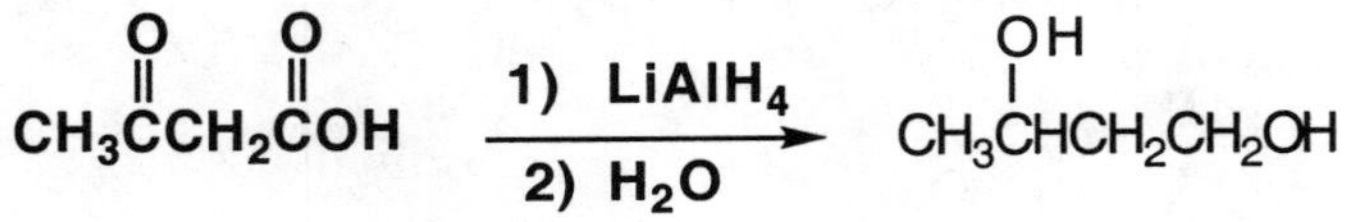 CH_3CCH_2COH $\xrightarrow[\text{2) H}_2\text{O}]{\text{1) LiAlH}_4}$ $CH_3CHCH_2CH_2OH$

(c) 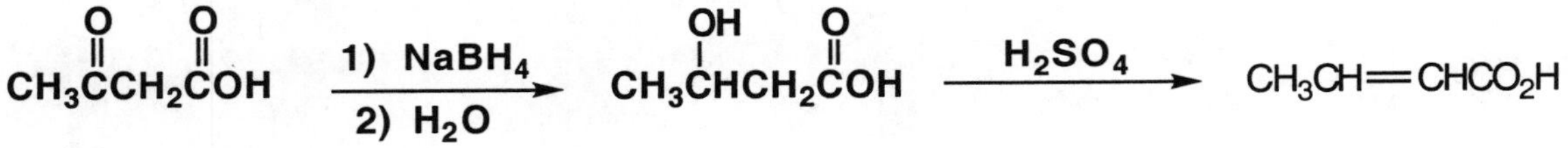 CH_3CCH_2COH $\xrightarrow[\text{2) H}_2\text{O}]{\text{1) NaBH}_4}$ CH_3CHCH_2COH $\xrightarrow{\text{H}_2\text{SO}_4}$ $CH_3CH{=}CHCO_2H$

Problem 12.30 Complete these examples of Fischer esterification. Assume an excess of the alcohol.

(a) 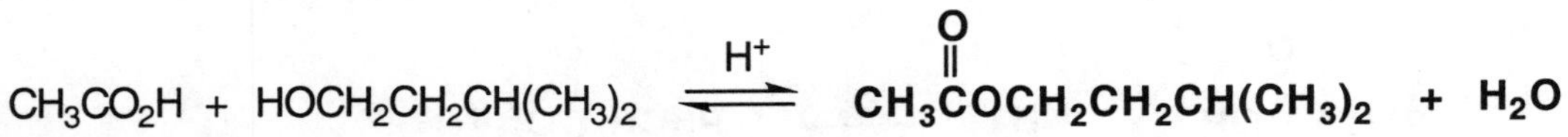 $CH_3CO_2H + HOCH_2CH_2CH(CH_3)_2 \underset{\text{H}^+}{\rightleftharpoons} CH_3COCH_2CH_2CH(CH_3)_2 + H_2O$

(b) 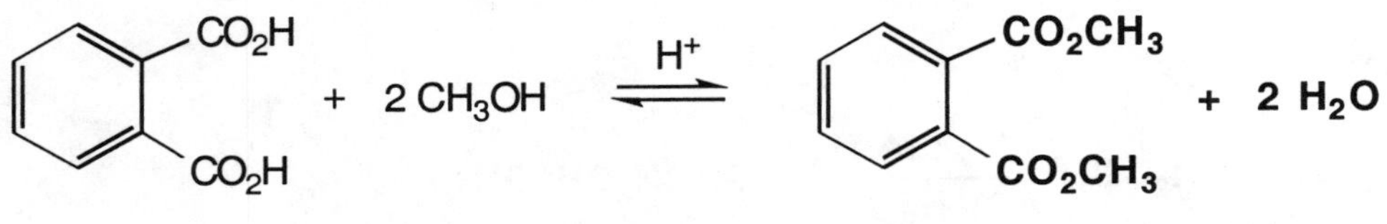 $+ \ 2\,CH_3OH \underset{\text{H}^+}{\rightleftharpoons}$ $+ \ 2\,H_2O$

(c) 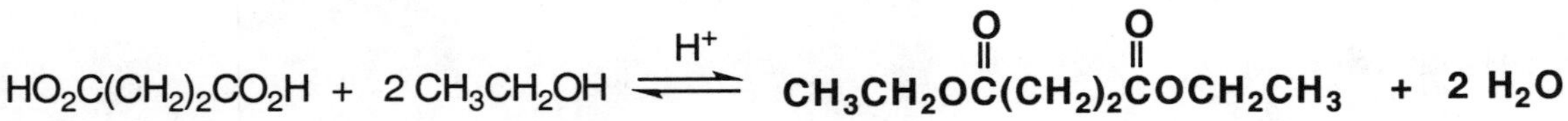 $HO_2C(CH_2)_2CO_2H + 2\,CH_3CH_2OH \underset{\text{H}^+}{\rightleftharpoons} CH_3CH_2OC(CH_2)_2COCH_2CH_3 + 2\,H_2O$

Problem 12.31 Methyl 2-hydroxybenzoate (methyl salicylate) has the odor of oil of wintergreen. This ester is prepared by Fischer esterification of 2-hydroxybenzoic acid (salicylic acid) with methanol. Draw the structural formula of methyl 2-hydroxybenzoic acid.

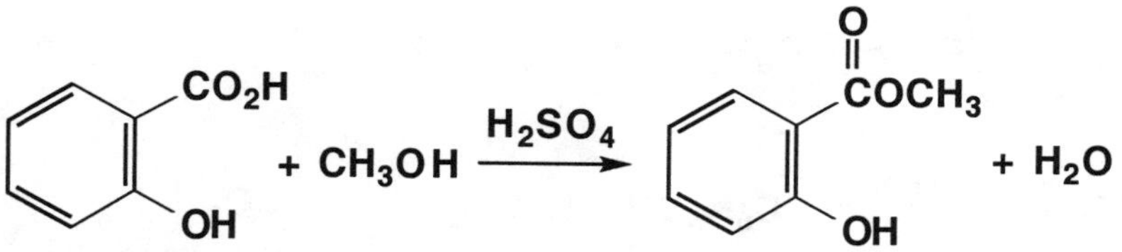

2-Hydroxybenzoic acid
(Salicylic acid)

Methyl 2-hydroxybenzoate
(Oil of wintergreen)

Problem 12.32 Benzocaine, a topical anesthetic, is prepared by treatment of 4-aminobenzoic acid with ethanol in the presence of an acid catalyst, followed by neutralization. Draw the structural formula of benzocaine.

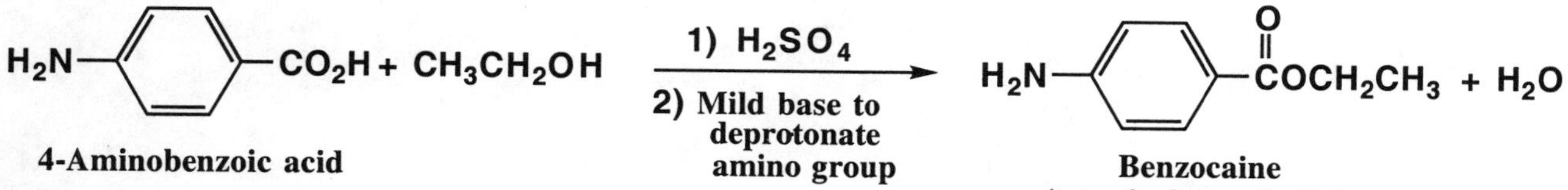

4-Aminobenzoic acid

1) H_2SO_4
2) Mild base to deprotonate amino group

Benzocaine
(a topical anesthetic)

Problem 12.33 From what carboxylic acid and alcohol is each ester derived?

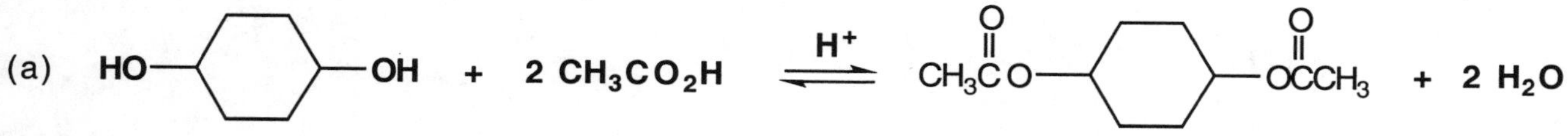

(a)

(b)

(c)

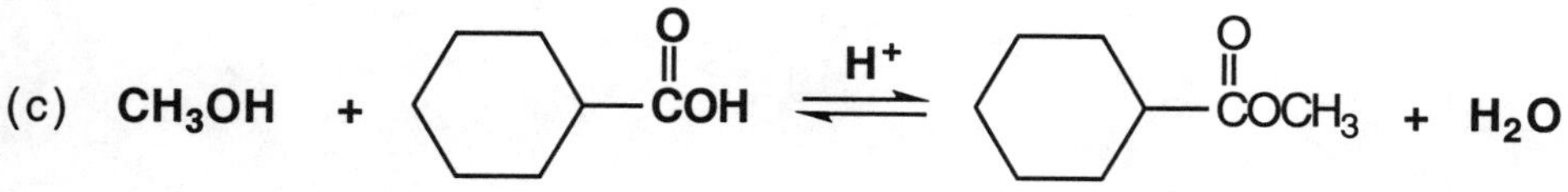

(d)

Problem 12.34 When 4-hydroxybutanoic acid is treated with an acid catalyst, it forms a lactone (a cyclic ester). Draw the structural formula of this lactone.

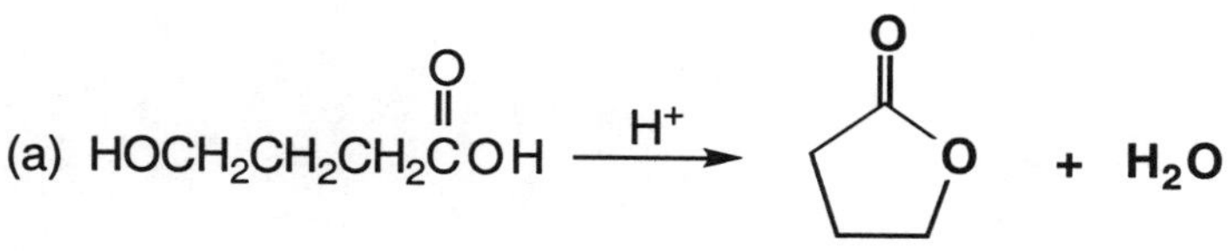

(a)

Problem 12.35 Draw the product formed on thermal decarboxylation of these compounds.

(a) $\underset{\displaystyle \overset{\displaystyle O}{\parallel}}{C_6H_5C}CH_2CO_2H$ $\xrightarrow{\text{Heat}}$ $\underset{\displaystyle \overset{\displaystyle O}{\parallel}}{C_6H_5C}CH_3$ + CO_2

(b) $\underset{\displaystyle \overset{\displaystyle CO_2H}{\vert}}{C_6H_5CH_2CH}CO_2H$ $\xrightarrow{\text{Heat}}$ $C_6H_5CH_2CH_2CO_2H$ + CO_2

(c) $\xrightarrow{\text{Heat}}$ + CO_2

CHAPTER 13
Solutions to Problems

Problem 13.1 Draw a structural formula for each compound:

(a) *N*-Cyclohexylacetamide.

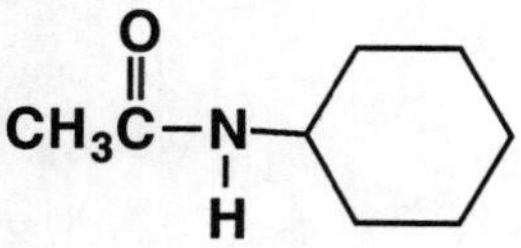

(b) *sec*-Butyl acetate

$CH_3\overset{O}{\overset{\|}{C}}O\underset{}{\overset{CH_3}{\overset{|}{C}}}HCH_2CH_3$

(c) Cyclobutyl butanoate

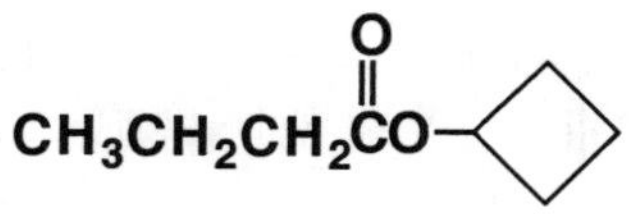

(d) *N*-(2-Octyl)succinimide

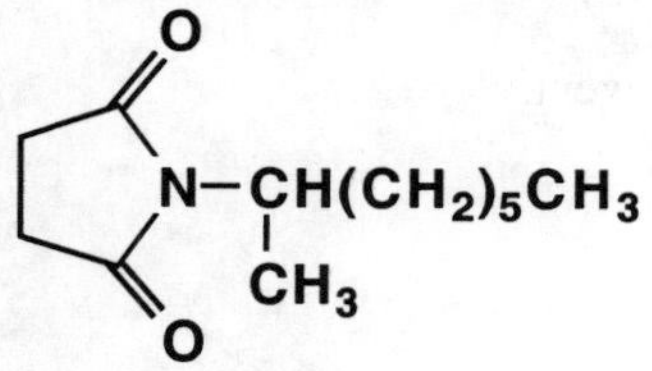

(e) Diethyl adipate

$C_2H_5O\overset{O}{\overset{\|}{C}}(CH_2)_4\overset{O}{\overset{\|}{C}}OC_2H_5$

(f) Propanoic anhydride

$CH_3CH_2\overset{O}{\overset{\|}{C}}O\overset{O}{\overset{\|}{C}}CH_2CH_3$

Problem 13.2 Complete and balance equations for hydrolysis of each ester in aqueous solution. Show each product as it is ionized under the given experimental conditions.

(a)

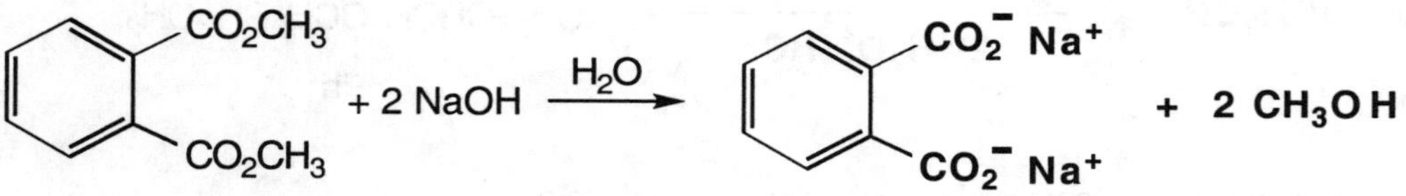

(b) $CH_3\overset{O}{\overset{\|}{C}}CH_2CH_2CH_2\overset{O}{\overset{\|}{C}}OCH_2CH_3 + H_2O \xrightarrow{HCl} CH_3\overset{O}{\overset{\|}{C}}CH_2CH_2CH_2\overset{O}{\overset{\|}{C}}OH + CH_3CH_2OH$

Problem 13.3 Complete equations for the hydrolysis of the amides in Example 13.3 in concentrated aqueous NaOH. Show all products as they exist in aqueous NaOH, and show the number of moles of NaOH required for hydrolysis of each amide.

(a) $CH_3\overset{O}{\overset{\|}{C}}-N(CH_3)_2 + NaOH \xrightarrow{H_2O} CH_3\overset{O}{\overset{\|}{C}}O^-Na^+ + (CH_3)_2NH$

(b)

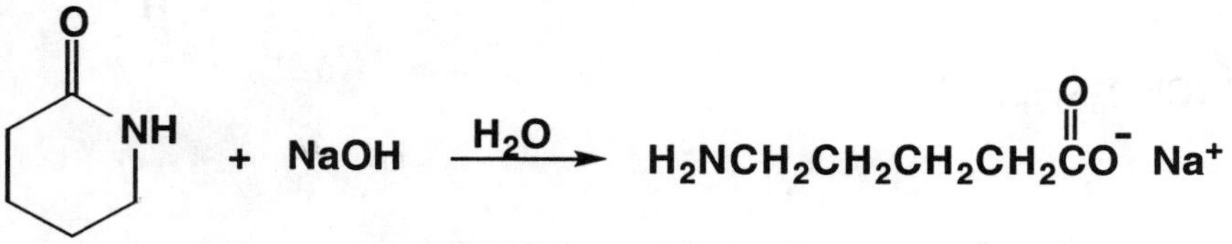

Problem 13.4 Complete these equations. The stoichiometry of each is given in the equation.

Each is an example of amminolysis of an ester.

(a)

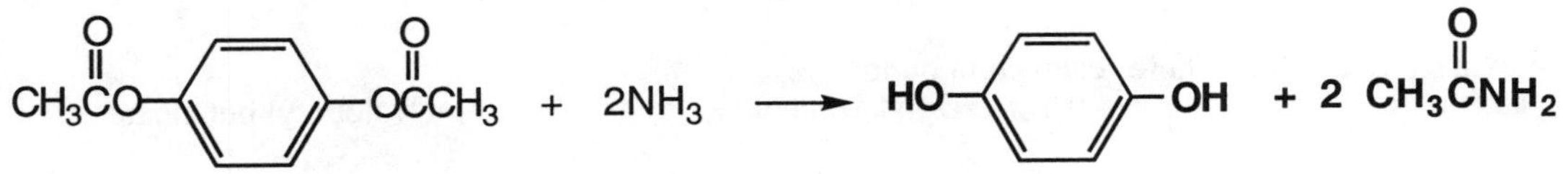

(b)

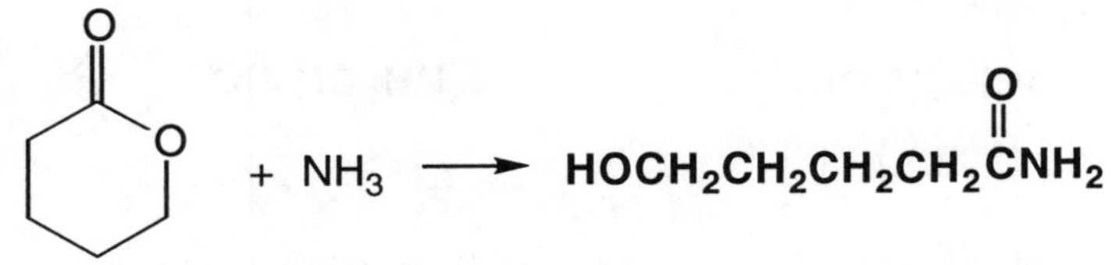

Problem 13.5 Show how to prepare each alcohol by treatment of an ester with a Grignard reagent.

(a)

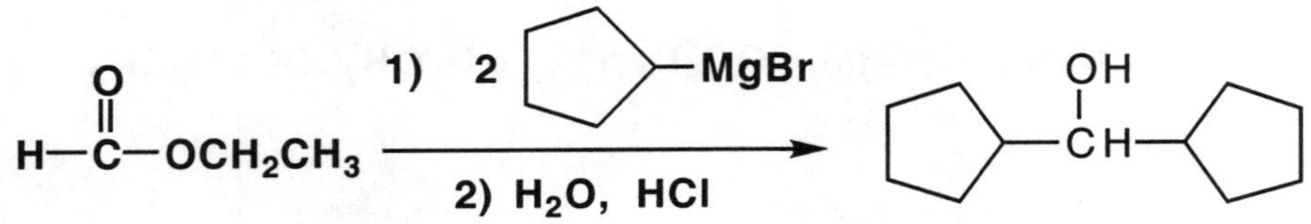

(b)

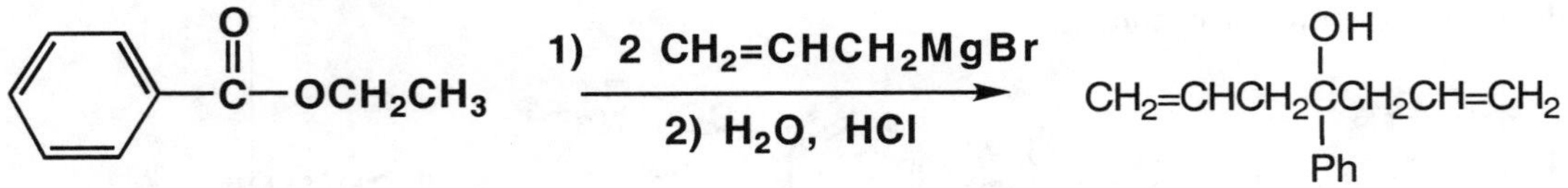

Problem 13.6 Show how to convert hexanoic acid to each amine in good yield.

(a)

(b)

Problem 13.7 Show how to convert (R)-2-phenylpropanoic acid to these compounds:

(a)

(R)-2-Phenyl-
propanoic acid

(R)-2-Phenyl-1-propanol

(b) (R)-PhCHCO$_2$H + SOCl$_2$ $\longrightarrow$ (R)-PhCHCCl $\xrightarrow{\text{NH}_3}$ (R)-PhCHCNH$_2$
     | || ||
     CH$_3$ O   O
                                                    | |
                                                    CH$_3$ CH$_3$

(R)-2-Phenyl-
propanoic acid

$\xrightarrow[\text{2) H}_2\text{O}]{\text{1) LiAlH}_4, \text{ ether}}$ (R)-PhCHCH$_2$NH$_2$
                                                    |
                                                    CH$_3$

(R)-2-Phenyl-1-propanamine

Structure and Nomenclature

Problem 13.8 Draw a structural formula for each compound.

(a) Dimethyl carbonate

$$\underset{\text{CH}_3\text{OCOCH}_3}{\overset{\text{O}}{\|}}$$

(b) *p*-Nitrobenzamide

O$_2$N—⟨benzene ring⟩—C(=O)NH$_2$

(c) Octanoyl chloride

$$\underset{\text{CH}_3(\text{CH}_2)_6\text{CCl}}{\overset{\text{O}}{\|}}$$

(d) Diethyl oxalate

$$\underset{\text{CH}_3\text{CH}_2\text{OC-COCH}_2\text{CH}_3}{\overset{\text{O O}}{\| \|}}$$

(e) Ethyl *cis*-2-pentenoate

CH$_3$CH$_2$ and COCH$_2$CH$_3$ on C=C with H, H (cis)

(f) Butanoic anhydride

(CH$_3$CH$_2$CH$_2$CO)$_2$O

(g) Dodecanamide

$$\underset{\text{CH}_3(\text{CH}_2)_{10}\text{CNH}_2}{\overset{\text{O}}{\|}}$$

(h) Ethyl 3-hydroxybutanoate

$$\underset{\text{CH}_3\text{CHCH}_2\text{COCH}_2\text{CH}_3}{\overset{\text{OH   O}}{|\|}}$$

Problem 13.9 Write the IUPAC name for each compound.

(a) ⟨Ph⟩—C(=O)—O—C(=O)—⟨Ph⟩

Benzoic anhydride

(b) CH$_3$(CH$_2$)$_{14}$COCH$_3$ (with C=O)

Methyl hexadecanoate

(c) CH$_3$(CH$_2$)$_4$CNHCH$_3$ (with C=O)

N-**Methylhexanamide**

(d) H$_2$N—⟨benzene ring⟩—C(=O)NH$_2$

4-Aminobenzamide

(e) CH$_2$(CO$_2$CH$_2$CH$_3$)$_2$

**Diethyl propanedioate
(Diethyl malonate)**

(f) PhCH$_2$CCHCOCH$_3$ (two C=O)
                |
               CH$_3$

**Methyl 2-methyl-3-oxo-
4-phenylbutanoate**

<u>Problem 13.10</u> When oil from the head of the sperm whale is cooled, spermaceti, a translucent wax with a white, pearly luster, crystallizes from the mixture. Spermaceti, which makes up 11% of whale oil, is composed mainly of hexadecyl hexadecanoate (cetyl palmitate). At one time, spermaceti was widely used in the making of cosmetics, fragrant soaps, and candles. Draw the structural formula of cetyl palmitate.

$$CH_3(CH_2)_{14} \overset{\overset{\displaystyle O}{\|}}{C} OCH_2(CH_2)_{14}CH_3$$

Hexadecyl hexadecanoate
(Cetyl palmitate)

Physical Properties

<u>Problem 13.11</u> Acetic acid and methyl formate are constitutional isomers. Both are liquids at room temperature: one with a boiling point of 32°C, the other with a boiling point of 118°C. Which of the two has the higher boiling point?

$$CH_3\overset{\overset{\displaystyle O}{\|}}{C}OH \qquad\qquad HC\overset{\overset{\displaystyle O}{\|}}{}OCH_3$$

Acetic acid   **Methyl formate**

Because of the polar O-H bond in acetic acid that is not present in methyl formate, only acetic acid has the possibility for intermolecular association by hydrogen bonding. Thus, acetic acid has a higher boiling point than its constitutional isomer methyl formate.

<u>Problem 13.12</u> Acetic acid has a boiling point of 118°C, whereas its methyl ester has a boiling point of 57°C. Account for the fact that the boiling point of acetic acid is higher than that of its methyl ester, even though acetic acid has a lower molecular weight.

$$CH_3\overset{\overset{\displaystyle O}{\|}}{C}OH \qquad\qquad CH_3\overset{\overset{\displaystyle O}{\|}}{C}OCH_3$$

Acetic acid   **Methyl acetate**

As in Problem 13.11, the difference in boiling points is due to the polar O-H bond in acetic acid that is not present in methyl acetate. Thus, only acetic acid has the possibility for intermolecular association by hydrogen bonding and, as a result, its boiling point is much higher than that of methyl acetate.

Reactions

<u>Problem 13.13</u> Arrange these compounds in order of increasing reactivity toward nucleophilic acyl substitution.

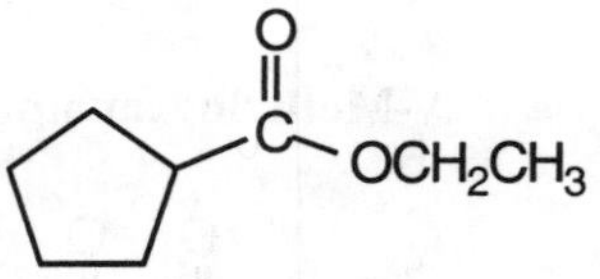

(1) 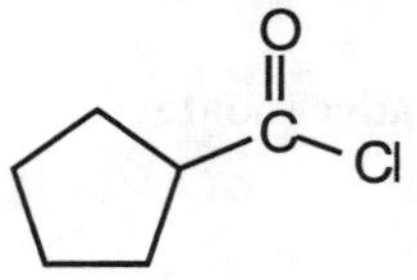   (2) 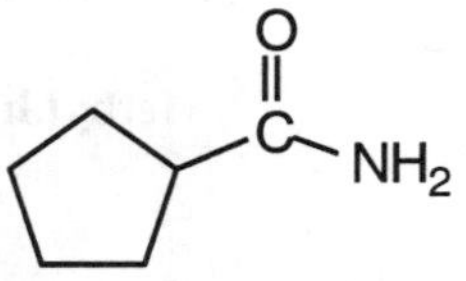   (3) 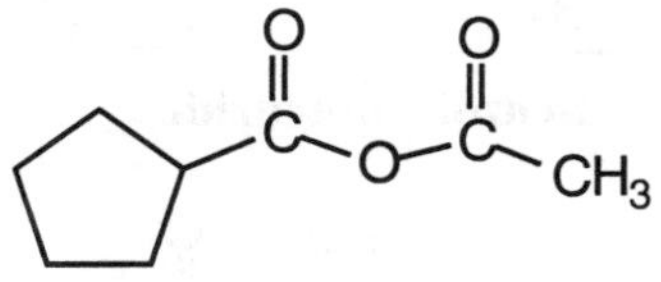   (4)

In general, the order of reactivity is acid chlorides > acid anhydrides > esters > amides. Therefore, the order for the molecules listed above is:

(ranked from least to most reactive) 3 < 1 < 4 < 2

Problem 13.14 A common method for preparing acid anhydrides is treatment of an acid chloride with the sodium salt of a carboxylic acid. For example, treatment of benzoyl chloride with sodium acetate gives acetic benzoic anhydride. Write a mechanism for this nucleophilic acyl substitution reaction.

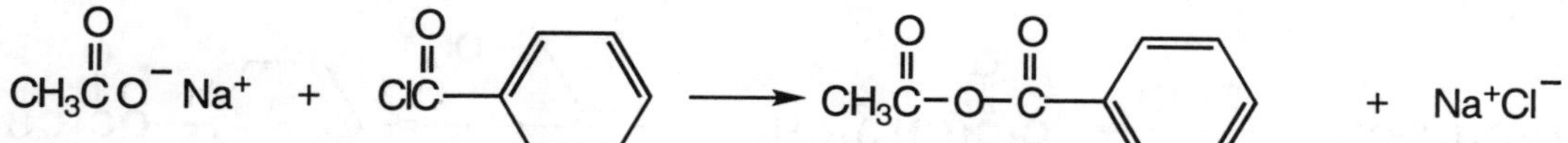

Sodium acetate   Benzoyl chloride   Acetic benzoic anhydride

Like the other nucleophilic acyl substitution reactions discussed throughout the chapter, the mechanism for this reaction involves attack of the nucleophilic acetate anion on the carbonyl carbon atom of the acid chloride. The resulting tetrahedral intermediate then decomposes with loss of chloride anion to give the final product.

Step 1:

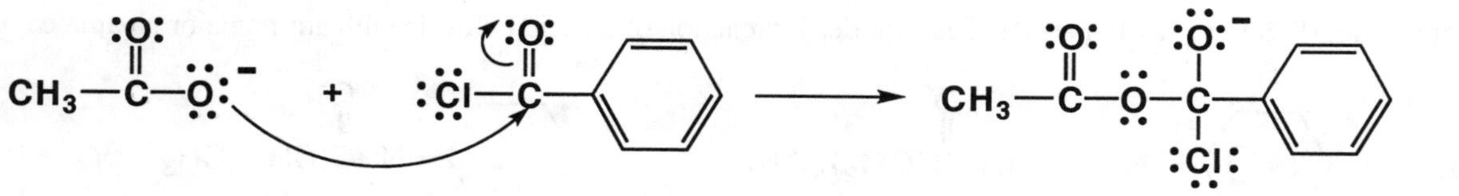

Step 2:

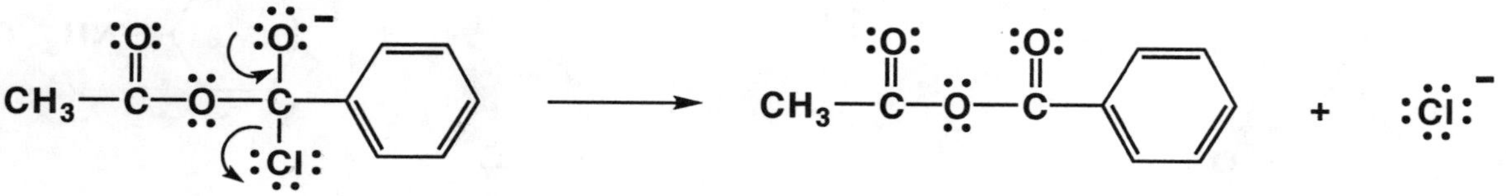

Problem 13.15 Show how to prepare these mixed anhydrides using the method described in Problem 13.14. (Hint: HCOCl is too unstable to use).

There is one combination of acid chloride and carboxylate ion that can be used to form the mixed anhydride in (a) and two combinations for the mixed anhydride in (b).

(a)
$$H-\overset{O}{\overset{||}{C}}-O^- \; Na^+ \quad + \quad Cl-\overset{O}{\overset{||}{C}}-CH_3 \longrightarrow H-\overset{O}{\overset{||}{C}}-O-\overset{O}{\overset{||}{C}}-CH_3$$

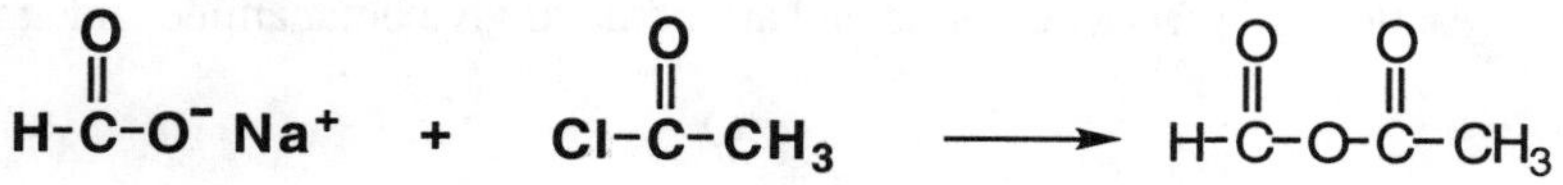

(b)
$$CH_3-\overset{O}{\overset{||}{C}}-Cl \; + \; Na^+ \; {}^-O-\overset{O}{\overset{||}{C}}CH_2C_6H_5$$

$$CH_3-\overset{O}{\overset{||}{C}}-O^- Na^+ \; + \; Cl-\overset{O}{\overset{||}{C}}CH_2C_6H_5$$

$$\longrightarrow CH_3-\overset{O}{\overset{||}{C}}-O-\overset{O}{\overset{||}{C}}CH_2C_6H_5$$

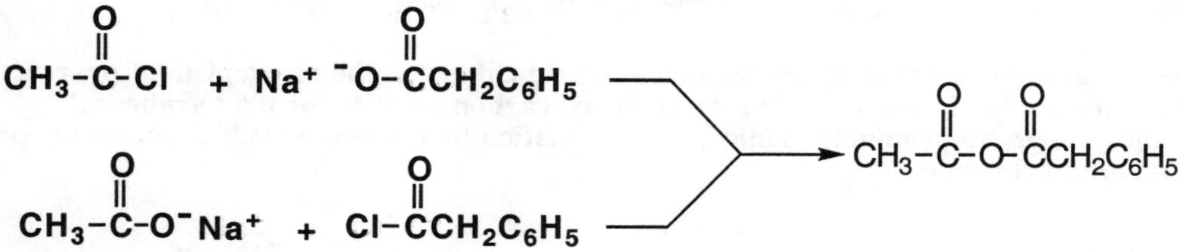

Problem 13.16 A carboxylic acid can be converted to an ester by Fischer esterification. Show how to synthesize each ester from a carboxylic acid and an alcohol by Fischer esterification.

(a)

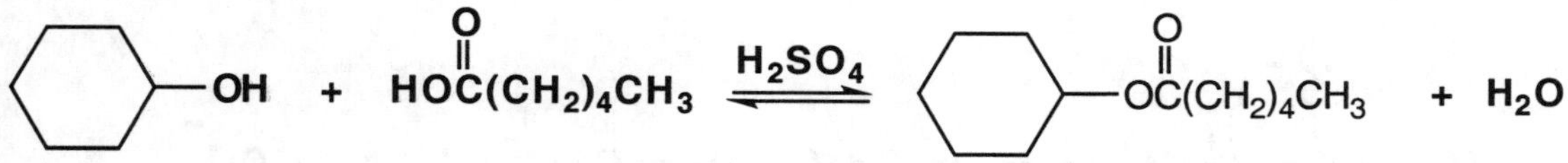

(b) $(CH_3)_2CH\overset{O}{\overset{||}{C}}OH \; + \; HOCH_2CH_3 \xrightarrow{H_2SO_4} (CH_3)_2CH\overset{O}{\overset{||}{C}}OCH_2CH_3 \; + \; H_2O$

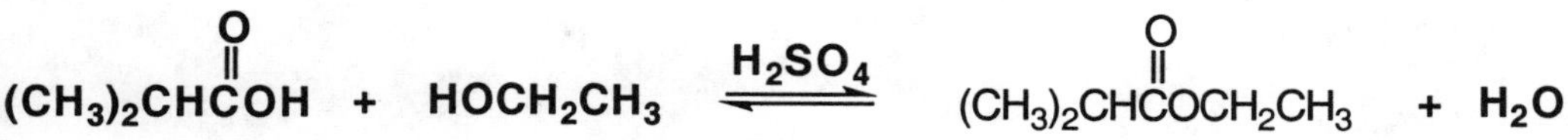

<u>Problem 13.17</u> A carboxylic acid can also be converted to an ester in two reactions by first converting the carboxylic acid to its acid chloride and then treating the acid chloride with an alcohol. Show how to prepare each ester in Problem 13.16 from a carboxylic acid and an alcohol by this two-step scheme.

(a) $HO\overset{O}{\overset{\|}{C}}(CH_2)_4CH_3 \xrightarrow{SOCl_2} Cl\overset{O}{\overset{\|}{C}}(CH_2)_4CH_3$ ⬡—OH → ⬡—$O\overset{O}{\overset{\|}{C}}(CH_2)_4CH_3$

(b) $(CH_3)_2CH\overset{O}{\overset{\|}{C}}OH \xrightarrow{SOCl_2} (CH_3)_2CH\overset{O}{\overset{\|}{C}}Cl \xrightarrow{HOCH_2CH_3} (CH_3)_2CH\overset{O}{\overset{\|}{C}}OCH_2CH_3$

<u>Problem 13.18</u> Show how to prepare these amides by reaction of an acid chloride with ammonia or an amine.

(a) 2 ⬡—NH_2 + $Cl-\overset{O}{\overset{\|}{C}}(CH_2)_4CH_3$ ⟶ ⬡—$NH\overset{O}{\overset{\|}{C}}(CH_2)_4CH_3$

+ ⬡—$NH_3^+ Cl^-$

(b) $(CH_3)_2CH\overset{O}{\overset{\|}{C}}-Cl$ + 2 $NH(CH_3)_2$ ⟶ $(CH_3)_2CH\overset{O}{\overset{\|}{C}}N(CH_3)_2$ + $(CH_3)_2NH_2^+ Cl^-$

(c) $Cl-\overset{O}{\overset{\|}{C}}(CH_2)_4\overset{O}{\overset{\|}{C}}-Cl$ + 4 NH_3 ⟶ $H_2N\overset{O}{\overset{\|}{C}}(CH_2)_4\overset{O}{\overset{\|}{C}}NH_2$ + 2 $NH_4^+Cl^-$

<u>Problem 13.19</u> Write a mechanism for the reaction of butanoyl chloride and ammonia to give butanamide and ammonium chloride.

$$CH_3(CH_2)_2\overset{O}{\overset{\|}{C}}Cl + 2 NH_3 \longrightarrow CH_3(CH_2)_2\overset{O}{\overset{\|}{C}}NH_2 + NH_4^+ Cl^-$$

Like the other reactions that involve attack by nucleophiles on acid chlorides, the mechanism of the reaction begins with the nucleophilic ammonia reacting with the carbonyl carbon atom to form a tetrahedral intermediate that collapses to give a protonated amide. Deprotonation by ammonia leads to the amide product and a molecule of ammonium chloride.

Step 1: $CH_3(CH_2)_2 -\overset{:O:}{\overset{\|}{C}}-\ddot{C}\ddot{l}:$ + $:\overset{H}{\underset{H}{N}}-H$ ⟶ $CH_3(CH_2)_2 -\overset{:\overset{-}{O}:}{\underset{:\ddot{C}\ddot{l}:}{C}}-\overset{H}{\underset{H}{\overset{+}{N}}}-H$

Step 2: $CH_3(CH_2)_2 -\overset{:\overset{-}{O}:}{\underset{:\ddot{C}\ddot{l}:}{C}}-\overset{H}{\underset{H}{\overset{+}{N}}}-H$ ⟶ $CH_3(CH_2)_2 -\overset{:O:}{\overset{\|}{C}}-\overset{H}{\underset{H}{\overset{+}{N}}}-H$ + $:\ddot{C}\ddot{l}:^-$

Step 3:

$$CH_3(CH_2)_2-\overset{\overset{\displaystyle :O:}{\|}}{C}-\overset{\overset{\displaystyle H}{|}}{\underset{\underset{\displaystyle H}{|}}{N^+}}-H \quad + \quad \overset{\overset{\displaystyle H}{|}}{\underset{\underset{\displaystyle H}{|}}{:N}}-H \quad \longrightarrow \quad CH_3(CH_2)_2-\overset{\overset{\displaystyle :O:}{\|}}{C}-\overset{\overset{\displaystyle H}{|}}{N:} \quad + \quad H-\overset{\overset{\displaystyle H}{|}}{\underset{\underset{\displaystyle H}{|}}{N^+}}-H$$

Problem 13.20 What product is formed when benzoyl chloride is treated with these reagents?
(a) C_6H_6, $AlCl_3$ (b) $CH_3CH_2CH_2CH_2OH$ (c) $CH_3CH_2CH_2CH_2SH$

(d) $CH_3CH_2CH_2CH_2NH_2$ (e) H_2O (f) piperidine N—H
     (two equivalents)   (two equivalents)

Problem 13.21 Write the product(s) of treatment of propanoic anhydride with each reagent.
(a) Ethanol (1 equivalent)

$$CH_3CH_2\overset{\overset{\displaystyle O}{\|}}{C}-O-\overset{\overset{\displaystyle O}{\|}}{C}CH_2CH_3 \;+\; CH_3CH_2OH \;\longrightarrow\; CH_3CH_2\overset{\overset{\displaystyle O}{\|}}{C}OCH_2CH_3 \;+\; CH_3CH_2\overset{\overset{\displaystyle O}{\|}}{C}OH$$

(b) Ammonia (2 equivalents)

$$CH_3CH_2\overset{\overset{\displaystyle O}{\|}}{C}-O-\overset{\overset{\displaystyle O}{\|}}{C}CH_2CH_3 \;+\; 2\ NH_3 \;\longrightarrow\; CH_3CH_2\overset{\overset{\displaystyle O}{\|}}{C}NH_2 \;+\; CH_3CH_2\overset{\overset{\displaystyle O}{\|}}{C}O^-\ NH_4^+$$

Problem 13.22 Write the product of treatment of succinic anhydride with each reagent.
(a) Ethanol (1 equivalent)

$+\ CH_3CH_2OH \longrightarrow CH_3CH_2O\overset{\overset{\displaystyle O}{\|}}{C}CH_2CH_2\overset{\overset{\displaystyle O}{\|}}{C}OH$

(b) Ammonia (2 equivalents)

$+\ 2\ NH_3 \longrightarrow H_2N\overset{\overset{\displaystyle O}{\|}}{C}CH_2CH_2\overset{\overset{\displaystyle O}{\|}}{C}O^-\ NH_4^+$

<u>Problem 13.23</u> The analgesic phenacetin is synthesized by treating 4-ethoxyaniline with acetic anhydride. Write an equation for the formation of phenacetin.

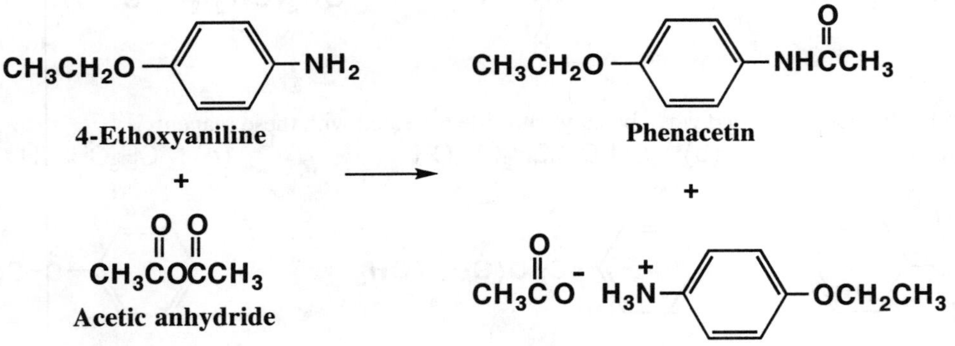

CH_3CH_2O—⬡—NH_2 CH_3CH_2O—⬡—$NHCCH_3$ (C=O above)

4-Ethoxyaniline   **Phenacetin**

+   +

CH_3COCCH_3 (O O above both C's) $CH_3\overset{O}{\overset{\|}{C}}O^-$ $H_3\overset{+}{N}$—⬡—OCH_2CH_3

Acetic anhydride

<u>Problem 13.24</u> The analgesic acetaminophen is synthesized by treatment of 4-aminophenol with one equivalent of acetic anhydride. Write an equation for the formation of acetaminophen. (Hint: An -NH_2 group is a better nucleophile than an -OH group.)

Note how, in the following reaction scheme, the acylation occurs at the more nucleophilic amino group rather than the less nucleophilic hydroxyl group of 4-aminophenol.

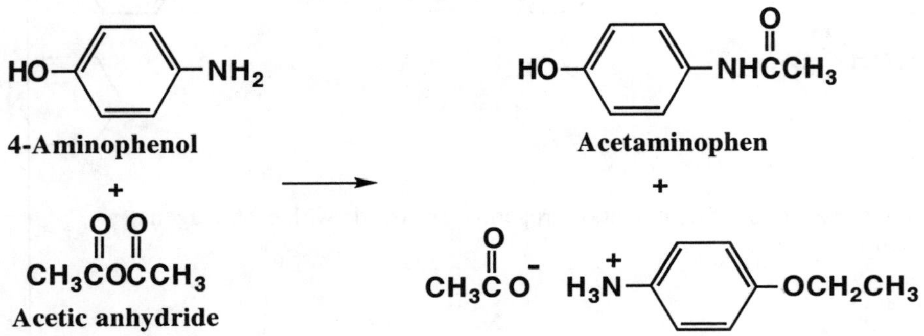

HO—⬡—NH_2 HO—⬡—$NHCCH_3$ (C=O above)

4-Aminophenol   **Acetaminophen**

+   +

CH_3COCCH_3 (O O above both C's) $CH_3\overset{O}{\overset{\|}{C}}O^-$ $H_3\overset{+}{N}$—⬡—OCH_2CH_3

Acetic anhydride

<u>Problem 13.25</u> Nicotinic acid, more commonly named niacin, is one of the B vitamins. Show how nicotinic acid can be converted to ethyl nicotinate and then to nicotinamide.

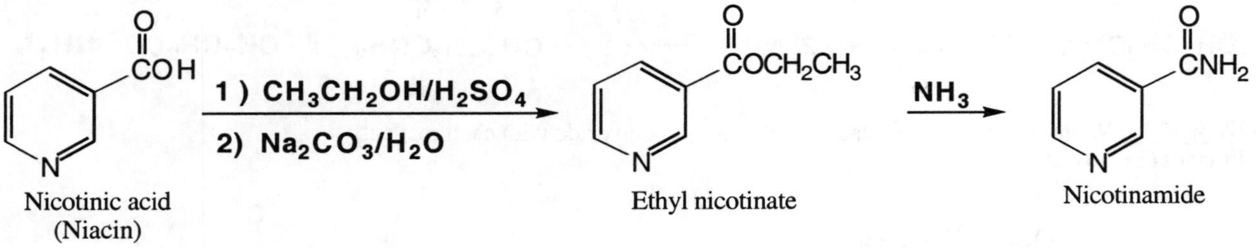

(pyridine ring)—$\overset{O}{\overset{\|}{C}}OH$ $\xrightarrow[\text{2) } Na_2CO_3/H_2O]{\text{1) } CH_3CH_2OH/H_2SO_4}$ (pyridine ring)—$\overset{O}{\overset{\|}{C}}OCH_2CH_3$ $\xrightarrow{NH_3}$ (pyridine ring)—$\overset{O}{\overset{\|}{C}}NH_2$

Nicotinic acid
(Niacin)   Ethyl nicotinate   Nicotinamide

Problem 13.26 Complete these reactions.

(a)

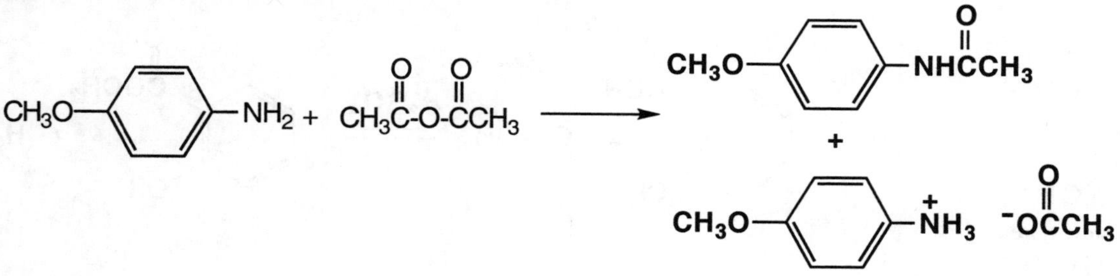

(b)

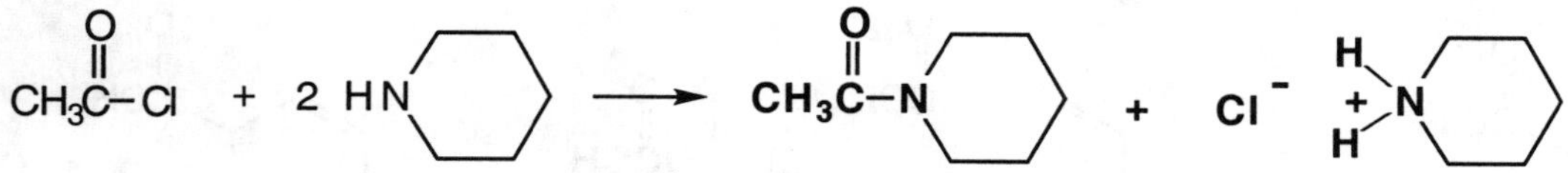

(c)

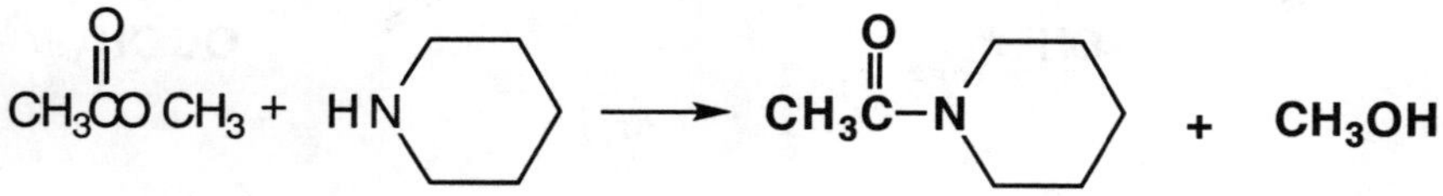

(d)

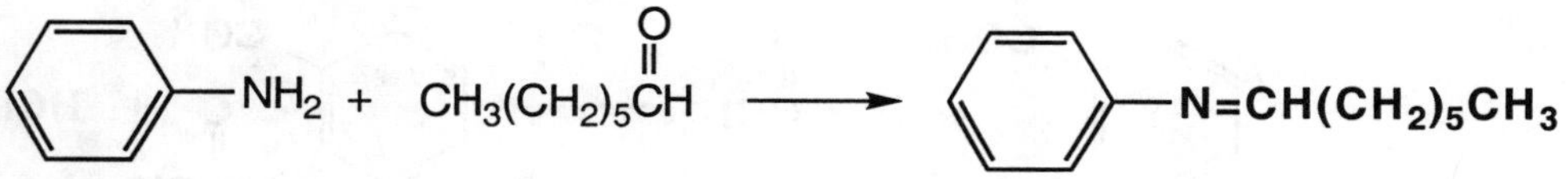

Problem 13.27 What product is formed when ethyl benzoate is treated with these reagents?

(a) H$_2$O, NaOH, heat

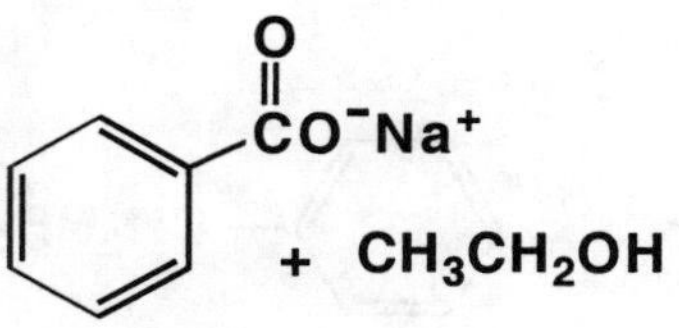

(b) LiAlH$_4$, then H$_2$O

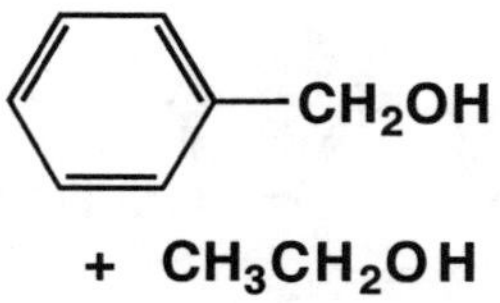

(c) H$_2$O, H$_2$SO$_4$, heat

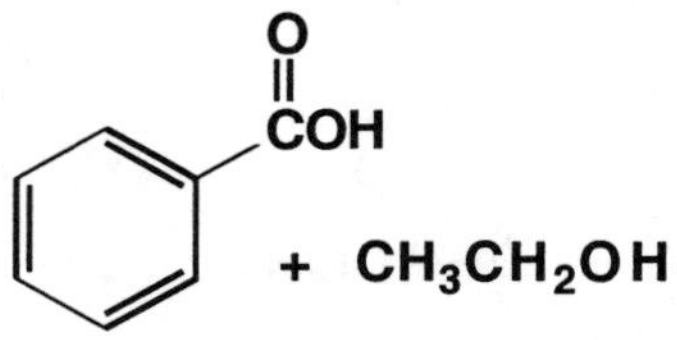

+ CH$_3$CH$_2$OH

+ CH$_3$CH$_2$OH

+ CH$_3$CH$_2$OH

(d) CH$_3$CH$_2$CH$_2$CH$_2$NH$_2$

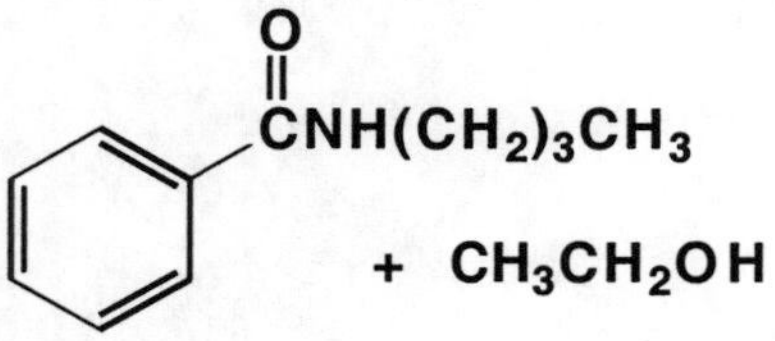

(e) C$_6$H$_5$MgBr (two moles), then H$_2$O/HCl

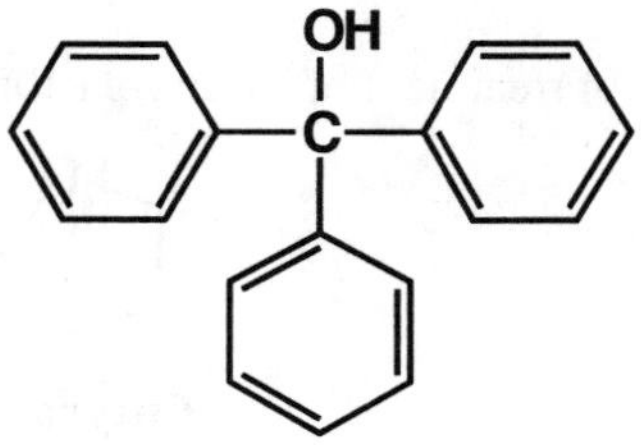

Problem 13.28 Show how to convert 2-hydroxybenzoic acid (salicylic acid) to these compounds.

(a)

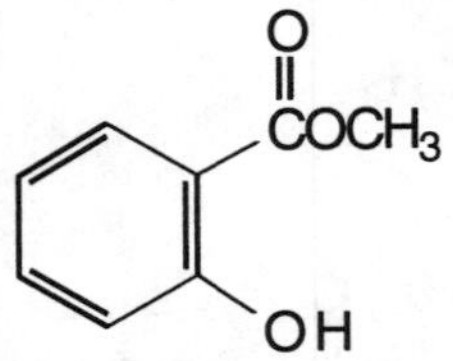

Methyl salicylate
(Oil of wintergreen)

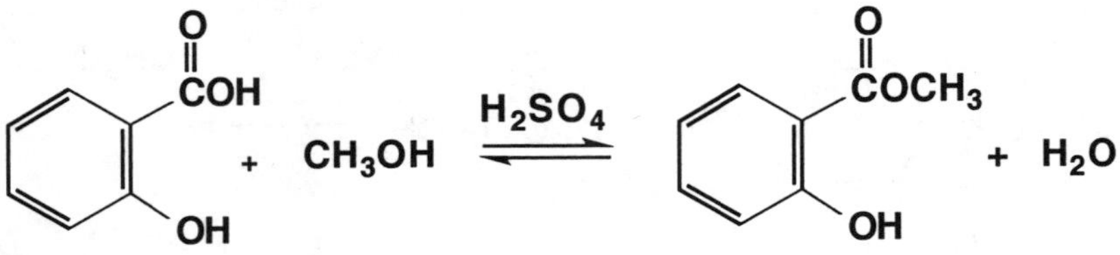

$$\text{salicylic acid} + CH_3OH \underset{}{\overset{H_2SO_4}{\rightleftharpoons}} \text{methyl salicylate} + H_2O$$

(b)

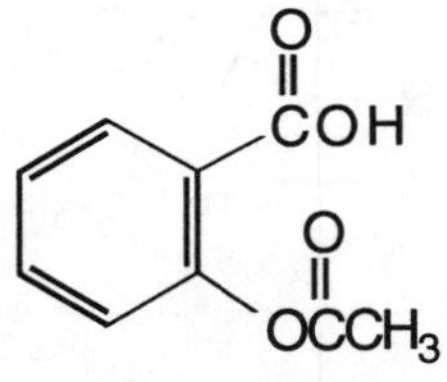

Acetyl salicylic acid
(Aspirin)

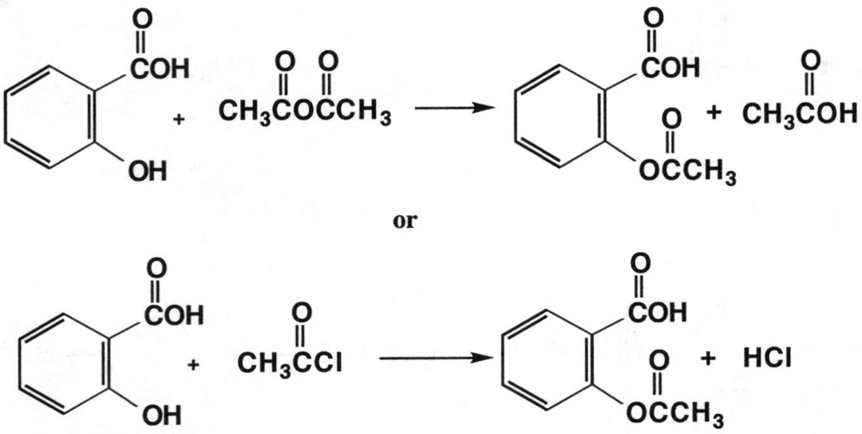

$$\text{salicylic acid} + CH_3COCCH_3 \longrightarrow \text{aspirin} + CH_3COH$$

or

$$\text{salicylic acid} + CH_3CCl \longrightarrow \text{aspirin} + HCl$$

Problem 13.29 What product is formed when benzamide is treated with these reagents?
(a) H_2O, HCl, heat   (b) NaOH, H_2O, heat   (c) LiAlH$_4$, then H_2O

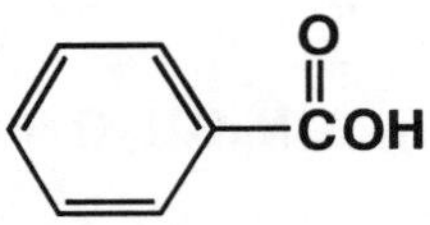

benzoic acid

+ NH$_4$Cl

sodium benzoate

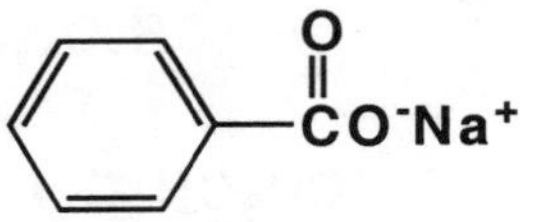

+ NH$_3$

benzylamine CH$_2$NH$_2$

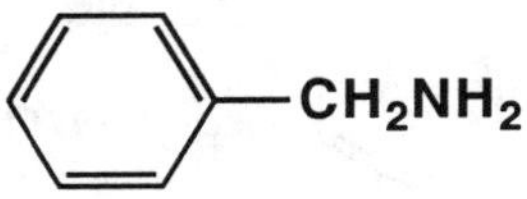

Problem 13.30 Show the product of treatment of γ-butyrolactone with each reagent.

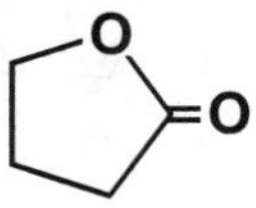

γ-Butyrolactone

(a)   NH$_3$ $\longrightarrow$ HOCH$_2$CH$_2$CH$_2$CNH$_2$

(b) LiAlH$_4$, then H$_2$O $\longrightarrow$ HOCH$_2$CH$_2$CH$_2$CH$_2$OH

(c) NaOH, H$_2$O, heat $\longrightarrow$ HOCH$_2$CH$_2$CH$_2$CO$^-$Na$^+$

Problem 13.31 Show the product of treatment of *N*-methyl-γ-butyrolactam with each reagent.

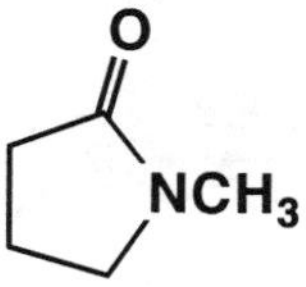

N-Methyl-γ-butyrolactam

(a) H_2O, HCl, heat $\longrightarrow$

$$\underset{\underset{CH_3}{|}}{\overset{\overset{H}{|}}{H-\overset{+}{N}}}-CH_2CH_2CH_2\overset{O}{\overset{\|}{C}}OH \quad Cl^-$$

(b) NaOH, H_2O, heat $\longrightarrow$

$$\underset{\underset{CH_3}{|}}{H-N}-CH_2CH_2CH_2\overset{O}{\overset{\|}{C}}O^- \; Na^+$$

(c) $LiAlH_4$, then H_2O $\longrightarrow$ NCH₃

Problem 13.32 Complete these reactions:

(a) $\overset{O}{\overset{\|}{-COC_2H_5}}$

1) 2 $CH_2=CHCH_2MgBr$
2) H_2O/HCl

$$\underset{\underset{CH_2CH=CH_2}{|}}{\overset{\overset{OH}{|}}{C}}-CH_2CH=CH_2$$

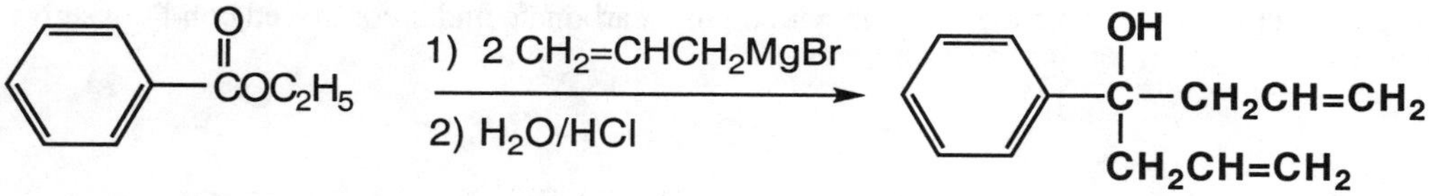

(b) $\overset{O}{\overset{\|}{C}}$

1) CH_3MgBr
2) H_2O/HCl

$$\underset{\underset{CH_3}{|}}{\overset{\overset{OH}{|}}{C}}$$

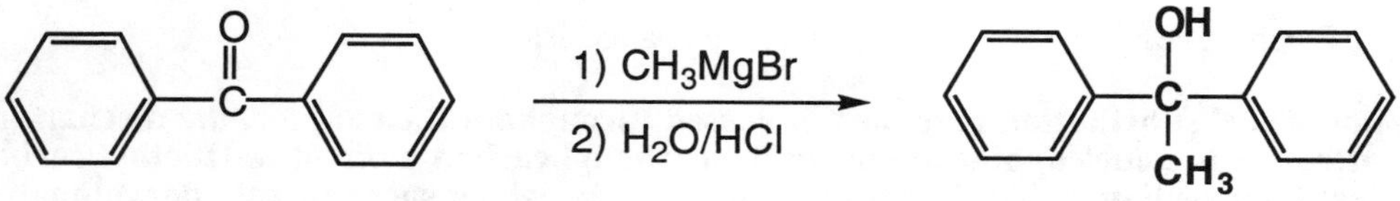

(c) $CH_3CH_2CH_2CH_2\overset{O}{\overset{\|}{C}}OCH_2CH_3$

1) 2 CH_3MgBr
2) H_2O/HCl

$$CH_3CH_2CH_2CH_2\underset{\underset{CH_3}{|}}{\overset{\overset{OH}{|}}{C}}CH_3$$

Problem 13.33 What combination of ester and Grignard reagent can be used to prepare each alcohol?

(a) 2-Methyl-2-butanol

$$CH_3CH_2\overset{O}{\overset{\|}{C}}OCH_2CH_3 \quad \xrightarrow[\text{2) } H_2O/HCl]{\text{1) } 2 \; CH_3MgBr} \quad \underset{\underset{CH_3}{|}}{\overset{\overset{OH}{|}}{CH_3C}}CH_2CH_3$$

(b) 3-Phenyl-3-pentanol

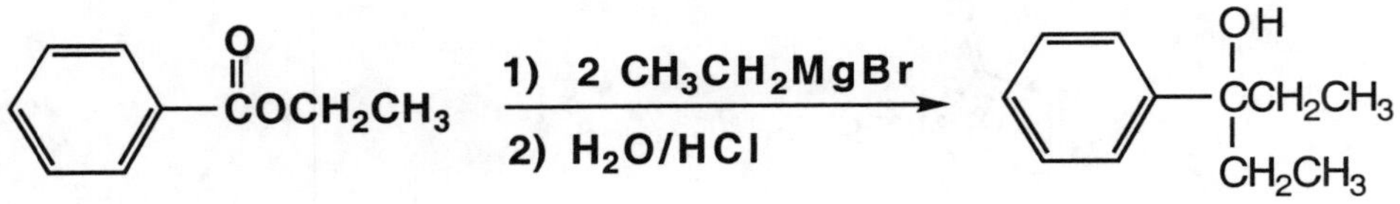

(c) 1,1-Diphenylethanol

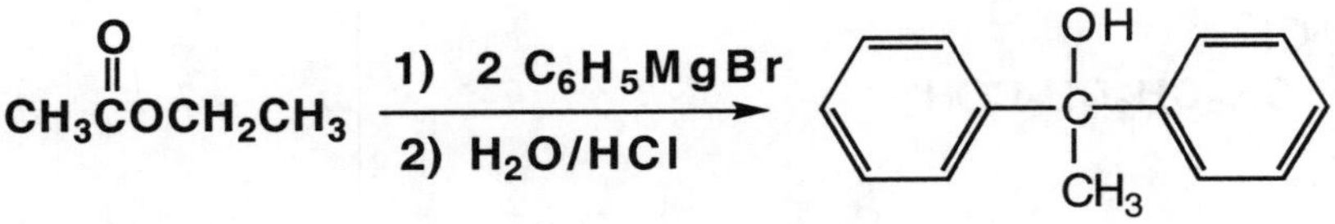

<u>Problem 13.34</u> Treatment of γ–butyrolactone with two equivalents of methylmagnesium bromide followed by hydrolysis in aqueous acid gives a compound of molecular formula $C_6H_{14}O_2$. Propose a structural formula for this compound.

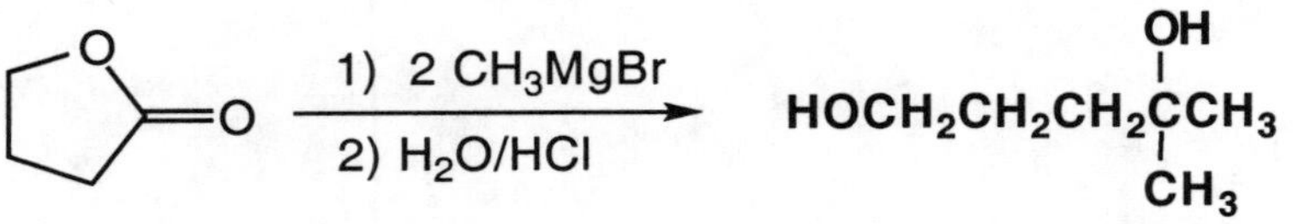

$$HOCH_2CH_2CH_2\underset{\underset{CH_3}{|}}{\overset{\overset{OH}{|}}{C}}CH_3$$

4-Methyl-1,4-pentanediol
$C_6H_{14}O_2$

<u>Problem 13.35</u> Reaction of a primary or secondary amine with diethyl carbonate under controlled conditions gives a carbamic ester. Propose a mechanism for this reaction.

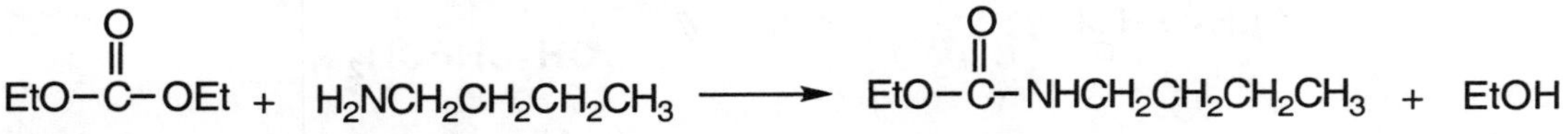

Diethyl carbonate   Butanamine   A carbamic ester

Like the other nucleophilic acyl substitution reactions discussed throughout the chapter, the mechanism for this reaction involves attack of the nucleophilic amine on the carbonyl carbon atom of the diethyl carbonate. The resulting tetrahedral intermediate collapses with loss of ethoxide, which subsequently deprotonates the protonated amide nitrogen to give the final products.

Step 1:

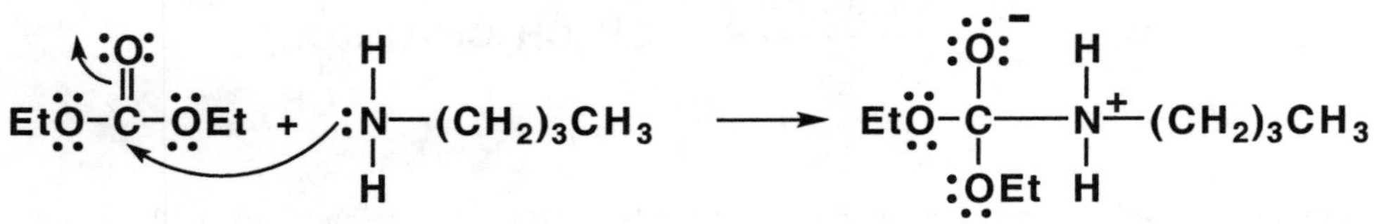

Step 2:

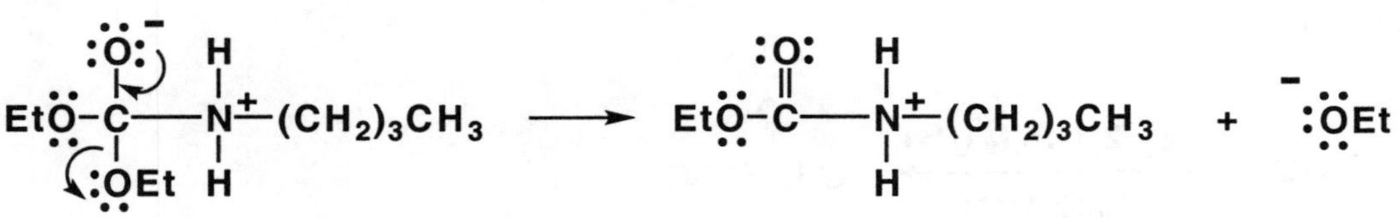

Step 3:

Problem 13.36 Barbiturates are prepared by treatment of diethyl malonate or a derivative of diethyl malonate with urea in the presence of sodium ethoxide as a catalyst. Following is an equation for the preparation of barbital, a long-duration hypnotic and sedative, from diethyl diethylmalonate and urea. Barbital is prescribed under one of a dozen or more trade names.

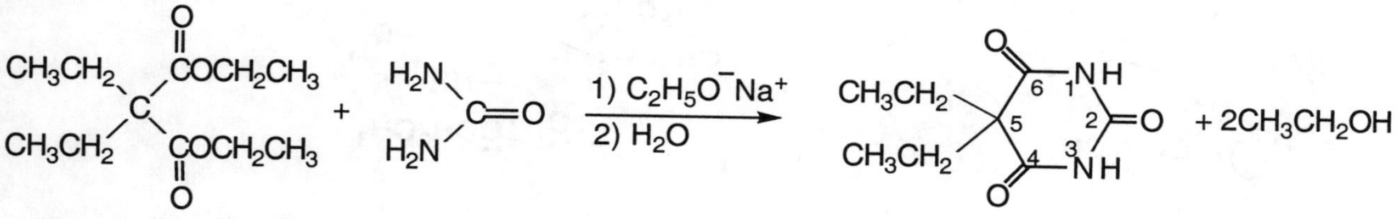

Diethyl diethylmalonate   Urea   5,5-Diethylbarbituric acid
(Barbital)

(a) Propose a mechanism for this reaction.

Step 1:

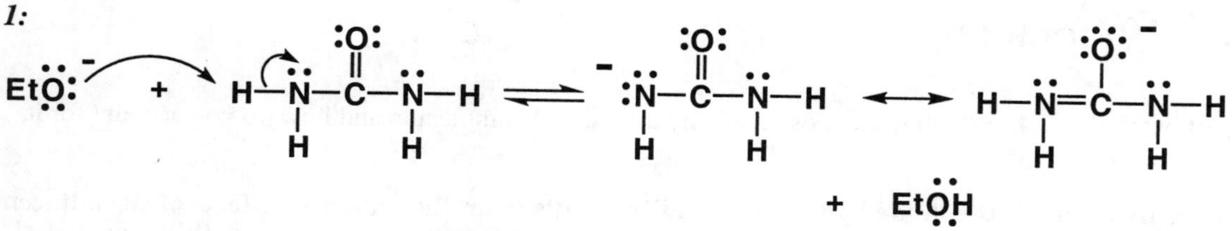

Step 2:

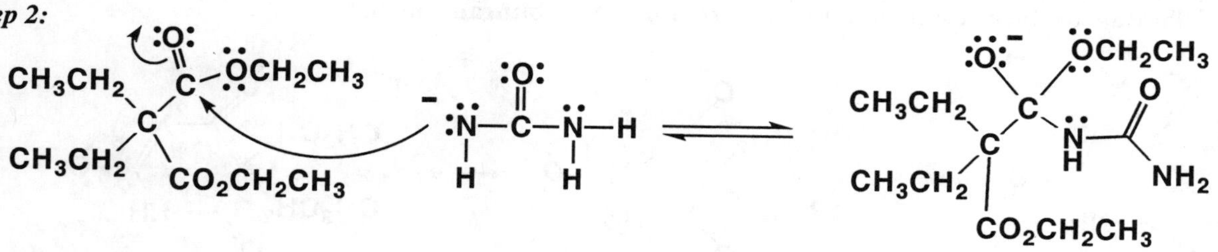

Step 3:

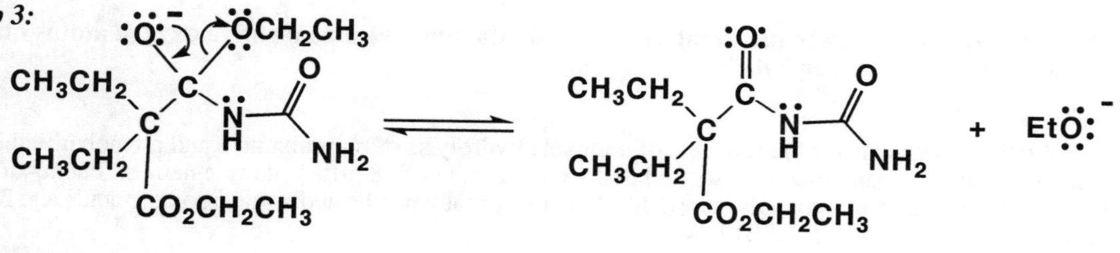

Step 4:

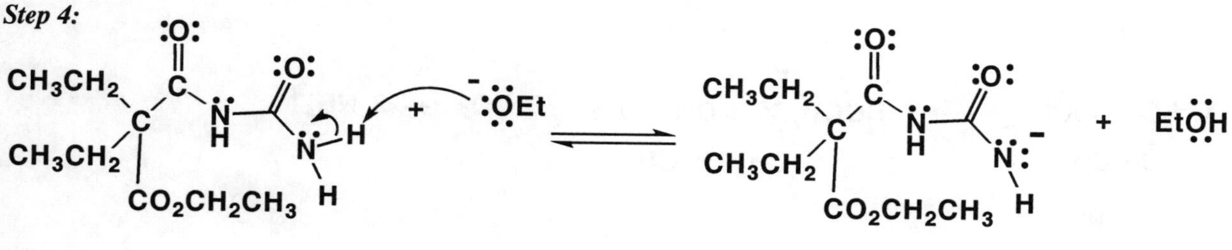

Step 5:

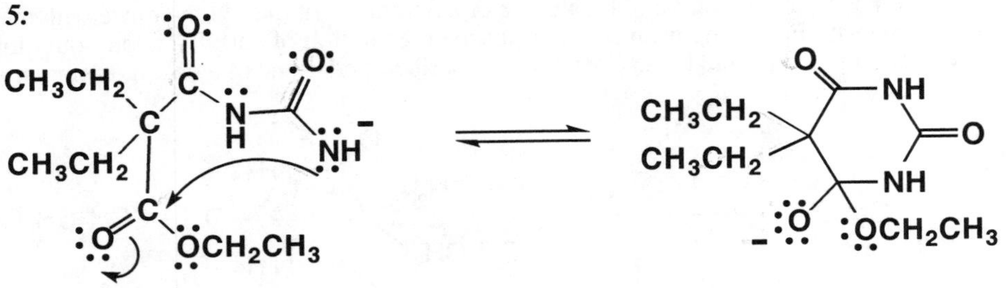

Step 6:

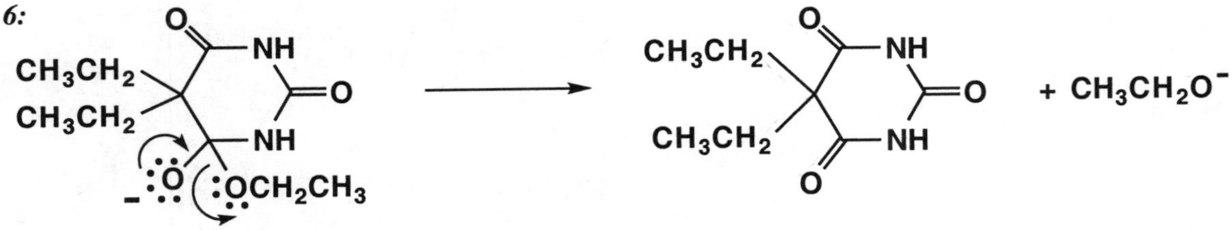

(b) The pK$_a$ of barbital is 7.4. Which is the most acidic hydrogen in this molecule and how do you account for its acidity?

The most acidic hydrogen is the imide hydrogen. Acidity results from the inductive effects of the adjacent carbonyl groups and stabilization of the deprotonated anion by resonance interaction with the carbonyl groups. Following are three contributing structures for the barbiturate anion.

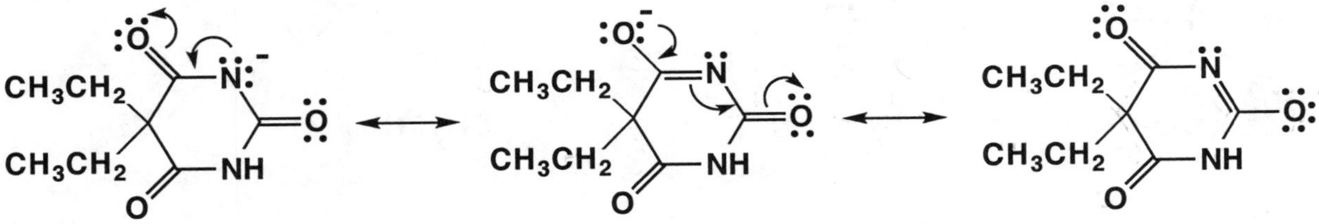

The two contributing structures that place the negative charge on the more electronegative oxygen atoms make the greatest contribution to the resonance hybrid.

Problem 13.37 Draw structural formulas for the products of complete hydrolysis of meprobamate and phenobarbital in hot aqueous acid. Meprobamate is a tranquilizer prescribed under one or more of 58 different trade names. Phenobarbital is a long-acting sedative, hypnotic, and anti-convulsant. [Hint: Remember that when heated, β-dicarboxylic acids and β-ketoacids undergo decarboxylation (Section 12.8B).]

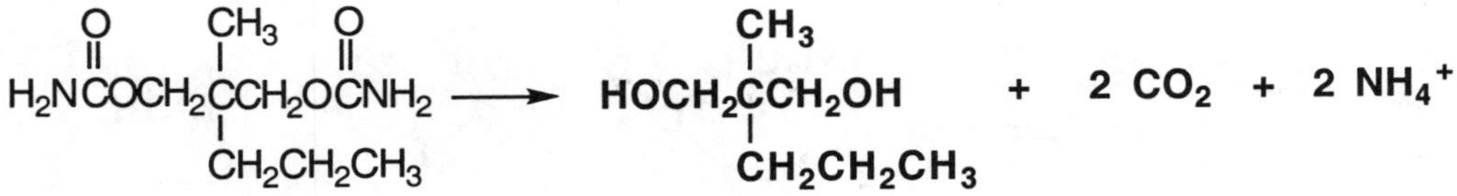

Meprobamate

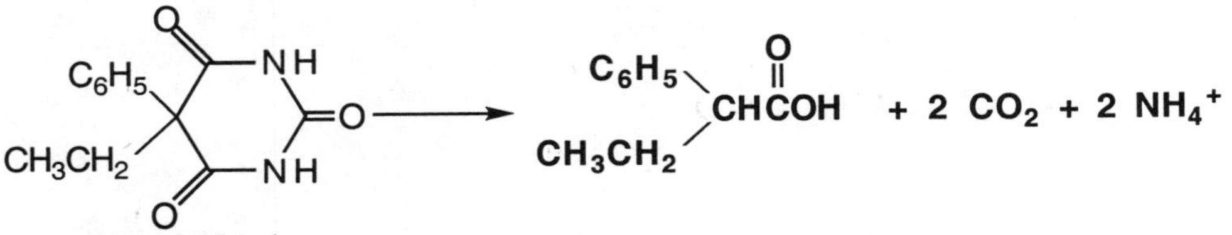

Phenobarbital

Problem 13.38 *N,N*-Diethyl-*m*-toluamide (Deet), the active ingredient in several common insect repellents, is synthesized from 3-methylbenzoic acid (*m*-toluic acid) and diethylamine. Show how this synthesis can be accomplished.

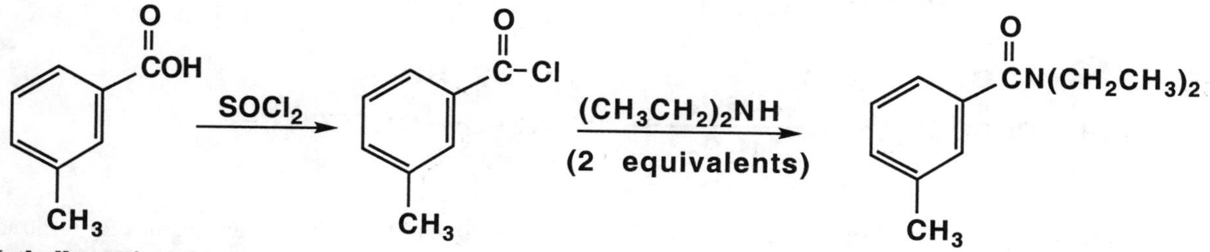

3-Methylbenzoic acid
(*m*-Toluic acid)

$(CH_3CH_2)_2NH$
(2 equivalents)

N,N-Diethyltoluamide
(DEET)

Problem 13.39 Show reagents for the synthesis of the following tertiary amine.

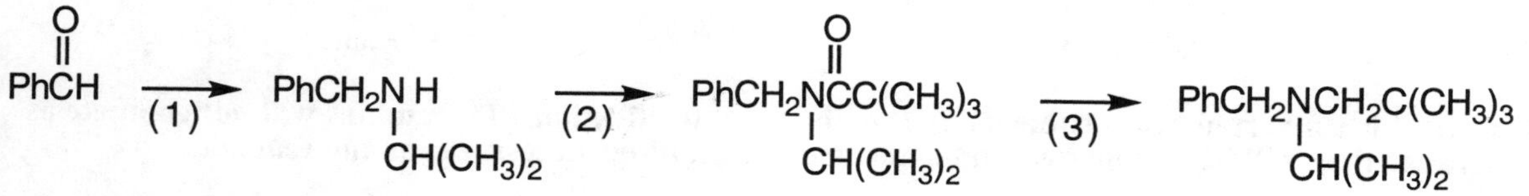

(1) Reductive amination of benzaldehyde with isopropyl amine.

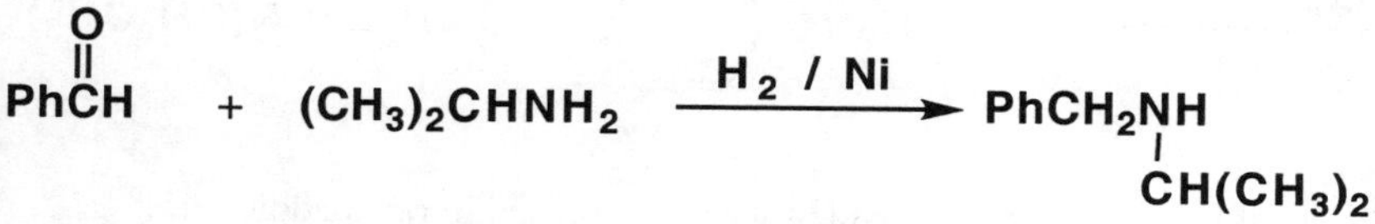

(2) Formation of an amide with 2,2-dimethylpropanoyl chloride (pivaloyl chloride)

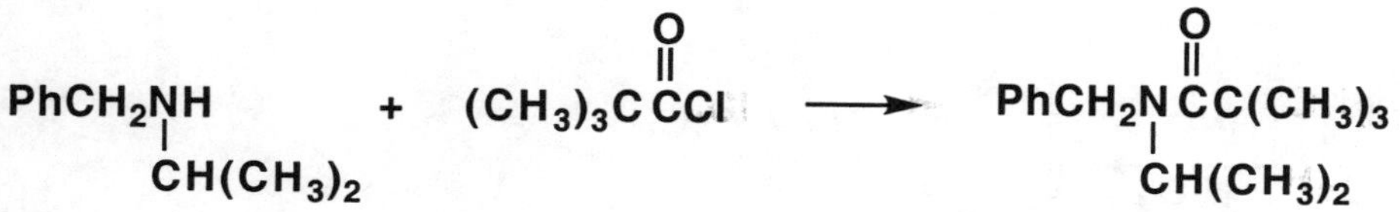

(3) Reduction of the amide with lithium aluminum hydride or with hydrogen in the presence of a transition metal catalyst.

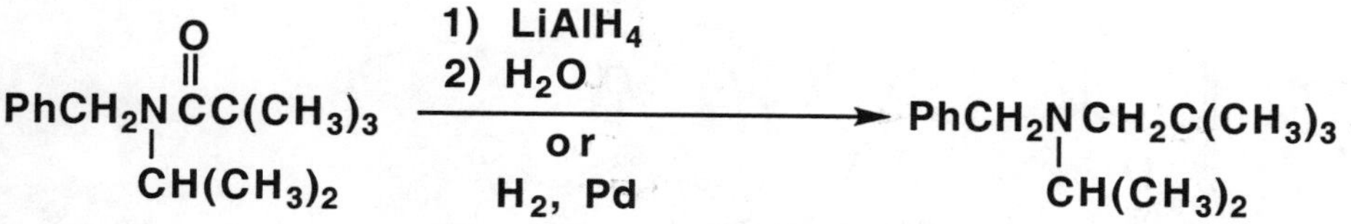

Problem 13.40 Show how to convert ethyl 2-pentenoate to these compounds.

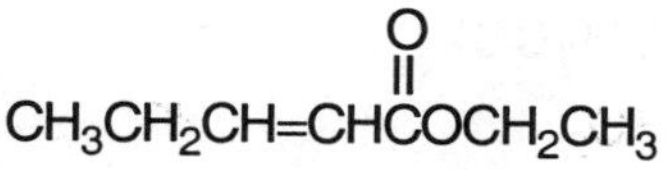

$$CH_3CH_2CH=CHCOCH_2CH_3$$

Ethyl 2-pentenoate

(a) $CH_3CH_2CH=CHCOCH_2CH_3$ $\xrightarrow{H_2/Pd}$ $CH_3CH_2CH_2CH_2COCH_2CH_3$

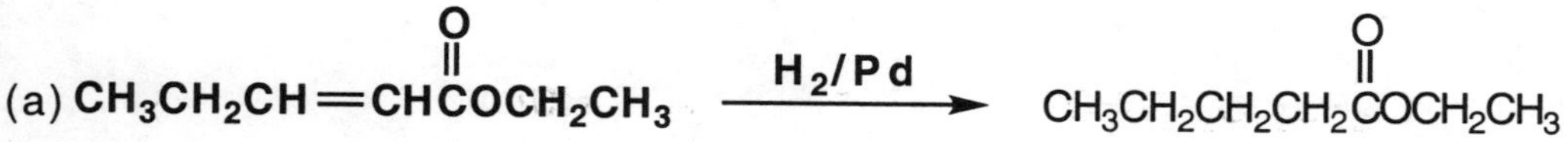

(b) $CH_3CH_2CH=CHCOCH_2CH_3$ $\xrightarrow[\text{2) } H_2O]{\text{1) } LiAlH_4}$ $CH_3CH_2CH=CHCH_2OH$

(c) $CH_3CH_2CH=CHCOCH_2CH_3$ $\xrightarrow[H_2O_2]{OsO_4}$ $CH_3CH_2CH-CHCOCH_2CH_3$ (with HO HO groups)

Problem 13.41 Procaine (its hydrochloride is marketed as Novocaine) was one of the first local anesthetics for infiltration and regional anesthesia. Show how to synthesize procaine using the given reagents as sources of carbon atoms.

H_2N—⬡—$\overset{O}{\overset{\|}{C}}OH$ + $HOCH_2CH_2N(CH_2CH_3)_2$ $\xrightarrow{?}$ H_2N—⬡—$\overset{O}{\overset{\|}{C}}OCH_2CH_2N(CH_2CH_3)_2$

p-Aminobenzoic acid   2-Diethylaminoethanol   Procaine

p-Aminobenzoic acid is converted to procaine using the Fischer esterification. The amines will not compete as nucleophiles, because they will be protonated under the acidic conditions necessary for the reaction.

H_2N—⬡—$\overset{O}{\overset{\|}{C}}OH$ $\xrightarrow[\text{2) } NaOH/H_2O]{\substack{\text{1) } HOCH_2CH_2N(CH_2CH_3)_2 \\ H_2SO_4}}$ H_2N—⬡—$\overset{O}{\overset{\|}{C}}OCH_2CH_2N(CH_2CH_3)_2$

Problem 13.42 Starting materials for the synthesis of the herbicide propranil, a weed killer used in rice paddies, are benzene and propanoic acid. Show reagents to bring about this synthesis.

⬡ $\xrightarrow{Cl_2 \text{ / } FeCl_3}$ ⬡Cl $\xrightarrow[H_2SO_4]{HNO_3}$ ⬡ (Cl, NO_2) $\xrightarrow{Cl_2 \text{ / } FeCl_3}$

⬡ (Cl, Cl, NO_2) $\xrightarrow[\substack{\text{1) Fe, HCl, } H_2O \\ \text{2) NaOH}}]{\substack{H_2 \text{ / } Ni \\ or}}$ ⬡ (Cl, Cl, NH_2)

⬡ (Cl, Cl, N HCCH_2CH_3 with O)
Propranil

$CH_3CH_2\overset{O}{\overset{\|}{C}}OH$ $\xrightarrow{SOCl_2}$ $CH_3CH_2\overset{O}{\overset{\|}{C}}Cl$

Notice how in the second chlorination reaction, the new Cl atom is directed to the correct position by the groups already present on the ring.

Problem 13.43 Following are structural formulas for three local anesthetics. Lidocaine was introduced in 1948 and is now the most widely used local anesthetic for infiltration and regional anesthesia. Its hydrochloride is marketed under the name Xylocaine. Etidocaine (hydrochloride marketed as Duranest) is comparable to lidocaine in onset, but its analgesic action lasts two to three times longer. Anesthetic action from mepivacaine (hydrochloride marketed as Carbocaine) is faster and somewhat longer in duration than lidocaine.

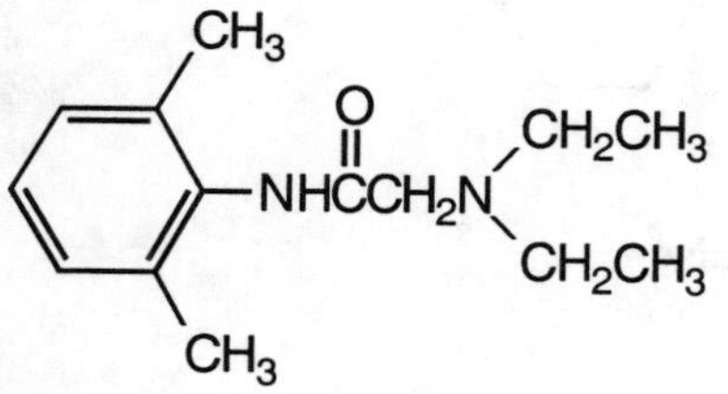

Lidocaine
(Xylocaine)

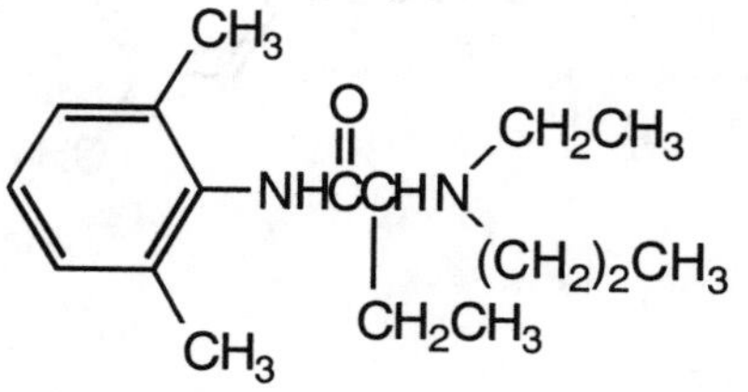

Etidocaine
(Duranest)

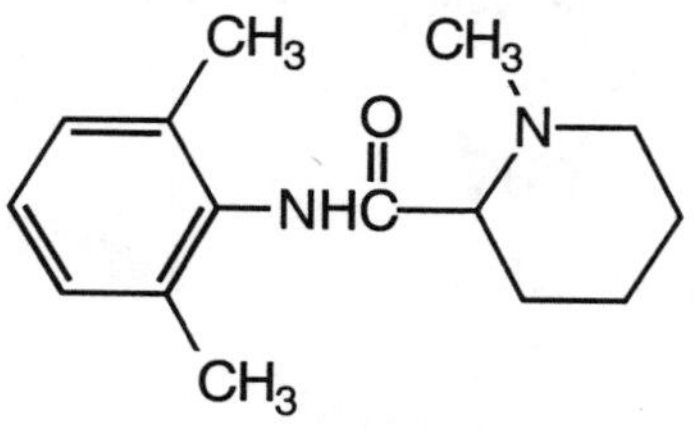

Mepivacaine
(Carbocaine)

(a) Propose a synthesis of lidocaine from 2,6-dimethylaniline, chloroacetyl chloride, and diethylamine.

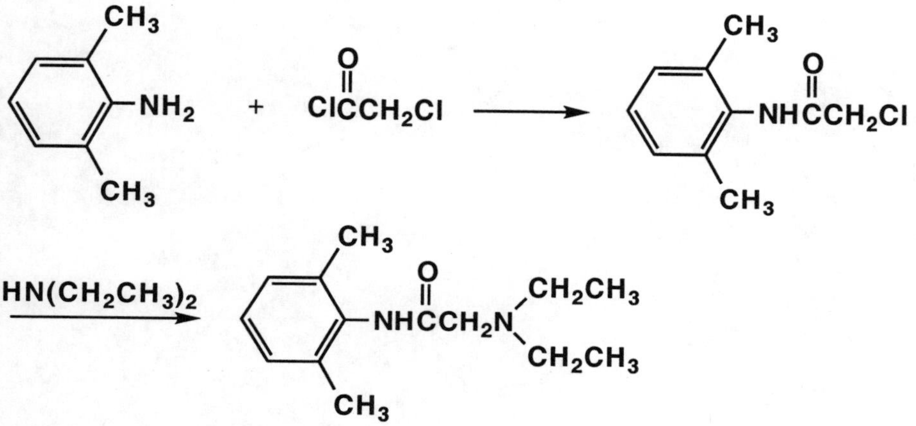

The chloroacetyl chloride reacts with 2,6-dimethylaniline to form the amide, because the acid chloride function is more reactive than the chloromethyl moiety. The amide is reacted with diethyl amine (S$_N$2) to complete the synthesis.

(b) Propose a synthesis of etidocaine form 2,6-dimethylaniline, 2-chlorobutanoyl chloride, and ethylpropylamine.

This synthesis is the same as above, except 2-chlorobutanoyl chloride and N-ethylaminopropane (ethylpropylamine) are used.

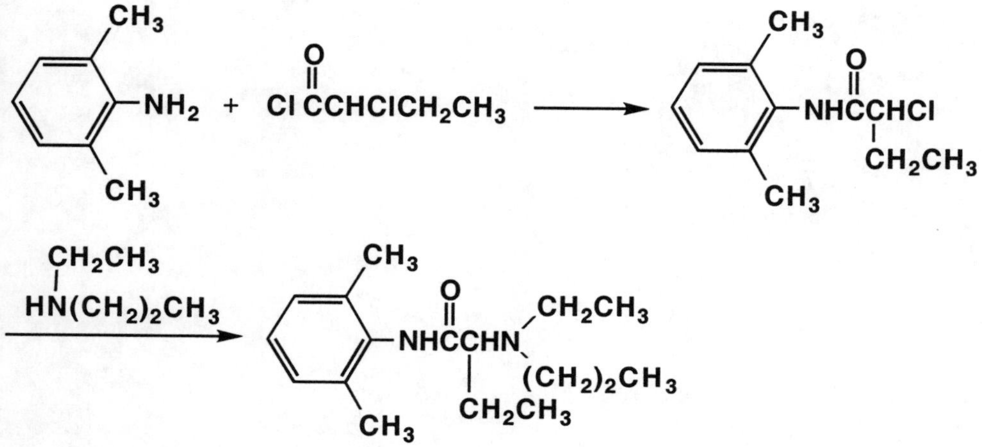

(c) What amine and acid chloride can be reacted to give mepivacaine?

Mepivacaine
(Carbocaine)

CHAPTER 14
Solutions to Problems

Problem 14.1 Identify the acidic hydrogens in each compound.

In the following structures, the acidic hydrogens are on carbon atoms that are adjacent to the carbonyl group. These hydrogens are marked by arrows.

(a) Cyclohexanone

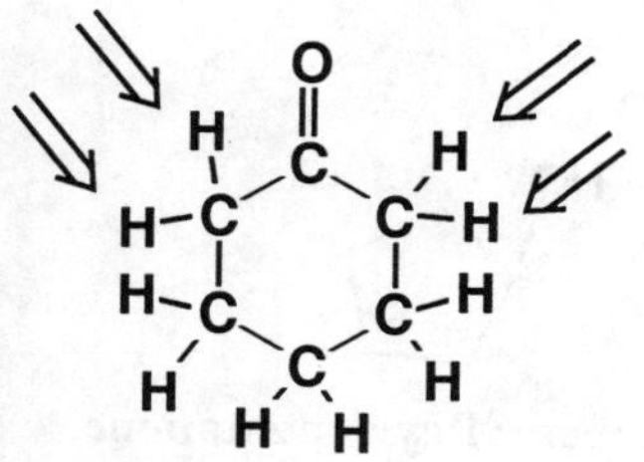

(b) Acetophenone

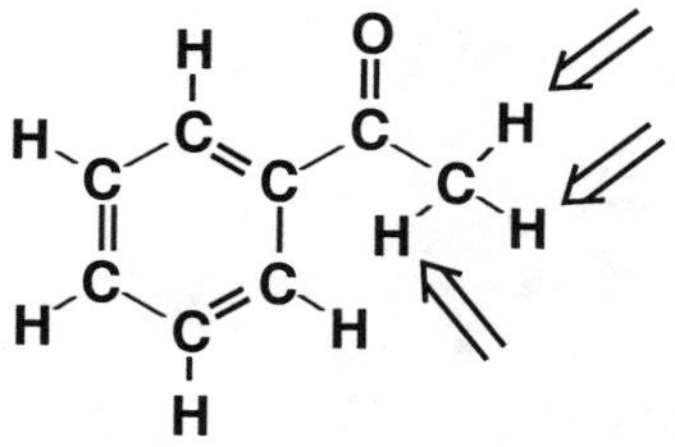

Problem 14.2 Treatment of 2-butanone with base gives two enolate anions. Draw each enolate anion as a hybrid of two contributing structures.

One enolate is derived from each of the two carbon atoms adjacent to the carbonyl group.

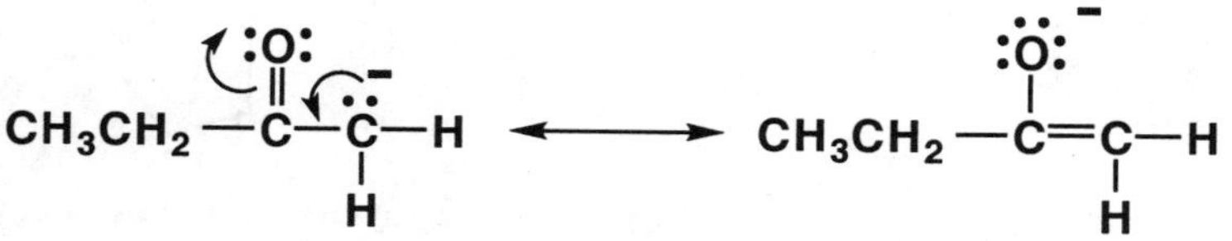

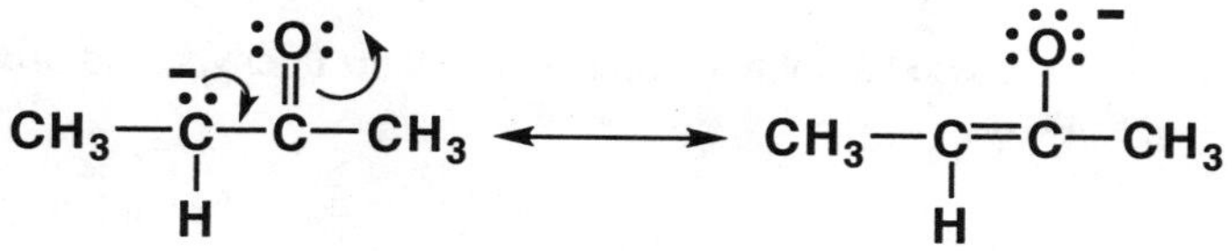

Problem 14.3 Predict the position of equilibrium for this reaction. Refer to Table 2.1 for the pK$_a$ of NH$_3$.

$$CH_3CCH_3 \quad + \quad NaNH_2 \rightleftharpoons CH_3C=CH_2 \quad + \quad NH_3$$

Acetone is a much stronger acid than ammonia. Therefore, the equilibrium lies far to the right.

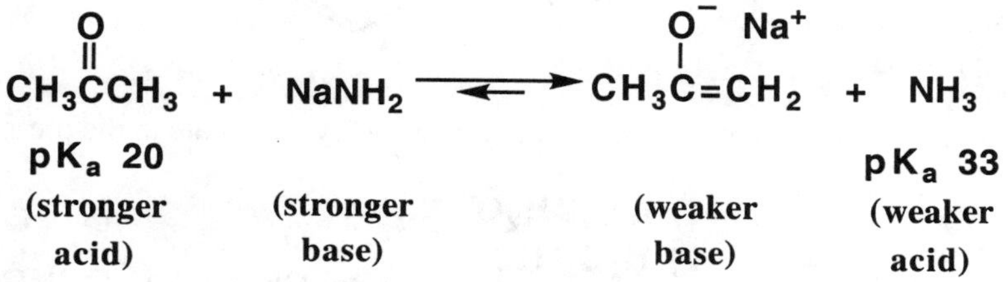

$$CH_3CCH_3 \quad + \quad NaNH_2 \rightleftharpoons CH_3C=CH_2 \quad + \quad NH_3$$

pK$_a$ 20	(stronger base)	(weaker base)	pK$_a$ 33
(stronger acid)			(weaker acid)

Problem 14.4 Draw the enolate anion formed by treatment of methyl phenylacetate with sodium methoxide.

In this case, there is only one enolate possible.

Problem 14.5 Draw the product of the base-catalyzed aldol reaction of each compound.
(a) Acetophenone   (b) Cyclopentanone

3-Hydroxy-1,3-diphenyl-1-butanone

2-(1-Hydroxycyclopentyl)cyclopentanone

Problem 14.6 Draw the product of acid-catalyzed dehydration of each aldol product from Problem 14.5.
(a) Acetophenone   (b) Cyclopentanone

1,3-Diphenyl-2-buten-1-one

2-Cyclopentenylidenecyclopentanone

Problem 14.7 Draw the product of the crossed aldol reaction between benzaldehyde and 3-pentanone and the product formed by its base-catalyzed dehydration.

Benzaldehyde   **3-Pentanone**

Problem 14.8 Show the product of Claisen condensation of ethyl 3-methylbutanoate in the presence of sodium ethoxide.

Ethyl 3-methylbutanoate

Problem 14.9 Complete the equation for this crossed Claisen condensation:

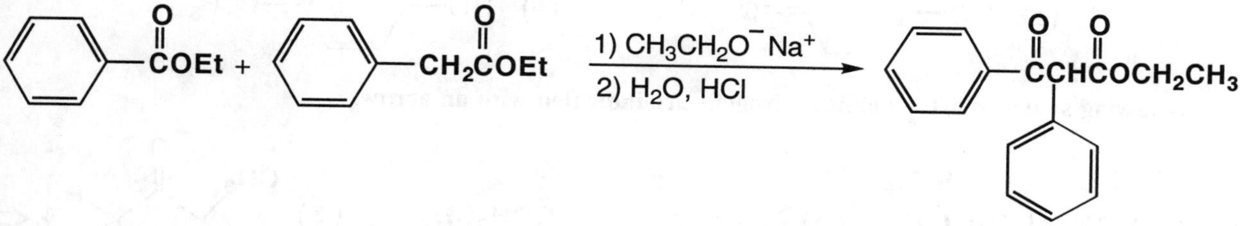

Problem 14.10 Show how to convert ethyl benzoate to 3-methyl-1-phenyl-1-butanone (isobutyl phenyl ketone) using a Claisen condensation at some stage in the synthesis.

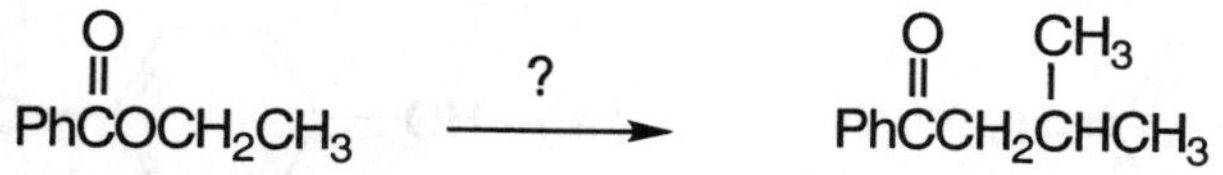

Ethyl benzoate   3-Methyl-1-phenyl-1-butanone

A crossed Claisen condensation of ethyl benzoate and ethyl 3-methylbutanoate gives a β-ketoester. Saponification of the ester followed by acidification gives the β-ketoacid. Heating causes decarboxylation and gives the desired product.

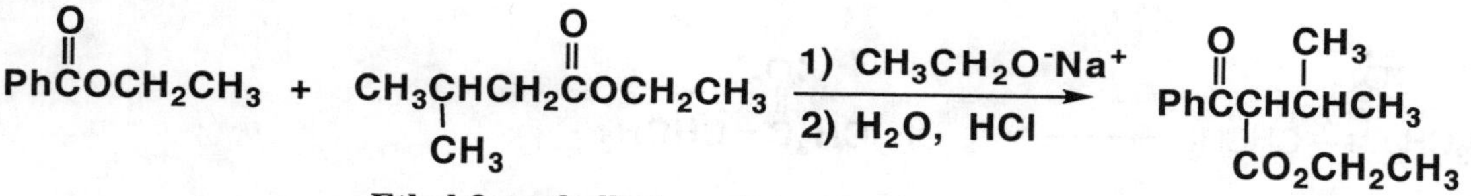

Ethyl 3-methylbutanoate

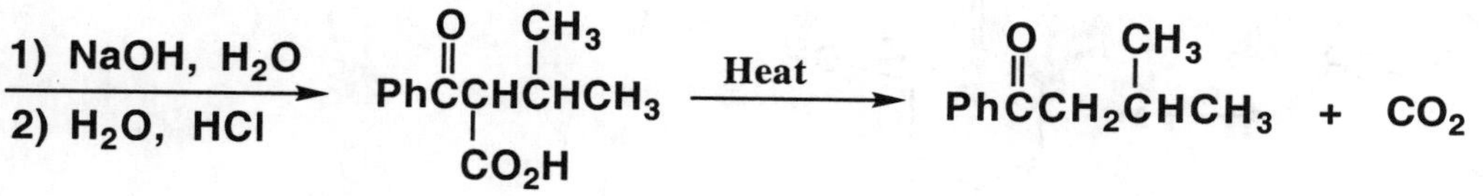

The Aldol Reaction
Problem 14.11 Estimate the pK_a of each compound and then arrange them in order of increasing acidity.

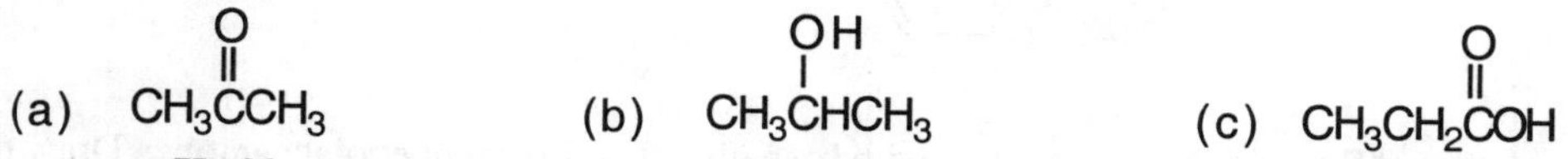

(a) CH_3CCH_3 (b) CH_3CHCH_3 (c) CH_3CH_2COH
     pK_a 20 pK_a 17 pK_a 5

The order of acidity ranked from least to most acidic is:

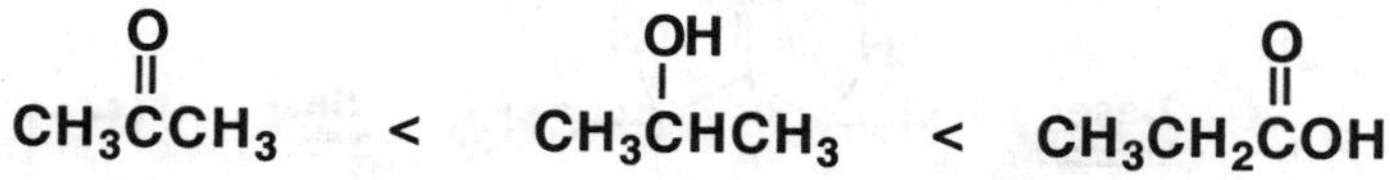

$$CH_3CCH_3 \quad < \quad CH_3CHCH_3 \quad < \quad CH_3CH_2COH$$

Problem 14.12 Identify the most acidic hydrogen(s) in each compound.

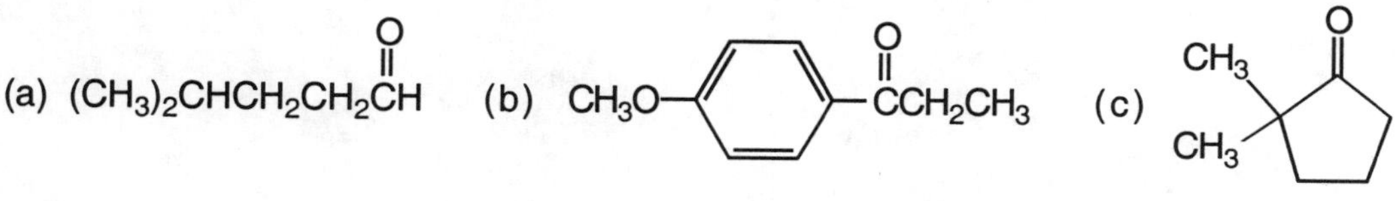

(a) $(CH_3)_2CHCH_2CH_2CH$ (b) $CH_3O-\!\!\!\!\bigcirc\!\!\!\!-CCH_2CH_3$ (c)

(d) 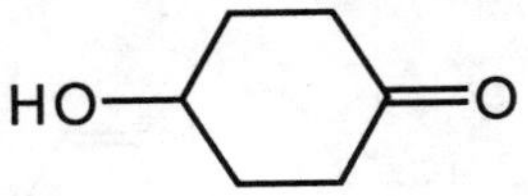 $HO-$⬡$=O$ (e) 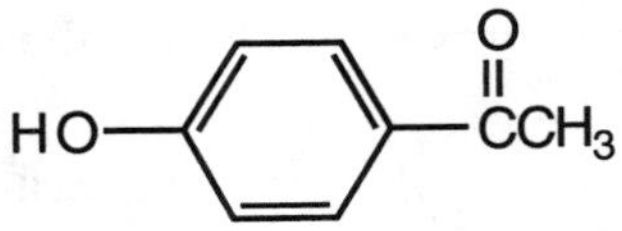 $HO-$⬡$-\overset{\overset{O}{\|}}{C}CH_3$

On the following structures, the acidic hydrogens are indicated with an arrow.

(a) 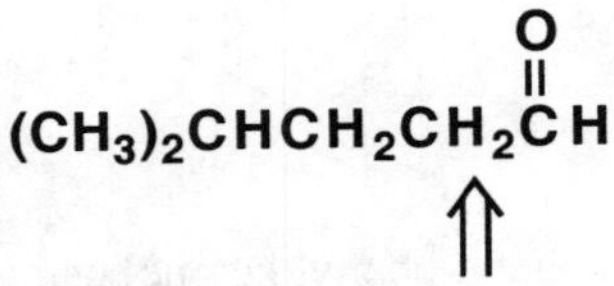 $(CH_3)_2CHCH_2CH_2\overset{\overset{O}{\|}}{C}H$ ⇑

(b) CH_3O-⬡$-\overset{\overset{O}{\|}}{C}CH_2CH_3$ ⇑

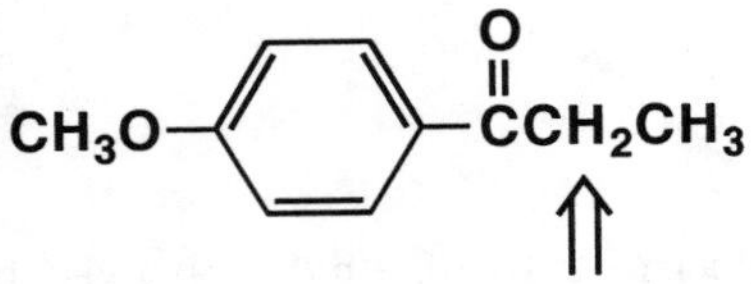

(c) cyclopentanone with two CH_3 groups and two H's   ⇐

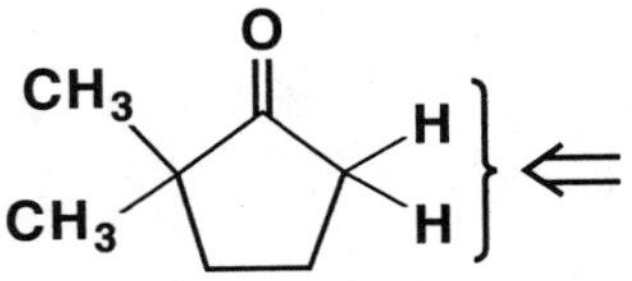

(d) 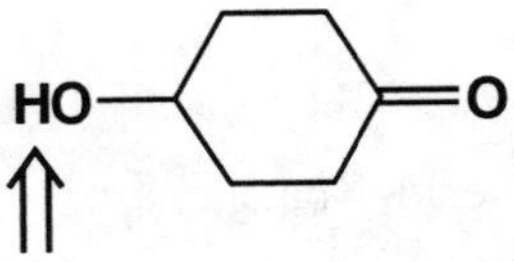 $HO-$⬡$=O$ ⇑

(e) 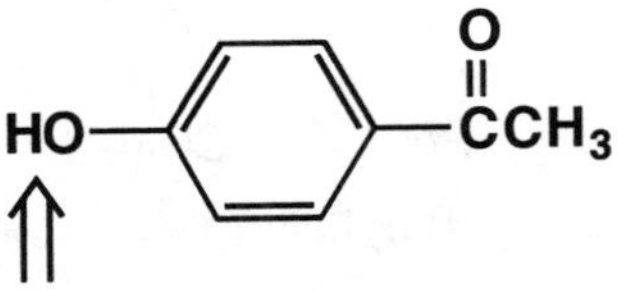 $HO-$⬡$-\overset{\overset{O}{\|}}{C}CH_3$ ⇑

Problem 14.13 Write a second contributing structure of each anion and use curved arrows to show the redistribution of electrons to give your second structure.

(a) 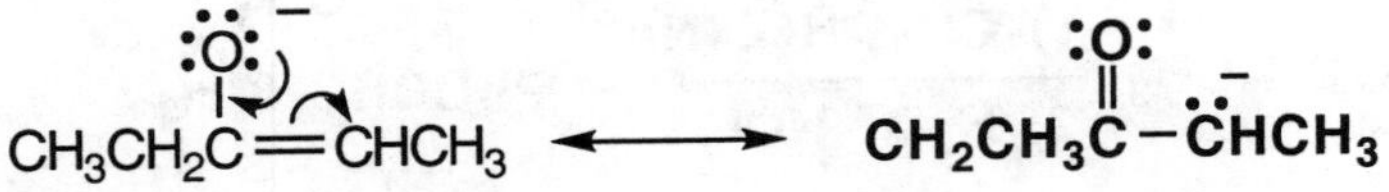 $CH_3CH_2\overset{\overset{:\ddot{O}:^-}{\|}}{C}=CHCH_3$ ⟷ $CH_2CH_3\overset{\overset{:O:}{\|}}{C}-\overset{-}{\ddot{C}}HCH_3$

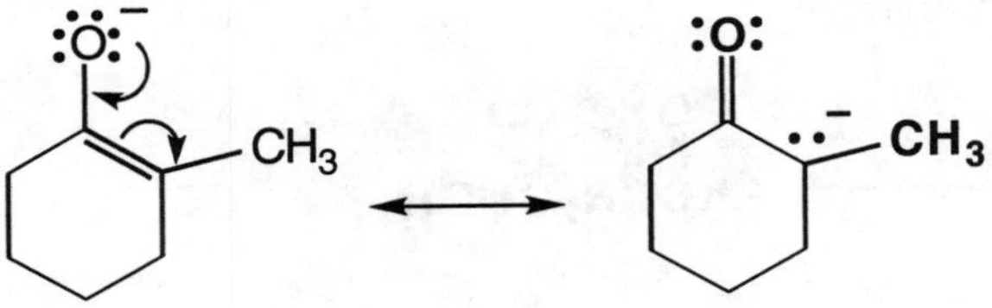

(b) cyclohexene enolate ⟷ cyclohexanone carbanion with CH₃

(c) $C_6H_5-\overset{\overset{:O:}{\|}}{C}-\overset{-}{\ddot{C}}H_2$ ⟷ $C_6H_5-\overset{\overset{:\ddot{O}:^-}{|}}{C}=CH_2$

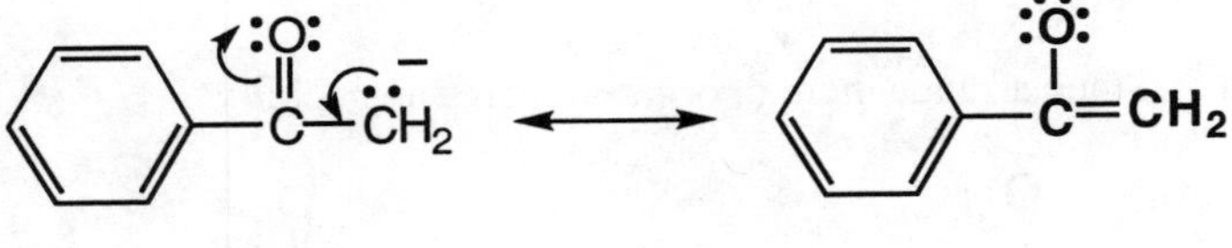

Problem 14.14 Treatment of 2-methylcyclohexanone with base gives two different enolate anions. Draw the contributing structure for each that places the negative charge on carbon.

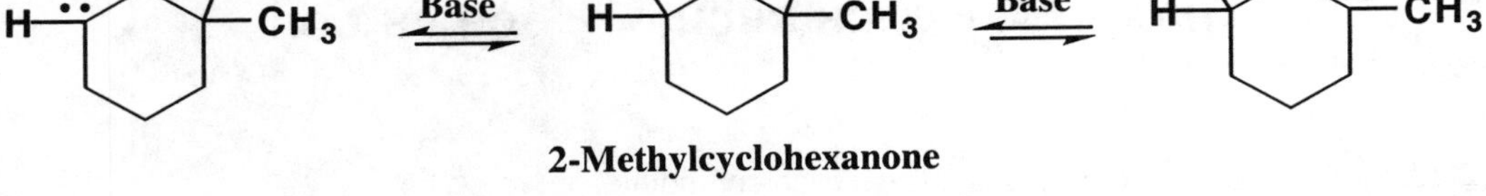

2-Methylcyclohexanone

Problem 14.15 Draw a structural formula for the product of the aldol reaction of each compound and for the α,β-unsaturated aldehyde or ketone formed from dehydration of each aldol product.

(a)

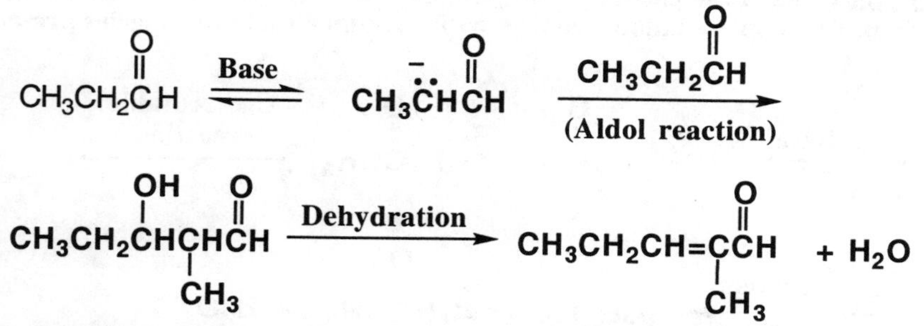

3-Hydroxy-2-methylpentanal   2-Methyl-2-pentenal

(b)

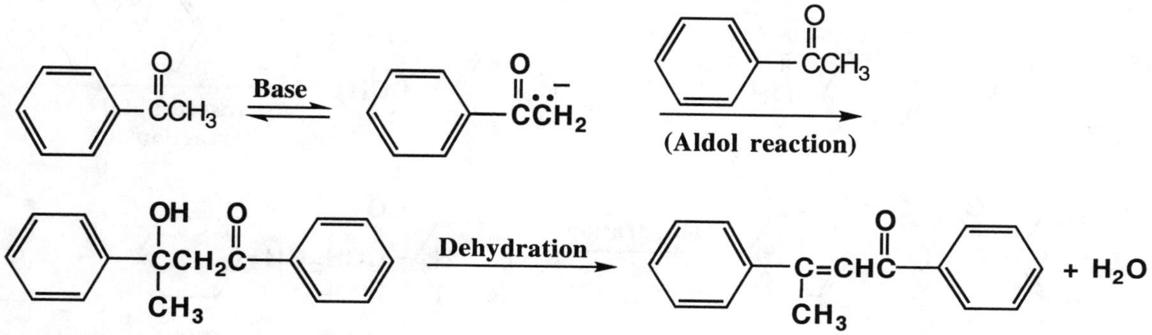

3-Hydroxy-1,3-diphenyl-1-butanone   1,3-Diphenyl-2-buten-1-one

(c)

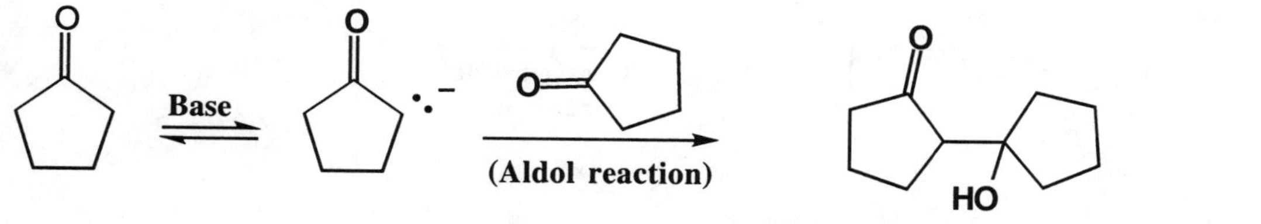

2-(1-Hydroxycyclopentyl)cyclopentanone

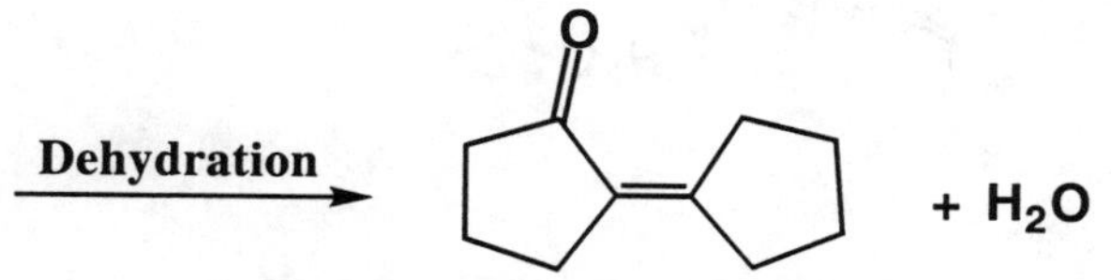

2-Cyclopentylidenecyclopentanone

(d)

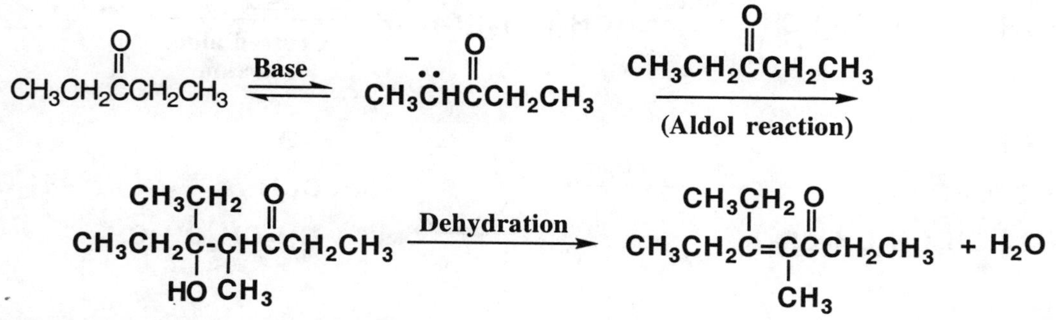

5-Ethyl-5-hydroxy-4-methyl-3-heptanone   5-Ethyl-4-methyl-4-hepten-3-one

<u>Problem 14.16</u> Draw a structural formula for the product of each crossed aldol reaction and for the compound formed by dehydration of each aldol product.

Note that in the following reactions, only one of the carbonyl compounds can form an enolate anion. The aldol product shown is the one derived from that enolate anion reacting with the other carbonyl species present in the reaction.

(a) $(CH_3)_3CCH$ + CH_3CCH_3 $\xrightarrow{\text{Base}}$ $(CH_3)_3CCH$ CH_2CCH_3 $\xrightarrow{\text{Crossed aldol reaction}}$

$(CH_3)_3CCHCH_2CCH_3$ $\xrightarrow{\text{Dehydration}}$ $(CH_3)_3CCH=CHCCH_3$ + H_2O

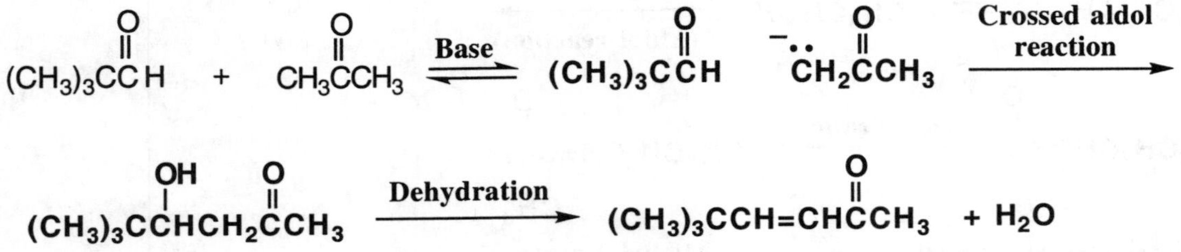

(b) [benzene ring]$-CCH_3$ + [benzene ring]$-CH$ $\xrightarrow{\text{Base}}$ [benzene ring]$-CCH_2^-$ [benzene ring]$-HC=O$ $\xrightarrow{\text{Crossed aldol reaction}}$

[benzene ring]$-CCH_2CH-$[benzene ring] $\xrightarrow{\text{Dehydration}}$ [benzene ring]$-CCH=CH-$[benzene ring] + H_2O

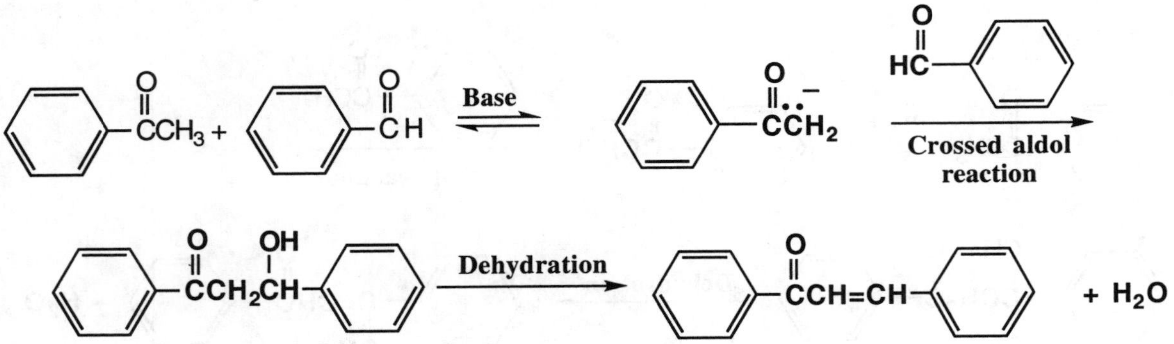

(c) [cyclohexanone] + HCH $\xrightarrow{\text{Base}}$ [cyclohexanone enolate]$^-$ HCH $\xrightarrow{\text{Crossed aldol reaction}}$

[cyclohexanone with CH_2OH] $\xrightarrow{\text{Dehydration}}$ [cyclohexanone with $=CH_2$] + H_2O

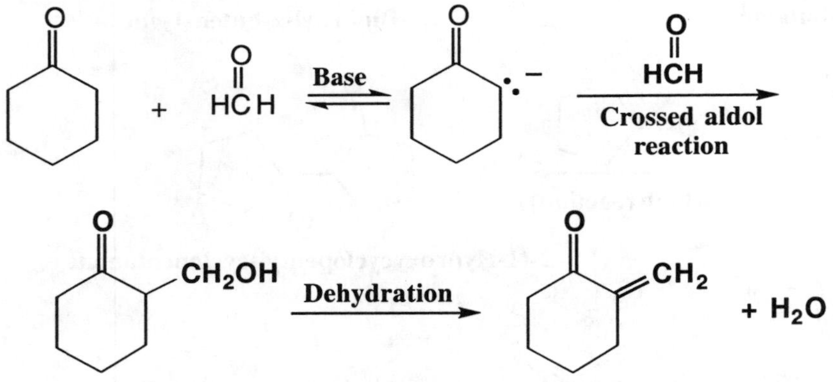

(d) [benzene ring]$-CH$ + $CH_3(CH_2)_4CH$ $\xrightarrow{\text{Base}}$ $CH_3(CH_2)_3CHCH$ [benzene ring]$-CH=O$ $\xrightarrow{\text{Crossed aldol reaction}}$

[benzene ring]$-CHCHCH$ with $CH_2CH_2CH_2CH_3$ $\xrightarrow{\text{Dehydration}}$ [benzene ring]$-CH=CCH$ with $CH_2CH_2CH_2CH_3$ + H_2O

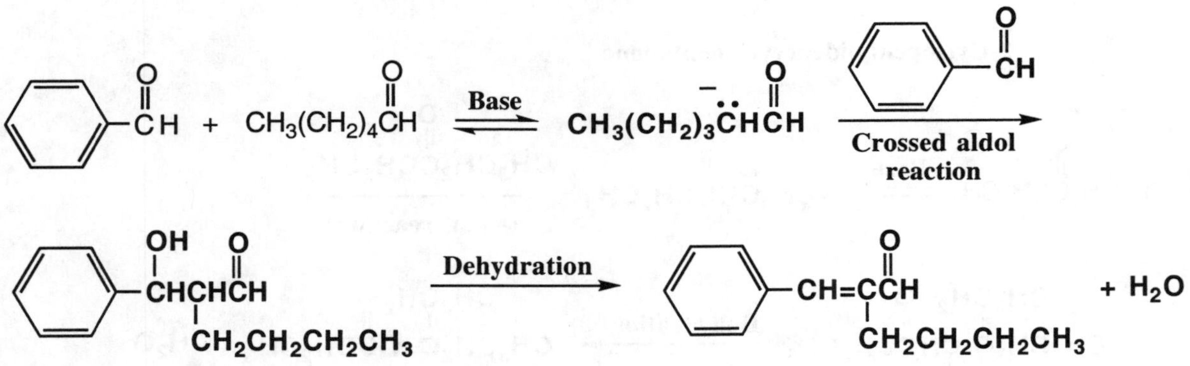

Problem 14.17 When a 1:1 mixture of acetone and 2-butanone is treated with base, six aldol products are possible. Draw a structural formula for each aldol product.

$$CH_3CH_2\overset{\overset{\displaystyle O}{\|}}{C}CH_3 \ + \ CH_3\overset{\overset{\displaystyle O}{\|}}{C}CH_3 \ \xrightarrow{\ \text{NaOH}\ } \ \begin{array}{c}\text{A mixture of six}\\ \text{aldol products}\end{array}$$

Three different enolate anions can be formed and each of these can react with either of the two ketones as shown.

$$CH_3CH_2\overset{\overset{\displaystyle O}{\|}}{C}CH_3 \ \xrightarrow{\ \text{NaOH}\ } \ \left[CH_3CH_2\overset{\overset{\displaystyle O}{\|}}{C}\overset{..}{C}H_2\right]^{-} Na^+ \ \xrightarrow[]{CH_3\overset{\overset{\displaystyle O}{\|}}{C}CH_3}$$

$$CH_3CH_2\overset{\overset{\displaystyle O}{\|}}{C}CH_2\underset{\underset{\displaystyle CH_3}{|}}{\overset{\overset{\displaystyle OH}{|}}{C}}CH_3$$

$$\xrightarrow[]{CH_3CH_2\overset{\overset{\displaystyle O}{\|}}{C}CH_3} \quad CH_3CH_2\overset{\overset{\displaystyle O}{\|}}{C}CH_2\underset{\underset{\displaystyle CH_2CH_3}{|}}{\overset{\overset{\displaystyle OH}{|}}{C}}CH_3$$

$$CH_3CH_2\overset{\overset{\displaystyle O}{\|}}{C}CH_3 \ \xrightarrow{\ \text{NaOH}\ } \ \left[CH_3\overset{..}{C}HC\overset{\overset{\displaystyle O}{\|}}{}CH_3\right]^{-} Na^+ \ \xrightarrow[]{CH_3\overset{\overset{\displaystyle O}{\|}}{C}CH_3}$$

$$CH_3\underset{\underset{\displaystyle H_3C}{|}}{\overset{\overset{\displaystyle OH}{|}}{C}}-CH\overset{\overset{\displaystyle O}{\|}}{C}CH_3 \quad \underset{CH_3}{}$$

$$\xrightarrow[]{CH_3CH_2\overset{\overset{\displaystyle O}{\|}}{C}CH_3} \quad CH_3CH_2\underset{\underset{\displaystyle H_3C}{|}}{\overset{\overset{\displaystyle OH}{|}}{C}}-CH\overset{\overset{\displaystyle O}{\|}}{C}CH_3 \quad \underset{CH_3}{}$$

$$CH_3\overset{\overset{\displaystyle O}{\|}}{C}CH_3 \ \xrightarrow{\ \text{NaOH}\ } \ \left[CH_3\overset{\overset{\displaystyle O}{\|}}{C}\overset{..}{C}H_2\right]^{-} Na^+ \ \xrightarrow[]{CH_3\overset{\overset{\displaystyle O}{\|}}{C}CH_3}$$

$$CH_3\overset{\overset{\displaystyle O}{\|}}{C}CH_2\underset{\underset{\displaystyle CH_3}{|}}{\overset{\overset{\displaystyle OH}{|}}{C}}CH_3$$

$$\xrightarrow[]{CH_3CH_2\overset{\overset{\displaystyle O}{\|}}{C}CH_3} \quad CH_3\overset{\overset{\displaystyle O}{\|}}{C}CH_2\underset{\underset{\displaystyle CH_3}{|}}{\overset{\overset{\displaystyle OH}{|}}{C}}CH_2CH_3$$

Problem 14.18 Show how to prepare each α,β-unsaturated ketone by an aldol reaction followed by dehydration of the aldol product.

(a)

Only one product is formed in this crossed aldol reaction because benzaldehyde cannot form an enolate ion.

(b) $CH_3\underset{\underset{CH_3}{|}}{C}{=}CHCCH_3$ ($\overset{O}{\overset{||}{}}$)

This compound can be produced from an aldol reaction utilizing only acetone followed by dehydration.

Problem 14.19 Show how to prepare each α,β-unsaturated aldehyde by an aldol reaction followed by dehydration of the aldol product.

(a)

Benzaldehyde   Acetaldehyde

Only one product is formed in this crossed aldol reaction because only acetaldehyde can form an enolate ion.

(b) $C_7H_{15}CH{=}\underset{\underset{C_6H_{13}}{|}}{C}CH$ ($\overset{O}{\overset{||}{}}$)

This compound can be produced from an aldol reaction of octanal followed by dehydration.

<u>Problem 14.20</u> When treated with base, the following compound undergoes an intramolecular aldol reaction to give a product containing a ring (yield 78%). Propose a structural formula for this product.

$$CH_3CH_2CH{=}CHCH_2CH_2\overset{\overset{\displaystyle O}{\|}}{C}CH_2CH_2\overset{\overset{\displaystyle O}{\|}}{C}H \xrightarrow[\text{reaction}]{\substack{\text{Base} \\ \text{Aldol}}} C_{10}H_{14}O + H_2O$$

Analyze this problem in the following way. There are three α-carbons which might form an anion and then condense with one of the carbonyl groups. Two of these condensations lead to three-membered rings and, therefore, are not feasible. The third anion leads to formation of the five-membered ring product.

$$CH_3CH_2CH{=}CHCH_2CH_2\overset{\overset{\displaystyle O}{\|}}{C}CH_2CH_2\overset{\overset{\displaystyle O}{\|}}{C}H \longrightarrow CH_3CH_2CH{=}CHCH_2{-}$$

Enolate forms   Enolate carbon attacks
at this carbon   this carbonyl carbon

<u>Problem 14.21</u> Propose a structural formula for the compound of molecular formula $C_6H_{10}O_2$ that undergoes aldol reaction followed by dehydration to give this α,β-unsaturated aldehyde.

$$C_6H_{10}O_2 \xrightarrow{\text{base}} \text{CHO} + H_2O$$

1-Cyclopentenecarbaldehyde

The compound in question is hexanedial.

$$\text{CHO, CHO} \xrightarrow{\text{base}} \text{CHO} + H_2O$$

Hexanedial   1-Cyclopentenecarbaldehyde

<u>Problem 14.22</u> Show how to bring about this conversion.

This is actually nothing more than an intramolecular aldol reaction followed by dehydration. Using the numbering scheme shown on the diketone, the aldol reaction involves an enolate at carbon 2 reacting with the carbonyl carbon atom 6.

$$\xrightarrow{\substack{\text{Aldol} \\ \text{reaction}}} \xrightarrow{\text{Dehydration}}$$

Problem 14.23 Oxanamide, a mild sedative, is synthesized from butanal in these five steps.

Butanal   2-Ethyl-2-hexenal   2-Ethyl-2-hexenoic acid

2-Ethyl-2-hexenoyl chloride   2-Ethyl-2-hexenamide   2-Ethyl-2,3-epoxyhexanamide
                                                                        (Oxanamide)

(a) Show reagents and experimental conditions to bring about each step in the synthesis.

**Step 1: Base-catalyzed aldol condensation followed by dehydration to give an
α, β-unsaturated ketone.**

Butanal   **2-Ethyl-3-hydroxyhexanal**   **2-Ethyl-2-hexenal**

**Step 2: Oxidation of the aldehyde to a carboxylic acid can be accomplished using Tollens' reagent. In the
industrial process, the oxidizing agent is oxygen, O_2.**

2-Ethyl-2-hexenal   **2-Ethyl-2-hexanoic acid**

Step 3: Reaction of the carboxylic acid with thionyl chloride gives the acid chloride.

2-Ethyl-2-hexenoic acid   **2-Ethyl-2-hexenoyl chloride**

Step 4: Reaction with ammonia makes the amide.

2-Ethyl-2-hexenoyl chloride   **2-Ethyl-2-hexenamide**

Step 5: Oxidation of the alkene to an epoxide can be brought about using a peroxy acid, RCO$_3$H.

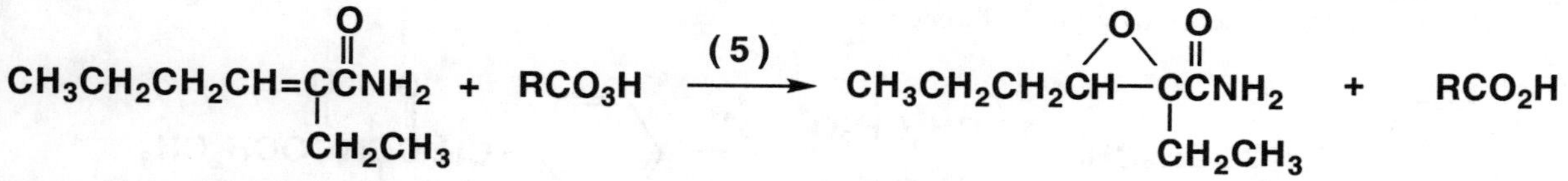

2-Ethyl-2-hexenamide

2-Ethyl-2,3-epoxyhexanamide
(Oxanamide)

(b) How many stereocenters are in oxanamide? How many stereoisomers are possible for this compound?

There are two stereocenters in oxanamide and these are marked with an asterisk in the structure. Four stereoisomers are possible.

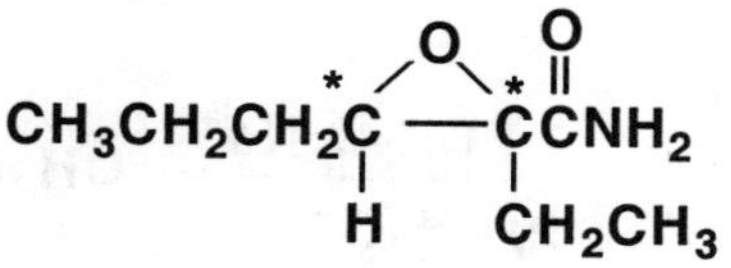

<u>Problem 14.24</u> This reaction is one of the 10 steps in glycolysis (Section 20.6), a series of enzyme-catalyzed reactions by which glucose is oxidized to two molecules of pyruvate. Show that this step is the reverse of an aldol reaction.

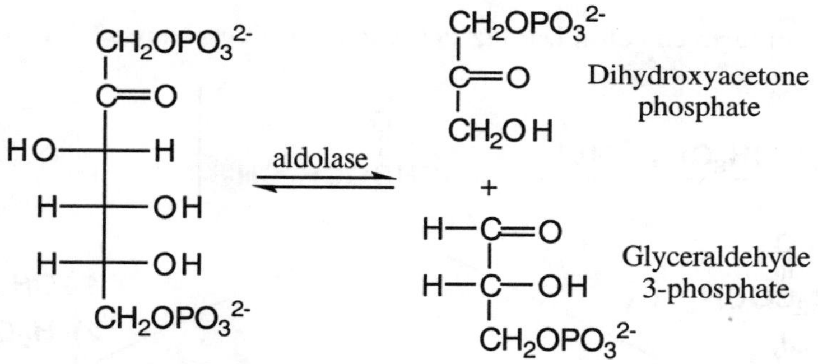

Fructose 1,6-bisphosphate

As shown in the following scheme, this reaction is functionally the reverse of an aldol reaction between the enolate of dihydroxyacetone phosphate and glyceraldehyde 3-phosphate.

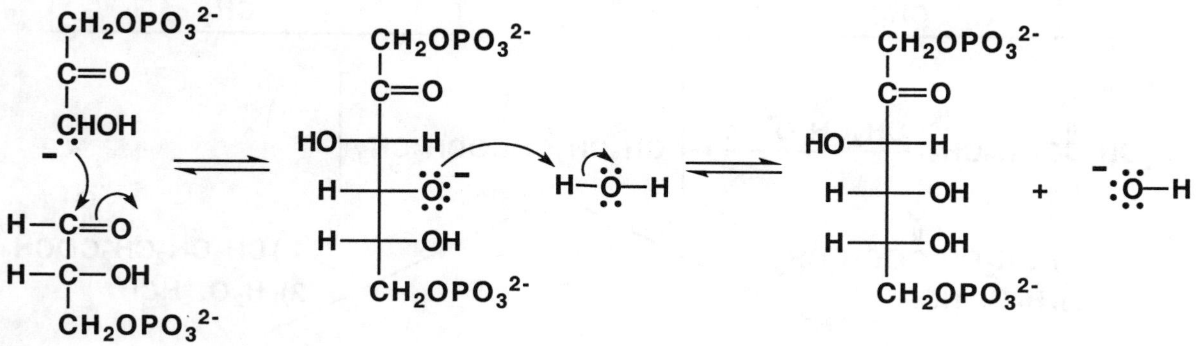

The Claisen and Dieckmann Condensations

Problem 14.25 Show the product of Claisen condensation of each ester.
(a) Ethyl phenylacetate in the presence of sodium ethoxide.

$$2 \ C_6H_5-CH_2COCH_2CH_3 \xrightarrow[\ 2)\ H_3O^+\ ]{1)\ CH_3CH_2O^-Na^+} C_6H_5-CH_2CCHCOCH_2CH_3 \ (\text{with } C_6H_5 \text{ substituent})$$

(b) Methyl hexanoate in the presence of sodium methoxide.

$$2 \ CH_3(CH_2)_4COCH_3 \xrightarrow[\ 2)\ H_3O^+\ ]{1)\ CH_3O^-Na^+} CH_3(CH_2)_4CCHCOCH_2CH_3 \quad \underset{(CH_2)_3CH_3}{|}$$

Problem 14.26 When a 1:1 mixture of ethyl propanoate and ethyl butanoate is treated with sodium ethoxide, four Claisen condensation products are possible. Draw a structural formula for each product.

In this case, both esters can form enolates, leading to the four products shown.

$$CH_3CH_2COCH_2CH_3 \xrightarrow{CH_3CH_2O^-Na^+} \left[CH_3\overset{..}{C}HCOCH_2CH_3 \right]$$

1) $CH_3CH_2COCH_2CH_3$
2) H_2O, HCl

1) $CH_3CH_2CH_2COCH_2CH_3$
2) H_2O, HCl

$$\boxed{CH_3CH_2CCHCOCH_2CH_3 \underset{CH_3}{|}}$$

$$\boxed{CH_3CH_2CH_2CCHCOCH_2CH_3 \underset{CH_3}{|}}$$

$$CH_3CH_2CH_2COCH_2CH_3 \xrightarrow{CH_3CH_2O^-Na^+} \left[CH_3CH_2\overset{..}{C}HCOCH_2CH_3 \right]$$

1) $CH_3CH_2COCH_2CH_3$
2) H_2O, HCl

1) $CH_3CH_2CH_2COCH_2CH_3$
2) H_2O, HCl

$$\boxed{CH_3CH_2CCHCOCH_2CH_3 \underset{CH_2CH_3}{|}}$$

$$\boxed{CH_3CH_2CH_2CCHCOCH_2CH_3 \underset{CH_2CH_3}{|}}$$

Problem 14.27 Draw a structural formula for the β-ketoester formed in the crossed Claisen condensation of ethyl propanoate with each ester:

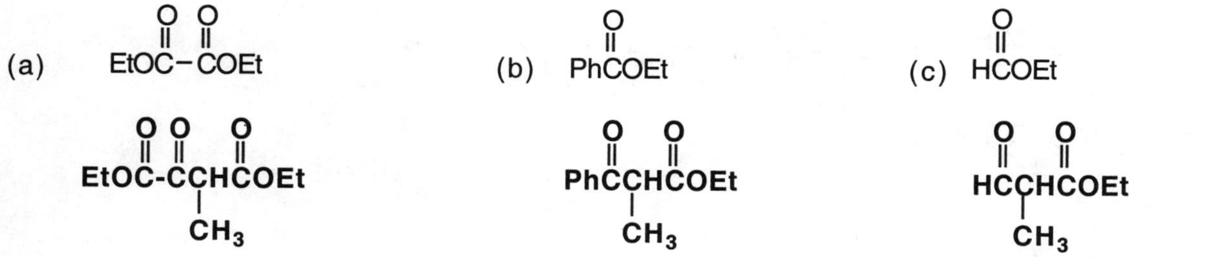

(a) $\overset{O}{\underset{\|}{EtOC}} - \overset{O}{\underset{\|}{COEt}}$ (b) $\overset{O}{\underset{\|}{PhCOEt}}$ (c) $\overset{O}{\underset{\|}{HCOEt}}$

$\overset{O}{\underset{\|}{EtOC}} - \overset{O}{\underset{\|}{C}} \overset{}{\underset{\underset{CH_3}{|}}{CH}} \overset{O}{\underset{\|}{COEt}}$ $\overset{O}{\underset{\|}{PhC}} \overset{}{\underset{\underset{CH_3}{|}}{CH}} \overset{O}{\underset{\|}{COEt}}$ $\overset{O}{\underset{\|}{HC}} \overset{}{\underset{\underset{CH_3}{|}}{CH}} \overset{O}{\underset{\|}{COEt}}$

Problem 14.28 Draw a structural formula for the product of saponification, acidification, and decarboxylation of each β-ketoester formed in Problem 14.27.

Following are the structures for the products of saponification and decarboxylation of each β-ketoester from the previous problem.

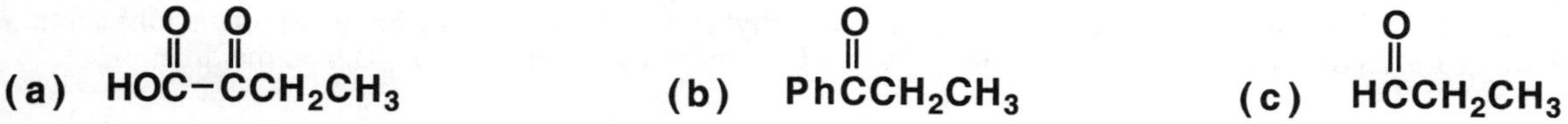

(a) $\overset{O}{\underset{\|}{HOC}} - \overset{O}{\underset{\|}{CCH_2CH_3}}$ (b) $\overset{O}{\underset{\|}{PhCCH_2CH_3}}$ (c) $\overset{O}{\underset{\|}{HCCH_2CH_3}}$

Problem 14.29 Complete the equation for this crossed Claisen condensation.

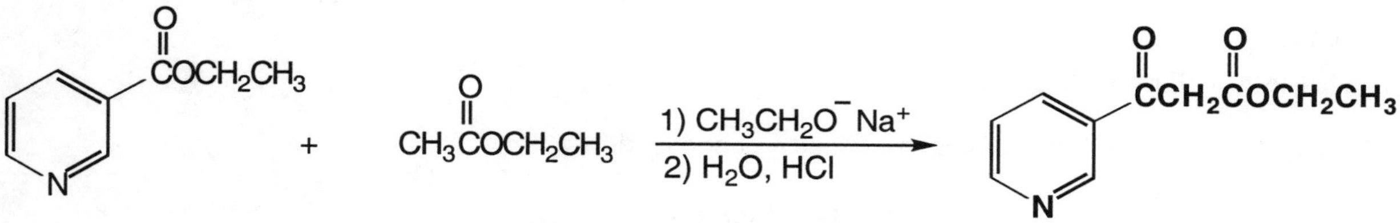

Problem 14.30 The Claisen condensation can be used as one step in the synthesis of ketones, as illustrated by this reaction sequence. Propose structural formulas for compounds A, B, and the ketone formed in this sequence.

$$2\ CH_3CH_2CH_2CH_2\overset{O}{\underset{\|}{C}}OEt \xrightarrow[\text{2) HCl, } H_2O]{\text{1) EtO}^-\text{Na}^+} A \xrightarrow[\text{heat}]{\text{NaOH, } H_2O} B \xrightarrow[\text{heat}]{\text{HCl, } H_2O} C_9H_{18}O$$

Compound (A) is a β-ketoester, compound (B) is the sodium salt of a β-ketoacid, and the final ketone is 5-nonanone.

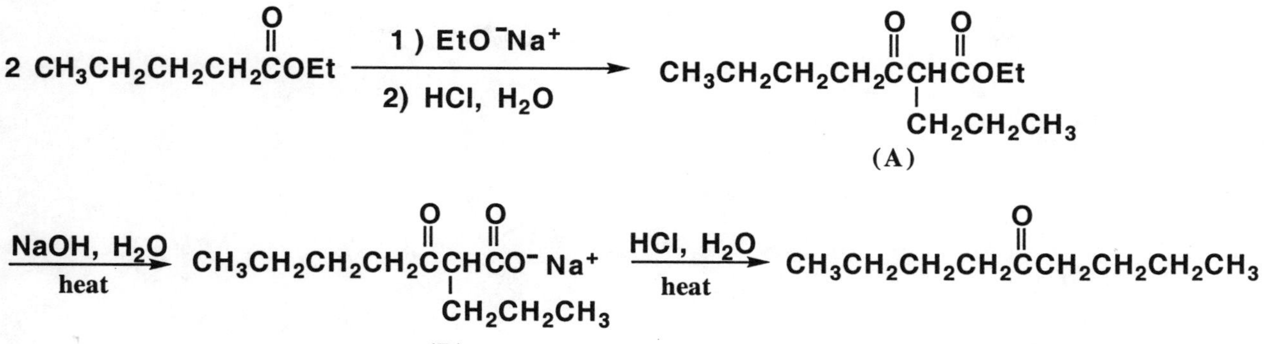

$$2\ CH_3CH_2CH_2CH_2\overset{O}{\underset{\|}{C}}OEt \xrightarrow[\text{2) HCl, } H_2O]{\text{1) EtO}^-\text{Na}^+} CH_3CH_2CH_2CH_2\overset{O}{\underset{\|}{C}}\overset{}{\underset{\underset{CH_2CH_2CH_3}{|}}{CH}}\overset{O}{\underset{\|}{C}}OEt$$

(A)

$$\xrightarrow[\text{heat}]{\text{NaOH, } H_2O} CH_3CH_2CH_2CH_2\overset{O}{\underset{\|}{C}}\overset{}{\underset{\underset{CH_2CH_2CH_3}{|}}{CH}}\overset{O}{\underset{\|}{C}}O^-\,Na^+ \xrightarrow[\text{heat}]{\text{HCl, } H_2O} CH_3CH_2CH_2CH_2\overset{O}{\underset{\|}{C}}CH_2CH_2CH_2CH_3$$

(B)   **5-Nonanone**

Problem 14.31 Draw a structural formula for the ketone formed by treating each diester with sodium ethoxide followed by acidification with HCl. (Hint: These are Dieckmann condensations.)

(a)

(b) $CH_3CH_2O\overset{O}{\overset{\|}{C}}(CH_2)_5\overset{O}{\overset{\|}{C}}OCH_2CH_3$

Problem 14.32 Claisen condensation between diethyl phthalate and ethyl acetate followed by saponification, acidification, and decarboxylation forms a diketone, $C_9H_6O_2$. Propose structural formulas for compounds A, B, and the diketone.

Diethyl phthalate   Ethyl acetate

Compound (A) is formed by two consecutive Claisen condensations.

<u>Problem 14.33</u> The rodenticide pindone is synthesized by the following sequence of reactions. Propose a structural formula for pindone.

Pindone

<u>Problem 14.34</u> This reaction is the fourth in the set of four enzyme-catalyzed steps by which the hydrocarbon chain of a fatty acid (Section 20.3) is oxidized, two carbons at a time, to acetyl-coenzyme A. Show that this reaction is the reverse of a Claisen condensation.

The enzyme-catalyzed process is functionally the reverse of a crossed Claisen condensation that takes place between the enolate anion of acetyl-CoA and an acyl-CoA.

CHAPTER 15
Solutions to Problems

<u>Problem 15.1</u> Given the following structure, determine the polymer's repeat unit, redraw the structure using the simplified parenthetical notation, and name the polymer.

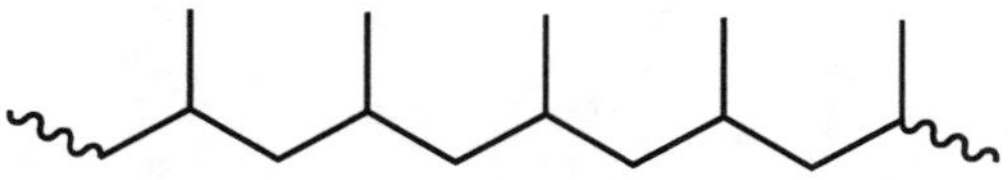

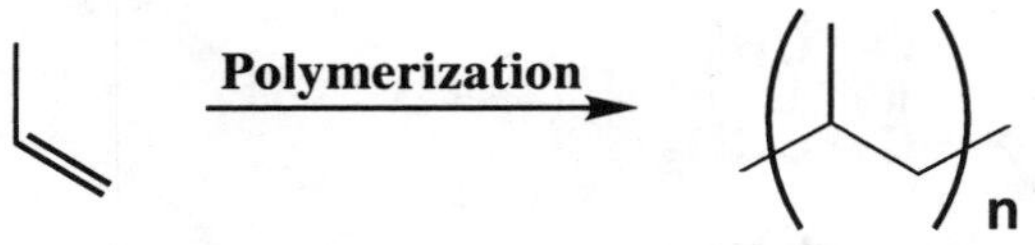

Polymerization

Monomer   **Repeat unit**

This polymer is derived from propylene and, therefore, is called polypropylene.

<u>Problem 15.2</u> Write the repeating unit of the epoxy resin formed from the following reaction.

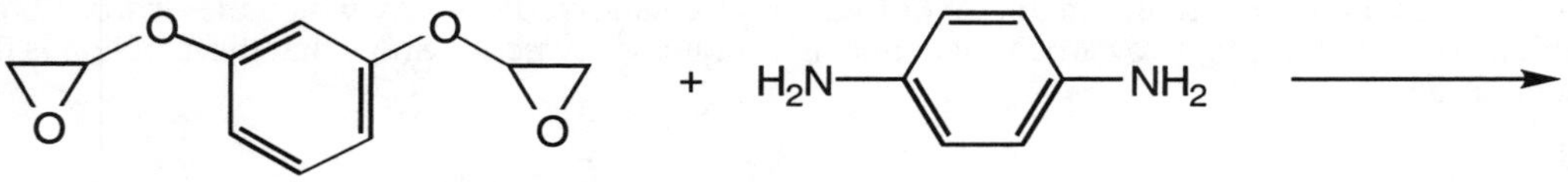

A diepoxide   A diamine

Polymerization occurs by attack of the amine on the less substituted carbon of the epoxide to give the following repeat unit:

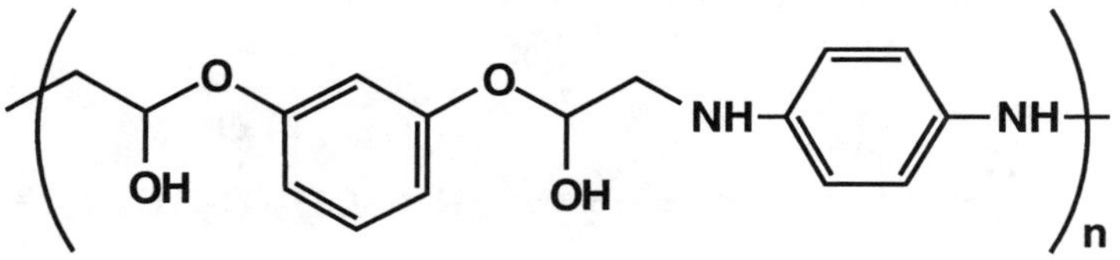

Step-Growth Polymers

<u>Problem 15.3</u> Identify the monomers required for the synthesis of each step-growth polymer.

(a)

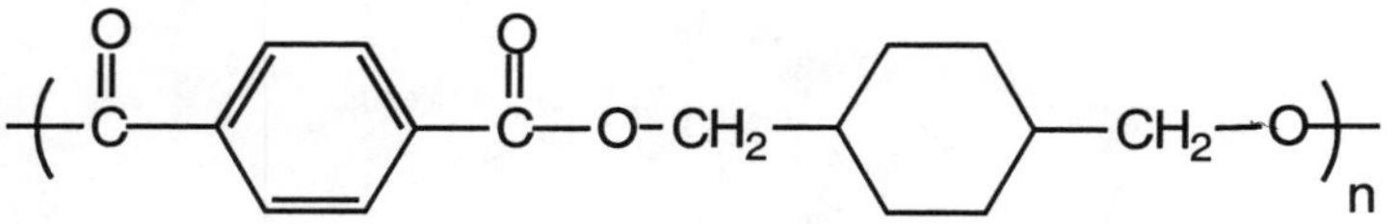

Kodel (a polyester)

The following two monomers can be used to produce Kodel:

(b)

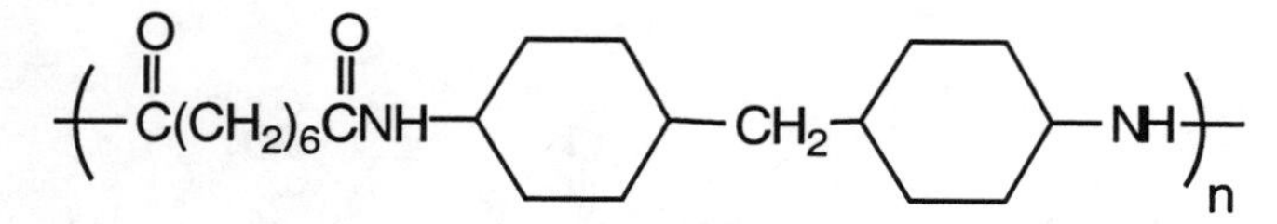

Quiana (a polyamide)

The following two monomers can be used to produce Quiana:

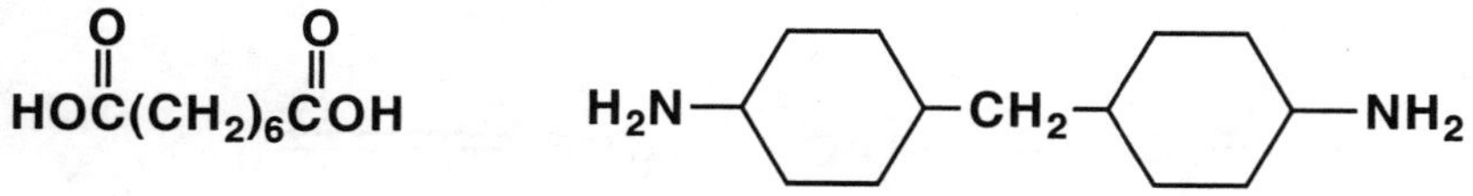

$HO\overset{O}{\overset{\|}{C}}(CH_2)_6\overset{O}{\overset{\|}{C}}OH$

(c)

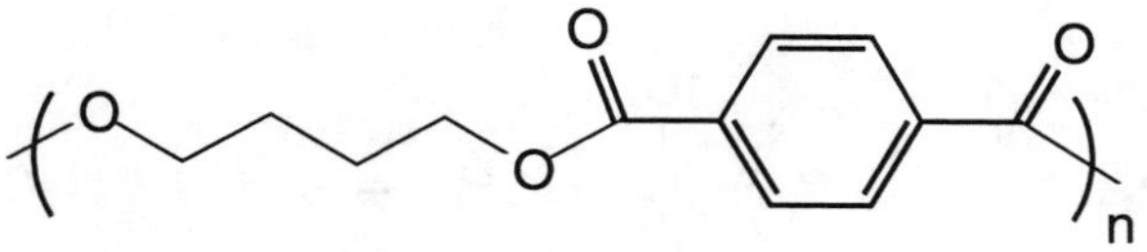

(a polyester)

The following two monomers can be used to produce this polyester:

(d)

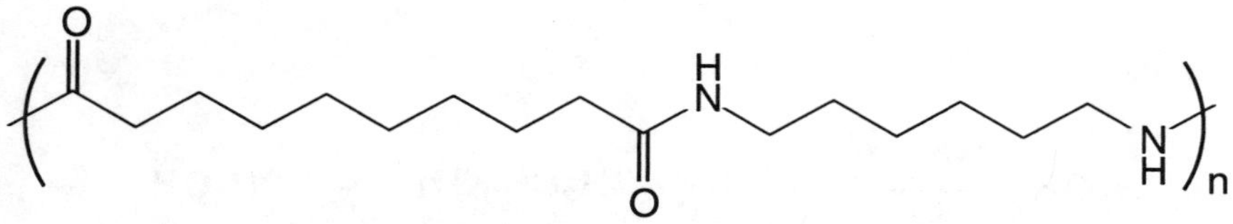

Nylon 6,10 (a polyamide)

The following two monomers can be used to produce nylon 6,10:

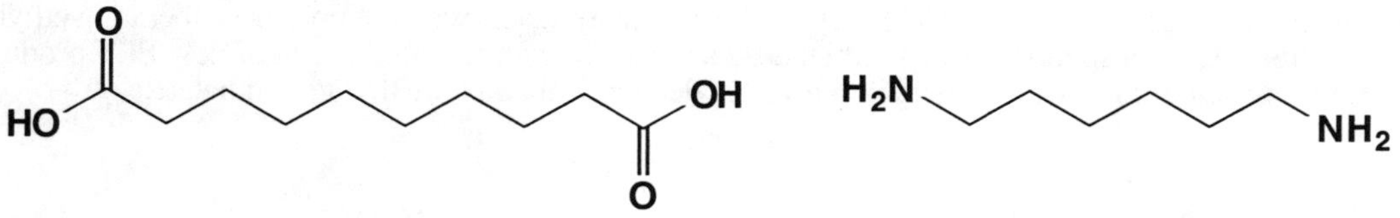

<u>Problem 15.4</u> Poly(ethylene terephthalate) (PET) can be prepared by this reaction. Propose a mechanism for the step-growth reaction in this polymerization.

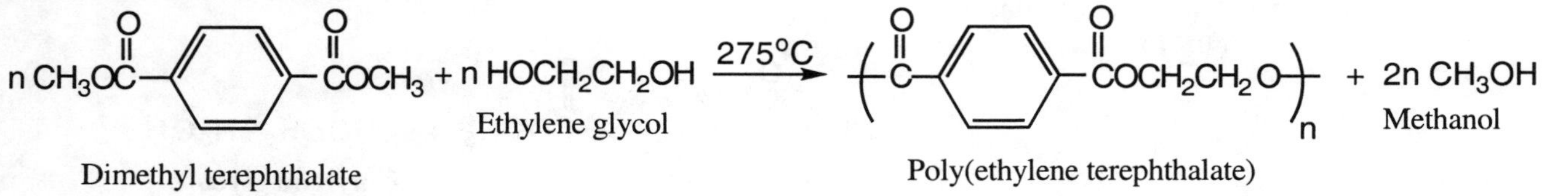

Dimethyl terephthalate   Poly(ethylene terephthalate)

The polymer grows by a series of transesterification reactions. Because of the high temperatures employed, catalysis by acid or base is unnecessary. Following is a mechanism for the first step of the polymerization.

Step 1:

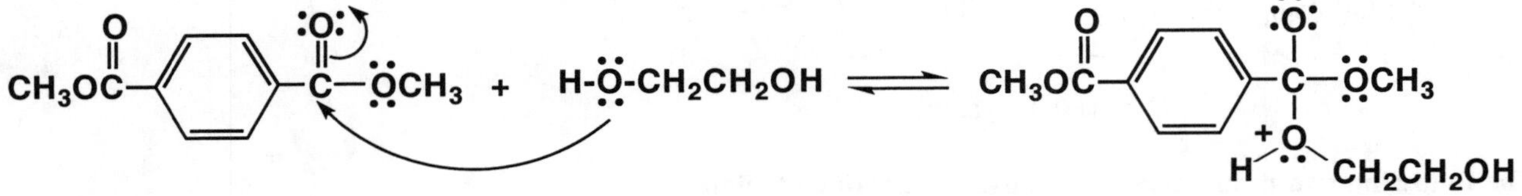

Step 2:

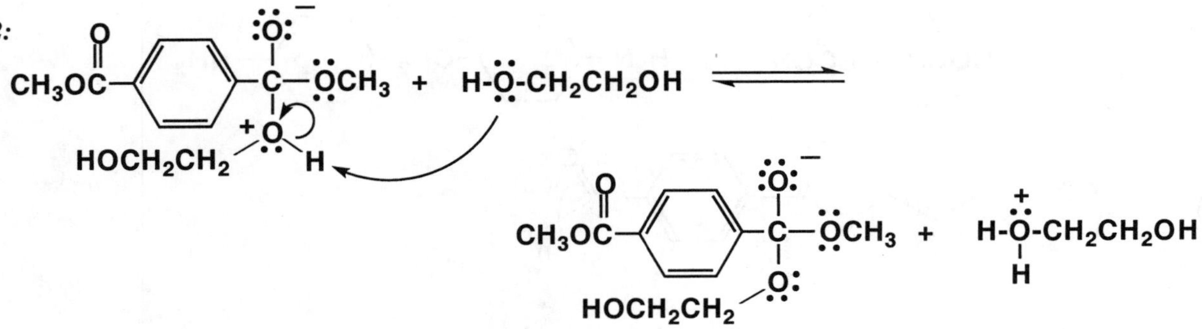

Step 3:

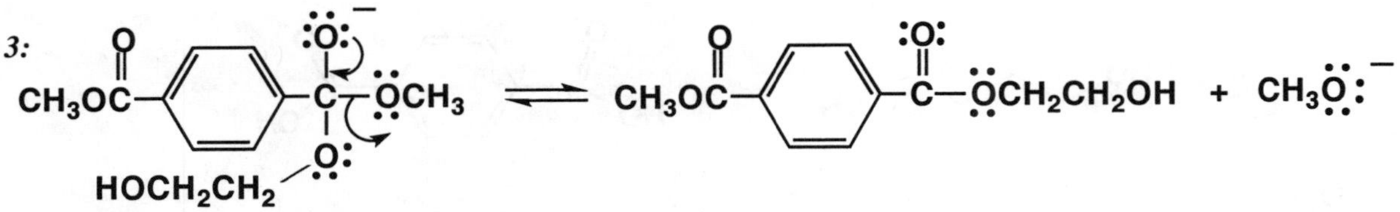

Step 4:

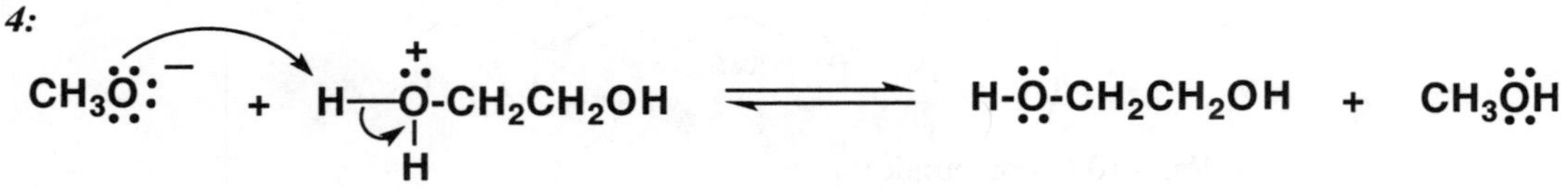

<u>Problem 15.5</u> Currently about 30% of PET soft drink bottles are being recycled. In one recycling process, scrap PET is heated with methanol in the presence of an acid catalyst. The methanol reacts with the polymer, liberating ethylene glycol and dimethyl terephthalate. These monomers are then used as feedstock for the production of new PET products. Write an equation for the reaction of PET with methanol to give ethylene glycol and dimethyl terephthalate.

Poly(ethylene terephthalate)
(PET)   **Methanol**

Dimethyl terephthalate

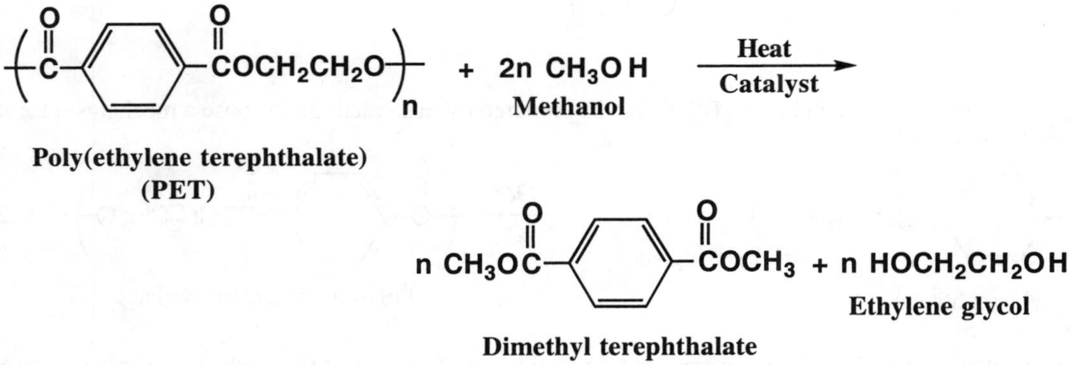

Problem 15.6 Nomex is an aromatic polyamide (aramid) prepared from polymerization of 1,3-benzenediamine and the acid chloride of 1,3-benzenedicarboxylic acid. The physical properties of the polymer make it suitable for high strength, high temperature applications such as parachute cords and jet aircraft tires. Draw a structural formula for the repeating unit of Nomex.

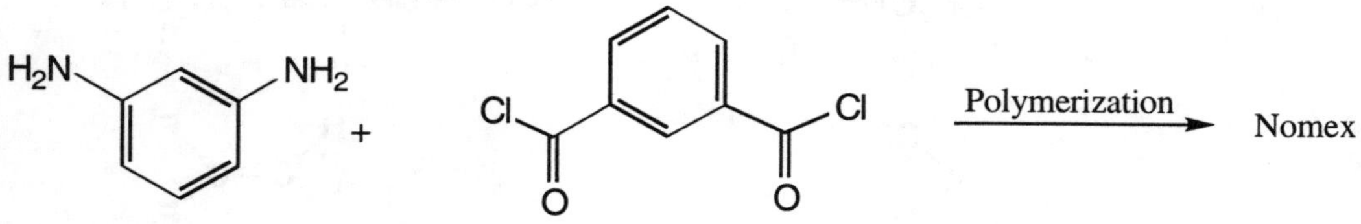

1,3-Benzenediamine   1,3-Benzenedicarboxylic
                                      acid chloride

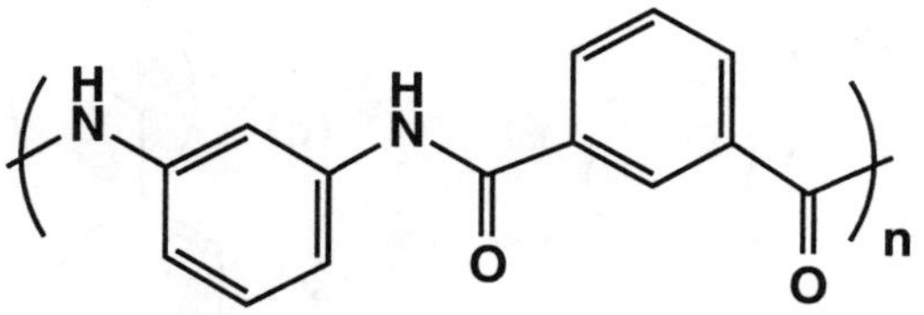

Nomex

Problem 15.7 Nylon 6,10 [Problem 15.3 (d)] can be prepared by reaction of a diamine and a diacid chloride. Draw the structural formula of each reactant.

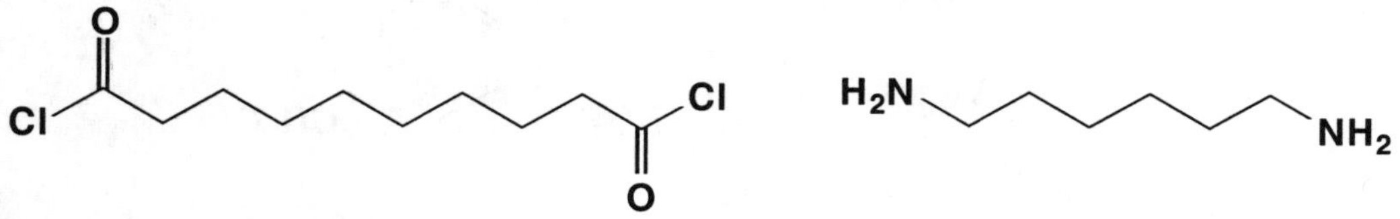

Chain-Growth Polymerization
Problem 15.8 Following is the structural formula of a section of polypropylene derived from three units of propylene monomer.

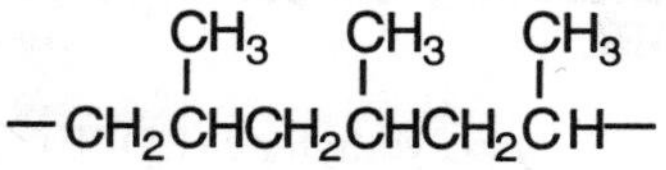

$$\begin{array}{ccc} CH_3 & CH_3 & CH_3 \\ | & | & | \end{array}$$
$$-CH_2CHCH_2CHCH_2CH-$$

Draw a structural formula for a comparable section of:

(a) Poly(vinyl chloride)   (b) Poly(tetrafluoroethylene)

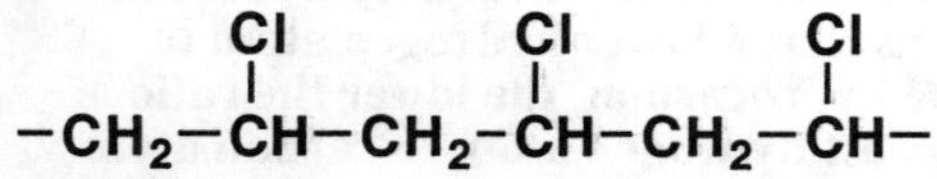

$$\begin{array}{ccc} Cl & Cl & Cl \\ | & | & | \end{array}$$
$$-CH_2-CH-CH_2-CH-CH_2-CH-$$

$$-CF_2-CF_2-CF_2-CF_2-CF_2-CF_2-$$

(c) Poly(methyl methacrylate)

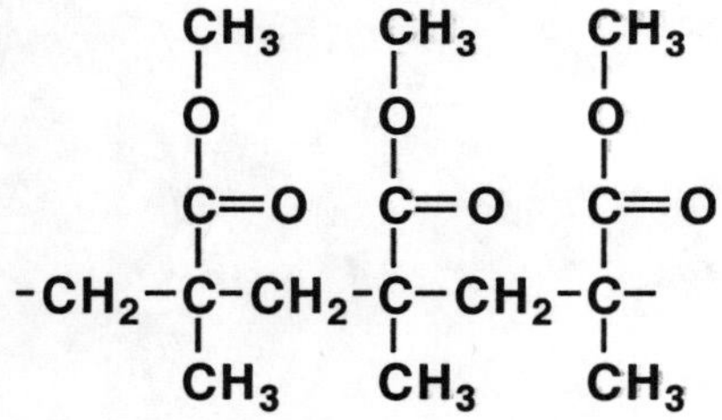

$$\begin{array}{ccc} CH_3 & CH_3 & CH_3 \\ | & | & | \\ O & O & O \\ | & | & | \\ C=O & C=O & C=O \\ | & | & | \end{array}$$
$$-CH_2-C-CH_2-C-CH_2-C-$$
$$\begin{array}{ccc} | & | & | \\ CH_3 & CH_3 & CH_3 \end{array}$$

Problem 15.9 Following are structural formulas for sections of two polymers. From what alkene monomer is each derived?

(a) $-CH_2CCH_2CCH_2CH-$ with Cl substituents

(b) $-CH_2CCH_2CCH_2CH-$ with F substituents

(a) monomer:

$$H_2C=CCl_2 \text{ (drawn as } \begin{array}{c} H \\ \diagdown \\ C=C \\ \diagup \quad \diagdown \\ H \quad \quad Cl \end{array} \begin{array}{c} Cl \\ \\ \\ Cl \end{array})$$

(b) monomer:

$$\begin{array}{c} H \\ \diagdown \\ C=C \\ \diagup \quad \diagdown \\ H \quad \quad F \end{array} \begin{array}{c} F \\ \\ \\ F \end{array}$$

Problem 15.10 Draw the structure of the alkene monomer used to make each chain-growth polymer.

(a) $\left(\text{...} \right)_n$

(b) $\left(\text{...}O \text{...} \right)_n$

(c) $\left(\text{...} \right)_n$ with benzene ring bearing CH_2Cl

(d) $\left(\begin{array}{c} CF-CF_2 \\ | \\ CF_3 \end{array} \right)_n$

monomers:

(a) 1-butene structure

(b) ethyl vinyl ether O structure

(c) styrene with CH_2Cl substituent

(d) $\begin{array}{c} F \\ \diagdown \\ C=CF_2 \\ \diagup \\ CF_3 \end{array}$

Problem 15.11 Low-density polyethylene (LDPE) has a higher degree of chain branching than high-density polyethylene (HDPE). Explain the relationship between chain branching and density.

The branches of LDPE prevent the polymer chains from packing together as tightly as the less-branched HDPE. Thus, LDPE is less dense than HDPE.

Problem 15.12 Compare the densities of low-density polyethylene (LDPE) and high-density polyethylene (HDPE) with the densities of the liquid alkanes listed in Table 2.4. How might you account for the differences between them?

As stated in the chapter, the density of LDPE is between 0.91 and 0.94 g/cm^3, while the density of HDPE is 0.96 g/cm^3. These values are considerably higher than the values of 0.626 to 0.730 g/cm^3 for pentane through decane listed in Table 2.4. A key parameter here is the ratio of hydrogen to carbon atoms in a hydrocarbon. Hydrogen atoms have such a low atomic weight compared to carbon atoms that a lower hydrogen atom to carbon atom ratio increases the density of a hydrocarbon. The longer the hydrocarbon, the lower the ratio of hydrogen atoms to carbon atoms. Thus, the polymers, which have the lowest hydrogen atom to carbon atom ratio, are significantly more dense than the shorter hydrocarbons.

Problem 15.13 Polymerization of vinyl acetate gives poly(vinyl acetate). Hydrolysis of this polymer in aqueous sodium hydroxide gives poly(vinyl alcohol). Draw the repeat units of both poly(vinyl acetate) and poly(vinyl alcohol).

$$CH_3-\overset{\overset{O}{\|}}{C}-O-CH=CH_2$$

Vinyl acetate

Poly(vinyl acetate)   **Poly(vinyl alcohol)**

Problem 15.14 As seen in the previous problem, poly(vinyl alcohol) is made by polymerization of vinyl acetate followed by hydrolysis in aqueous sodium hydroxide. Why is poly(vinyl alcohol) not made instead by polymerization of vinyl alcohol, $CH_2=CHOH$?

Vinyl alcohol is the enol form of acetaldehyde (CH_3CHO). Poly(vinyl alcohol) cannot be made from vinyl alcohol because only a very small fraction of acetaldehyde exists as vinyl alcohol at equilibrium.

CHAPTER 16
Solutions to Problems

Problem 16.1
(a) Draw Fischer projections for all 2-ketopentoses.
(b) Show which are D-ketopentoses, which are L-ketopentoses, and which are enantiomers.
(c) Refer to Table 16.2, and write names of the ketopentoses you have drawn.

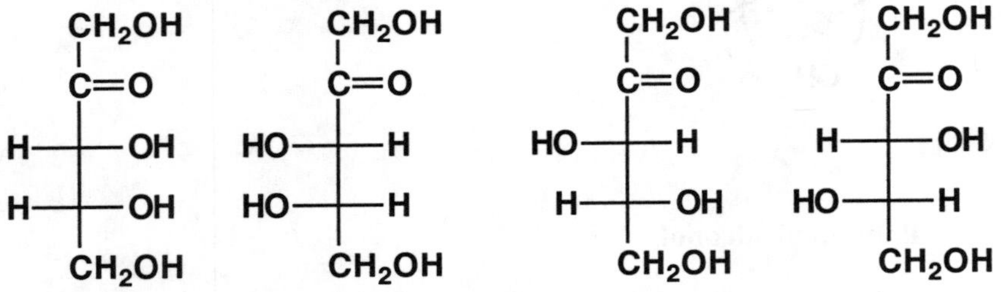

D-Ribulose   L-Ribulose   D-Xylulose   L-Xylulose

A pair of enantiomers   A pair of enantiomers

Problem 16.2 Mannose exists in aqueous solution as a mixture of α-D-mannopyranose and β-D mannopyranose. Draw Haworth projections for these molecules.

D-Mannose differs in configuration from D-glucose at carbon 2. Therefore, the alpha and beta forms of D-mannopyranose differ from those of alpha and beta D-glucopyranoses only in the orientation of the -OH on carbon-2. Following are Haworth projections for these compounds.

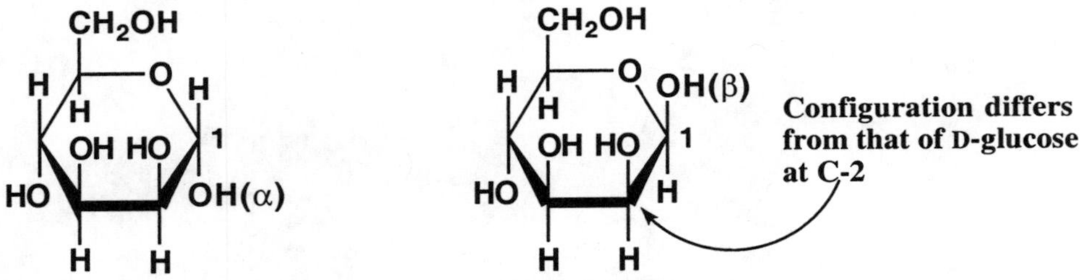

Configuration differs
from that of D-glucose
at C-2

α-D-Mannopyranose   β-D-Mannopyranose
(α-D-Mannose)   (β-D-Mannose)

Problem 16.3 Draw chair conformations for α-D-mannopyranose and β-D-mannopyranose. Label the anomeric carbon atom in each.

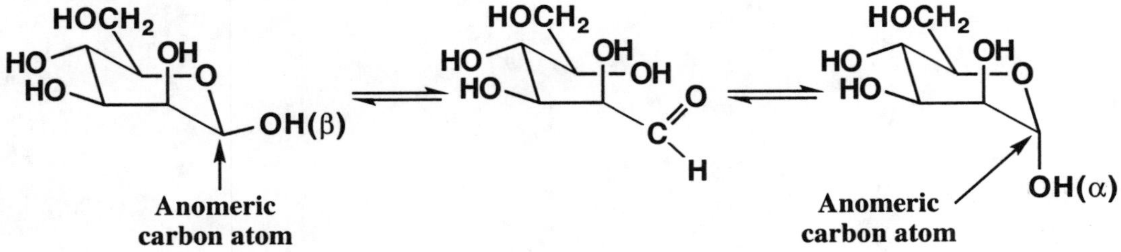

Anomeric   Anomeric
carbon atom   carbon atom

β-D-Mannopyranose   D-Mannose   α-D-Mannopyranose
(β-D-Mannose)   (open form)   (α-D-Mannose)

Problem 16.4 Draw structural formulas for these glycosides. Label each anomeric carbon and glycoside bond.
(a) Methyl β-D-fructofuranoside (methyl β-D-fructoside).

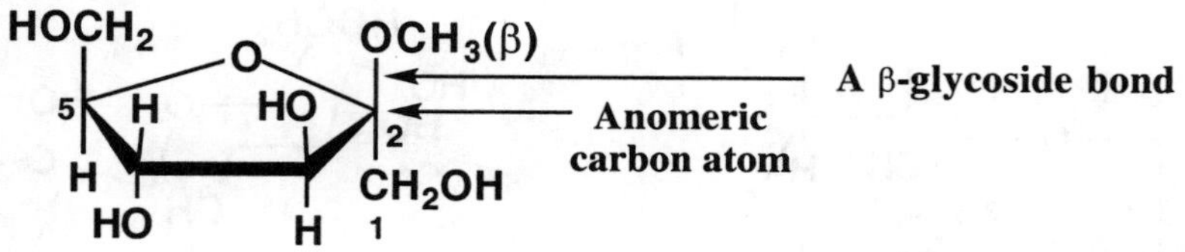

A β-glycoside bond

Anomeric carbon atom

Methyl β-D-fructofuranoside

(b) Methyl α-D-mannopyranoside (methyl α-D-mannoside).

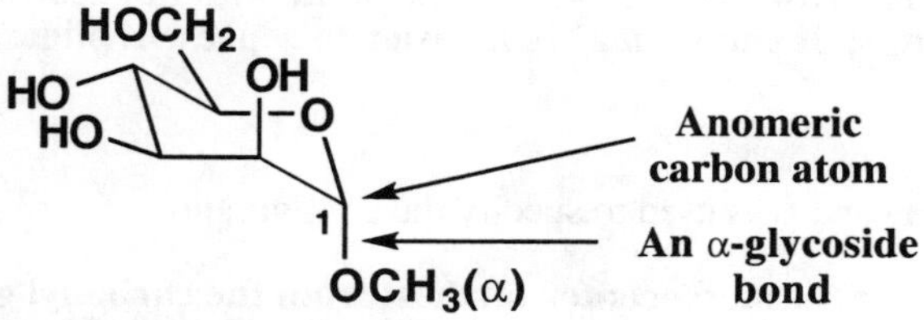

Anomeric carbon atom

An α-glycoside bond

**Methyl α-D-mannopyranoside
(Chair conformation)**

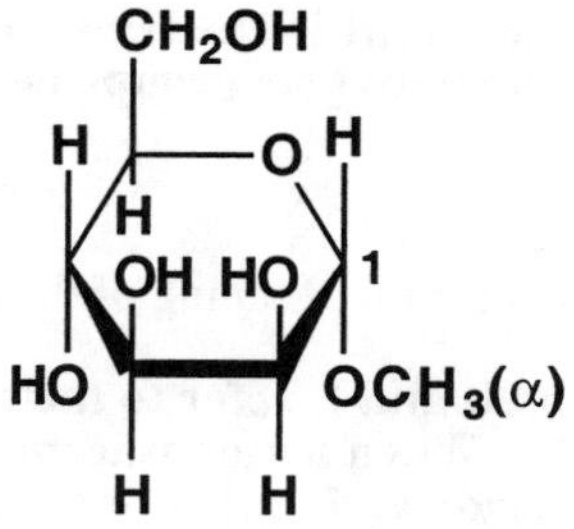

**Methyl α-D-mannopyranoside
(Haworth projection)**

Problem 16.5 Draw a structural formula for the β-N-glycoside formed between β-D-ribofuranose and adenine.

Following are structural formulas for adenine, the monosaccharide hemiacetal, and the N-glycoside.

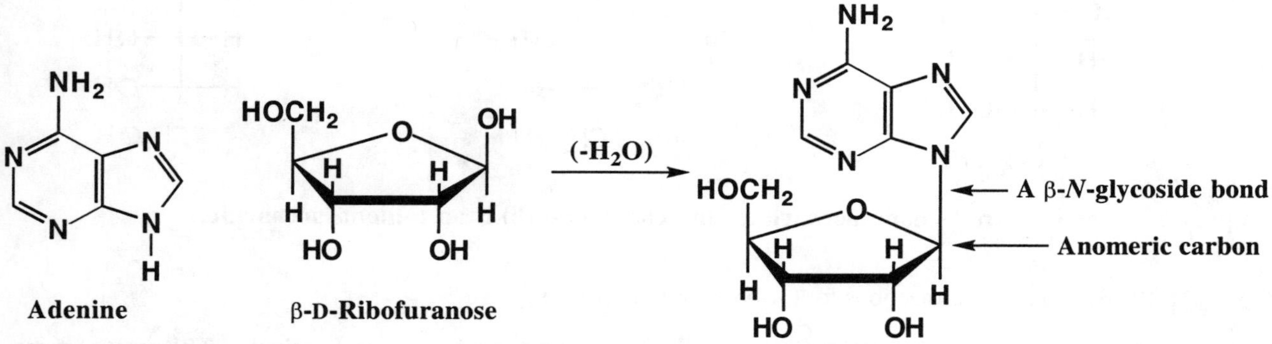

Adenine   **β-D-Ribofuranose**

$(-H_2O)$

A β-N-glycoside bond

Anomeric carbon

Problem 16.6 D-Erythrose is reduced by $NaBH_4$ to erythritol. Do you expect the alditol formed under these conditions to be optically active or optically inactive? Explain.

Erythritol is a meso compound. Thus, it is achiral and will be optically inactive.

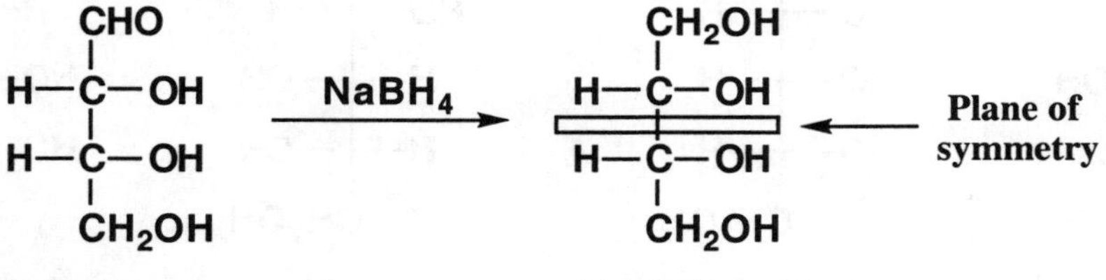

D-Erythrose   **Erythritol
(meso)**

Plane of symmetry

Problem 16.7 Draw Haworth and chair formulas for the α form of a disaccharide in which two units of D-glucopyranose are joined by a β-1,3-glycoside bond.

β-1,3-Glycoside bond

Note that the bonds between carbon one of the glucose unit on the left, the glycosidic oxygen, and carbon three of the other glucose unit in the Haworth formula are drawn with a bend. The bend does not indicate the presence of extra methylene groups between the rings; it simply makes it easier to represent oligosaccharides

Monosaccharides

Problem 16.8 Explain the meaning of the designations D and L as used to specify the configuration of carbohydrates.

The designations D and L refer to the configuration of the stereocenter farthest from the carbonyl group of the monosaccharide. When a monosaccharide is drawn in a Fischer projection, the reference -OH is on the right in a D-monosaccharide and on the left in an L-monosaccharide. Note that the conventions D and L specify the configuration at one and only one stereocenter of the molecule, no matter how many others are present.

Problem 16.9 Which compounds are D-monosaccharides and which are L-monosaccharides?

Compounds (a) and (c) are D-monosaccharides, and compound (b) is an L-monosaccharide.

Problem 16.10 Write Fischer projections for L-ribose for L-arabinose.

L-Ribose and L-arabinose are the mirror images of D-ribose and D-arabinose, respectively. The most common error in answering this question is to start with the Fischer projection for the D sugar and then invert the configuration of carbon 4 only. While the monosaccharide thus drawn is an L-sugar, it is not the correct one. All of the stereocenters must be changed to draw the true enantiomers.

| D-Ribose | L-Ribose | D-Arabinose | L-Arabinose |

Problem 16.11 2,6-Dideoxy-D-altrose, known alternatively as D-digitoxose, is a monosaccharide obtained on hydrolysis of digitoxin, a natural product extracted from purple foxglove (*Digitalis purpurea*). Digitoxin has found wide use in cardiology because it reduces pulse rate, regularizes heart rhythm and strengthens heart beat. Draw the structural formula of 2,6-dideoxy-D-altrose.

CHO
H——H
H——OH
H——OH
H——OH
CH₃

2,6-Dideoxy-D-altose
(D-Digitoxose)

The Cyclic Structure of Monosaccharides
Problem 16.12 Define the term anomeric carbon.

The anomeric carbon is the hemiacetal carbon of the cyclic form of a monosaccharide.

Problem 16.13 Explain the conventions for using α and β to designate the configuration of cyclic forms of monosaccharides.

If the anomeric hydroxyl group is cis to the -CH₂OH group, the cyclic monosaccharide is designated as beta (β). When the anomeric hydroxyl group and the -CH₂OH group are trans, the cyclic monosaccharide is designated as alpha (α).

Problem 16.14 Draw α-D-glucopyranose (α-D-glucose) in a Haworth projection. Now, using only the information given here, draw Haworth projections for these monosaccharides.

(a) α-D-Mannopyranose (α-D-mannose). The configuration of D-mannose differs from the configuration of D-glucose only at carbon 2.

(b) α-D-Gulopyranose (α-D-gulose). The configuration of D-gulose differs from the configuration of D-glucose at carbons 3 and 4.

α-D-Glucopyranose
(α-D-Glucose)

α-D-Mannopyranose
(α-D-Mannose)

α-D-Gulopyranose
(α-D-Gulose)

Problem 16.15 Convert each Haworth projection to an open-chain form and then to a Fischer projection. Name the monosaccharide you have drawn.

(a)

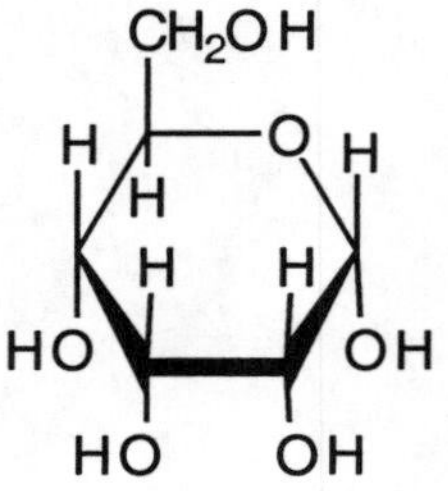

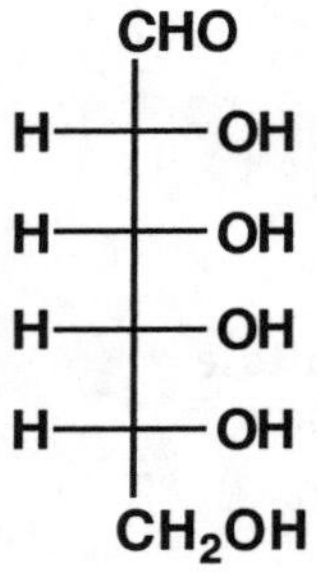

D-Allose

(b)

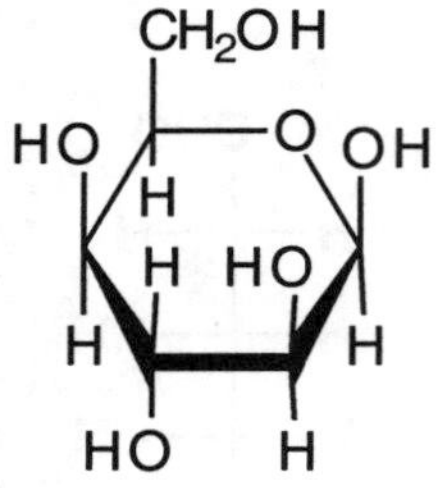

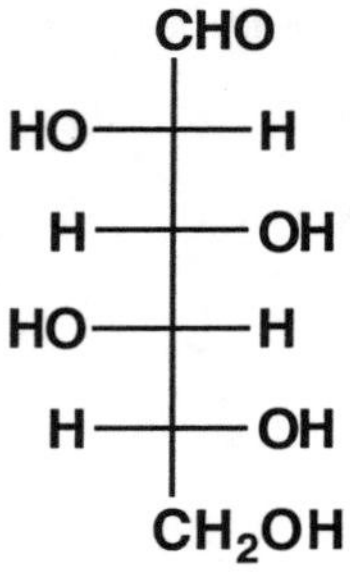

D-Idose

Problem 16.16 Convert each chair conformation to an open-chain form and then to a Fischer projection. Name the monosaccharide you have drawn.

(a)

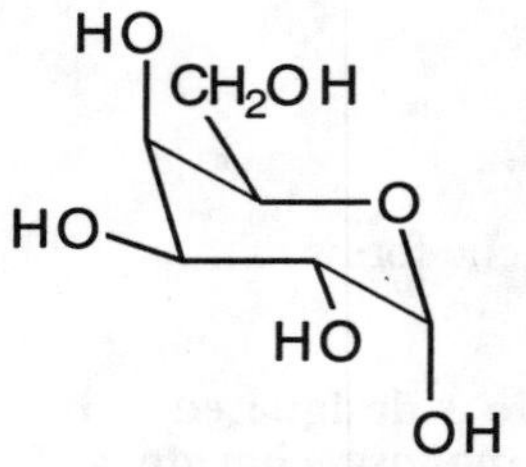

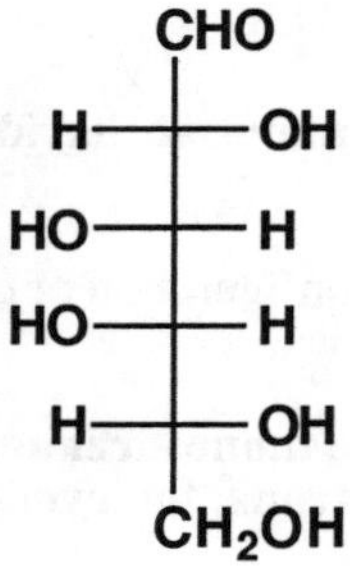

D-Galactose

(b)

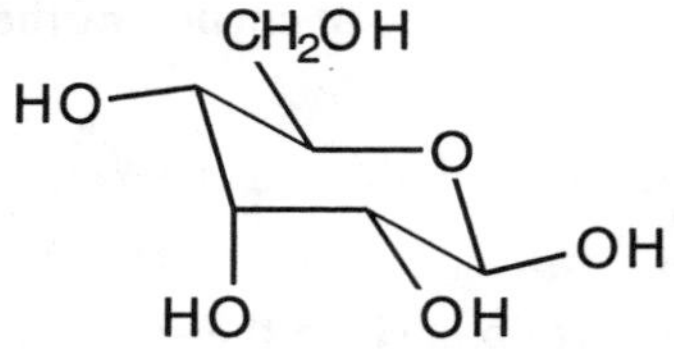

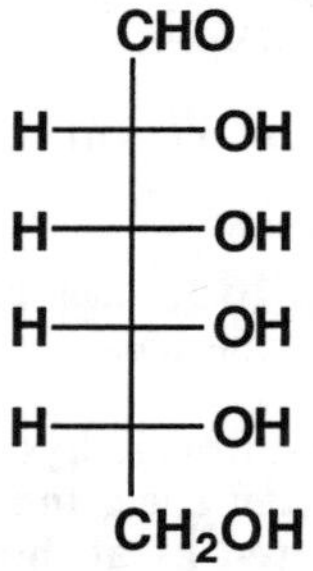

D-Allose

Problem 16.17 The configuration of D-arabinose differs from the configuration of D-ribose only at carbon 2. Using this information, draw a Haworth projection for α-D-arabinofuranose (α-D-arabinose).

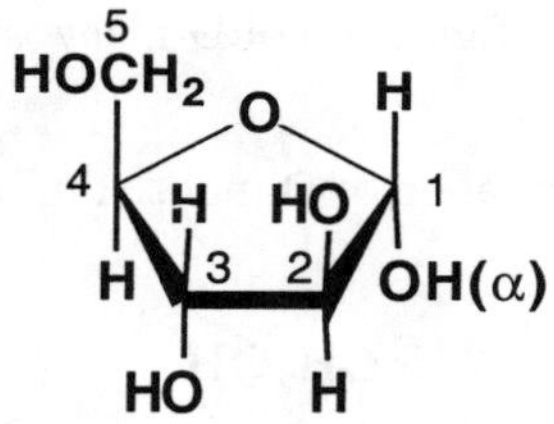

α-D-Arabinofuranose
(α-D-arabinose)

Problem 16.18 Explain the phenomenon of mutarotation with reference to carbohydrates. By what means is it detected?

Any monosaccharide of four or more carbons can exist in an open-chain form and two or more cyclic hemiacetal (i.e. furanose or pyranose) forms, each having a different specific rotation. The specific rotation of an aqueous solution, measured with a polarimeter, of any one form changes until an equilibrium value is reached, representing an equilibrium concentration of the different forms. Mutarotation is the change in specific rotation toward the equilibrium value.

Reactions of Monosaccharides

Problem 16.19 Draw Fischer projections for the product(s) formed by reaction of D-galactose with the following. In addition state whether each product is optically active or optically inactive.

(a) $NaBH_4$ in H_2O (b) H_2/Pt (c) $AgNO_3$ in NH_3, H_2O

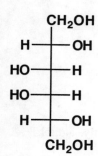

Galactitol	**Galactitol**
(Meso; inactive)	**(Meso; inactive)**

D-Galactonic acid
(Chiral; optically active)

Problem 16.20 Repeat problem 16.19 using D-ribose.

(a) $NaBH_4$ in H_2O (b) H_2/Pt (c) $AgNO_3$ in NH_3, H_2O

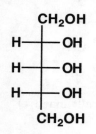 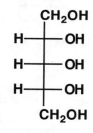 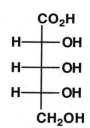

Ribitol	**Ribitol**	**D-Ribonic acid**
(Meso; inactive)	**(Meso; inactive)**	**(Chiral; optically active)**

Problem 16.21 There are four D-aldopentoses (Table 16.1). If each is reduced with $NaBH_4$, which yield optically active alditols? Which yield optically inactive alditols?

D-Ribose and D-xylose yield different achiral (meso) alditols. D-Arabinose and D-lyxose yield the same chiral alditol.

CHO —NaBH₄→ CH₂OH   CHO —NaBH₄→ CH₂OH

D-Ribose   **Ribitol (meso)**   **D-Xylose**   **Xylitol (meso)**

```
        CHO                        CH2OH                       CHO
  HO ──┼── H               HO ──┼── H                 HO ──┼── H
   H ──┼── OH      NaBH4     H ──┼── OH     NaBH4      HO ──┼── H
   H ──┼── OH      ────▶     H ──┼── OH     ◀────       H ──┼── OH
       CH2OH                    CH2OH                       CH2OH
     D-Arabinose             D-Arabinitol               D-Lyxose
                         [α]²⁵_D = -32°
```

$[\alpha]_D^{25} = -32°$

Problem 16.22 Account for the observation that reduction of D-glucose with NaBH$_4$ gives an optically active alditol, whereas reduction of D-galactose with NaBH$_4$ gives an optically inactive alditol.

Reduction of D-glucose gives an optically active alditol product, while reduction of D-galactose gives an optically inactive meso alditol.

```
        CHO                      CH2OH                    CHO                     CH2OH
   H ──┼── OH              H ──┼── OH              H ──┼── OH              H ──┼── OH
  HO ──┼── H     NaBH4    HO ──┼── H             HO ──┼── H     NaBH4    HO ──┼── H
   H ──┼── OH    ────▶     H ──┼── OH            HO ──┼── H     ────▶    HO ──┼── H
   H ──┼── OH              H ──┼── OH             H ──┼── OH              H ──┼── OH
       CH2OH                  CH2OH                   CH2OH                  CH2OH
    D-Glucose          (Optically active)        D-Galactose       (Meso: optically inactive)
```

Problem 16.23 Which two D-aldohexoses give optically inactive (meso) alditols on reduction with NaBH$_4$?

D-Allose and D-galactose both give optically inactive (meso) alditols.

```
        CHO                      CH2OH                    CHO                     CH2OH
   H ──┼── OH              H ──┼── OH              H ──┼── OH              H ──┼── OH
   H ──┼── OH     NaBH4    H ──┼── OH             HO ──┼── H     NaBH4    HO ──┼── H
   H ──┼── OH    ────▶     H ──┼── OH             HO ──┼── H     ────▶    HO ──┼── H
   H ──┼── OH              H ──┼── OH              H ──┼── OH              H ──┼── OH
       CH2OH                  CH2OH                   CH2OH                  CH2OH
    D-Allose        (Meso; optically inactive)     D-Galactose       (Meso: optically inactive)
```

<u>Problem 16.24</u> Name the two alditols formed by NaBH4 reduction of D-fructose.

D-Fructose   **D-Glucitol**   **D-Mannitol**

Reduction of D-fructose gives D-glucitol and D-mannitol. Each differs in configuration only at carbon-2.

<u>Problem 16.25</u> One pathway for the metabolism of glucose-6-phosphate is its enzyme-catalyzed conversion to fructose-6-phosphate. Show that this transformation can be regarded as two enzyme-catalyzed keto-enol tautomerizations.

D-Glucose-6-phosphate   D-Fructose-6-phosphate

Enol form

Problem 16.26 L-Fucose, one of several monosaccharides commonly found in the surface polysaccharides of animal cells (Section 16.7D), is synthesized biochemically from D-mannose in the following eight steps:

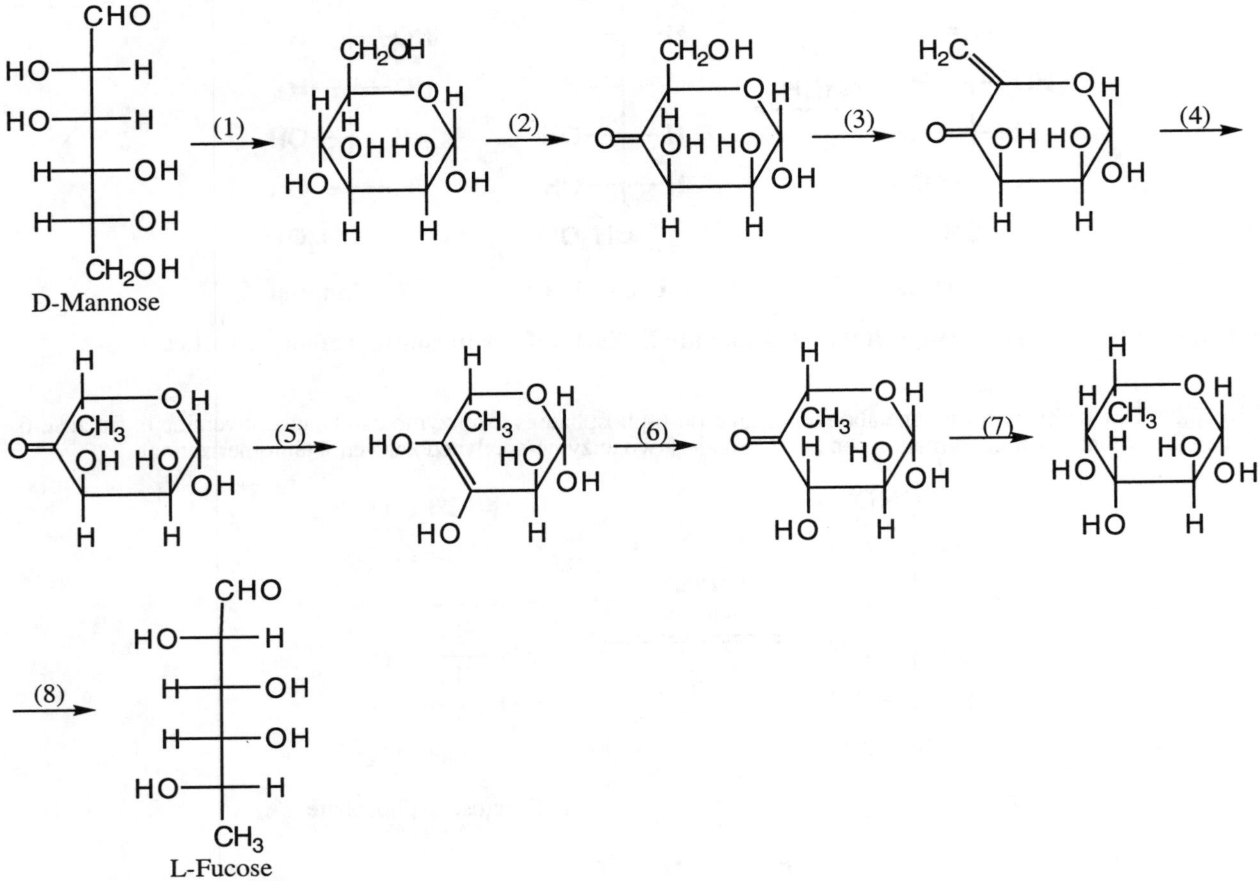

D-Mannose

L-Fucose

(a). Describe the type of reaction (i.e., oxidation, reduction, hydration, dehydration, etc.) involved in each step.

Following is the type of reaction in each step.
(1) Formation of a hemiacetal from a carbonyl group and a secondary alcohol.
(2) A two-electron oxidation of a secondary alcohol to a ketone.
(3) Dehydration of a β-hydroxyketone to an α,β-unsaturated ketone.
(4) A two-electron reduction of a carbon-carbon double bond to a carbon-carbon single bond.
(5) Keto-enol tautomerism of an α-hydroxyketone to form an enediol.
(6) Keto-enol tautomerism of an enediol to form an α-hydroxyketone.
(7) A two-electron reduction of a ketone to a secondary alcohol.
(8) Opening of a cyclic hemiacetal to form an aldehyde and an alcohol.

(b) Explain why this monosaccharide, which is derived from D-mannose, now belongs to the L series.

It is the configuration at carbon-5 of this aldohexose that determines whether it is of the D-series or of the L-series. The result of steps 3 and 4 is inversion of configuration at carbon 5 and, therefore, conversion of a D-aldohexose to an L-aldohexose.

Problem 16.27 Draw structural formulas for the products formed by hydrolysis at pH 7.4 (the pH of blood plasma) of all ester, thioester, amide, anhydride, and glycoside bonds in acetyl coenzyme A (acetyl-CoA). Name as many of the hydrolysis products as you can.

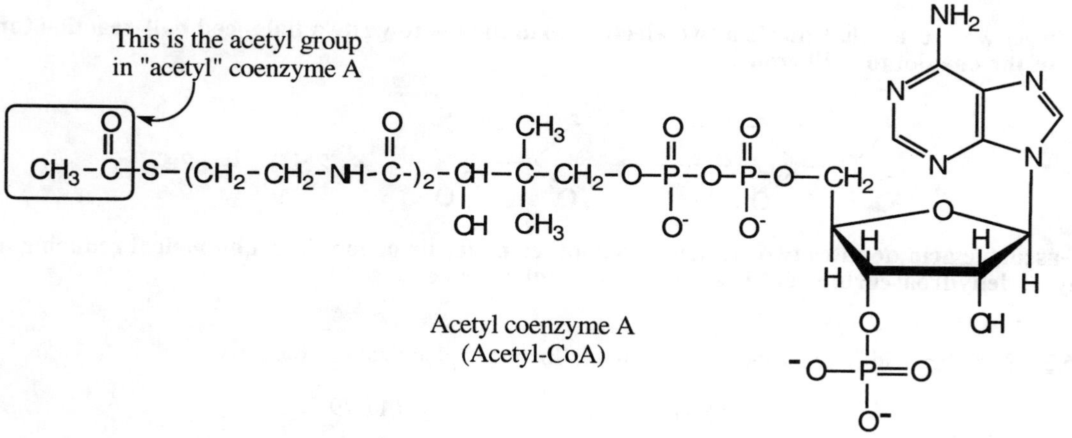

Acetyl coenzyme A
(Acetyl-CoA)

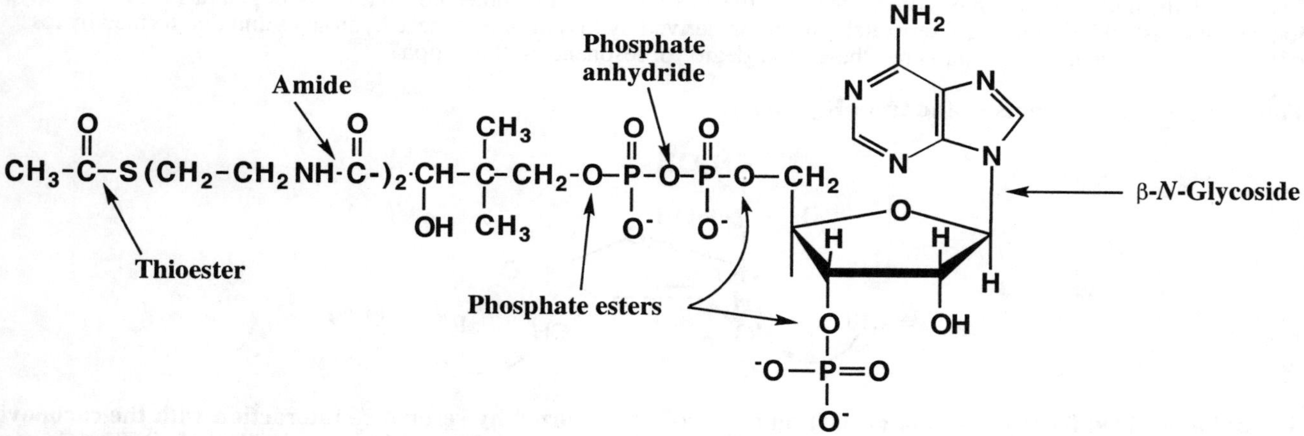

Following are the smaller molecules formed by hydrolysis of each ester, thioester, amide, anhydride and glycoside bond. They are arranged to correspond roughly to their location from left to right in acetyl CoA.

CH_3COH

Acetic acid

$HOCCH_2CH_2NH_2$

3-Aminopropanoic acid

$2\ HPO_4^{2-}$

Phosphate

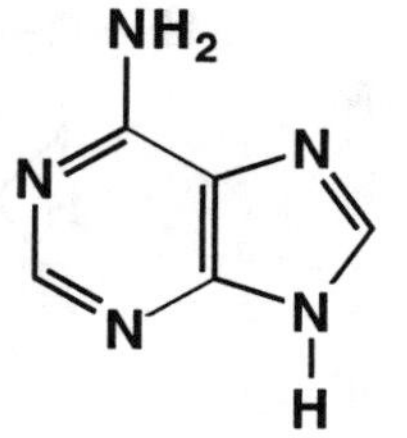

Adenine

$HSCH_2CH_2NH_2$

2-Aminoethanethiol

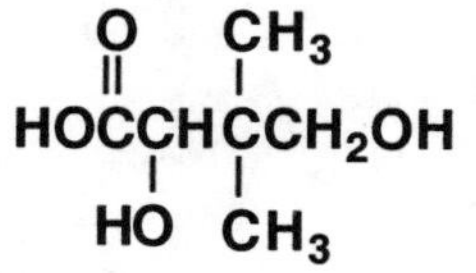

$HOCCHCCH_2OH$

2,4-Dihydroxy-3,3-dimethylbutanoic acid

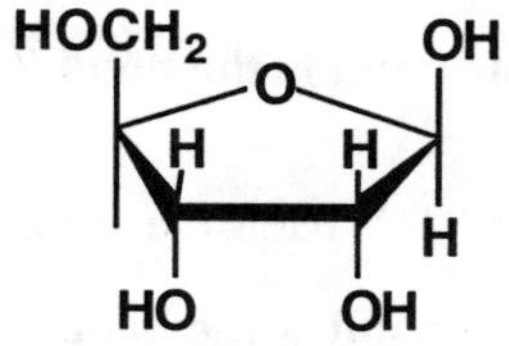

β-D-Ribofuranose

HPO_4^{2-}
Phosphate

Ascorbic Acid

Problem 16.28 Write a balanced half-reaction to show that conversion of L-ascorbic acid to L-dehydroascorbic acid is an oxidation. How many electrons are involved in this oxidation? Is ascorbic acid a biological oxidizing agent or a biological reducing agent?

The most direct way to see that this is a two-electron oxidation is to write a balanced half-reaction for the conversion of the enediol to a diketone.

Because L-ascorbic acid donates two electrons to another molecule or ion, it is a biological reducing agent. Conversely, L-dehydroascorbic acid is a biological oxidizing agent.

Problem 16.29 Ascorbic acid is a diprotic acid with the following acid ionization constants.

$$pK_{a1} = 4.10 \qquad pK_{a2} = 11.79$$

The two acidic hydrogens are those connected with the enediol part of the molecule. Which hydrogen has which ionization constant? (Hint: Draw separately the anion derived by loss of one of these hydrogens and that formed by loss of the other hydrogen. Which anion has the greater degree of resonance stabilization?)

Following are assignments of the two pK$_a$ values.

The anion derived from ionization of -OH on carbon-3 is stabilized by resonance interaction with the carbonyl oxygen. There is no comparable resonance stabilization of the anion derived from ionization of -OH on carbon-2.

Disaccharides and Oligosaccharides

Problem 16.30 Define the term glycoside bond.

A glycoside bond is the bond from the anomeric carbon of a glycoside to an -OR group.

Problem 16.31 What is the difference in meaning between the terms glycoside bond and glucoside bond?

A glycoside bond is the bond from the anomeric carbon of a glycoside to an -OR group. A glucoside bond is a glycoside bond that yields glucose upon hydrolysis.

<u>Problem 16.32</u> In making candy or sugar syrups, sucrose is boiled in water with a little acid, such as lemon juice. Why does the product mixture taste sweeter than the starting sucrose solution?

Sucrose is a disaccharide composed of the monosaccharides glucose and fructose linked through an α−1,2-glycoside bond. The acid catalyzes hydrolysis of the glycoside bond, and the monomeric glucose and fructose are more soluble than sucrose itself. Because of this, the syrups and candy have higher concentrations of these sugars than is possible with sucrose. As a result, the candy and syrup taste sweeter. Furthermore, fructose actually tastes sweeter than sucrose, having a relative sweetness of 174, compared with 100 for sucrose. Thus, converting the sucrose into fructose increases the sweetness of the mixture.

<u>Problem 16.33</u> Which disaccharides are reduced by NaBH₄?
(a) Maltose   (b) Lactose   (c) Sucrose

In order for these disaccharides to react with NaBH₄, they must contain at least one carbonyl group that is equilibrium with the open chain form. The first two have such a carbonyl group, but the last one, sucrose does not. Therefore, (a) and (b) will be reduced by NaBH₄, but (c) will not.

<u>Problem 16.34</u> Trehalose is found in young mushrooms and is the chief carbohydrate in the blood of certain insects. Trehalose is a disaccharide consisting of two D-monosaccharide units, each joined to the other by an α-1,1-glycoside bond.

Trehalose

(a) Is trehalose a reducing sugar?

Trehalose is not a reducing sugar because each anomeric carbon is involved in formation of the glycoside bond.

(b) Does trehalose undergo mutarotation?

It will not undergo mutarotation because there is no open chain form possible for either monosaccharide. Both anomeric carbons are involved in the glycoside bond.

(c) Name the two monosaccharides units of which trehalose is composed.

Trehalose is composed of two molecules of D-glucose.

<u>Problem 16.35</u> Hot water extracts of ground willow bark are an effective pain reliever. Unfortunately, the liquid is so bitter that most persons refuse it. The pain reliever in these infusions is salicin. Name the monosaccharide unit in salicin.

Salicin

The monosaccharide unit of salicin is β-D-glucose.

Polysaccharides

Problem 16.36 A Fischer projection of *N*-acetyl-D-glucosamine is given in Section 16.1F.
(a) Draw Haworth and chair structures for the α- and β-pyranose forms of this monosaccharide.

Following are Haworth and chair formulas for the β-pyranose form of this monosaccharide. To draw the α-pyranose form, invert configuration at carbon 1.

(b) Draw Haworth and chair structures for the disaccharide formed by joining two units of the pyranose form of *N*-acetyl-D-glucosamine by a β-1,4-glycoside bond. If you drew this correctly, you have the structural formula for the repeating dimer of chitin, the structural polysaccharide component of the shell of lobster and other crustaceans.

Following are Haworth and chair formulas for the β-anomer of this disaccharide.

Problem 16.37 Propose structural formulas for the repeating disaccharide unit in these polysaccharides:
(a) Alginic acid, isolated from seaweed, is used as a thickening agent in ice cream and other foods. Alginic acid is a polymer of D-mannuronic acid in the pyranose form joined by β-1,4-glycoside bonds.

D-Mannuronic acid   D-Galacturonic acid

Following is the chair conformation for repeating disaccharide units of alginic acid.

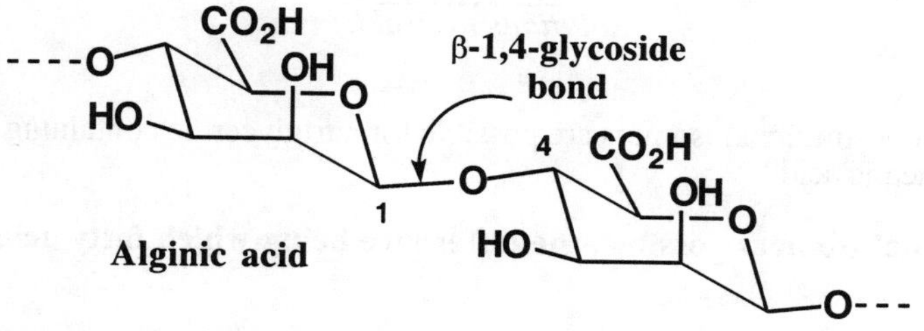

Alginic acid

β-1,4-glycoside bond

(b) Pectic acid is the main component of pectin, which is responsible for the formation of jellies from fruits and berries. Pectic acid is a polymer of D-galacturonic acid in the pyranose form joined by α-1,4-glycoside bonds.

Following is the chair conformation for repeating disaccharide units of pectic acid.

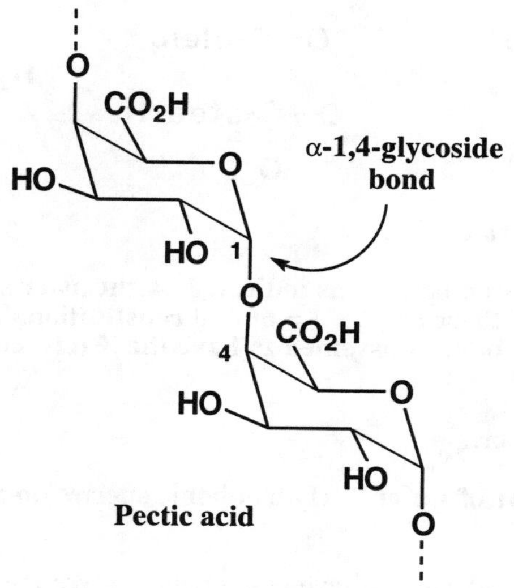

α-1,4-glycoside bond

Pectic acid

CHAPTER 17
Solutions to Problems

<u>Problem 17.1</u> (a) How many constitutional isomers are possible for a triglyceride containing one molecule each of palmitic acid, oleic acid, and stearic acid?

There are three constitutional isomers possible, the difference being which fatty acid is in the middle of the molecule:

(b) Which of these constitutional isomers are chiral?

Each of the molecules in (a) has one stereocenter as indicated by the asterisk, so each constitutional isomer can exist as a pair of enantiomers. Thus, there are 2 x 3 = 6 total constitutional isomers possible. Note that for oleic acid, the carbon-carbon double bond is assumed to have the Z (cis) configuration only.

<u>Problem 17.2</u> Define the term hydrophobic.

The term literally means "having fear of water." Hydrophobic species do not dissolve in water.

<u>Problem 17.3</u> Identify the hydrophobic and hydrophilic region(s) of a triglyceride.

Note that the vast majority of the molecule is hydrophobic, thus explaining why triglycerides are so hydrophobic overall.

<u>Problem 17.4</u> Explain why the melting points of unsaturated fatty acids are lower than those of saturated fatty acids.

Fatty acids in the solid state are attracted to one another via a combination of dispersion forces, involving the long hydrocarbon chains, and hydrogen bonding and dipole-dipole interactions between the carboxylic acid groups. Saturated fatty acids are able to pack very close to one another and adopt a very compact structure in the solid state, which maximizes dispersion forces. The kink in unsaturated fatty acids, resulting from the cis double bond, prevents them from packing together as well. Thus, there are fewer attractive dispersion forces between unsaturated fatty acid molecules in the solid state and, as a result, their melting points are lower.

Problem 17.5 Which would you expect to have the higher melting point, glyceryl trioleate or glyceryl trilinoleate?

The triglyceride with fewer cis double bonds will have the higher melting point. Each oleic acid unit has only one cis double bond, while each linoleic acid unit has two (See Table 17.1). Glycerol trioleate will have the higher melting point.

Problem 17.6 Draw a structural formula for methyl linoleate. Be certain to show the correct configuration of groups about each carbon-carbon double bond.

Methyl linoleate

Problem 17.7 Explain why coconut oil is a liquid triglyceride, even though most of its fatty acid components are saturated.

Triglycerides having fatty acids with shorter chains have lower melting points. From Table 17.2, we see that coconut oil is composed of 45% lauric acid. Lauric acid is a C12 fatty acid, so coconut oil has a melting point that is low enough to make it a liquid near room temperature.

Problem 17.8 It is common now to see "contains no tropical oils" on cooking oil labels, meaning that the oil contains no palm or coconut oil. What is the difference between the composition of tropical oils and that of vegetable oils, such as corn oil, soybean oil, and peanut oil?

The tropical oils contain mostly lower-molecular-weight saturated fatty acids, while the vegetable oils contain mostly unsaturated fatty acids.

Problem 17.9 What is meant by the term hardening as applied to vegetable oils?

The term hardening refers to the process of catalytic reduction of polyunsaturated vegetable oils using H_2 and a transition metal catalyst. By removing the (Z) double bonds, the reduction reaction allows the fatty acids to pack together better and, thus, the triacylglycerols become more solid.

Problem 17.10 How many moles of H_2 are used in the catalytic hydrogenation of 1 mole of a triglyceride derived from glycerol, stearic acid, linoleic acid, and arachidonic acid?

One molecule of H_2 is used per double bond in the triglyceride. Stearic acid does not have any double bonds, linoleic acid has 2 double bonds, and arachidonic acid has 4 double bonds. Thus, 2 + 4 = 6 moles of H_2 will be used per mole of the triglyceride.

Problem 17.11 Characterize the structural features necessary to make a good synthetic detergent.

A good synthetic detergent should have a long hydrocarbon tail and a very polar head group at one end. This combination will allow for the production of micelle structures in aqueous solution that will dissolve hydrophobic dirt, such as grease and oil. The polar head group should not form insoluble salts with the ions normally found in hard water, such as Ca(II), Mg(II), and Fe(III).

Problem 17.12 Following are structural formulas for a cationic detergent and a neutral detergent. Account for the detergent properties of each.

$$CH_3(CH_2)_6CH_2\overset{\overset{\displaystyle CH_3}{\overset{\displaystyle |}{+}}}{\underset{\underset{\displaystyle CH_2C_6H_5}{\displaystyle |}}{N}}CH_3 \quad Cl^-$$

Benzyldimethyloctylammonium chloride
(a cationic detergent)

$$HOCH_2CCH_2O\overset{O}{\overset{||}{C}}(CH_2)_{14}CH_3$$

with $HOCH_2$ and $HOCH_2$ groups

Pentaerythrityl palmitate
(a neutral detergent)

In each case there is a long hydrocarbon tail attached to a very polar group. This combination will allow for the production of micelle structures in aqueous solution that will dissolve nonpolar, hydrophobic dirt, such as grease and oil. In the case of benzyldimethyloctylammonium chloride, the polar group is the positively-charged ammonium group, while for the pentaerythrityl palmitate the polar group is composed of a triol function.

Problem 17.13 Identify some of the detergents used in shampoos and dish washing liquids. Are they primarily anionic, neutral, or cationic detergents.

Most detergents in shampoos and dish washing detergents are anionic detergents, such as the following alkyl benzene solfonate.

$$CH_3(CH_2)_{10}CH_2\text{—}\underset{O}{\overset{O}{\underset{||}{\overset{||}{S}}}}\text{—}O^-\ Na^+$$

Problem 17.14 Show how to convert palmitic acid (hexadecanoic acid) into the following:
(a) Ethyl palmitate

$$CH_3(CH_2)_{14}\overset{O}{\overset{||}{C}}OH \ + \ CH_3CH_2OH \ \xrightarrow{H^+} \ CH_3(CH_2)_{14}\overset{O}{\overset{||}{C}}OCH_2CH_3$$

Ethyl palmitate

(b) Palmitoyl chloride

$$CH_3(CH_2)_{14}\overset{O}{\overset{||}{C}}OH \ + \ SOCl_2 \ \longrightarrow \ CH_3(CH_2)_{14}\overset{O}{\overset{||}{C}}Cl$$

Palmitoyl chloride

(c) 1-Hexadecanol (cetyl alcohol)

$$CH_3(CH_2)_{14}\overset{O}{\overset{||}{C}}OH \ \xrightarrow[\text{2) } H_2O]{\text{1) } LiAlH_4,\ \text{ether or THF}} \ CH_3(CH_2)_{14}CH_2OH$$

1-Hexadecanol
(Cetyl alcohol)

(d) 1-Hexadecanamine

$$CH_3(CH_2)_{14}\overset{\overset{\displaystyle O}{\|}}{C}OH \;+\; SOCl_2 \longrightarrow CH_3(CH_2)_{14}\overset{\overset{\displaystyle O}{\|}}{C}Cl \xrightarrow{\;NH_3\;}$$

$$CH_3(CH_2)_{14}\overset{\overset{\displaystyle O}{\|}}{C}NH_2 \xrightarrow[\text{2) } H_2O]{\text{1) } LiAlH_4,\ ether\ or\ THF} CH_3(CH_2)_{14}CH_2NH_2$$

1-Hexadecanamine

(e) *N,N*-Dimethylhexadecanamide

$$CH_3(CH_2)_{14}\overset{\overset{\displaystyle O}{\|}}{C}OH \;+\; SOCl_2 \longrightarrow CH_3(CH_2)_{14}\overset{\overset{\displaystyle O}{\|}}{C}Cl \xrightarrow{\;HN(CH_3)_2\;} CH_3(CH_2)_{14}\overset{\overset{\displaystyle O}{\|}}{C}N(CH_3)_2$$

***N,N*-Dimethylhexadecanamide**

Problem 17.15 Palmitic acid (hexadecanoic acid) is the source of the hexadecyl (cetyl) group in the following compounds. Each is a mild surface-acting germicide and fungicide and is used as a topical antiseptic and disinfectant.

Cetylpyridinium chloride   Benzylcetyldimethylammonium chloride

(a) Cetylpyridinium chloride is prepared by treating pyridine with 1-chlorohexadecane (cetyl chloride). Show how to convert palmitic acid to cetyl chloride.

$$CH_3(CH_2)_{14}\overset{\overset{\displaystyle O}{\|}}{C}OH \xrightarrow[\text{2) } H_2O]{\text{1) } LiAlH_4\ in\ ether\ or\ THF} CH_3(CH_2)_{14}CH_2OH \xrightarrow{\;SOCl_2\;} CH_3(CH_2)_{14}CH_2Cl$$

**1-Chlorohexadecane
(Cetyl chloride)**

(b) Benzylcetyldimethylammonium chloride is prepared by treating benzyl chloride with *N,N*-dimethyl-1-hexadecanamine. Show how this tertiary amine can be prepared from palmitic acid.

$$CH_3(CH_2)_{14}\overset{\overset{\displaystyle O}{\|}}{C}OH \;+\; SOCl_2 \longrightarrow CH_3(CH_2)_{14}\overset{\overset{\displaystyle O}{\|}}{C}Cl \xrightarrow{\;HN(CH_3)_2\;} CH_3(CH_2)_{14}\overset{\overset{\displaystyle O}{\|}}{C}N(CH_3)_2$$

$$\xrightarrow[\text{2) } H_2O]{\text{1) } LiAlH_4,\ ether\ or\ THF} CH_3(CH_2)_{14}CH_2N(CH_3)_2$$

***N,N*-Dimethyl-1-hexadecanamine**

Prostaglandins

Problem 17.16 Examine the structure of PGF$_{2\alpha}$ and
(a) Identify all stereocenters

Each stereocenter is indicated with an asterisk.

(b) Identify all double bonds about which cis,trans isomerism is possible.

These double bonds are indicated by arrows.

(c) State the number of stereoisomers possible for a molecule of this structure.

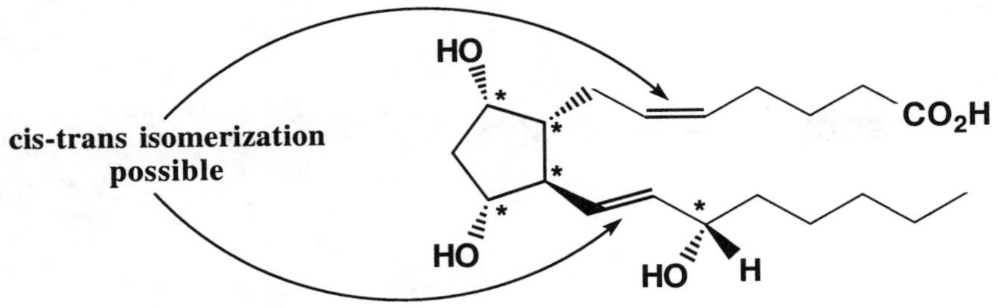

PGF$_{2\alpha}$

There are 2^5 x 2^2 = 128 stereoisomers possible for a molecule of this structure.

Problem 17.17 Following is the structure of unoprostone, a compound patterned after the natural prostaglandins (Section 17.3). Rescula, the isopropyl ester of unoprostone, is an antiglaucoma drug used to treat ocular hypertension. Compare the structural formula of this synthetic prostaglandin with that of PGF2α.

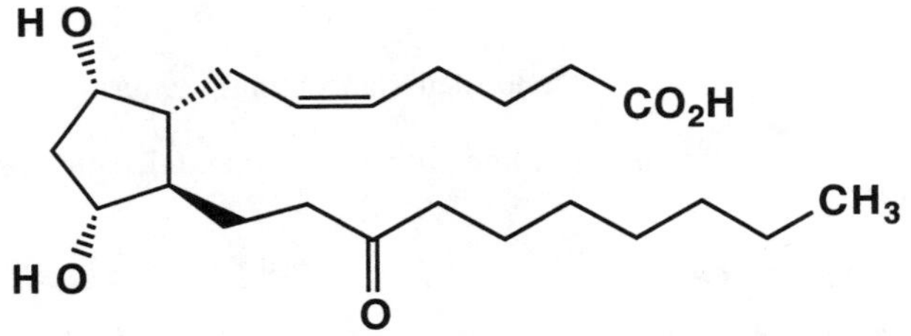

Unoprostone
(antiglaucoma)

Both unoprostone and PGF$_{2\alpha}$ consist of five-membered rings connected to alkyl side chains, which are trans to one another. The chain terminating in the carboxylic acid functionality is identical in both compounds. The other chain, however, is two carbons longer in unoprostone than in PGF$_{2\alpha}$ and is missing the trans double bond. In addition, unoprostone contains a carbonyl group in the position where PGF$_{2\alpha}$ contains a hydroxyl group on this chain. Both compounds contain two cis hydroxyl groups on the five-membered ring.

Steroids

<u>Problem 17.18</u> Draw the structural formula for the product formed by treatment of cholesterol with H_2/Pd; with Br_2.

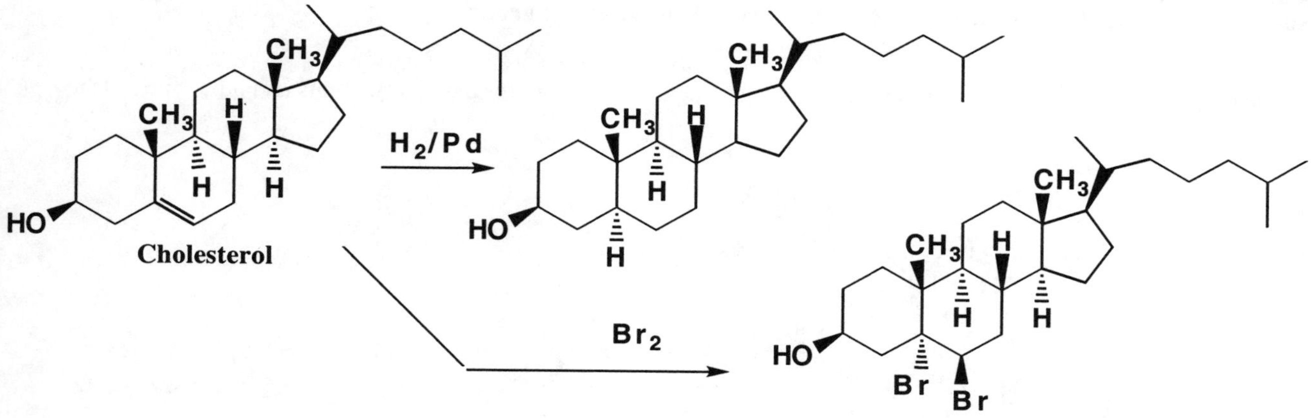

Cholesterol

H_2/Pd

Br_2

<u>Problem 17.19</u> List several ways in which cholesterol is necessary for human life. Why do so many people find it necessary to restrict their dietary intake of cholesterol?

Cholesterol is an important component of biological membranes where it serves to modulate membrane fluidity. In addition, cholesterol is an important precursor to a variety of steroid hormones. Cholesteryl esters are a major component of atherosclerotic plaque, so restricting the dietary intake of cholesterol is helpful for limiting atherosclerosis.

<u>Problem 17.20</u> Both low-density lipoproteins (LDL) and high-density lipoproteins (HDL) consist of a core of triacylglycerols and cholesterol esters surrounded by a single phospholipid layer. Draw the structural formula of cholesteryl linoleate, one of the cholesterol esters found in this core.

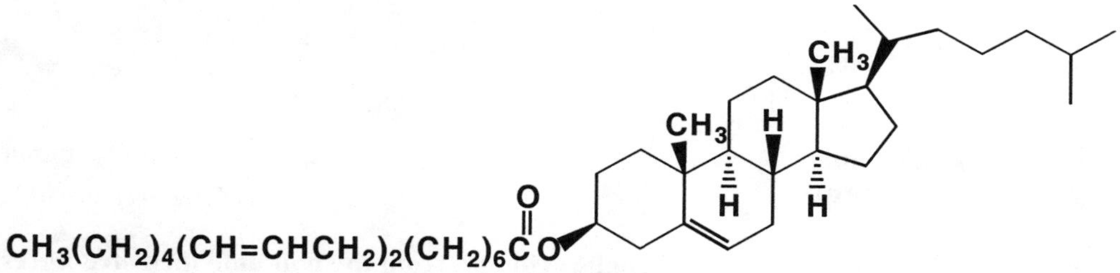

$CH_3(CH_2)_4(CH=CHCH_2)_2(CH_2)_6\overset{O}{\overset{\|}{C}}O$

Cholesteryl linoleate

<u>Problem 17.21</u> Examine the structural formulas of testosterone (a male sex hormone) and progesterone (a female sex hormone). What are the similarities in structure between the two? What are the differences?

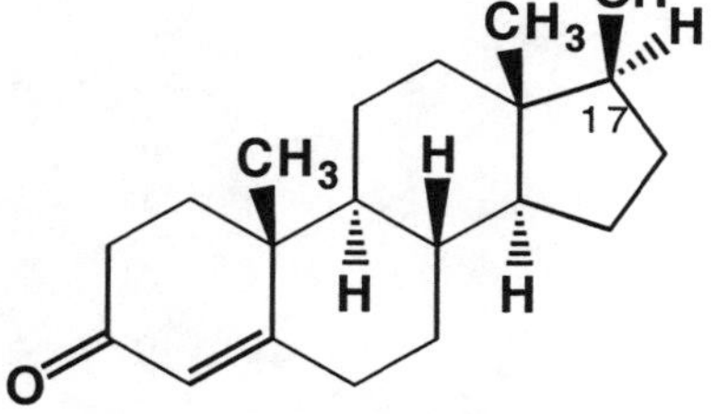

Testosterone

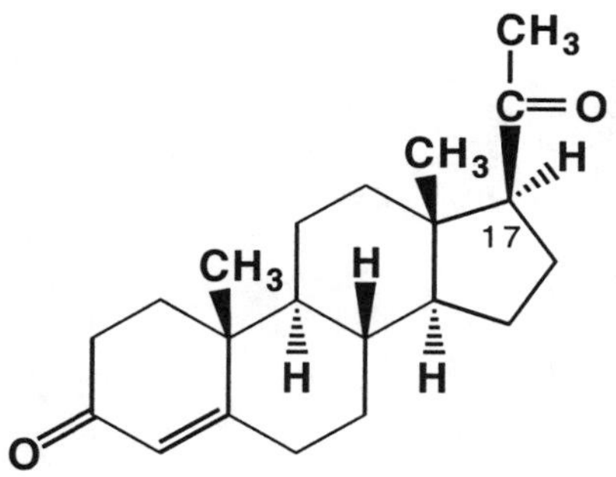

Progesterone

These structures are remarkably similar. Both contain the standard four-ring steroid structure with axial methyl groups at C10 and C13. In addition, both structures contain an ene-one group in the A ring. On the other hand, the two structures differ in the nature of the D ring substituent at C17. In testosterone, the substituent is a hydroxy group and in progesterone it is an acetyl group.

Problem 17.22 Examine the structural formula of cholic acid and account for the ability of this and other bile salts to emulsify fats and oils and thus aid in their digestion.

Cholic acid as CO₂⁻ anion
Structural formula

Cholic acid as CO₂⁻ anion
Conformational formula

Cholic acid as CO₂⁻ anion
(hydrogens not shown)

Cholic acid as CO₂⁻ anion
(all hydrogens shown)

As can be seen in the conformational formula of cholic acid and from the ball-and-stick structures, all three of the hydroxyl groups are pointing in the same direction. Thus, one "face" of the cholic acid molecule is hydrophilic. Similarly, all of the methyl groups are pointing in the same direction, so the other face is hydrophobic. Several molecules of cholic acid can surround hydrophobic molecules, such as fats and oils, interacting with them via their hydrophobic faces. The hydrophilic faces of the cholic acid molecules in these micelle-like assemblies are exposed to water, thus enabling the entire assembly to be soluble, as in a micelle derived from detergents. The highly polar carboxylate group at the end of the molecule further aids in solubility.

Problem 17.23 Following is a structural formula for cortisol (hydrocortisone). Draw a stereorepresentation of this molecule showing the conformations of the five- and six-membered rings.

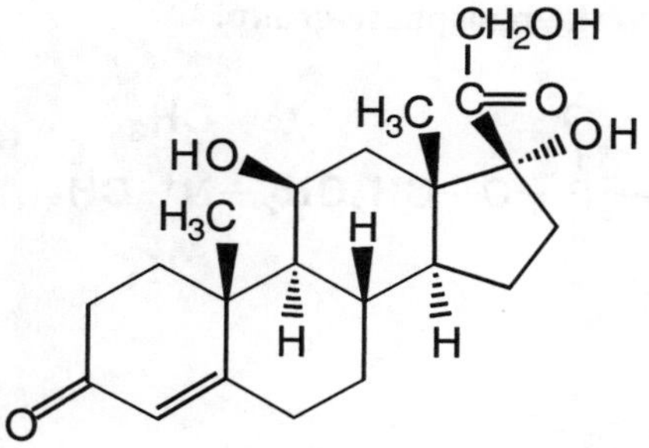

Cortisol (Hydrocortisone)
Structural formula

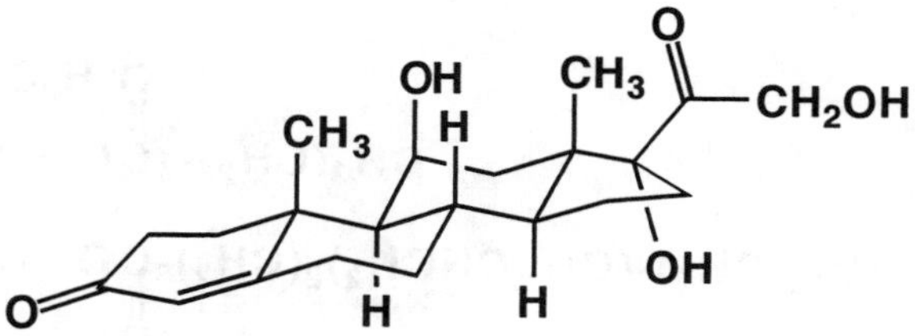

Cortisol (Hydrocortisone)
Conformational formula

Problem 17.24 Because some types of tumors need an estrogen to survive, compounds that compete with the estrogen receptor on tumor cells are useful anticancer drugs. The compound tamoxifen is one such drug. To what part of the estrone molecule is the shape of tamoxifen similar?

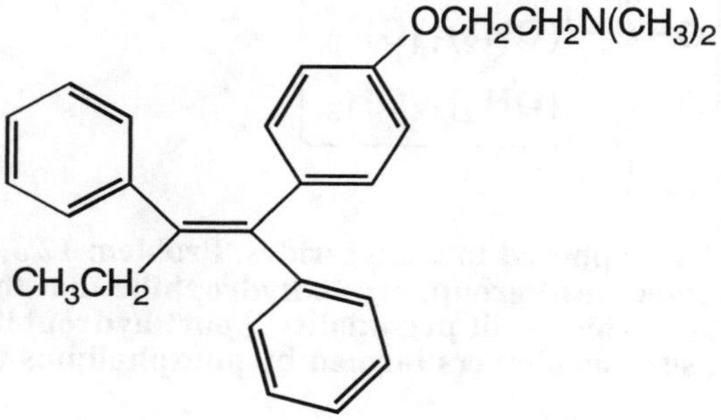

Tamoxifen

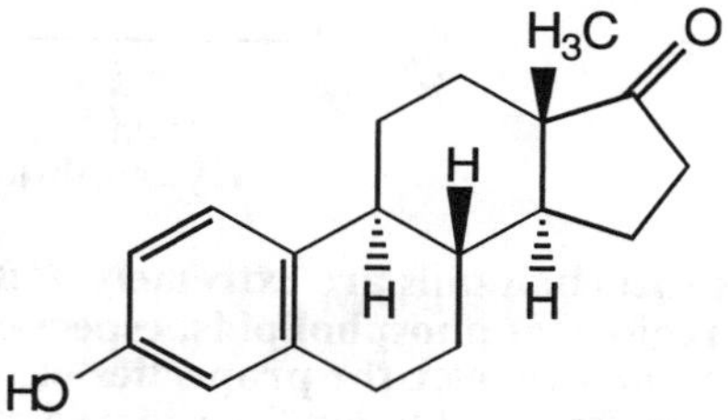

Estrone

Both tamoxifen and estrone are very hydrophobic. Drawn below are highlighted regions of tamoxifen and estrone that emphasize structural similarity. It should be pointed out that some liberties are taken in the following structures when it comes to certain bond angles in tetrahedral and trigonal carbon atoms.

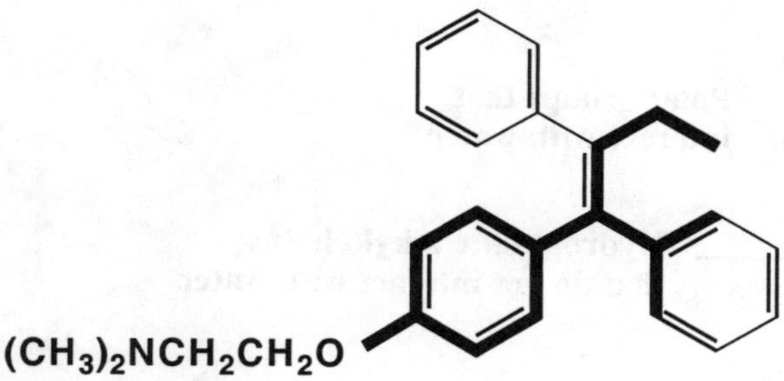

Tamoxifen

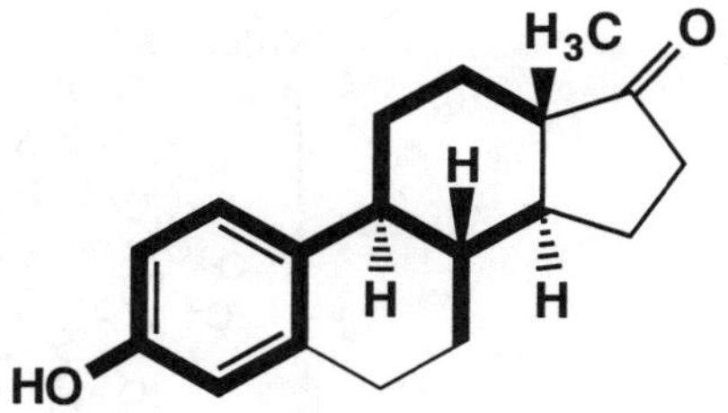

Estrone

Phospholipids

Problem 17.25 Draw the structural formula of a lecithin containing one molecule each of palmitic acid and linoleic acid.

Lecithins are phosphoacylglycerols in which choline is attached to the phosphate group.

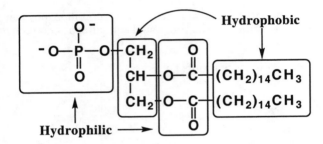

Problem 17.26 Identify the hydrophobic and hydrophilic region(s) of a phospholipid.

The long hydrocarbon tails are extremely hydrophobic. As opposed to triglycerides, Problem 17.3, the hydrophilic regions of phospholipids, especially the charged head group, are so hydrophilic that they exert a tremendous influence over the properties of the molecule. This "split personality", part hydrophilic and part hydrophobic, is responsible for the ordered structures, such as bilayers formed by phospholipids in aqueous solution.

Problem 17.27 The hydrophobic effect is one of the most important noncovalent forces directing the self-assembly of biomolecules in aqueous solution. The hydrophobic effect arises from tendencies (1) to arrange polar groups so that they interact with the aqueous environment by hydrogen bonding and (2) to arrange nonpolar groups so that they are shielded from the aqueous environment. Show how the hydrophobic effect is involved in directing:

(a) Formation of micelles by soaps and detergents.

In micelles, the hydrophobic hydrocarbon tails are associated with each other to form the hydrophobic interior, while the polar groups are located on the outside surface where they interact with water.

(b) Formation of lipid bilayers by phospholipids.

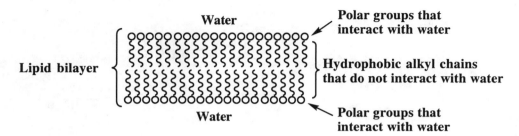

In lipid bilayers, the hydrophobic hydrocarbon tails are associated with each other to form the hydrophobic inner layer, while the polar head groups are grouped together on both outside surfaces where they interact with water.

Problem 17.28 How does the presence of unsaturated fatty acids contribute to the fluidity of biological membranes?

As seen previously (Problem 17.4), unsaturated fatty acids have lower melting points than saturated fatty acids. This results from the kink induced by the cis double bonds, which prevents close packing of unsaturated fatty acids in the solid state. In biological membranes, the cis double bonds in unsaturated fatty acids will likewise prevent dense packing within the hydrophobic inner layer, resulting in greater membrane fluidity.

Problem 17.29 Lecithins can act as emulsifying agents. The lecithin of egg yolk, for example, is used to make mayonnaise. Identify the hydrophobic part(s) and the hydrophilic part(s) of a lecithin. Which parts interact with the oils used in making mayonnaise? Which parts interact with the water?

Fat-soluble Vitamins
Problem 17.30 Examine the structural formula of vitamin A, and state the number of cis,trans isomers possible for this molecule.

As shown in the structure above, vitamin A has four double bonds that can be either cis or trans. Thus, there are 2^4 or 16 possible cis,trans isomers. Note that the double bond in the ring cannot have cis,trans isomers.

<u>Problem 17.31</u> The form of vitamin A present in many food supplements is vitamin A palmitate. Draw the structural formula of this molecule.

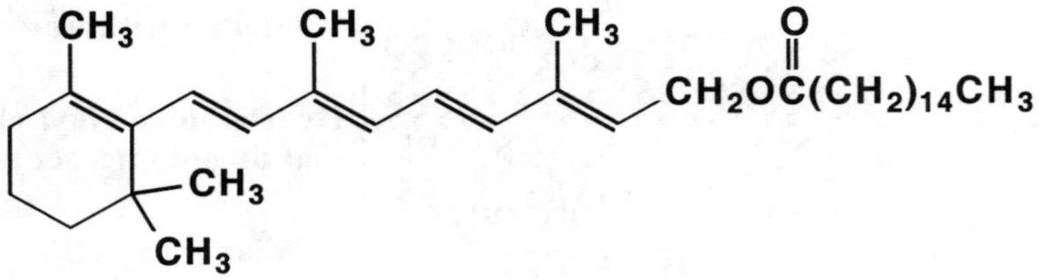

<u>Problem 17.32</u> Examine the structural formulas of vitamin A, 1,25-dihydroxyvitamin D_3, vitamin E, and vitamin K_1 (Section 17.6). Do you expect them to be more soluble in water or in dichloromethane? Do you expect them to be soluble in blood plasma?

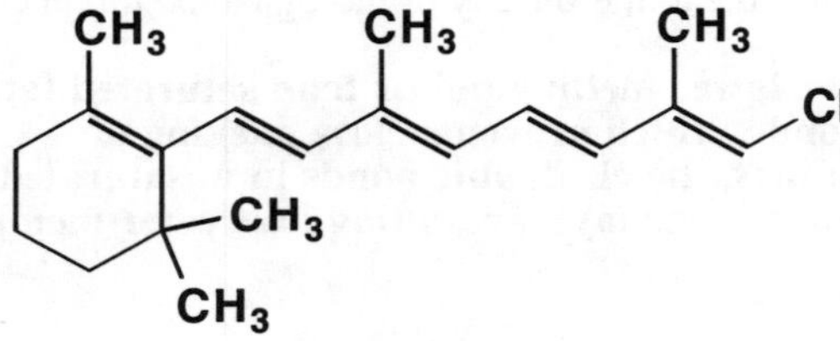

Vitamin A
(Retinol)

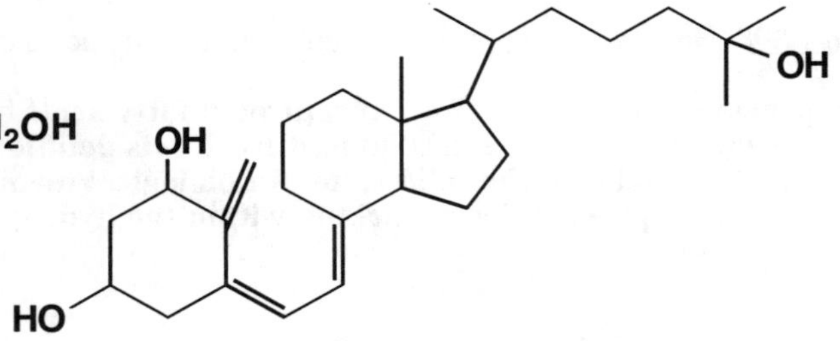

1,25-Dihydroxyvitamin D_3

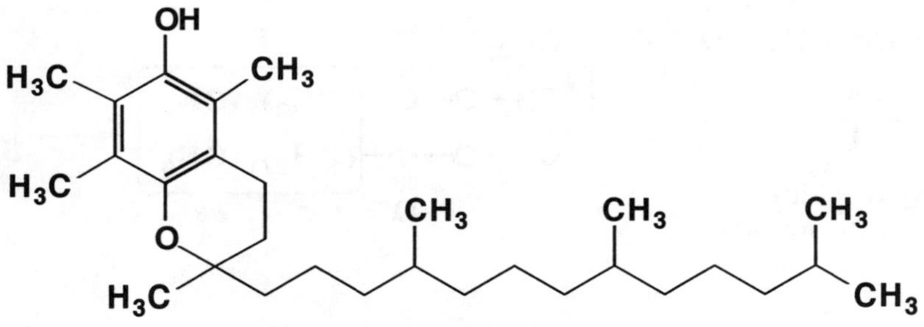

Vitamin E
(α-Tocopherol)

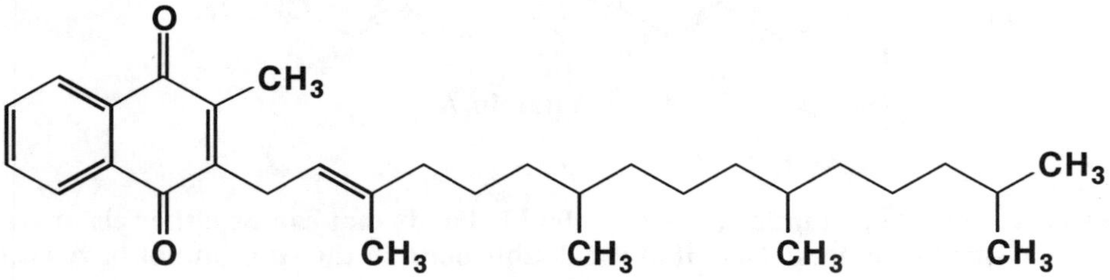

Vitamin K_1

All of these structures are extremely hydrophobic, so they will be more soluble in organic solvents, such as dichloromethane, than in polar solvents such as water. Since blood plasma is an aqueous solution, these vitamins will only be sparingly soluble in blood plasma.

CHAPTER 18
Solutions to Problems

Problem 18.1 Of the 20 protein-derived amino acids shown in Table 18.1, which contain (a) no stereocenter, (b) two stereocenters?

The only amino acid with no stereocenters is glycine (Gly, G). Both isoleucine (Ile, I) and threonine (Thr, T) have two stereocenters, shown with asterisks in the structures below.

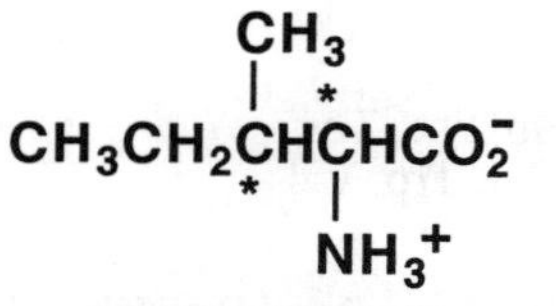

Isoleucine (Ile, I) 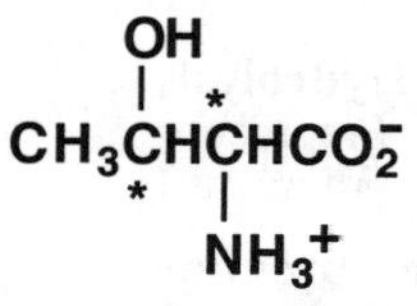   Threonine (Thr, T)

Problem 18.2 The isoelectric point of histidine is 7.64. Toward which electrode does histidine migrate on paper electrophoresis at pH 7.0?

An amino acid will have at least a partial positive charge at any pH that is below its isoelectric point. Because the isoelectric point of histidine is 7.64, it will have a partial positive charge at pH 7.0. Therefore, at pH 7.0 histidine migrates toward the negative electrode.

Problem 18.3 Describe the behavior of a mixture of glutamic acid, arginine, and valine on paper electrophoresis at pH 6.0.

The pI's for glutamic acid, arginine, and valine are 3.08, 10.76, and 6.00, respectively. Therefore, at pH 6.0 glutamic acid is negatively charged, arginine is positively charged, and valine is neutral. Thus, on paper electrophoresis, glutamic acid will migrate toward the positive electrode, arginine will migrate toward the negative electrode, and valine will not move.

Problem 18.4 Draw a structural formula for Lys-Phe-Ala. Label the *N*-terminal amino acid and the *C*-terminal amino acid. What is the net charge on this tripeptide at pH 6.0?

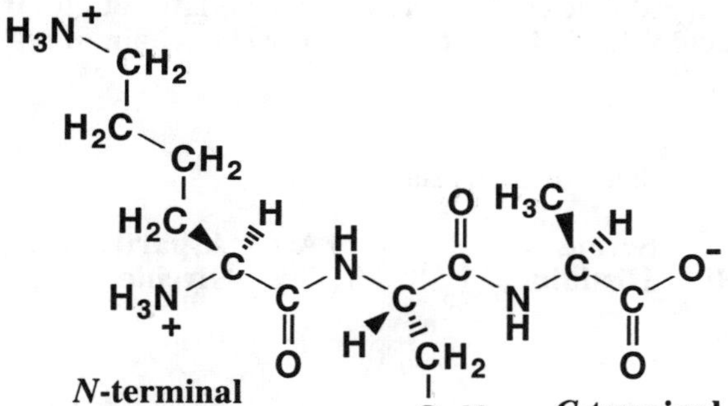

N-terminal amino acid   *C*-terminal amino acid

Due to the presence of the basic lysine residue, this tripeptide will have a net positive charge at pH 6.0

Problem 18.5 Which of these tripeptides are hydrolyzed by trypsin? By chymotrypsin?
(a) Tyr-Gln-Val   (b) Thr-Phe-Ser   (c) Thr-Ser-Phe

Based on the substrate specificities listed in Table 18.3, trypsin will not cleave any of these tripeptides because there are no arginine or lysine residues present. Chymotrypsin will cleave between the Tyr and Gln residues in (a) and between the Phe and Ser residues of (b).

<u>Problem 18.6</u> Deduce the amino acid sequence of an undecapeptide (11 amino acids) from the experimental results shown in the table.

Experimental Procedure	Amino Acid Composition
Undecapeptide	Ala, Arg, Glu, Lys$_2$, Met, Phe, Ser, Thr, Trp, Val
Edman degradation	Ala
Trypsin-Catalyzed Hydrolysis	
Fragment E	Ala, Glu, Arg
Fragment F	Thr, Phe, Lys
Fragment G	Lys
Fragment H	Met, Ser, Trp, Val
Chymotrypsin-Catalyzed Hydrolysis	
Fragment I	Ala, Arg, Glu, Phe, Thr
Fragment J	Lys$_2$, Met, Ser, Trp, Val
Treatment with Cyanogen Bromide	
Fragment K	Ala, Arg, Glu, Lys$_2$, Met, Phe, Thr, Val
Fragment L	Trp, Ser

Based on the Edman degradation result, alanine (Ala) is the *N*-terminal residue of the peptide. Fragment E must have Arg on the *C*-terminal end because it is a peptide produced by trypsin cleavage. Since we know Ala is the *N*-terminal residue, fragment E must have the sequence Ala-Glu-Arg. There must be two lysine residues or an arginine and a lysine residue adjacent to each other based on the appearance of a single lysine residue as Fragment G. Since Fragment J has two lysines and no arginine residues, the two lysine residues must be adjacent to each other. From the chymotrypsin cleavage, we know that the *C*-terminal residues of Fragments I and J must be Phe and Trp, respectively. The sequence of Fragment L must therefore be Ser-Trp. Generation of Fragment L by CNBr cleavage indicates that the last three amino acids in the peptide are Met-Ser-Trp. Thus, the sequence of Fragment H must be Val-Met-Ser-Trp. This information, combined with the knowledge that there are two lysine residues adjacent to each other, indicates that Fragment J has the sequence Lys-Lys-Val-Met-Ser-Trp. Because we know that the C-terminal residue of Fragment I must be Phe and that Fragment I must start with Ala-Glu-Arg, the entire sequence of Fragment I must be Ala-Glu-Arg-Thr-Phe.
Putting Fragments I and J together gives the following sequence for the entire peptide:

Ala-Glu-Arg-Thr-Phe-Lys-Lys-Val-Met-Ser-Trp

<u>Problem 18.7</u> At pH 7.4, with what amino acid side chains can the side chain of lysine form salt linkages?

At pH 7.4, the only negatively charged side chains are the carboxylates of glutamic acid and aspartic acid. Therefore, these are the amino acid side chains with which the side chain of lysine can form a salt linkage.

<u>Amino Acids</u>
<u>Problem 18.8</u> What amino acid does each abbreviation stand for?

(a) Phe **Phenylalanine**   (b) Ser **Serine**   (c) Asp **Aspartic acid**
(d) Gln **Glutamine**   (e) His **Histidine**   (f) Gly **Glycine**
(g) Tyr **Tyrosine**

<u>Problem 18.9</u> Configuration of the stereocenter in α-amino acids is most commonly specified using the D,L convention. It can also be identified using the R,S convention (Section 3.3) Does the stereocenter in L-serine have the R or the S configuration?

$$\text{HOCH}_2$$
$$\underset{\overset{|}{\underset{H_3\overset{+}{N}}{}}}{C}\cdots H \quad CO_2^-$$

L-Serine

L-serine has the S configuration.

Problem 18.10 Assign an R or S configuration to the stereocenter in each amino acid.
(a) L-Phenylalanine   (b) L-Glutamic acid   (c) L-Methionine

The configuration of the stereocenter in each amino acid is S.

Problem 18.11 The amino acid threonine has two stereocenters. The stereoisomer found in proteins has the configuration 2S, 3R about the two stereocenters. Draw a Fischer projection of this stereoisomer and also a three-dimensional representation.

L-Threonine

Problem 18.12 Define the term zwitterion.

A zwitterion is a molecule that has a full positive charge and a full negative charge. The two charges cancel one another and, thus, a zwitterion has no net charge.

Problem 18.13 Draw zwitterion forms of these amino acids.
(a) Valine   (b) Phenylalanine   (c) Glutamine

Problem 18.14 Why are Glu and Asp often referred to as acidic amino acids?

Glutamic acid (Glu) and aspartic acid (Asp) are referred to as acidic amino acids because their side chains contain carboxylic acid functions. Note that both Glu and Asp are negatively charged at neutral pH.

Problem 18.15 Why is Arg often referred to as a basic amino acid? Which two other amino acids are also basic amino acids?

The guanidine function of arginine (Arg) is strongly basic, so this amino acid is referred to as a basic amino acid. Note that this means arginine is positively charged at neutral pH. Lysine (Lys) and histidine (His) are also referred to as basic amino acids because their side chains contain a basic primary amine and an imidazole function, respectively.

Problem 18.16 What is the meaning of the alpha as it is used in α-amino acid?

The alpha in α-amino acid indicates that the amino group is on the carbon atom that is α to the carboxyl group.

Problem 18.17 Several β-amino acids exist. There is a unit of β-alanine, for example, contained within the structure of coenzyme A (Problem 13.27). Write the structural formula of β-alanine.

$$\overset{+}{H_3N}CH_2CH_2CO_2^-$$

β-Alanine

Problem 18.18 Although only L-amino acids occur in proteins, D-amino acids are often a part of the metabolism of lower organisms. The antibiotic actinomycin D, for example, contains a unit of D-valine, and the antibiotic bacitracin A contains units of D-asparagine and D-glutamic acid. Draw Fischer projections and three-dimensional representations for these three D-amino acids.

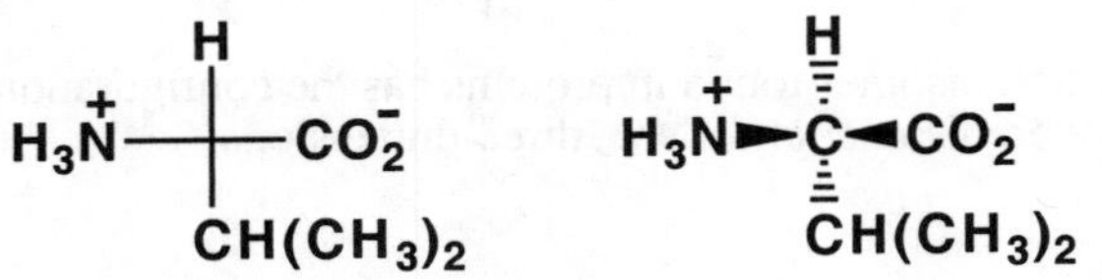

$$H_3\overset{+}{N}\text{---}\overset{\overset{H}{|}}{\underset{\underset{CH(CH_3)_2}{|}}{C}}\text{---}CO_2^-$$

D-Valine

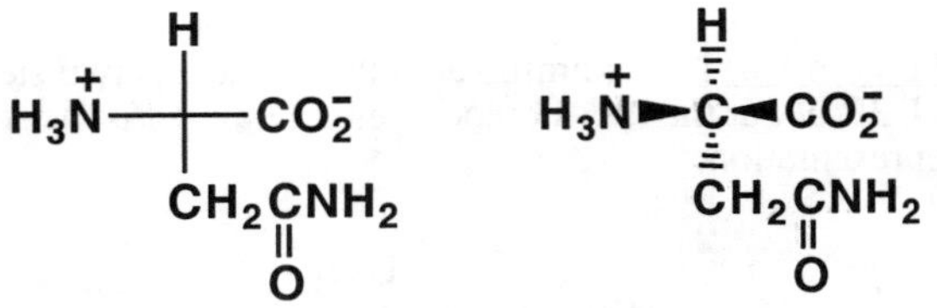

D-Asparagine

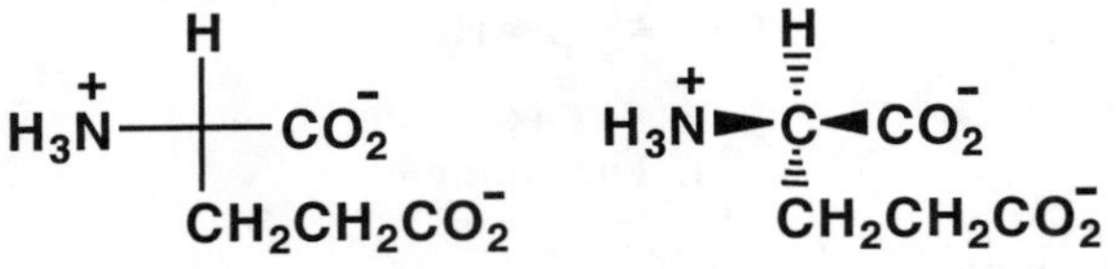

D-Glutamic acid

Problem 18.19 Histamine is synthesized from one of the 20 protein-derived amino acids. Suggest which amino acid is its biochemical precursor and the type of organic reaction(s) involved in its biosynthesis (e.g., oxidation, reduction, decarboxylation, nucleophilic substitution).

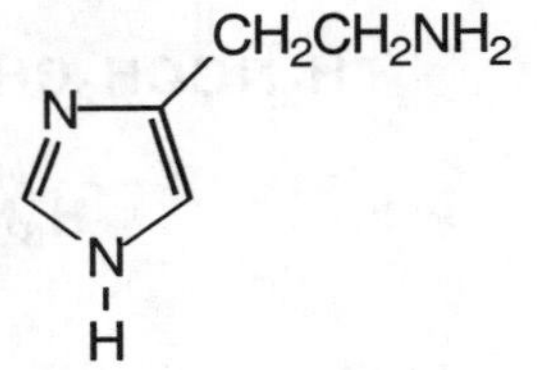

Histamine

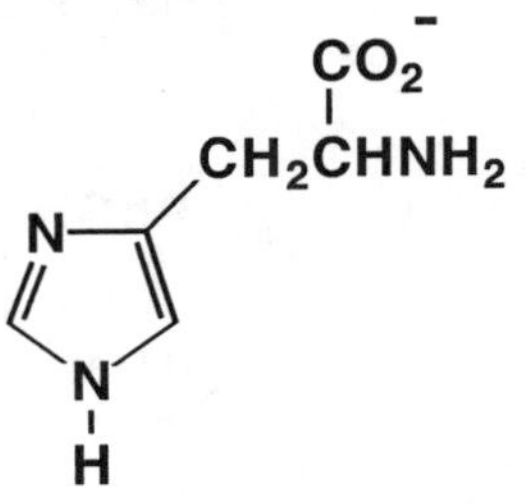

Histidine

Histamine is derived from the amino acid histidine and is the result of a biosynthetic decarboxylation reaction. Both the histamine and histidine are drawn in the form present at basic pH.

Problem 18.20 Both norepinephrine and epinephrine are synthesized from the same protein-derived amino acid. From which amino acid are they synthesized and what types of reactions are involved in their biosynthesis?

(a)

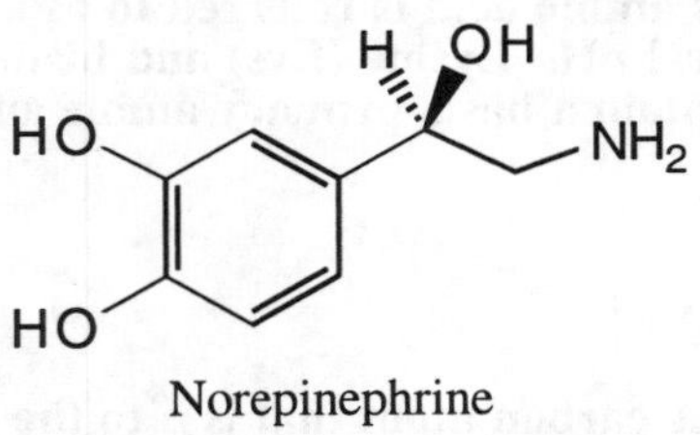

Norepinephrine

(b)

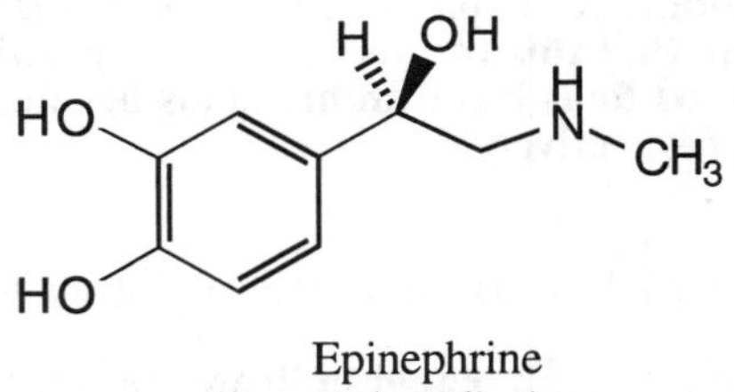

Epinephrine
(Adrenaline)

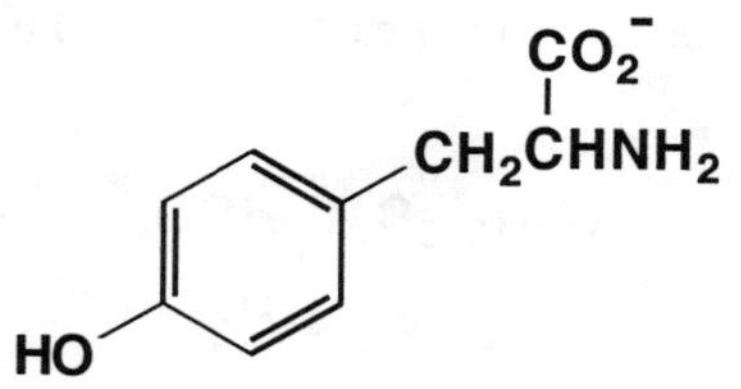

Tyrosine

Norepinephrine and epinephrine are derived from the amino acid tyrosine. In both cases, biosynthesis of these molecules involves decarboxylation, aromatic hydroxylation ortho to the original phenolic -OH group, and hydroxylation of the benzylic methylene group. Epinephrine is also methylated on the α-amino group. All of the molecules in this problem are drawn in the form present at basic pH.

Problem 18.21 From which amino acid are serotonin and melatonin synthesized and what types of reactions are involved in their biosynthesis?

(a) 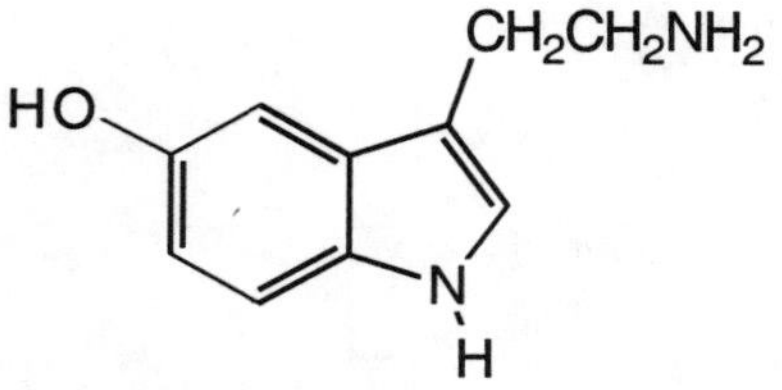   Serotonin

(b) 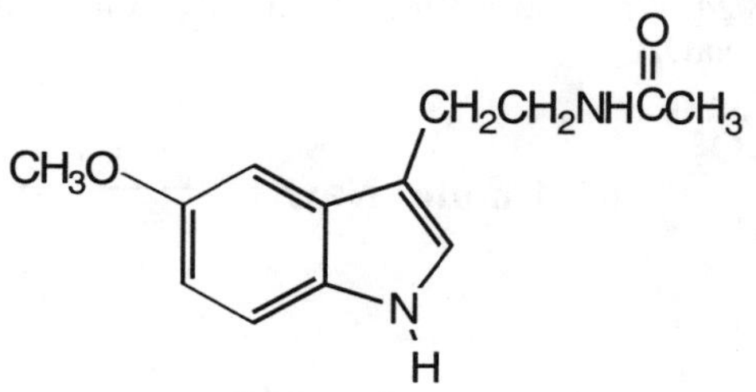   Melatonin

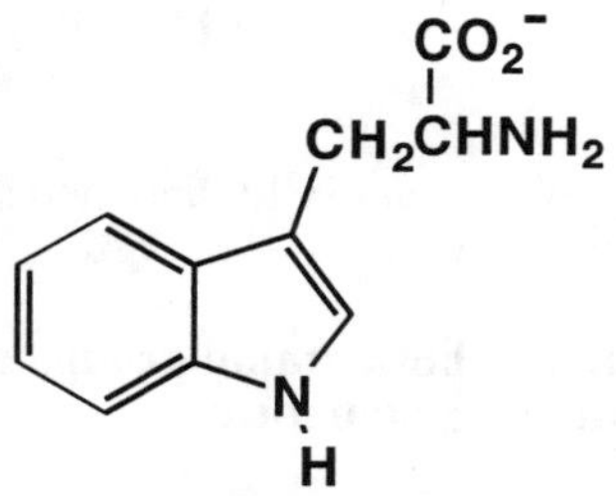

Tryptophan

Serotonin and melatonin are derived from the amino acid tryptophan. In both cases, biosynthesis of these molecules involves decarboxylation. In the case of serotonin there is also an aromatic hydroxylation. For melatonin there is an aromatic methoxy group added, and the amine is acetylated. Note that all of the molecules in the problem are drawn in the form present at basic pH.

Acid-Base Behavior of Amino Acids

Problem 18.22 Draw the structural formula for the form of each amino acid most prevalent at pH 1.0.
(a) Threonine   (b) Arginine

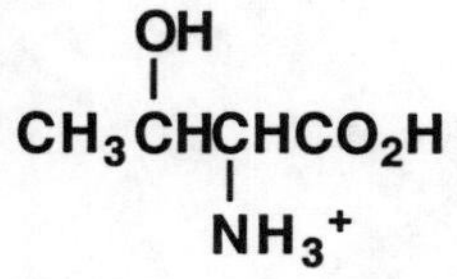

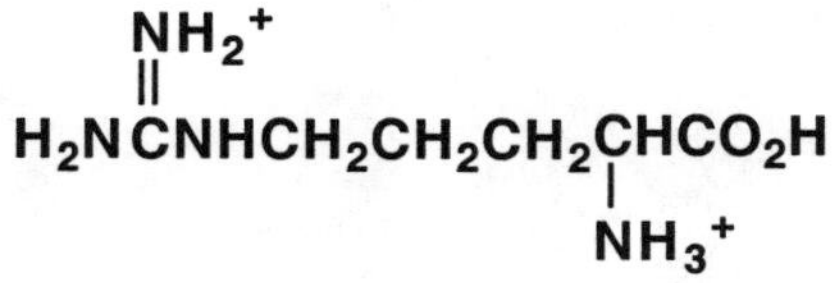

(c) Methionine 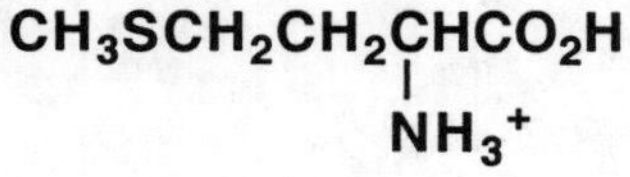   (d) Tyrosine

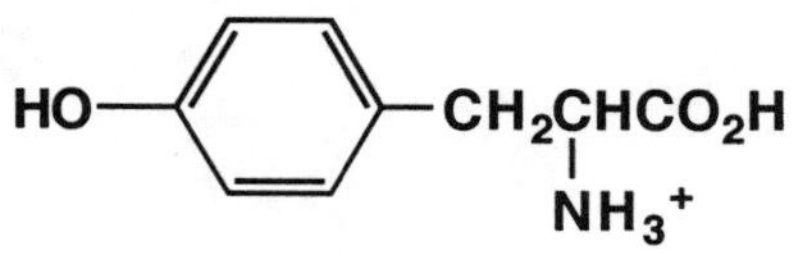

Problem 18.23 Draw the structural formula for the form of each amino acid most prevalent at pH 10.0.
(a) Leucine   (b) Valine

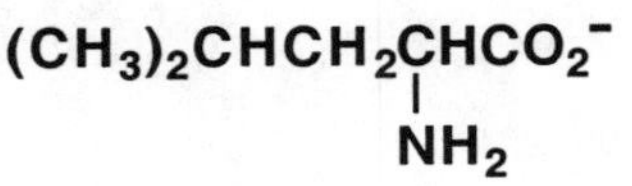

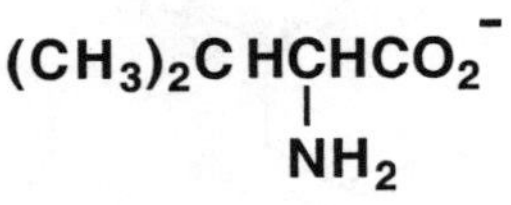

$(CH_3)_2CHCH_2CHCO_2^-$
                   |
                  NH_2

$(CH_3)_2CHCHCO_2^-$
                |
               NH_2

(c) Proline 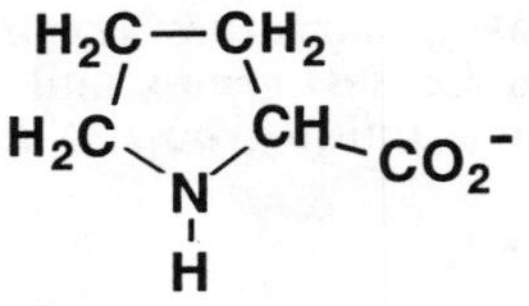   (d) Aspartic acid

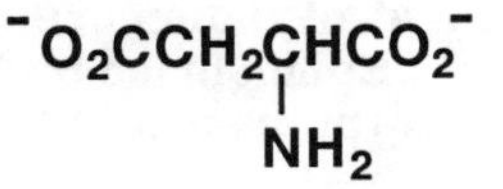

H_2C-CH_2
$H_2C\diagdown \quad CH-CO_2^-$
$\quad \diagdown N \diagup$
$\quad\quad |$
$\quad\quad H$

$^-O_2CCH_2CHCO_2^-$
                    |
                   NH_2

Problem 18.24 Write the zwitterion form of alanine and show its reaction with
(a) 1 mole NaOH

$CH_3CHCO_2^-$
        | $+$ 1 mole NaOH $\longrightarrow$
       NH_3^+

$CH_3CHCO_2^-$
        |
       NH_2

(b) 1 mole HCl

$CH_3CHCO_2^-$
        | $+$ 1 mole HCl $\longrightarrow$
       NH_3^+

CH_3CHCO_2H
        |
       NH_3^+

Problem 18.25 Write the form of lysine most prevalent at pH 1.0 and then show its reaction with the following. Consult Table 18.2 for pK_a values of the ionizable groups in lysine.

At pH 1.0, the most prevalent form of lysine has both amino groups as well as the carboxyl group protonated and a total charge of +2 as shown in the following structure.

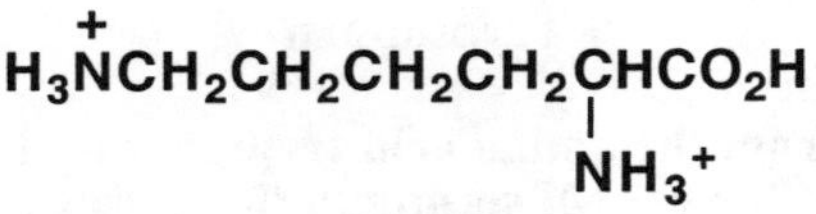

$\overset{+}{H_3}NCH_2CH_2CH_2CH_2CHCO_2H$
                                    |
                                   NH_3^+

Lysine

(a) 1 mole NaOH   (b) 2 moles NaOH

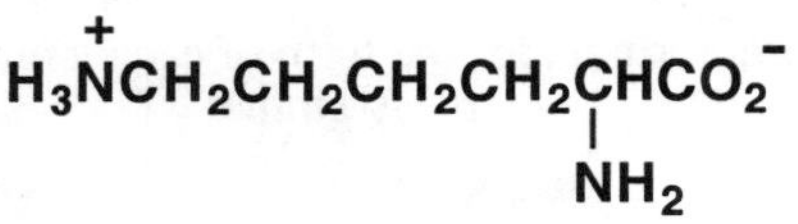

$\overset{+}{H_3}NCH_2CH_2CH_2CH_2CHCO_2^-$
                                    |
                                   NH_3^+

$\overset{+}{H_3}NCH_2CH_2CH_2CH_2CHCO_2^-$
                                    |
                                   NH_2

(c) 3 mole NaOH

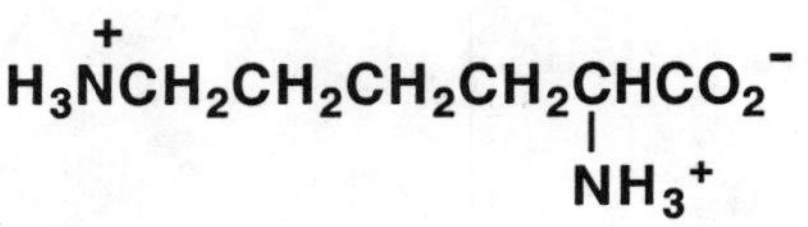

$H_2NCH_2CH_2CH_2CH_2CHCO_2^-$
                             |
                            NH_2

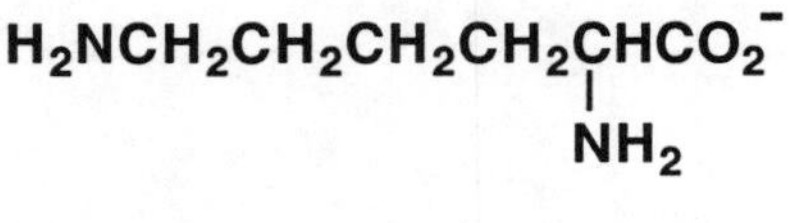

<u>Problem 18.26</u> Write the form of aspartic acid most prevalent at pH 1.0 and then show its reaction with the following. Consult Table 18.2 for pK_a values of the ionizable groups in aspartic acid.

At pH 1.0, the most prevalent form of aspartic acid has both carboxylic acid groups as well as the amino group protonated and a total charge of +1 as shown in the following structure.

$$HO_2CCH_2CHCO_2H$$
$$|$$
$$NH_3^+$$

Aspartic acid

(a) 1 mol NaOH   (b) 2 mol NaOH   (c) 3 mol NaOH

$$HO_2CCH_2CHCO_2^-$$ $$^-O_2CCH_2CHCO_2^-$$ $$^-O_2CCH_2CHCO_2^-$$
$$|$$ $$|$$ $$|$$
$$NH_3^+$$ $$NH_3^+$$ $$NH_2$$

<u>Problem 18.27</u> Given pK_a values for ionizable groups from Table 18.2, sketch curves for the titration of (a) glutamic acid with NaOH, and (b) histidine with NaOH.

Glutamic acid has pK_a values of 2.1, 4.07, and 9.47 so the titration curve would look something like the following:

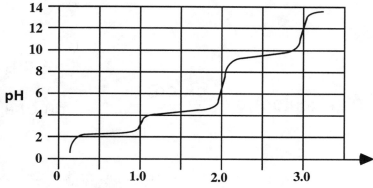

Histidine has pK_a values of 1.77, 6.10, and 9.18 so the titration curve would look something like the following:

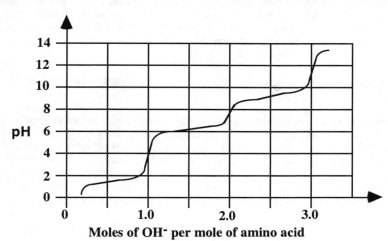

Problem 18.28 Draw a structural formula for the product formed when alanine is treated with the following reagents:
(a) Aqueous NaOH

(b) Aqueous HCl

(c) CH_3CH_2OH, H_2SO_4

(d) $(CH_3CO)_2O$, CH_3CO_2Na

Problem 18.29 Account for the fact that the isoelectric point of glutamine (pI 5.65) is higher than the isoelectric point of glutamic acid (pI 3.08).

Amino acids have no net charge at their pI. For this to happen with glutamic acid, the net charge on the α-carboxyl and side chain carboxyl groups must be -1 to balance the +1 charge of the α-amino group. This will occur at a pI = (1/2)(2.10 + 4.07) = 3.08. The amide side chain of glutamine is already neutral near neutral pH, so the pI of the amino acid is determined by the values for the only ionizable groups, namely, the α-carboxyl and α-amino groups, according to the equation pI = (1/2)(2.17 + 9.03) = 5.6. This value is near that of the other amino acids with non-ionizable functional groups on their side chains.

Problem 18.30 Enzyme-catalyzed decarboxylation of glutamic acid gives 4-aminobutanoic acid (Section 18.1D). Estimate the pI of 4-aminobutanoic acid.

There is little, if any, inductive effect operating between the amino and carboxyl groups of 4-aminobutanoic acid because there are three methylene groups between them. Thus, the pK$_a$ of the amino group of 4-aminobutanoic acid is like that of a simple amino group, near 10.0. Similarly, the pK$_a$ of the carboxyl group is like that of a simple carboxyl group, near 4.5. Given these estimates for the pK$_a$ values, the pI would be:

$$\mathbf{pI} = \frac{1}{2}(\mathbf{p}K_{\mathbf{a}}\ \alpha - \mathbf{CO_2H} + \mathbf{p}K_{\mathbf{a}}\ \alpha - \mathbf{NH_3}^+) = \frac{1}{2}(4.5 + 10.0) = 7.25$$

Problem 18.31 Guanidine and the guanidino group present in arginine are two of the strongest organic bases known. Account for this basicity.

The guanidino group is strongly basic because of resonance stabilization of the protonated guanidinium ion as shown below:

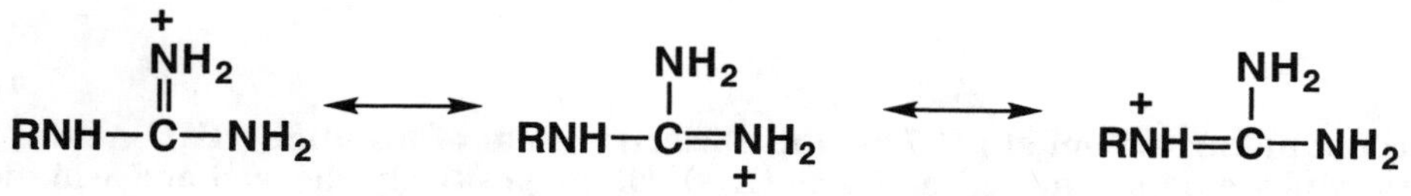

R = H or alkyl group

Problem 18.32 At pH 7.4, the pH of blood plasma, do the majority of protein-derived amino acids bear a net negative charge or a net positive charge?

The majority of amino acids have a pI near 5 or 6, so they will bear a net negative charge at pH 7.4. The only exceptions are arginine, histidine and lysine (the three basic amino acids).

Problem 18.33 Do the following compounds migrate to the cathode or to the anode on electrophoresis at the specified pH?

The key to determining which way a molecule migrates is to estimate its net charge at the given pH. Molecules with a net positive charge will migrate toward the negative electrode (cathode) and molecules with a net negative charge will migrate toward the positive electrode (anode). Molecules at a pH below their isoelectric point (Table 18.2) have a net positive charge, molecules at a pH above their isoelectric point have a net negative charge, and molecules at a pH that equals their isoelectric point have no net charge.

(a) Histidine at pH 6.8

pI = 7.64, so at pH 6.8 histidine has a net positive charge and migrates toward the negative electrode (cathode).

(b) Lysine at pH 6.8

pI = 9.74, so at pH 6.8 lysine has a net positive charge and migrates toward the negative electrode (cathode).

(c) Glutamic acid at pH 4.0

pI = 3.08, so at pH 4.0 glutamic acid has a net negative charge and migrates toward the positive electrode (anode).

(d) Glutamine at pH 4.0

pI = 5.65, so at pH 4.0 glutamine has a net positive charge and migrates toward the negative electrode (cathode).

(e) Glu-Ile-Val at pH 6.0

The carboxyl group on the glutamic acid side chain will be largely deprotonated at pH 6.0, as will the α-carboxyl group (on valine). The α-amino group (on glutamic acid) will be mostly protonated at pH 6.0. Thus, the molecule will have a net negative charge and will migrate toward the positive electrode (anode).

(f) Lys-Gln-Tyr at pH 6.0

The α-amino group and the side chain amino group of lysine will both be protonated at pH 6.0, while the α-carboxyl group of tyrosine will be deprotonated. Thus, the molecule will have a net positive charge at pH 6.0 and will migrate toward the negative electrode (cathode).

<u>Problem 18.34</u> At what pH would you carry out an electrophoresis to separate the amino acids in each mixture of amino acids?

Recall that an amino acid below its isoelectric point will have some degree of positive charge, an amino acid above its isoelectric point will have some degree of negative charge, and an amino acid at its isoelectric point will have no net charge.

(a) Ala, His, Lys

Electrophoresis could be carried out at pH 7.64, the isoelectric point of histidine (His). At this pH, the histidine is neutral and would not move, the lysine (Lys) will be positively charged and will move toward the negative electrode, and the alanine (Ala) will be slightly negatively charged and will move toward the positive electrode.

(b) Glu, Gln, Asp

Electrophoresis could be carried out at pH 3.08, the isoelectric point of glutamic acid (Glu). At this pH, the glutamic acid is neutral and would not move, the glutamine (Gln) will be positively charged and will move toward the negative electrode, and the aspartic acid (Asp) will be slightly negatively charged and will move toward the positive electrode.

(c) Lys, Leu, Tyr

Electrophoresis could be carried out at pH 6.04, the isoelectric point of leucine (Leu). At this pH, the leucine is neutral and would not move, the lysine (Lys) will be positively charged and will move toward the negative electrode, and the tyrosine (Tyr) will be slightly negatively charged and will move toward the positive electrode.

<u>Problem 18.35</u> Examine the amino acid sequence of human insulin (Figure 18.14) and list each Asp, Glu, His, Lys, and Arg in this molecule. Do you expect human insulin to have an isoelectric point nearer that of the acidic amino acids (pI 2.0-3.0), the neutral amino acids (pI 5.5-6.5), or the basic amino acids (pI 9.5-11.0)?

A listing of the amino acids present are shown below:

aspartic acid (Asp)	**0**
glutamic acid (Glu)	**4**
histidine (His)	**2**
lysine (Lys)	**1**
arginine (Arg)	**1**

The charge will be neutral only when there are four positively charged residues to neutralize the four negative charges of the carboxylates from the four Glu residues. For this to happen, the Lys, Arg, and both His residues must be positively charged. Since the imidazole of His is not protonated until the pH is below 6 or so, the entire molecule will be neutral only around this pH. Thus, insulin is expected to have an isoelectric point nearer to that of the neutral amino acids.

Primary Structure of Polypeptides and Proteins
<u>Problem 18.36</u> If a protein contains four different SH groups, how many different disulfide bonds are possible if only a single disulfide bond is formed? How many different disulfides are possible if two disulfide bonds are formed?

If only one disulfide bond were to be formed from the four different cysteine residues, then there would be a total of 6 different disulfide bonds possible. There would be three possibilities if two disulfide bonds were formed.

<u>Problem 18.37</u> How many different tetrapeptides can be made if
(a) The tetrapeptide contains one unit each of Asp, Glu, Pro, and Phe?

There could be any of the four residues in the first position, any of the remaining three amino acids in the second position, and so on. Thus, there are 4 x 3 x 2 x 1 = 24 possible tetrapeptides.

(b) All 20 amino acids can be used, but each only once?

Using the same logic as in (a), there are 20 x 19 x 18 x 17 = 116,280 possible tetrapeptides.

<u>Problem 18.38</u> A decapeptide has the following amino acid composition:

$$Ala_2, Arg, Cys, Glu, Gly, Leu, Lys, Phe, Val$$

Partial hydrolysis yields the following tripeptides:

$$Cys-Glu-Leu + Gly-Arg-Cys + Leu-Ala-Ala + Lys-Val-Phe + Val-Phe-Gly$$

One round of Edman degradation yields a lysine phenylthiohydantoin. From this information, deduce the primary structure of this decapeptide.

The Edman degradation result indicates that the Lys residue must be at the *N*-terminus. Given this information, the rest of the peptide sequence can be deduced from overlap among the tripeptide sequences as shown below.

The complete peptide is:
        Lys-Val-Phe-Gly-Arg-Cys-Glu-Leu-Ala-Ala

The peptides fit as follows:
        Lys-Val-Phe
            Val-Phe-Gly
                Gly-Arg-Cys
                    Cys-Glu-Leu
                        Leu-Ala-Ala

<u>Problem 18.39</u> Following is the primary structure of glucagon, a polypeptide hormone of 29 amino acids. Glucagon is produced in the α-cells of the pancreas and helps maintain blood glucose levels in a normal concentration range.

1	5	10	15

His-Ser-Glu-Gly-Thr-Phe-Thr-Ser-Asp-Tyr-Ser-Lys-Tyr-Leu-Asp-Ser-Arg-Arg-

20	25	29

Ala-Gln-Asp-Phe-Val-Gln-Trp-Leu-Met-Asn-Thr

Glucagon

Which peptide bonds are hydrolyzed when this polypeptide is treated with
(a) Phenyl isothiocyanate

This reagent hydrolyzes only the *N*-terminal amino acid, so the His-Ser bond would be hydrolyzed. The site of cleavage is indicated by the ✳.

1	5	10	15

His]✳[Ser-Glu-Gly-Thr-Phe-Thr-Ser-Asp-Tyr-Ser-Lys-Tyr-Leu-Asp-Ser-Arg-Arg-

20	25	29

Ala-Gln-Asp-Phe-Val-Gln-Trp-Leu-Met-Asn-Thr

(b) Chymotrypsin

Chymotrypsin catalyzes hydrolysis of peptide bonds that are located on the carboxyl side of phenylalanine, tyrosine, and tryptophan residues. The sites of cleavage are indicated by the ✳.

1   5   10   15
His-Ser-Glu-Gly-Thr-Phe]✳[Thr-Ser-Asp-Tyr]✳[Ser-Lys-Tyr]✳[Leu-Asp-Ser-Arg-Arg-

                 20   25   29
Ala-Gln-Asp-Phe]✳[Val-Gln-Trp]✳[Leu-Met-Asn-Thr

(c) Trypsin

Trypsin catalyzes hydrolysis of the peptide bonds that are located on the carboxyl side of arginine and lysine residues. The sites of cleavage are indicated by the ✳.

1   5   10   15
His-Ser-Glu-Gly-Thr-Phe-Thr-Ser-Asp-Tyr-Ser-Lys]✳[Tyr-Leu-Asp-Ser-

                   20   25   29
Arg]✳[Arg]✳[Ala-Gln-Asp-Phe-Val-Gln-Trp-Leu-Met-Asn-Thr

(d) Br-CN

Cyanogen bromide cleaves on the C-terminal side of methionine residues. The site of cleavage is indicated by the ✳.

1   5   10   15
His-Ser-Glu-Gly-Thr-Phe-Thr-Ser-Asp-Tyr-Ser-Lys-Tyr-Leu-Asp-Ser-Arg-Arg-

                 20   25   29
Ala-Gln-Asp-Phe-Val-Gln-Trp-Leu-Met]✳[Asn-Thr

Problem 18.40 A tetradecapeptide (14 amino acid residues) gives the following peptide fragments on partial hydrolysis. From this information, deduce the primary structure of this polypeptide. Fragments are grouped according to size.

Pentapeptide Fragments	Tetrapeptide Fragments
Phe-Val-Asn-Gln-His	Gln-His-Leu-Cys
His-Leu-Cys-Gly-Ser	His-Leu-Val-Glu
Gly-Ser-His-Leu-Val	Leu-Val-Glu-Ala

The complete peptide is:
Phe-Val-Asn-Gln-His-Leu-Cys-Gly-Ser-His-Leu-Val-Glu-Ala

The peptides fit as follows:
Phe-Val-Asn-Gln-His
             Gln-His-Leu-Cys
                   His-Leu-Cys-Gly-Ser
                             Gly-Ser-His-Leu-Val
                                     His-Leu-Val-Glu
                                           Leu-Val-Glu-Ala

Problem 18.41 Draw a structural formula of these tripeptides. Mark each peptide bond, the *N*-terminal amino acid, and the *C*-terminal amino acid.
(a) Phe-Val-Asn   (b) Leu-Val-Gln

Problem 18.42 Estimate the pI of each tripeptide in Problem 18.41.

These pI values can be estimated by using the pK$_a$ of the amino group for the *N*-terminal amino acid, and the pK$_a$ of the carboxylic acid group for the *C*-terminal amino acid. Using the values for the appropriate amino groups and carboxylic acid groups listed in table 18.2 leads to the values of pI = 1/2(9.24 + 2.02) = 5.63 and pI = 1/2(9.76 + 2.17) = 5.96 for (a) Phe-Val-Asn and (b) Leu-Val-Gln, respectively.

Problem 18.43 Glutathione (G-SH), one of the most common tripeptides in animals, plants, and bacteria, is a scavenger of oxidizing agents. In reacting with oxidizing agents, glutathione is converted to G-S-S-G.

Glutathione

(a) Name the amino acids in this tripeptide.

The amino acids in glutathione are glutamic acid (Glu), cysteine (Cys), and glycine (Gly).

(b) What is unusual about the peptide bond formed by the N-terminal amino acid?

The *N*-terminal glutamic acid is linked to the next residue by an amide bond with the carboxyl group of the side chain, not the α-carboxyl group.

(c) Write a balanced half-reaction for the reaction of two molecules of glutathione to form a disulfide bond. Is glutathione a biological oxidizing agent or a biological reducing agent?

$$2G\text{-}SH \rightarrow G\text{-}S\text{-}S\text{-}G + 2H^+ + 2e^-$$

The glutathione is oxidized in this process, so it is a biological reducing agent.

(d) Write a balanced equation for reaction of glutathione with molecular oxygen, O$_2$, to form G-S-S-G and H$_2$O. Is molecular oxygen oxidized or reduced in this process?

$$2G\text{-}SH + 1/2\ O_2 \rightarrow G\text{-}S\text{-}S\text{-}G + H_2O$$

The molecular oxygen is reduced in this process.

Problem 18.44 Following is a structural formula and ball-and-stick model for the artificial sweetener aspartame. Each amino acid has the L configuration.

$$\underset{\text{Aspartame}}{\overset{+}{H_3N}CHCNHCHCOCH_3}$$

Aspartame

(a) Name the two amino acids in this molecule.

Aspartame is composed of aspartic acid (Asp) attached via a peptide bond to the methyl ester of phenylalanine (Phe).

(b) Estimate the isoelectric point of Aspartame.

Using the values in Table 18.2 for the amino group and side chain carboxylic acid group of aspartic acid leads to pI = 1/2(9.82 + 3.86) = 6.84.

(c) Write the structural formula for the products of hydrolysis of aspartame in 1M HCl.

$$\overset{+}{H_3N}CHCNHCHCOCH_3 \quad \xrightarrow{\text{1M HCl}} \quad \overset{+}{H_3N}CHCO_2H \; + \; \overset{+}{H_3N}CHCO_2H \; + \; CH_3OH$$

Three-Dimensional Shapes of Polypeptides and Proteins
Problem 18.45 Examine the α-helix conformation. Are amino acid side chains arranged all inside the helix, all outside the helix, or randomly?

All of the amino acid side chains extend outside the helix.

Problem 18.46 Distinguish between intermolecular and intramolecular hydrogen bonding between the backbone groups on polypeptide chains. In what type of secondary structure do you find intermolecular hydrogen bonds? In what type do you find intramolecular hydrogen bonding?

Intermolecular hydrogen bonding is possible with β-sheet secondary structures, while only intramolecular hydrogen bonding is possible with α-helix secondary structures.

Problem 18.47 Many plasma proteins found in aqueous environment are globular in shape. Which amino acid side chains would you expect to find on the surface of a globular protein and in contact with the aqueous environment? Which would you expect to find inside, shielded from the aqueous environment? Explain.
(a) Leu   (b) Arg   (c) Ser   (d) Lys   (e) Phe

In general, charged or hydrophilic side chains are exposed to the aqueous solution on the surface of a globular protein. Thus, (b) Arg, (c) Ser, and (d) Lys will be on the surface. The hydrophobic amino acids (a) Leu and (e) Phe will generally be inside the protein, shielded from the aqueous environment.

CHAPTER 19
Solutions to Problems

<u>Problem 19.1</u> Draw a structural formula for each nucleotide.
(a) 2'-Deoxythymidine 5'-monophosphate

(b) 2'-Deoxythymidine 3'-monophosphate

<u>Problem 19.2</u> Draw a structural formula for the section of DNA that contains the base sequence CTG and is phosphorylated at the 3' end only.

<u>Problem 19.3</u> Write the complementary DNA base sequence for 5'-CCGTACGA-3'.

The complementary sequence would be 3'-GGCATGCT-5'.

<u>Problem 19.4</u> Here is a portion of the nucleotide sequence in phenylalanine tRNA.

3'-ACCACCUGCUCAGGCCUU-5'

Write the nucleotide sequence of its DNA complement.

Remember that the base uracil (U) in RNA is complementary to adenine (A) in DNA. The complement DNA sequence of the above RNA sequence would be:

5'-TGGTGGACGAGTCCGGAA-3'.

Problem 19.5 The following section of DNA codes for oxytocin, a polypeptide hormone:

3'-ACG-ATA-TAA-GTT-TTA-ACG-GGA-GAA-CCA-ACT-5'

(a) Write the base sequence of the mRNA synthesized from this section of DNA.

5'-UGC-UAU-AUU-CAA-AAU-UGC-CCU-CUU-GGU-UGA-3'

(b) Given the sequence of bases in part (a), write the primary structure of oxytocin.

The amino acid sequence of oxytocin would be:

Amino terminus- Cys-Tyr-Ile-Gln-Asn-Cys-Pro-Leu-Gly -Carboxyl terminus

Note how the last codon, UGA, does not code for an amino acid, but rather is the stop signal.

Problem 19.6 The following is another section of the bovine rhodopsin gene. Which of the endonucleases given in Example 19.6 will catalyze cleavage of this section?

FnuDII   **HpaII**

5'-ACGTCGGGTCGTCGTCCTCT CGCG GTGGTGAGTCTT CCGG CTCTTCT-3'

Only FnuDII and HpaII can cleave this sequence.

Nucleosides and Nucleotides
Problem 19.7 Two drugs used in the treatment of acute leukemia are 6-mercaptopurine and 6-thioguanine. In each drug, the oxygen at carbon 6 of the parent molecule is replaced by divalent sulfur. Draw structural formulas for the enethiol (the sulfur equivalent of an enol) forms of 6-mercaptopurine and 6-thioguanine.

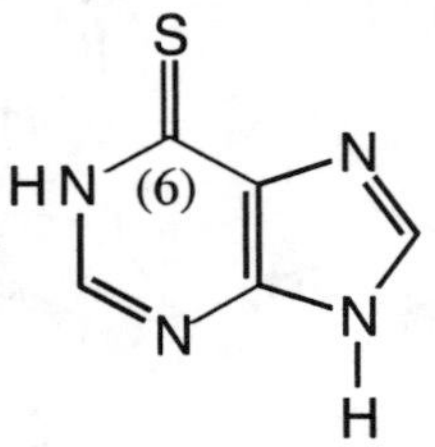

6-Mercaptopurine 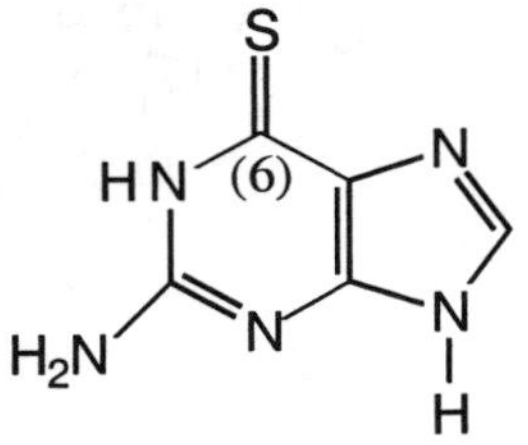   6-Thioguanine

The enethiol forms are as follows:

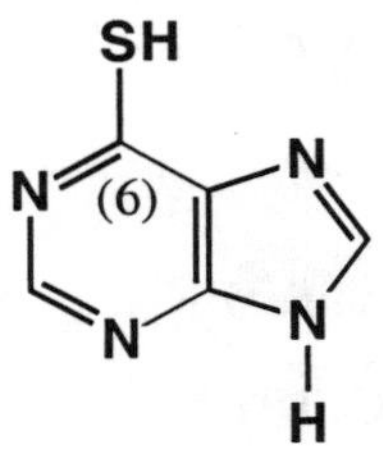

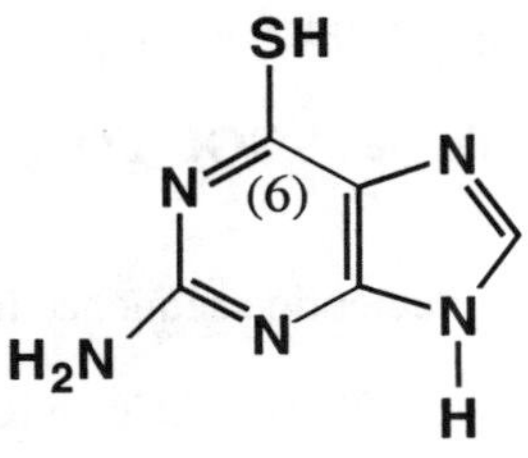

Problem 19.8 Following are structural formulas for cytosine and thymine. Draw two additional tautomeric forms for cytosine and three additional tautomeric forms for thymine.

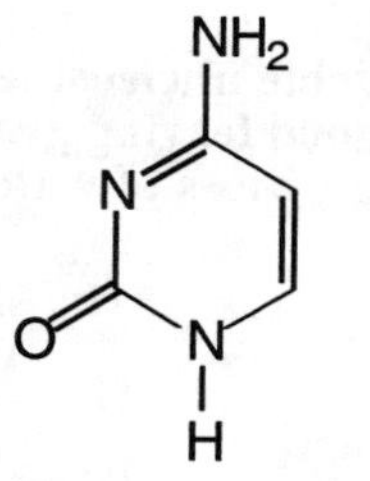

Cytosine (C) 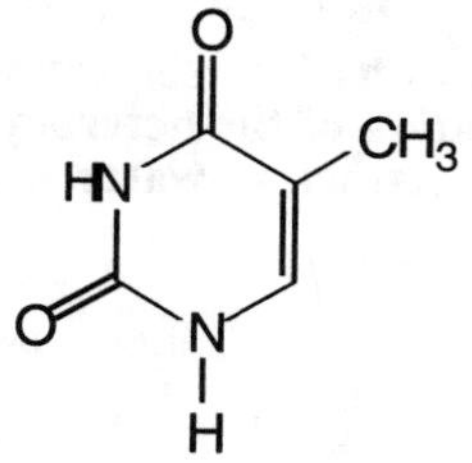   Thymine (T)

There are four additional tautomeric forms for cytosine:

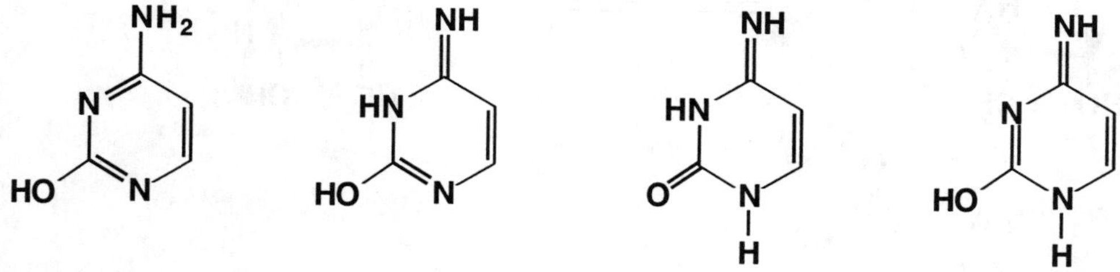

There are four additional tautomeric forms for thymine:

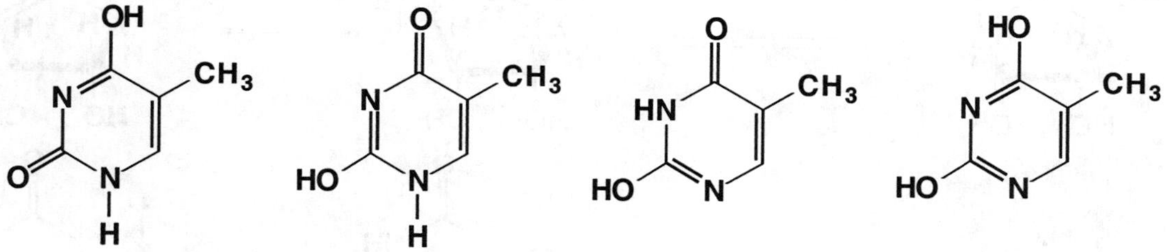

Problem 19.9 Draw a structural formula for a nucleoside composed of
(a) β-D-Ribose and adenine   (b) β-2-Deoxy-D-ribose and cytosine

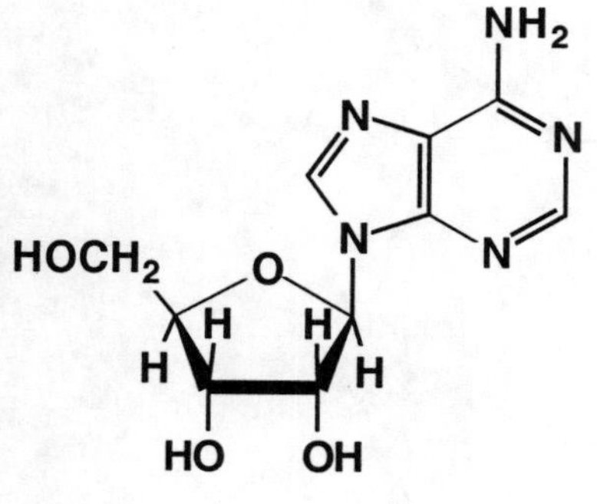

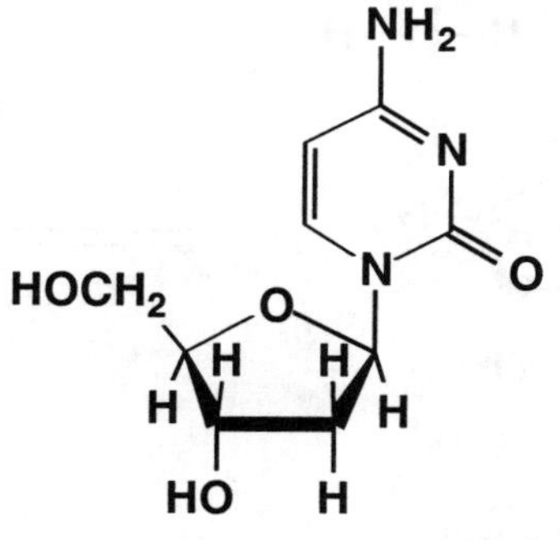

Problem 19.10 Nucleosides are stable in water and in dilute base. In dilute acid, however, the glycoside bond of a nucleoside undergoes hydrolysis to give a pentose and a heterocyclic aromatic amine base. Propose a mechanism for this acid-catalyzed hydrolysis.

Acid-catalyzed glycoside bond hydrolysis is most pronounced for purine nucleosides. A reasonable mechanism involves protonation of the heterocyclic base to create a good leaving group that departs to form a resonance stabilized cation. Attack of water with the cation followed by loss of water generates a pentose. This is an S_N1 mechanism.

Step 1:

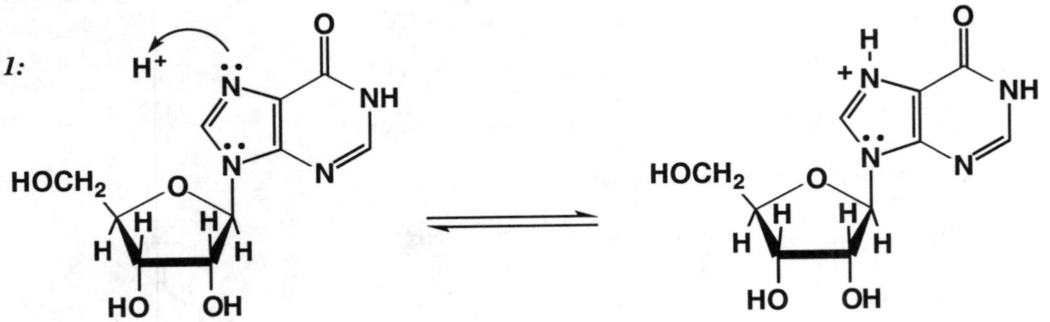

Step 2:

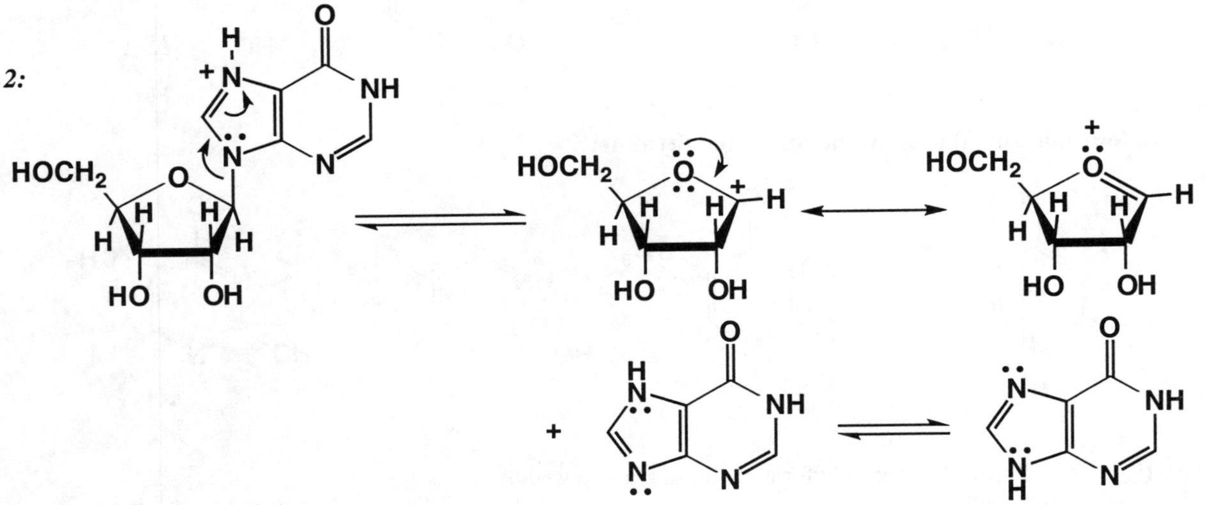

Step 3:

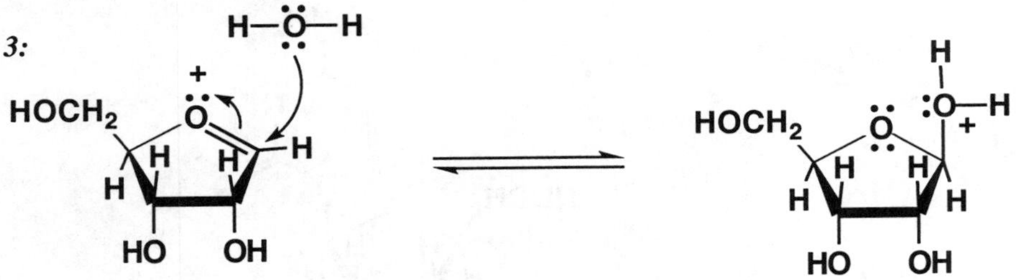

Step 4:

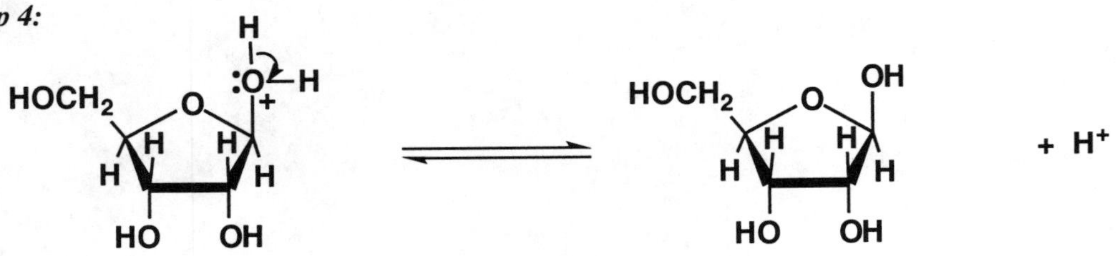

Problem 19.11 Explain the difference in structure between a nucleoside and a nucleotide.

A nucleoside consists of a D-ribose or 2'-deoxy-D-ribose bonded to an heterocylic aromatic amine base by a β-*N*-glycoside bond. A nucleotide is a nucleoside that has one or more molecules of phosphoric acid esterified at an -OH group of the monosaccharide, usually at the 3' and/or 5' -OH group.

Problem 19.12 Draw a structural formula for each nucleotide and estimate its net charge at pH 7.4, the pH of blood plasma.
(a) 2'-Deoxyadenosine 5'-triphosphate (dATP)

The values for the first three pK$_a$'s of dATP are all below 5.0, so these are fully deprotonated at pH 7.4. The fourth pK$_a$ of dATP is 7.0, so that at pH 7.4 the overall charge will be about midway between -3 and -4.

(b) Guanosine 3'-monophosphate (GMP)

The two pK$_a$ values for GMP are well below 7.4, so both are fully deprotonated giving an overall charge of -2.

(c) 2'-Deoxyguanosine 5'-diphosphate (dGDP)

The values for the first two pK$_a$'s of dGDP are both below 5.0, so these are fully deprotonated at pH 7.4. The third pK$_a$ of dGDP is 6.7, so it will be mostly deprotonated at pH 7.4, giving the molecule the net charge close to -3.

Problem 19.13 Cyclic-AMP, first isolated in 1959, is involved in many diverse biological processes as a regulator of metabolic and physiological activity. In it, a single phosphate group is esterified with both the 3' and 5' hydroxyls of adenosine. Draw a structural formula of cyclic-AMP.

Cyclic-AMP

The Structure of DNA

Problem 19.14 Why are deoxyribonucleic acids called acids? What are the acidic groups in their structure?

Deoxyribonucleic acids are called acids because the phosphodiester groups of the backbone are acidic. At neutral pH, they are fully deprotonated, leading to the anionic nature of DNA.

Problem 19.15 Human DNA contains approximately 30.4% A. Estimate the percentages of G, C, and T and compare them with the values presented in Table 19.1.

The A residues must be paired with T residues, so one would predict that there must also be 30.4% T. A and T must therefore account for 30.4% + 30.4% = 60.8% of the bases. That leaves (100% - 60.8%) / 2 = 19.6% each for G and C. In Table 19.1, there is actually slightly less T than expected, so there is also slightly more G and C than expected.

Problem 19.16 Draw a structural formula of the DNA tetranucleotide 5'-A-G-C-T-3'. Estimate the net charge on this tetranucleotide at pH 7.0. What is the complementary tetranucleotide to this sequence?

As shown in the structure, there is a net charge of -5 on this tetranucleotide at pH 7.0. This oligonucleotide is self-complementary, that is, the complementary oligonucleotide also has the sequence 5'-A-G-C-T-3'.

Problem 19.17 List the postulates of the Watson-Crick model of DNA secondary structure.

Major postulates of the Watson-Crick model are that:
1) A molecule of DNA consists of two antiparallel strands coiled in a right handed manner about the same axis, thereby creating a double helix.
2) The bases project inward toward the helix axis.
3) The bases are paired through hydrogen bonding, with a purine paired to a pyrimidine, so that each base pair is of the same size and shape.
4) In particular, A pairs with T and G pairs with C.
5) The paired bases are stacked one on top of another in the interior of the double helix.
6) There is a distance of 0.34 nm between adjacent stacked paired bases.
7) There are ten paired bases per turn of the helix, and these are slightly offset from each other. The slight offset provides two grooves of different dimensions along the helix, the so-called major and minor grooves.

Problem 19.18 The Watson-Crick model is based on certain experimental observations of base composition and molecular dimensions. Describe these observations and show how the Watson-Crick model accounts for each.

Chargaff found that in different organisms, the amount of A is very close to the amount of T and the amount of G always is very close to the amount of C, even though different organisms have different ratios of A to G. The base-pairing postulates of the Watson-Crick model fully explain the observed ratios of bases. The geometry of the Watson-Crick model also accounts perfectly for the periodicity and thickness observed in the X-ray diffraction data of Franklin and Wilkins.

Problem 19.19 Compare the α-helix of proteins and the double helix of DNA in these ways:
(a) The units that repeat in the backbone of the polymer chain.

The α-helix of a protein is composed of amino acids, so the repeating unit of the backbone is a carboxyl group bonded to a tetrahedral carbon atom and a nitrogen atom. The carboxyl group and nitrogen atoms are linked via amide bonds. The repeating unit of the double helix in DNA is a 2'-deoxy-D-ribose unit linked via 3'-5' phosphodiester bonds.

(b) The projection in space of substituents along the backbone (the R groups in the case of amino acids; purine and pyrimidine bases in the case of double-stranded DNA) relative to the axis of the helix.

The α-helix of a protein has the R groups pointed out away from the helix axis. The DNA bases of the double helix are pointed inward, toward the helix axis.

Problem 19.20 Discuss the role of the hydrophobic interactions in stabilizing:
(a) Double-stranded DNA

In the DNA double helix, the relatively hydrophobic bases are stacked on the inside, surrounded by the relatively hydrophilic sugar-phosphate backbone that is on the outside of the structure. The stacking of the hydrophobic bases minimizes contact with water.

(b) Lipid bilayers

In lipid bilayers, the hydrophobic hydrocarbon tails are associated with each other to form the hydrophobic inner layer, while the polar head groups are associated with each other on both outside surfaces.

(c) Soap micelles

In micelles, the hydrophobic hydrocarbon tails are associated with each other to form the hydrophobic interior, while the polar groups are associated with each other on the outside surface.

Problem 19.21 Name the type of covalent bond(s) joining monomers in these biopolymers.
(a) Polysaccharides   (b) Polypeptides   (c) Nucleic acids

Polysaccharides have glycosidic linkages, polypeptides have amide linkages and nucleic acids have phosphodiester linkages between the monomers.

<u>Problem 19.22</u> In terms of hydrogen bonding, which is more stable, an A-T base pair or a G-C base pair?

A G-C base pair is held together by three hydrogen bonds, while an A-T base pair is held together by only two hydrogen bonds. Thus, a G-C base pair is more stable than an A-T base pair.

<u>Problem 19.23</u> At elevated temperatures, nucleic acids become denatured, that is, they unwind into single-stranded DNA. Account for the observation that the higher the G-C content of a nucleic acid, the higher the temperature required for its thermal denaturation.

G-C base pairs have three hydrogen bonds between them, while A-T base pairs have only two. Thus, the G-C base pairs are held together with stronger overall attractive forces and require higher temperatures to denature.

<u>Problem 19.24</u> Write the DNA complement for 5'-ACCGTTAAT-3'. Be certain to label which is the 5' end and which is the 3' end of the complement strand.

The complementary sequence is 3'-TGGCAATTA-5'.

<u>Problem 19.25</u> Write the DNA complement for 5'-TCAACGAT-3'.

The complementary sequence is 3'-AGTTGCTA-5'.

Ribonucleic Acids
<u>Problem 19.26</u> Compare the degree of hydrogen bonding in the base pair A-T found in DNA with that in the base pair A-U found in RNA.

The only difference between uracil (U) and thymine (T) is the methyl group at the 5 position of thymine. The presence or absence of this methyl group has little, if any, influence on hydrogen bonding.

<u>Problem 19.27</u> Compare DNA and RNA in these ways.
(a) Monosaccharide units

DNA contains 2'-deoxy-D-ribose units, while RNA contains D-ribose units.

(b) Principal purine and pyrimidine bases

DNA		RNA	
Purines	**Pyrimidines**	**Purines**	**Pyrimidines**
Adenine	**Thymine**	**Adenine**	**Uracil**
Guanine	**Cytosine**	**Guanine**	**Cytosine**

(c) Primary structure

The monosaccharide unit in DNA is 2'-deoxy-D-ribose while the monosaccharide unit in RNA is D-ribose. The bases are the same in the two types of nucleic acids, except that thymine is found in DNA while uracil is found in RNA. DNA is usually double stranded and RNA is primarily single stranded. In both DNA and RNA, the primary sequence consists of linear chains of the nucleic acids linked by phosphodiester bonds involving the 3' and 5' hydroxyl groups of the monosaccharide units.

(d) Location in the cell

DNA is found in cell nuclei, while the bulk of RNA occurs as ribosome particles in the cytoplasm.

(e) Function in the cell

DNA serves to store and transmit genetic information, and RNA is primarily involved with the transcription and translation of that genetic information during the synthesis of proteins.

Problem 19.28 What type of RNA has the shortest lifetime in cells?

Messenger RNA has the shortest lifetime in cells, usually on the order of a few minutes or less. This short lifetime is thought to allow for very tight control over how much protein is synthesized in the cell at any one time.

Problem 19.29 Write the mRNA complement for 5'-ACCGTTAAT-3'. Be certain to label which is the 5' end and which is the 3' end of the mRNA strand.

The mRNA complement would be 3'-UGGCAAUUA-5'.

Problem 19.30 Write the mRNA complement for 5'-TCAACGAT-3'.

The mRNA complement would be 3'-AGUUGCUA-5'.

The Genetic Code
Problem 19.31 What does it mean to say that the genetic code is degenerate?

The genetic code is referred to as degenerate because more than one codon can code for the same amino acid. There are 64 different codons, but only twenty amino acids. Three of the codons represent stop signals.

Problem 19.32 Write the mRNA codons for
(a) Valine **GUU, GUC, GUA, GUG** (b) Histidine **CAU, CAC**
(c) Glycine **GGU, GGC, GGA, GGG**

Problem 19.33 Aspartic acid and glutamic acid have carboxyl groups on their side chains and are called acidic amino acids. Compare the codons for these two amino acids.

All of the codons for these two acidic amino acids begin with GA. The codons for aspartic acid are GAU and GAC, while the codons for glutamic acid are GAA and GAG. This suggests that earlier in evolution, there may have been only one acidic amino acid, coded for by all four of these codons.

Problem 19.34 Compare the structural formulas of the amino acids phenylalanine and tyrosine. Compare also the codons for these two amino acids.

$$\text{Phenylalanine}: \quad \text{C}_6\text{H}_5-\text{CH}_2\text{CHCO}_2^- \ (\text{NH}_3^+)$$

$$\text{Tyrosine}: \quad \text{HO}-\text{C}_6\text{H}_4-\text{CH}_2\text{CHCO}_2^- \ (\text{NH}_3^+)$$

Phenylalanine   **Tyrosine**

Phenylalanine has a phenyl side chain, while tyrosine has a phenol side chain. The mRNA codons for phenylalanine are UUU and UUC, while the mRNA codons for tyrosine are UAU and UAC. The single base difference for these two very similar amino acids suggests that tyrosine may have been introduced later in evolution than phenylalanine (Phe is the biological precursor of Tyr, so probably was present earlier).

Problem 19.35 Glycine, alanine, and valine are classified as nonpolar amino acids. Compare the codons for these three amino acids. What similarities do you find? What differences do you find?

Glycine	Alanine	Valine
GGU	GCU	GUU
GGC	GCC	GUC
GGA	GCA	GUA
GGG	GCG	GUG

All of these amino acids have four mRNA codons, all codons start with G, and in each case, the first two bases of the codon are identical for a given amino acid. This makes the last base irrelevant.

Problem 19.36 Codons in the set CUU, CUC, CUA, and CUG all code for the amino acid leucine. In this set, the first and second bases are identical, and the identity of the third base is irrelevant. For what other sets of codons is the third base also irrelevant, and for what amino acid(s) does each set code?

The third base is also irrelevant for GUX (valine), GCX (alanine), GGX (glycine), ACX (threonine), CCX (proline), CGX (arginine), and UCX (serine). In the preceding codons, X stands for any of the bases.

Problem 19.37 Compare the codons with a pyrimidine, either U or C, as the second base. Do the majority of the amino acids specified by these codons have hydrophobic or hydrophilic side chains?

The majority of amino acids with a pyrimidine in the second position of their codons are hydrophobic. This set contains phenylalanine, leucine, isoleucine, methionine, valine, proline, and alanine. Only serine and threonine have a pyrimidine in the second position and also have hydrophilic side chains.

Problem 19.38 Compare the codons with a purine, either A or G, as the second base. Do the majority of the amino acids specified by these codons have hydrophilic or hydrophobic side chains?

The majority of amino acids with a purine in the second position of their codons are hydrophilic. This set contains histidine, glutamine, asparagine, lysine, aspartic acid, glutamic acid, arginine, cysteine, serine and glycine (glycine can be considered a hydrophilic amino acid because it has only a hydrogen atom as a side chain). Only tryptophan is not hydrophilic. Tyrosine is a special case wherein the side chain is aromatic, but it bears a polar hydroxyl group.

Problem 19.39 What polypeptide is coded for by this mRNA sequence?

5'-GCU-GAA-GUC-GAG-GUG-UGG-3'

This mRNA codes for the following polypeptide:

Amino terminus- Ala-Glu-Val-Glu-Val-Trp -Carboxyl terminus.

Problem 19.40 The alpha chain of human hemoglobin has 141 amino acids in a single polypeptide chain. Calculate the minimum number of bases on DNA necessary to code for the alpha chain. Include in your calculation the bases necessary for specifying termination of polypeptide synthesis.

The minimum number of bases needed for the alpha chain of human hemoglobin must code for the 141 amino acids as well as three extra bases for the stop codon. Therefore, the minimum number of bases that will be required is (3 x 141) + (1 x 3) = 426 bases.

Problem 19.41 In HbS, the human hemoglobin found in individuals with sickle-cell anemia, glutamic acid at position 6 in the beta chain is replaced by valine.
(a) List the two codons for glutamic acid and the four codons for valine.

The two mRNA codons for glutamic acid are GAA and GAG, while the four mRNA codons for valine are GUU, GUC, GUA, and GUG.

(b) Show that one of the glutamic acid codons can be converted to a valine codon by a single substitution mutation, that is, by changing one letter in one codon.

Both of the glutamic acid codons can be converted to valine by replacing the central A with a U residue.

CHAPTER 20
Solutions to Problems

<u>Problem 20.1</u> Under anaerobic (without oxygen) conditions, glucose is converted to lactate by a metabolic pathway called anaerobic glycolysis or, alternatively, lactate fermentation. Is anaerobic glycolysis a net oxidation, a net reduction, or neither?

$$C_6H_{12}O_6 \xrightarrow[\text{glycolysis}]{\text{Anaerobic}} 2\ \underset{\text{Lactate}}{CH_3\overset{\underset{|}{OH}}{C}HCO_2^-} + 2\ H^+$$

Glucose

The overall process of anaerobic glycolysis that converts glucose to lactate is neither an oxidation nor a reduction, because the oxidative and reductive steps cancel each other. As seen in Example 20.1, converting one molecule of glucose to two molecules of pyruvate is an oxidation involving a total of four electrons. Two molecules of NAD+ are utilized as oxidizing agents, each being reduced to NADH by a hydride transfer (a two electron reduction). However, this is balanced exactly by the next step in which two molecules of pyruvate are converted into two molecules of lactate (reduction of a carbonyl group to a hydroxyl group). Each of these reactions is a two electron reduction. The four electrons come from two molecules of NADH to regenerate two molecules of NAD+.

<u>Problem 20.2</u> Does lactate fermentation result in an increase or decrease in blood pH?

Lactate fermentation leads to an increase of the H+ concentration in the bloodstream, therefore the bloodstream pH must decrease.

β-Oxidation
<u>Problem 20.3</u> Write structural formulas for palmitic, oleic, and stearic acids, the three most abundant fatty acids.

Palmitic and stearic acids are fully saturated, having 16 and 18 carbons in their chains, respectively. Oleic acid has 18 carbons and a single cis double bond.

Palmitic acid

Oleic acid

Stearic acid

Problem 20.4 A fatty acid must be activated before it can be metabolized in cells. Write a balanced equation for the activation of palmitic acid.

Activation of a fatty acid involves formation of a thioester with coenzyme A. The proton is derived from the thiol group of CoA-SH.

$$CH_3(CH_2)_{14}\overset{\overset{\displaystyle O}{\|}}{C}O^- \ + \ CoA\text{-}SH \ + \ ATP \longrightarrow \ CH_3(CH_2)_{14}\overset{\overset{\displaystyle O}{\|}}{C}SCoA$$

$$+ \ AMP \ + \ P_2O_7^{4-} \ + \ H^+$$

Palmitic acid
(as anion)

Problem 20.5 Name three coenzymes necessary for β-oxidation of fatty acids. From what vitamin is each derived?

The three coenzymes needed for β-oxidation are:
  1) Coenzyme A (CoA-SH), derived from the vitamin pantothenic acid.
  2) Nicotinamide adenine dinucleotide (NAD+), derived from the vitamin niacin.
  3) Flavin adenine dinucleotide (FAD), derived from the vitamin riboflavin (vitamin B$_2$).
All three coenzymes contain the heterocyclic aromatic amine base adenosine.

Problem 20.6 We have examined β-oxidation of saturated fatty acids, such as palmitic acid and stearic acid. Oleic acid, an unsaturated fatty acid, is also a common component of dietary fats and oils. This unsaturated fatty acid is degraded by β-oxidation but, at one stage in its degradation, requires an additional enzyme named enoyl-CoA isomerase. Why is this enzyme necessary, and what isomerization does it catalyze? (Hint: Consider both the configuration of the carbon-carbon double bond in oleic acid and its position in the carbon chain.)

If you count the carbon atoms in oleic acid carefully, you will see that after three rounds of β-oxidation you are left with a fragment having a cis double bond. Enzyme-catalyzed rotation of the cis double bond would give a trans double bond, but the double bond would not be in the correct position for β-oxidation to continue. Thus, the enoyl-CoA-isomerase must also change the position of the double bond, so that it is in conjugation with the carbonyl group, as shown in the following scheme.

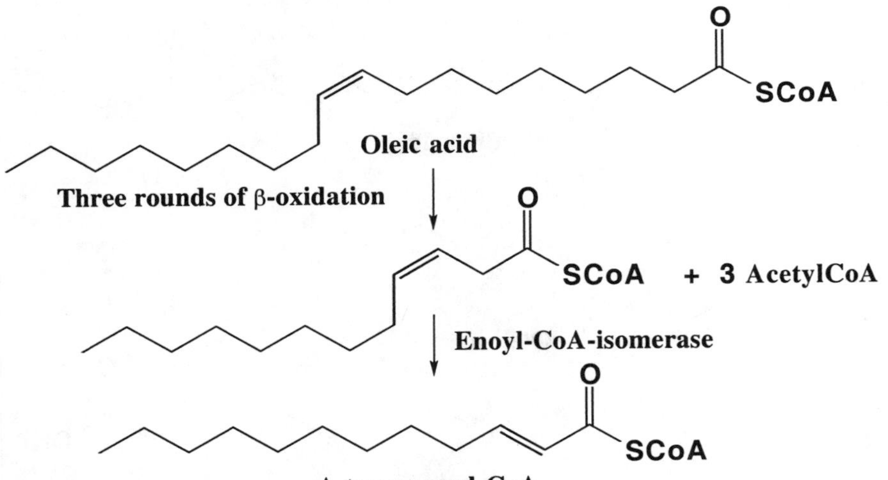

Oleic acid

Three rounds of β-oxidation

+ 3 AcetylCoA

Enoyl-CoA-isomerase

A *trans*-enoyl-CoA

Glycolysis
Problem 20.7 Name one coenzyme required for glycolysis. From what vitamin is it derived?

The one coenzyme required for glycolysis is NAD+, which is derived from the vitamin niacin.

Problem 20.8 Number the carbon atoms of glucose 1 through 6 and show from which carbon atom of glucose the carboxyl group of each molecule of pyruvate is derived.

By numbering the carbon atoms of glucose and following the different atoms through the pathway, it can be seen that the carboxyl group carbon atoms are derived from carbon atoms 3 and 4 of glucose.

$$^1\text{CHO}$$
$$\text{H}-^2\text{C}-\text{OH}$$
$$\text{HO}-^3\text{C}-\text{H}$$
$$\text{H}-^4\text{C}-\text{OH}$$
$$\text{H}-^5\text{C}-\text{OH}$$
$$^6\text{CH}_2\text{OH}$$

D-Glucose

$$\xrightarrow{\text{Glycolysis}}$$

$$\underset{1\quad2\quad3}{\text{CH}_3-\overset{\text{O}}{\overset{||}{\text{C}}}-\overset{\text{O}}{\overset{||}{\text{C}}}\text{O}^-} \quad + \quad \underset{6\quad5\quad4}{\text{CH}_3-\overset{\text{O}}{\overset{||}{\text{C}}}-\overset{\text{O}}{\overset{||}{\text{C}}}\text{O}^-}$$

Pyruvate

Problem 20.9 How many moles of lactate are produced from 3 moles of glucose?

During anaerobic glycolysis, 2 moles of lactate are produced for each mole of glucose used. Six moles of lactate will be produced from 3 moles of glucose.

Problem 20.10 Although glucose is the principal source of carbohydrates for glycolysis, fructose and galactose are also metabolized for energy.
(a) What is the main dietary source of fructose? Of galactose?

The main dietary source of D-fructose is the disaccharide sucrose (table sugar), in which D-fructose is combined with D-glucose. The main dietary source of D-galactose is the disaccharide lactose, from milk, in which D-galactose is combined with D-glucose.

(b) Propose a series of reactions by which fructose might enter glycolysis.

Fructose could be converted to fructose 6-phosphate and enter glycolysis at reaction 3, where it will be converted to fructose 1,6-bisphosphate.

$$\text{CH}_2\text{OH}$$
$$\text{C}=\text{O}$$
$$\text{HO}-\text{C}-\text{H}$$
$$\text{H}-\text{C}-\text{OH}$$
$$\text{H}-\text{C}-\text{OH}$$
$$\text{CH}_2\text{OH}$$

D-Fructose

$$\xrightarrow{\text{Phosphorylation}}$$

$$\text{CH}_2\text{OH}$$
$$\text{C}=\text{O}$$
$$\text{HO}-\text{C}-\text{H}$$
$$\text{H}-\text{C}-\text{OH}$$
$$\text{H}-\text{C}-\text{OH}$$
$$\text{CH}_2\text{OPO}_3^{2-}$$

Fructose 6-phosphate

(c) Propose a series of reactions by which galactose might enter glycolysis.

D-Galactose can be converted to D-glucose by enzyme-catalyzed inversion of configuration at C-4, and thereby enter glycolysis at step 1.

$$
\begin{array}{ccc}
\text{CHO} & & \text{CHO} \\
\text{H—C—OH} & & \text{H—C—OH} \\
\text{HO—C—H} & \xrightarrow{\textbf{Epimerization}} & \text{HO—C—H} \\
\text{HO—C—H} & & \text{H—C—OH} \\
\text{H—C—OH} & & \text{H—C—OH} \\
\text{CH}_2\text{OH} & & \text{CH}_2\text{OH}
\end{array}
$$

D-Galactose   **D-Glucose**

Problem 20.11 How many moles of ethanol are produced per mole of sucrose through the reactions of glycolysis and alcoholic fermentation? How many moles of CO_2 are produced?

A total of four moles of ethanol and four moles of carbon dioxide are produced from one mole of sucrose. This can be determined be remembering that one mole of the disaccharide sucrose is first hydrolyzed to one mole of glucose and one mole of fructose. Each of these 6-carbon monosaccharides enter glycolysis to give two moles of pyruvate, so a total of four moles of pyruvate are produced for each mole of sucrose used. Each mole of pyruvate is converted to one mole of ethanol and one mole of carbon dioxide, so a total of four moles of ethanol and four moles of carbon dioxide are produced for each mole of sucrose.

Problem 20.12 Glycerol derived from hydrolysis of triglycerides and phospholipids is also metabolized for energy. Propose a series of reactions by which the carbon skeleton of glycerol might enter glycolysis and be oxidized to pyruvate.

Glycerol enters glycolysis through the following enzyme catalyzed steps that lead to glyceraldehyde 3-phosphate, which is converted into pyruvate via the normal glycolysis pathway.

$$
\begin{array}{cccc}
\text{CH}_2\text{OH} & \text{CH}_2\text{OH} & \text{CH}_2\text{OH} & \text{CHO} \\
\text{HO—C—H} \longrightarrow & \text{HO—C—H} \longrightarrow & \text{O=C} \longrightarrow & \text{H—C—OH} \\
\text{CH}_2\text{OH} & \text{CH}_2\text{OPO}_3^{2-} & \text{CH}_2\text{OPO}_3^{2-} & \text{CH}_2\text{OPO}_3^{2-}
\end{array}
$$

Glycerol   **Glycerol**   **Dihydroxyacetone**   **Glyceraldehyde**
                    3-phosphate   **phosphate**   **3-phosphate**

Problem 20.13 Ethanol is oxidized in the liver to acetate ion by NAD^+.
(a) Write a balanced equation for this oxidation.

The production of acetate ion from ethanol is an overall four electron oxidation, so two moles of NAD^+ are required for every mole of ethanol. In addition, two protons are produced along with the proton that dissociates from acetic acid to give acetate.

$$
\text{CH}_3\text{CH}_2\text{OH} \;+\; 2\,\text{NAD}^+ \;+\; \text{H}_2\text{O} \longrightarrow \; \overset{\overset{\text{O}}{\|}}{\text{CH}_3\text{C}}\text{O}^- \;+\; 2\,\text{NADH} \;+\; 3\,\text{H}^+
$$

(b) Do you expect the pH of blood plasma to increase, decrease, or remain the same as a result of metabolism of a significant amount of ethanol?

The pH of blood plasma will drop due to the protons produced as the result of metabolism of a significant amount of ethanol.

Problem 20.14 Write a mechanism to show the role of NADH in the reduction of acetaldehyde to ethanol.

For this reduction, NADH delivers a hydride equivalent, and a group on the enzyme (denoted as A) delivers a proton to the oxygen atom. Note how the lone pair of electrons on the nitrogen in the ring is used as a source of electrons for the reaction.

NADH   NAD$^+$

Problem 20.15 When pyruvate is reduced to lactate by NADH, two hydrogens are added to pyruvate: one to the carbonyl carbon, the other to the carbonyl oxygen. Which of these hydrogens is derived from NADH?

As can be seen in the mechanism given in the answer to Problem 20.14, the NADH delivers a hydride equivalent, H-. This species is highly nucleophilic and reacts with the electrophilic carbonyl carbon atom.

Problem 20.16 Review the oxidation reactions of glycolysis and β-oxidation and compare the types of functional groups oxidized by NAD$^+$ with those oxidized by FAD.

NAD$^+$ oxidizes a secondary alcohol to a ketone (reaction 3 of β-oxidation), as well as an aldehyde to a carboxylic acid derivative (reaction 6 of glycolysis). FAD oxidizes a carbon-carbon single bond to a carbon-carbon double bond (reaction 1 of β-oxidation).

Problem 20.17 Why is glycolysis called an anaerobic pathway?

Glycolysis is called an anaerobic pathway because no oxygen is used. Glycolysis probably first evolved in organisms that appeared before there was oxygen in the environment.

Problem 20.18 Which carbons of glucose end up in CO_2 as a result of alcoholic fermentation?

As shown in the answer to Problem 20.8, carbons 3 and 4 of D-glucose end up as the carboxylic acid carbons of pyruvate. These same two carbon atoms, carbons 3 and 4, end up as CO_2 as a result of alcoholic fermentation.

D-Glucose   Pyruvate   Ethanol

<u>Problem 20.19</u> Which steps in glycolysis require ATP? Which steps produce ATP?

Reactions 1 and 3 of glycolysis require ATP, while reactions 7 and 10 produce ATP.

<u>Problem 20.20</u> The respiratory quotient (RQ) is used in studies of energy metabolism and exercise physiology. It is defined as the ratio of the volume of carbon dioxide produced to the volume of oxygen used:

$$RQ = \frac{\text{Volume } CO_2}{\text{Volume } O_2}$$

(a) Show that RQ for glucose is 1.00. (Hint: Look at the balanced equation for complete oxidation of glucose to carbon dioxide and water.)

In the balanced reaction for the complete oxidation of glucose into CO_2 and H_2O, 6 moles of O_2 are used and 6 moles of CO_2 are produced, so the RQ is 6/6 = 1.00.

$$C_6H_{12}O_6 \; + \; 6 \; O_2 \longrightarrow \; 6 \; CO_2 \; + \; 6 \; H_2O$$
$$\text{\textbf{D-Glucose}}$$

(b) Calculate RQ for triolein, a triglyceride of molecular formula $C_{57}H_{104}O_6$.

In the balanced equation for the complete oxidation of triolein, 80 moles of O_2 are used and 57 moles of CO_2 are produced for each mole of triolein consumed. The RQ = 57/80 = 0.71

$$C_{57}H_{104}O_6 \; + \; 80 \; O_2 \longrightarrow \; 57 \; CO_2 \; + \; 52 \; H_2O$$
$$\text{\textbf{Triolein}}$$

(c) For an individual on a normal diet, RQ is approximately 0.85. Would this value increase or decrease if ethanol were to supply an appreciable portion of caloric needs?

In the balanced equation for the complete oxidation of ethanol, C_2H_6O, 3 moles of O_2 are used and 2 moles of CO_2 are produced for each mole of ethanol consumed. The RQ = 2/3 = 0.67, so the individual's RQ would decrease if ethanol were to supply an appreciable portion of caloric needs.

$$C_2H_6O \; + \; 3 \; O_2 \longrightarrow \; 2 \; CO_2 \; + \; 3 \; H_2O$$
$$\text{\textbf{Ethanol}}$$

<u>Problem 20.21</u> Acetoacetate, β-hydroxybutyrate, and acetone are commonly known within the health sciences as ketone bodies, in spite of the fact that one of them is not a ketone at all. They are products of human metabolism and are always present in blood plasma. Most tissues, with the notable exception of the brain, have the enzyme systems necessary to use them as energy sources. Synthesis of ketone bodies occurs by the following enzyme-catalyzed reactions. Enzyme names are (1) thiolase, (2) β-hydroxy-β-methylglutaryl-CoA synthase, (3) β-hydroxy-β-methylglutaryl-CoA lyase, and (5) β-hydroxybutyrate dehydrogenase. Reaction (4) is spontaneous and uncatalyzed.

Describe the type of reaction involved in each step and the type of mechanism by which each occurs.

Reaction 1 is a Claisen condensation (Section 14.3) between two molecules of acetyl-CoA.
Reaction 2 is an aldol reaction (Section 14.2) that can be thought of as taking place between the enolate anion of acetyl-CoA and the ketone carbonyl of acetoacetyl-CoA.
Reaction 3 is a reverse aldol reaction (Section 14.2) that generates acetyl-CoA and acetoacetate.
Reaction 4 is a decarboxylation of a β-ketoacid (Section 12.8A) that generates CO_2 and acetone from acetoacetate.
Reaction 5 is a reduction of the ketone group of acetoacetate to a secondary alcohol (Section 11.10).

Problem 20.22 A connecting point between anaerobic glycolysis and β-oxidation is formation of acetyl-CoA. Which carbon atoms of glucose appear as methyl groups of acetyl-CoA? Which carbon atoms of palmitic acid appear as methyl groups of acetyl-CoA?

As shown in the answer to Problem 20.8, carbons 3 and 4 of D-glucose end up as the carboxylic acid carbons of pyruvate. These same two carbon atoms, carbons 3 and 4 end up as CO_2 as a result of oxidation and decarboxylation to acetyl CoA. This means that carbons 1 and 6 end up as the methyl groups of acetyl CoA.

Palmitic acid undergoes β-oxidation to produce acetyl-CoA, so the even number carbon atoms (2,4,6,8,10,12,14,16) end up being the methyl groups.

CHAPTER 21
Solutions to Problems

<u>Problem 21.1</u> Calculate the energy of red light (680 nm) in kilocalories per mole. Which form of radiation carries more energy, infrared radiation of wavelength 2.50 μm or red light of wavelength 680 nm?

Combining the two equations given in the text gives:

$$E = h\nu = h \left(\frac{c}{\lambda}\right)$$

Plugging in the appropriate values gives the desired answer:

$$E = \frac{(9.537 \times 10^{-14} \text{ kcal-sec-mol}^{-1})(3.00 \times 10^8 \text{ m-sec}^{-1})}{680 \times 10^{-9} \text{ m}} = \boxed{42.1 \text{ kcal-mol}^{-1}}$$

Notice how the units cancel to give the final answer in kcal-mol^{-1}. As can be seen from the equations, the longer the wavelength, the lower the energy, thus red light carries more energy.

<u>Problem 21.2</u> State the number of sets of equivalent hydrogens in each compound and the number of hydrogens in each set.

Numbers have been added to the carbon atoms of the structures to aid in referring to specific hydrogens. Use the "test atom" approach if you have trouble seeing the answers.

(a) 3-Methylpentane

There are four sets of equivalent hydrogens. <u>Set 1:</u> 6 hydrogens from the methyl groups of carbon atoms 1 and 6. <u>Set 2:</u> 4 hydrogens from the -CH$_2$- groups of carbon atoms 2 and 5. <u>Set 3:</u> 3 hydrogens from the methyl group of carbon atom 4. <u>Set 4:</u> 1 hydrogen from the -CH- group of carbon atom 3.

2,2,2-Trimethylpentane

There are four sets of equivalent hydrogens. <u>Set 1:</u> 9 hydrogens from the methyl groups of carbon atoms 6, 7 and 8. <u>Set 2:</u> 6 hydrogens from the methyl groups of carbon atoms 1 and 3. <u>Set 3:</u> 2 hydrogens from the -CH$_2$- group of carbon atom 4. <u>Set 4:</u> 1 hydrogen from the -CH- group of carbon atom 2.

<u>Problem 21.3</u> Each compound gives only one signal in its ^{1}H-NMR spectrum. Propose a structural formula for each.

In order for these molecules to give a single signal in their ^{1}H-NMR spectrum, each of the hydrogen nuclei must be in an identical environment. This will only occur in symmetrical molecules.

(a) C_3H_6O (b) C_5H_{10} (c) C_5H_{12} (d) $C_4H_6Cl_4$

Problem 21.4 The line of integration of the two signals in the ^{1}H-NMR spectrum of a ketone of molecular formula $C_7H_{14}O$ shows a vertical rise of 62 and 10 chart divisions, respectively. Calculate the number of hydrogens giving rise to each signal, and propose a structural formula for this ketone.

The ratio of signals is approximately 6:1, which corresponds to a 12:2 ratio of hydrogens. Thus, the larger signal represents 12 hydrogens and the smaller signal represents 2 hydrogens. A structure consistent with this assignment is 2,4-dimethyl-3-pentanone as shown below:

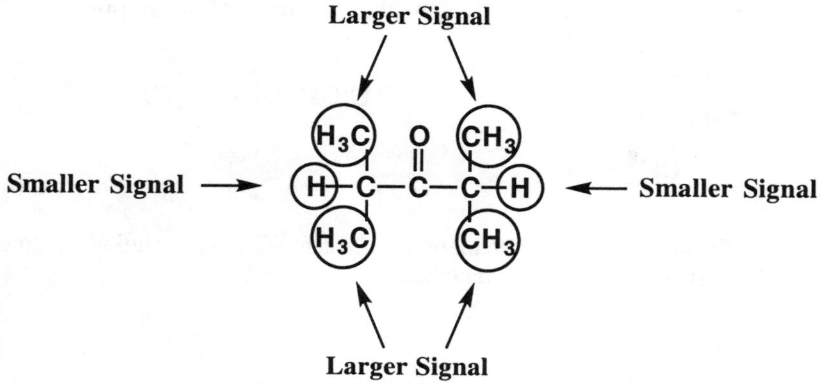

Problem 21.5 Following are two constitutional isomers of molecular formula $C_4H_8O_2$:

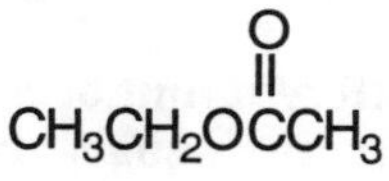

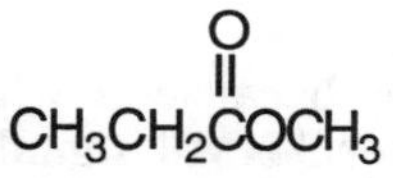

(a) Predict the number of signals in the ^{1}H-NMR spectrum of each isomer.

Each compound will exhibit three signals, one each for the two different -CH_3 groups and a third from the -CH_2- group.

(b) Predict the ratio of areas of the signals in each spectrum.

The ratio of signals in each case will be 3:3:2.

Problem 21.6 Following are pairs of constitutional isomers. Predict the number of signals and the splitting pattern of each signal in the ^{1}H-NMR spectrum of each isomer.

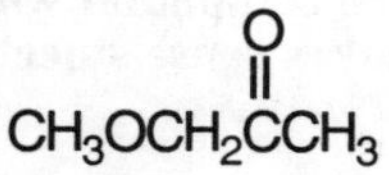

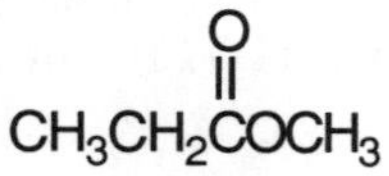

The ketone on the left and the ester on the right will each have three different signals, with the splitting patterns as indicated.

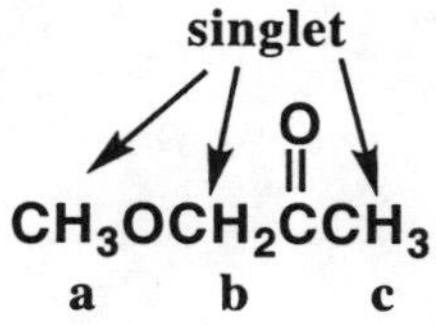

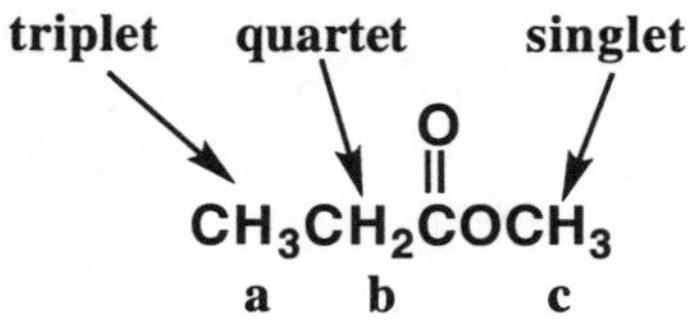

(b)   CH₃CCH₃   and   ClCH₂CH₂CH₂Cl

with Cl substituents (top and bottom Cl on central carbon)

The molecule on the left will have one signal and the molecule on the right will have two signals with splitting patterns as indicated.

singlet

$$CH_3CCH_3$$
a   Cl   a
Cl

triplet   quintet   triplet

$$ClCH_2CH_2CH_2Cl$$
a   b   a

Problem 21.7 Explain how to distinguish between the members of each pair of constitutional isomers based on the number of signals in the ^{13}C-NMR spectrum of each member.

(a) methylenecyclohexane with carbons labeled a-e   and   methylcyclohexene with carbons labeled a-g

These molecules can be distinguished because they have different numbers of nonequivalent carbon nuclei and thus will have different numbers of ^{13}C-NMR signals. Different signals are indicated by different letters on the structures. The molecule on the left has higher symmetry and will have 5 different signals corresponding to the carbon atoms labeled a - e, while the molecule on the right has less symmetry and will have 7 different signals corresponding to the carbon atoms labeled a - g.

(b) $\overset{a}{C}H_3\overset{b}{C}H=\overset{c}{C}H\overset{d}{C}H_2\overset{e}{C}H_2\overset{f}{C}H_3$ and $\overset{c}{C}H_3\overset{b}{C}H_2\overset{a}{C}H=\overset{a}{C}H\overset{b}{C}H_2\overset{c}{C}H_3$

The molecule on the left has lower symmetry and will have 6 different signals corresponding to the carbon atoms labeled a - f, while the molecule on the right has more symmetry and will only have 3 different signals corresponding to the carbon atoms labeled a - c.

Problem 21.8 Calculate the index of hydrogen deficiency of cyclohexene, and account for this deficiency by reference to its structural formula.

The molecular formula for cyclohexene is C_6H_{10}. The molecular formula for the reference compound with 6 carbon atoms is C_6H_{14}. Thus the index of hydrogen deficiency is (14-10)/2 or 2. This makes sense since cyclohexene has one ring and one pi bond.

Cyclohexene

Problem 21.9 The index of hydrogen deficiency of niacin is 5. Account for this index of hydrogen deficiency by reference to the structural formula of niacin.

Nicotinamide
(Niacin)

The index of hydrogen deficiency of niacin is 5 because there are four pi bonds and one ring in the structure.

Index of Hydrogen Deficiency
Problem 21.10 Complete the following table.

Class of Compound	Molecular Formula	Index of Hydrogen Deficiency	Reason for Hydrogen Deficiency
alkane	C_nH_{2n+2}	0	(reference hydrocarbon)
alkene	C_nH_{2n}	1	one pi bond
alkyne	C_nH_{2n-2}	**2**	**two pi bonds**
alkadiene	C_nH_{2n-2}	**2**	**two pi bonds**
cycloalkane	C_nH_{2n}	**1**	**one ring**
cycloalkene	C_nH_{2n-2}	**2**	**one ring and one pi bond**

Problem 21.11 Calculate the index of hydrogen deficiency of each compound.
(a) Aspirin, $C_9H_8O_4$ (b) Ascorbic acid (vitamin C), $C_6H_8O_6$

(20-8)/2 = 6   **(14-8)/2 = 3**

(c) Pyridine, C_5H_5N (d) Urea, CH_4N_2O

(13-5)/2 = 4 (nitrogen correction)   **(6-4)/2 = 1 (nitrogen correction)**

(e) Cholesterol, $C_{27}H_{46}O$ (f) Trichloroacetic acid, $C_2HCl_3O_2$

(56-46)/2 = 5   **(3-1)/2 = 1 (halogen correction)**

Interpretation of ^{1}H-NMR and ^{13}C-NMR Spectra
Problem 21.12 Following are structural formulas for the cis isomers of 1,2-, 1,3-, and 1,4-dimethylcyclohexanes and three sets of ^{13}C-NMR spectral data. Assign each constitutional isomer its correct spectral data.

Spectrum 1:	Spectrum 2:	Spectrum 3:
31.35	34.20	44.60
30.67	31.30	35.14
20.85	23.56	32.88
	15.97	26.54
		23.01

These constitutional isomers are most readily distinguished by the number of sets of nonequivalent carbon atoms and thus different ^{13}C signals. Compound (a) has 4 sets of nonequivalent carbon atoms corresponding to spectrum 2, compound (b) has 5 sets of nonequivalent carbon atoms corresponding to spectrum 3, and compound (c) has 3 sets of nonequivalent carbon atoms corresponding to spectrum 1. The different sets of equivalent carbon atoms are indicated by the letters.

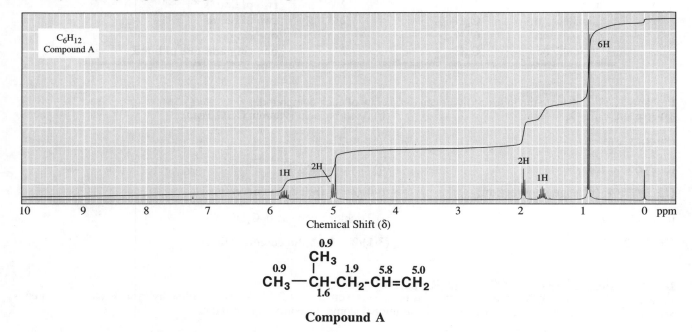

<u>Problem 21.13</u> Following is a ^1H-NMR spectrum for compound A, molecular formula C_6H_{12}. Compound A decolorizes a solution of bromine in carbon tetrachloride. Propose a structural formula of compound A.

Compound A has an index of hydrogen deficiency of 1, in the form of a double bond as evidenced by the reaction with Br_2. The rest of the detailed structure can be deduced from the spectrum.

For the spectral interpretations in this problem and the rest of the chapter, the chemical shift (δ) is given followed by the relative integration, the multiplicity of the peak (singlet, doublet, triplet, etc.) and finally the identity of the hydrogens giving rise to the signal are shown in bold.

Compound A

^1H-NMR δ 5.8 (1H, multiplet; this is more complex than expected because the adjacent vinylic hydrogens are not equivalent, -CH=), 5.0 (2H, multiplet; this is asymmetric because these two vinylic hydrogens are not equivalent, =CH₂), 1.9 (2H, multiplet; doublet of doublets, -CH₂-), 1.6 (1H, multiplet; a triplet of septets, -CH), 0.9 (6H, one doublet, -CH₃). The chemical shifts associated with each set of hydrogens are indicated on the structure.

Problem 21.14 Following is a ^{1}H-NMR spectrum of compound B, C_7H_{12}. Compound B decolorizes a solution of Br_2 in CCl_4. Propose a structural formula for compound B.

The molecular formula indicates that there is an index of hydrogen deficiency of 2, so there are two rings and/or pi bonds.

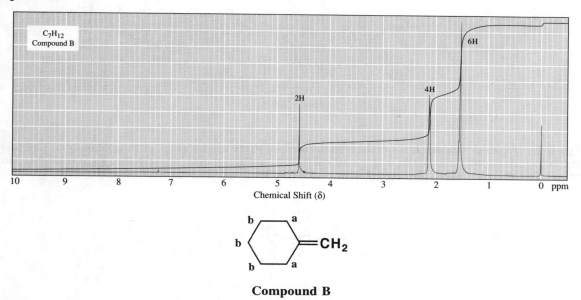

Compound B

^{1}**H-NMR δ 4.6 (2H, singlet, =CH$_2$), 2.1 (4H, broad peak, the -CH$_2$- groups adjacent to the sp^2 carbon atom on the ring labeled as "a" on the structure), 1.6 (6H, broad peak, the three -CH$_2$- groups labeled as "b" on the structure).**

Problem 21.15   Following are ^{1}H-NMR spectra for compounds C and D, each of molecular formula $C_5H_{12}O$. Each is a liquid at room temperature, is slightly soluble in water, and reacts with sodium metal with the evolution of a gas. Propose structural formulas for compounds C and D.

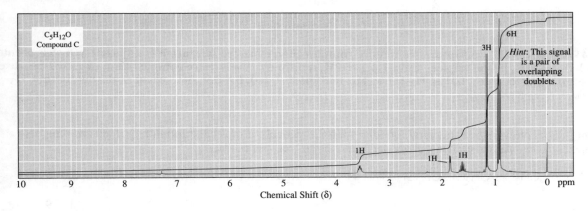

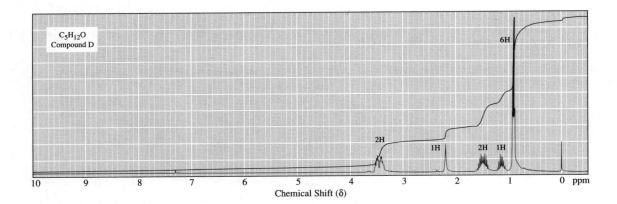

The index of hydrogen deficiency is 0 for these molecules, so there are no rings or double bonds. The fact that the compounds are slightly soluble in water and react with sodium metal indicate that each molecule has an -OH group. The chemical shifts associated with each set of hydrogens are indicated on the structures.

$$
\begin{array}{c}
\overset{0.9}{\underset{}{CH_3}} \\
\overset{0.9}{CH_3}-\underset{\underset{\underset{1.85}{OH}}{\overset{1.6}{|}}}{\overset{}{CH}}-\overset{3.5}{CH}-\overset{1.15}{CH_3}
\end{array}
$$

Compound C

¹H-NMR δ 3.5 (1H, multiplet, -CH-OH-), 1.85 (1H, doublet, -OH), 1.6 (1H, multiplet, -CH-(CH₃)₂), 1.15 (3H, doublet, -C(OH)-CH₃), 0.9 (6H, overlapping doublets, -CH-(CH₃)₂).

$$
\begin{array}{c}
\overset{0.8\text{-}0.9}{\underset{}{CH_3}} \\
\overset{0.8\text{-}0.9}{CH_3}\overset{}{CH_2}\underset{\underset{1.4\text{-}1.6}{}}{C}\overset{1.1}{H}\underset{3.4\text{-}3.5}{CH_2}\overset{}{OH}
\end{array}
$$
1.4-1.6   3.4-3.5   2.2

Compound D

¹H-NMR δ 3.4-3.5 (2H, multiplet; this is more complex than expected because it is adjacent to a stereocenter, -CH₂-OH), 2.2 (1H, broad triplet, -OH), 1.4-1.6 (2H, multiplet; this is more complex than expected because it is adjacent to a stereocenter, CH₃-CH₂-), 1.1 (1H, multiplet, -CH-), 0.8-0.9 (6H, broad multiplet, both -CH₃ groups).

Problem 21.16 Following are structural formulas for three alcohols of molecular formula C₇H₁₆O and three sets of ¹³C-NMR spectral data. Assign each constitutional isomer its correct spectral data.

(a) CH₃CH₂CH₂CH₂CH₂CH₂CH₂OH

(b) CH₃ĊCH₂CH₂CH₂CH₃ with OH above and CH₃ below the central carbon

(c) CH₃CH₂ĊCH₂CH₃ with OH above and CH₂CH₃ below the central carbon

Spectrum 1:	Spectrum 2:	Spectrum 3:
74.66	70.97	62.93
30.54	43.74	32.79
7.73	29.21	31.86
	26.60	29.14
	23.27	25.75
	14.09	22.63
		14.08

These constitutional isomers are most readily distinguished by the number of sets of nonequivalent carbon atoms and thus different ^{13}C signals. Using the following analysis, it can be seen that compound (a) has 7 sets of nonequivalent carbon atoms corresponding to spectrum 3, compound (b) has 6 sets of nonequivalent carbon atoms corresponding to spectrum 2, and compound (c) has 3 sets of nonequivalent carbon atoms corresponding to spectrum 1.

$$\overset{g}{C}H_3\overset{f}{C}H_2\overset{e}{C}H_2\overset{d}{C}H_2\overset{c}{C}H_2\overset{b}{C}H_2\overset{a}{C}H_2OH$$

$$\begin{array}{c} \overset{OH}{|} \\ \overset{e}{C}H_3\overset{a}{\underset{|}{C}}\overset{b}{C}H_2\overset{c}{C}H_2\overset{d}{C}H_2\overset{f}{C}H_3 \\ \underset{e}{C}H_3 \end{array}$$

$$\begin{array}{c} \overset{OH}{|} \\ \overset{c}{C}H_3\overset{b}{C}H_2\overset{a}{\underset{|}{C}}\overset{b}{C}H_2\overset{c}{C}H_3 \\ \underset{b}{C}H_2\underset{c}{C}H_3 \end{array}$$

<u>Problem 21.17</u> Alcohol E, molecular formula $C_6H_{14}O$, undergoes acid-catalyzed dehydration when warmed with phosphoric acid to give compound F, molecular formula C_6H_{12}, as the major product. The ^{1}H-NMR spectrum of compound E shows peaks at δ 0.89 (t, 6H), 1.12 (s, 3H), 1.38 (s, 1H), and 1.48 (q, 4H). The ^{13}C-NMR spectrum of compound E shows peaks at δ 72.98, 33.72, 25.85, and 8.16. Propose structural formulas for compounds E and F.

From the molecular formula, there is a hydrogen deficiency index of 0, so there are no rings or pi bonds in compound E. From the ^{13}C-NMR peak at 72.98 we know there is a carbon bonded to an -OH group. The rest of the structure can be deduced from the ^{1}H-NMR spectrum. The chemical shifts associated with each set of hydrogens are indicated on the structure.

$$\begin{array}{c} \overset{1.38}{OH} \\ \overset{0.89\ 1.48}{C}H_3CH_2\overset{|}{\underset{|}{C}}\overset{1.48\ 0.89}{C}H_2CH_3 \\ \underset{1.12}{C}H_3 \end{array}$$

Compound E

^{1}H-NMR δ 1.48 (4H, quartet, -CH$_2$), 1.38 (1H, singlet, -OH), 1.12 (3H, singlet, -C(OH)-CH$_3$), 0.89 (6H, triplet, both -CH$_2$-CH$_3$ groups)

Dehydration of compound E gives the following alkene.

$$\begin{array}{c} CH_3\qquad\quad CH_3 \\ \diagdown\quad\ \diagup \\ C=C \\ \diagup\quad\ \diagdown \\ H\qquad\quad CH_2CH_3 \end{array}$$

Compound F

<u>Problem 21.18</u> Compound G, $C_6H_{14}O$, does not react with sodium metal and does not discharge the color of Br$_2$ in CCl$_4$. Its ^{1}H-NMR spectrum consists of only two signals, a 12H doublet at δ 1.1 and a 2H septet at δ 3.6. Propose a structural formula for compound G.

From the molecular formula, there is a hydrogen deficiency index of 0, so there are no rings or pi bonds in compound G. Since it does not react with sodium metal there cannot be an -OH group. The oxygen atom must therefore be contained within an ether group. The simplicity of the ^{1}H-NMR spectrum indicates a highly level of symmetry in the molecule, with each methyl group being attached to a carbon with a single hydrogen atom. The only structure consistent with all of this information is the following ether. The chemical shifts associated with each set of hydrogens are indicated on the structure.

$$\underset{1.1}{H_3C} \qquad \underset{1.1}{CH_3}$$
$$3.6\ HC-O-CH\ 3.6$$
$$\underset{1.1}{H_3C} \qquad \underset{1.1}{CH_3}$$

Compound G

Problem 21.19 Propose a structural formula for each haloalkane.
(a) $C_2H_4Br_2$ δ 2.5 (d, 3H) and 5.9 (q, 1H)

$$\underset{2.5}{CH_3}\text{-}\underset{5.9}{CHBr_2}$$

(b) $C_4H_8Cl_2$ δ 1.60 (d, 3H), 2.15 (q, 2H), 3.72 (t, 2H), and 4.27 (sextet, 1H)

$$\underset{1.6}{CH_3}\text{-}\underset{4.27}{CHCl}\text{-}\underset{2.15}{CH_2}\text{-}\underset{3.72}{CH_2Cl}$$

(c) $C_5H_8Br_4$ δ 3.6 (s, 8H)

$$\underset{3.6}{CH_2Br}$$
$$\underset{3.6}{BrCH_2}-C-\underset{3.6}{CH_2Br}$$
$$\underset{3.6}{CH_2Br}$$

(d) C_4H_9Br δ 1.1 (d, 6H), 1.9 (m, 1H), and 3.4 (d, 2H)

$$\underset{1.1}{CH_3}$$
$$\underset{1.1}{CH_3}-\underset{1.9}{CH}-\underset{3.4}{CH_2Br}$$

(e) $C_5H_{11}Br$ δ 1.1 (s, 9H) and 3.2 (s, 2H)

$$\underset{1.1}{CH_3}$$
$$\underset{1.1}{CH_3}-C-\underset{3.2}{CH_2Br}$$
$$\underset{1.1}{CH_3}$$

(f) $C_7H_{15}Cl$ δ 1.1 (s, 9H) and 1.6 (s, 6H)

$$\underset{1.1}{CH_3}\ \underset{1.6}{CH_3}$$
$$\underset{1.1}{CH_3}-C-C-Cl$$
$$\underset{1.1}{CH_3}\ \underset{1.6}{CH_3}$$

Problem 21.20 Following are structural formulas for esters (1), (2), and (3) and three ^{1}H-NMR spectra. Assign each compound its correct spectrum and assign all signals to their corresponding hydrogens.

$$\underset{(1)}{CH_3\overset{O}{\overset{\|}{C}}OCH_2CH_3} \qquad \underset{(2)}{H\overset{O}{\overset{\|}{C}}OCH_2CH_2CH_3} \qquad \underset{(3)}{CH_3O\overset{O}{\overset{\|}{C}}CH_2CH_3}$$

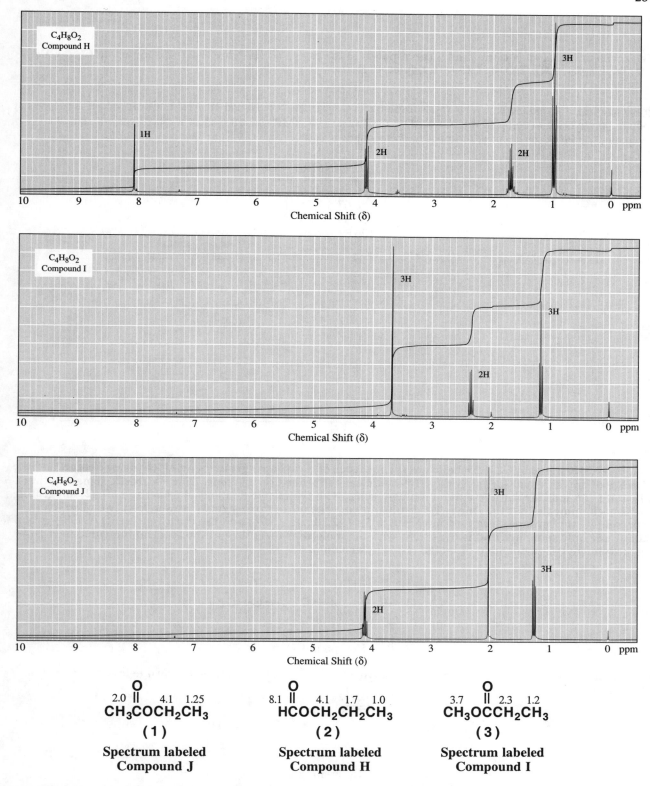

The spectrum labeled as compound H corresponds to the ester (2): ^{1}H-NMR δ 8.1 (1H, singlet, H-C(O)-), 4.1 (2H, triplet, -O-CH$_2$-), 1.7 (2H, multiplet; a doublet of triplets, -CH$_2$-), 1.0 (3H, triplet, -CH$_3$).

The spectrum labeled as compound I corresponds to the ester (3): ^{1}H-NMR δ 3.7 (3H, singlet, CH$_3$-O-), 2.3 (2H, quartet, -C(O)-CH$_2$-), 1.2 (3H, triplet, -CH$_3$).

The spectrum labeled as compound J corresponds to the ester (1): ^{1}**H-NMR δ 4.1 (2H, quartet, -C(O)-CH₂-),**
2.0 (3H, singlet, CH₃-C(O)-), 1.25 (3H, triplet, -CH₃).

<u>Problem 21.21</u> Compound K, $C_{10}H_{12}O_2$, is insoluble in water, 10% NaOH, and 10% HCl. Compound K is reduced by
sodium borohydride to compound L, $C_{10}H_{14}O_2$. Given are a ^{1}H-NMR spectrum and ^{13}C-NMR spectral data for
compound K. Propose structural formulas for compounds K and L.

^{13}C-NMR

206.51	114.17
158.67	55.21
130.33	50.07
126.31	29.03

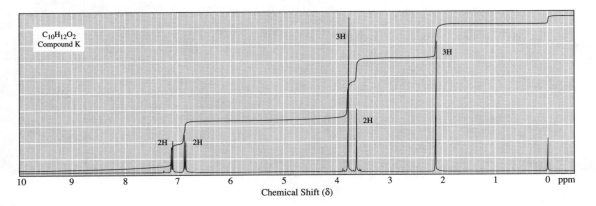

The ^{13}C-NMR signal at δ 206.51 indicates the presence of a carbonyl group of an aldehyde or ketone, and the
absence of any aldehyde signal in the ^{1}H-NMR means this must be a ketone. The ^{13}C-NMR signals between δ
114 and δ 158 indicate the presence of a phenyl ring. The symmetric doublets at δ 6.88 and δ 7.12, integrating
to 2H each, indicate that the ring is 1,4 disubstituted. The other three signals are singlets; two of which
represent methyl groups since they integrate to 3H (δ 2.10 and δ 3.76) and the third represents a -CH₂- group (δ
3.61). The only structure that is consistent with the molecular formula and the spectral information is 1-(4-
methoxyphenyl)-2-propanone (4-methoxy-phenylacetone).

3.76 CH₃O— ⟨benzene ring⟩ —CH₂—C(=O)—CH₃ 2.10
3.61

Compound H

Compound H will be reduced by sodium borohydride to the secondary alcohol shown below.

CH₃O— ⟨benzene ring⟩ —CH₂—CH(OH)—CH₃

Compound I

Problem 21.22 Propose a structural formula for each compound. Each contains an aromatic ring.
(a) $C_9H_{10}O$ δ 1.2 (t, 3H), 3.0 (q, 2H), 7.4-8.0 (m, 5H)

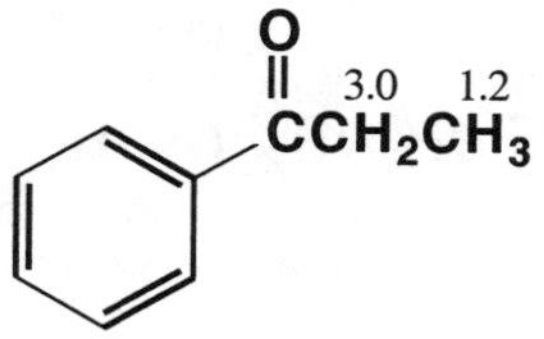

(b) $C_{10}H_{12}O_2$ δ 2.2 (s, 3H), 2.9 (t, 2H), 4.3 (t, 2H), and 7.3 (s, 5H)

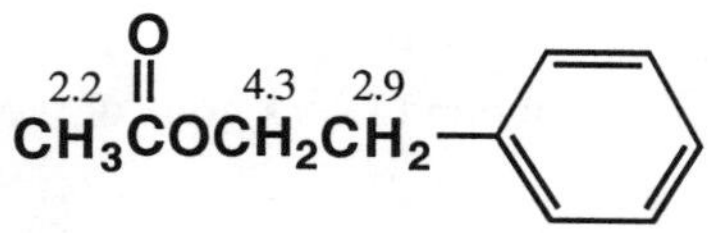

(c) $C_{10}H_{14}$ δ 1.2 (d, 6H), 2.3 (s, 3H), 2.9 (septet, 1H), and 7.0 (s, 4H)

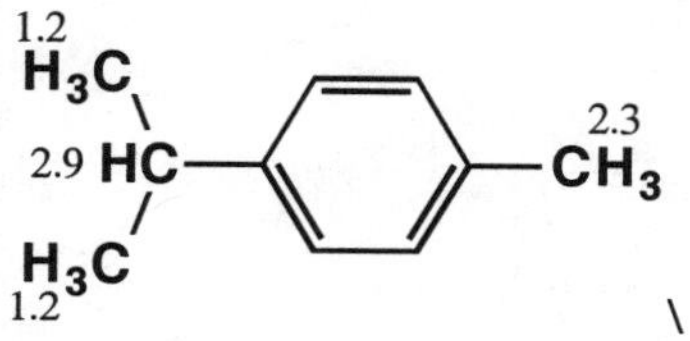

(d) C_8H_9Br δ 1.8 (d, 3H), 5.0 (q, 1H), 7.3 (s, 5H)

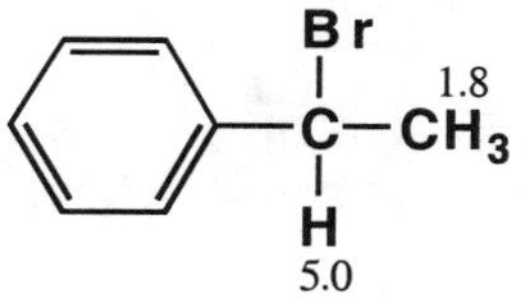

Problem 21.23 Compound M, molecular formula $C_9H_{12}O$, readily undergoes acid-catalyzed dehydration to give compound N, C_9H_{10}. A ^{1}H-NMR spectrum of compound M shows signals at δ 0.91 (t, 3H), 1.78 (m, 2H), 2.26 (d, 1H), 4.55 (m, 1H), and 7.31 (m 5H). From this information, propose structural formulas for compounds M and N.

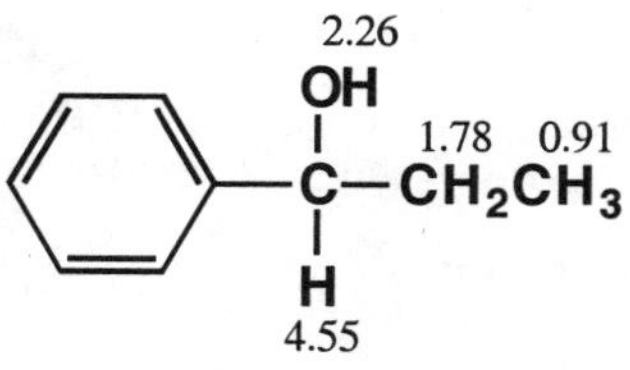

Compound J

This compound has the correct molecular formula of $C_9H_{12}O$ and is fully consistent with the ^{1}H-NMR spectrum. The chemical shift of each hydrogen is given on the structure. An alcohol function at the benzyl position explains the acid-catalyzed dehydration that gives the alkene Compound K.

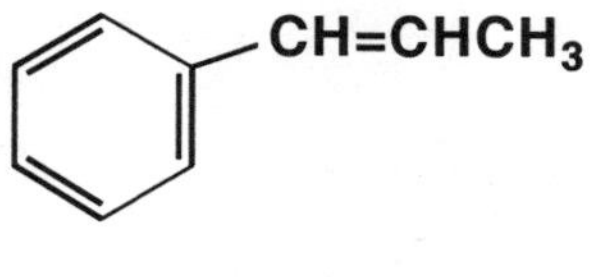

Compound K

<u>Problem 21.24</u> Propose a structural formula for each ketone.
(a) C_4H_8O δ 1.0 (t, 3H), 2.1 (s, 3H), and 2.4 (q, 2H)

$$\underset{\text{1.0}}{\text{CH}_3}\underset{\text{2.4}}{\text{CH}_2}-\overset{\displaystyle\overset{O}{\|}}{\text{C}}-\underset{\text{2.1}}{\text{CH}_3}$$

(b) $C_7H_{14}O$ δ 0.9 (t, 6H), 1.6 (sextet, 4H), and 2.4 (t, 4H)

$$\underset{\text{0.9}}{\text{CH}_3}-\underset{\text{1.6}}{\text{CH}_2}-\underset{\text{2.4}}{\text{CH}_2}-\overset{\displaystyle\overset{O}{\|}}{\text{C}}-\underset{\text{2.4}}{\text{CH}_2}-\underset{\text{1.6}}{\text{CH}_2}-\underset{\text{0.9}}{\text{CH}_3}$$

<u>Problem 21.25</u> Propose a structural formula for compound O, a ketone of molecular formula $C_{10}H_{12}O$.

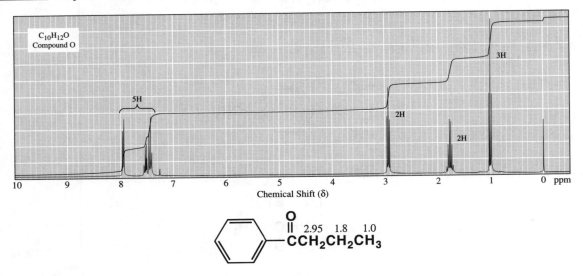

Compound O

This compound has the correct molecular formula of $C_{10}H_{12}O$, and is a ketone. The ^{1}H-NMR spectrum can be assigned as follows: δ 7.4-8 (5H, multiplet, aromatic hydrogens), 2.95 (2H, triplet, -C(O)CH$_2$-), 1.8 (2H, multiplet, -CH$_2$-CH$_3$), 1.0 (3H, triplet, -CH$_3$).

<u>Problem 21.26</u> Following is a ^{1}H-NMR spectrum for compound P, $C_6H_{12}O_2$. Compound P undergoes acid-catalyzed dehydration to give compound Q, $C_6H_{10}O$. Propose structural formulas for compounds P and Q.

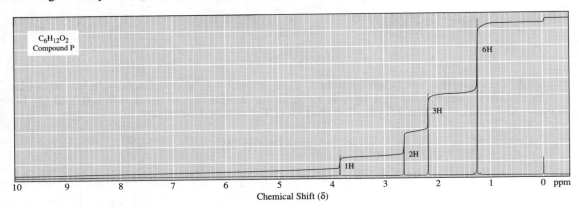

From the molecular formulas it is clear that compound P undergoes an acid-catalyzed dehydration reaction to create compound Q. Thus, compound P must have an -OH group. Furthermore, because its index of hydrogen deficiency is 1, compound P must have one pi bond or ring. Compound Q has an index of hydrogen deficiency of 2, so it must have two rings and or one pi bond and a ring, consistent with a dehydration to form an alkene. The ^{1}H-NMR spectrum of compound P shows all singlets. Especially helpful are the methyl group resonances; the singlet integrating to 6H at δ 1.22 and the singlet integrating to 3H at δ 2.18. This latter signal is assigned as a methyl ketone. The other two methyl groups are equivalent. There is the -OH hydrogen at δ 3.85 and a -CH$_2$- resonance at δ 2.62. The only structure consistent with these signals is 4-hydroxy-4-methyl-2-pentanone.

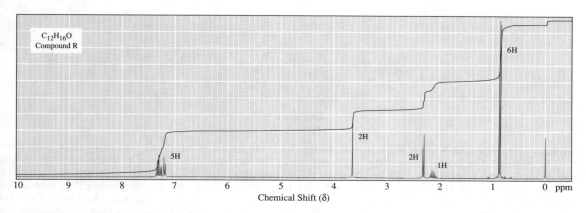

Compound P

Upon dehydration, Compound P would be turned into 4-methyl-3-pentene-2-one.

Compound Q

<u>Problem 21.27</u> Propose a structural formula for compound R, $C_{12}H_{16}O$. Following is its ^{1}H-NMR spectrum and the position of signals in its ^{13}C-NMR spectrum.

^{13}C-NMR	
207.82	50.88
134.24	50.57
129.36	24.43
128.60	22.48
126.86	

The signal in the ^{13}C-NMR spectrum at δ 207.82 indicates the presence of a carbonyl group. The signals around δ 130 indicate there is an aromatic ring. The doublet in the ^{1}H-NMR spectrum at δ 0.84 that integrates to 6H indicates two methyl groups adjacent to a -CH- group. There are also two -CH$_2$- groups; one that is not adjacent to other hydrogens (the singlet at δ 3.62) and one next to a -CH- group (the doublet at δ 2.30). The multiplet at δ 2.12 must be this -CH- group that is also adjacent to the two methyl groups. Five aromatic hydrogens are found in the complex set of signals at δ 7.3. The only structure that is consistent with all of these facts is 4-methyl-1-phenyl-2-pentanone.

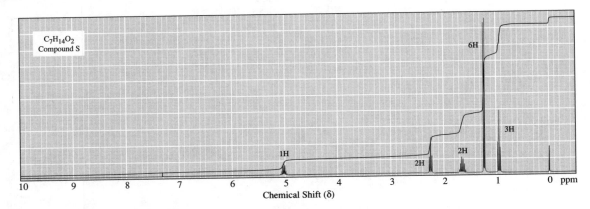

Compound R

Problem 21.28   Propose a structural formula for each carboxylic acid.

(a) $C_5H_{10}O_2$

^{1}H-NMR	^{13}C-NMR
0.94 (t, 3H)	180.71
1.39 (m, 2H)	33.89
1.62 (m, 2H)	26.76
2.35 (t, 2H)	22.21
12.0 (s, 1H)	13.69

$$\underset{0.94}{CH_3}\underset{1.39}{CH_2}\underset{1.62}{CH_2}\underset{2.35}{CH_2}\underset{12.0}{CO_2H}$$

(b) $C_6H_{12}O_2$

^{1}H-NMR	^{13}C-NMR
1.08 (s, 9H)	179.29
2.23 (s, 2H)	46.82
12.1 (s, 1H)	30.62
	29.57

$$\underset{1.08}{CH_3}\overset{\overset{1.08}{\overset{\displaystyle CH_3}{|}}}{\underset{\underset{1.08}{\underset{\displaystyle CH_3}{|}}}{\underset{2.23}{C}}}\underset{}{CH_2}\underset{12.1}{CO_2H}$$

(c) $C_5H_8O_4$

^{1}H-NMR	^{13}C-NMR
0.93 (t, 3H)	170.94
1.80 (m, 2H)	53.28
3.10 (t, 1H)	21.90
12.7 (s, 2H)	11.81

$$\underset{12.7}{HO_2C}\underset{3.10}{\overset{\overset{\displaystyle }{|}}{CH}}\underset{12.7}{CO_2H}$$
$$\underset{\underset{1.8\quad 0.93}{CH_2CH_3}}{|}$$

21.29 Following is the ^{1}H-NMR spectrum of compound S. Reduction of compound S using lithium aluminum hydride gives two alcohols, molecular formulas $C_4H_{10}O$ and C_3H_8O. Propose structural formulas for compound S and the two alcohols.

The reduction of compound S with lithium aluminum hydride into two alcohols indicates that compound S is an ester. The multiplet at δ 5.1 integrating to 1H indicates the carbon atom bound to the ester oxygen atom has a single hydrogen. The triplet at δ 2.25 integrating to 2H indicates the -CH$_2$- group adjacent to the carbonyl of the ester is also next to a -CH$_2$- group, presumably the multiplet at δ 1.65 that is the only other signal that integrates to 2H. The doublet at δ 1.3 integrating to 6H must be two methyl groups attached to a -CH- group, and the remaining triplet at δ 0.95 integrating to 3H must be another methyl group that is attached to a -CH$_2$- group. The only structure that is consistent with all of this information is isopropyl butyrate.

Compound S

Compound S will be reduced by lithium aluminum hydride to the two alcohols 1-butanol and 2-propanol.

$$CH_3CH_2CH_2CH_2OH$$

$$\overset{\displaystyle OH}{\underset{\displaystyle}{CH_3CHCH_3}}$$

1-Butanol
($C_4H_{10}O$)

2-Propanol
(C_3H_8O)

<u>Problem 21.30</u> Propose a structural formula for each ester.

(a) $C_6H_{12}O_2$

^{1}H-NMR	^{13}C-NMR
1.18 (d, 6H)	177.16
1.26 (t, 3H)	60.17
2.51 (m, 1H)	34.04
4.13 (q, 2H)	19.01
	14.25

$$\overset{1.18 \quad 2.51 \overset{O}{\overset{\|}{}} \quad 4.13 \quad 1.26}{(CH_3)_2CHCOCH_2CH_3}$$

(b) $C_7H_{12}O_4$

^{1}H-NMR	^{13}C-NMR
1.28 (t, 6H)	166.52
3.36 (s, 2H)	61.43
4.21 (q, 4H)	41.69
	14.07

$$\overset{1.28 \quad 4.21 \quad \overset{O}{\overset{\|}{}} \quad 3.36 \quad \overset{O}{\overset{\|}{}} \quad 4.21 \quad 1.28}{CH_3CH_2OCCH_2COCH_2CH_3}$$

(c) $C_7H_{14}O_2$

^{1}H-NMR	^{13}C-NMR
0.92 (d, 6H)	171.15
1.52 (m, 2H)	63.12
1.70 (m, 1H)	37.31
2.09 (s, 3H)	25.05
4.10 (t, 2H)	22.45
	21.06

$$\overset{2.09 \quad \overset{O}{\overset{\|}{}} \quad 4.10 \quad 1.52 \quad 1.7 \quad 0.92}{CH_3COCH_2CH_2CH(CH_3)_2}$$

<u>Problem 21.31</u> Following is the ^{1}H-NMR spectrum of compound T, $C_{11}H_{14}O_3$. Propose a structural formula for this compound. (Hint: The signal at δ 1.4 is actually two signals, each a closely spaced triplet.)

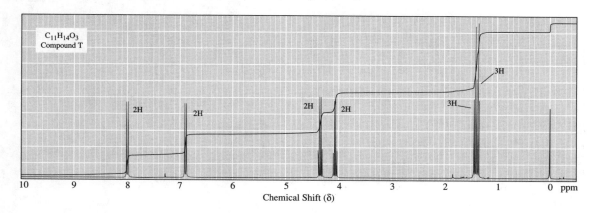

The two sets of doublets integrating to 2H each between δ 6.9 and 8.0 indicate the presence of a 1,4 disubstituted aromatic ring. The two quartets integrating to 2H each at δ 4.35 and δ 4.1 indicate there are two -CH$_2$- groups attached to oxygen atoms, we know one of which is part of an ester, and each attached to ethyl groups. The two triplets integrating to a total of 6H at δ 1.4 are the signals from the two methyl groups. The only structure that is consistent with the molecular formula $C_{11}H_{14}O_3$ is ethyl 4-ethoxybenzoate.

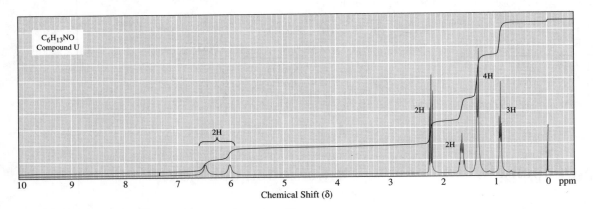

Compound T

Problem 21.32 Propose a structural formula for amide U, molecular formula $C_6H_{13}NO$.

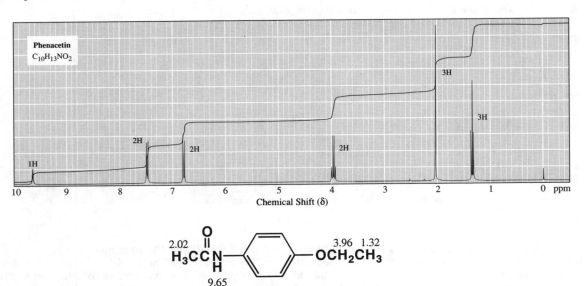

The signals at δ 0.9-2.2 are consistent with a $CH_3CH_2CH_2CH_2CH_2$- structure, with the last -CH_2- being adjacent to a carbonyl group indicated by a chemical shift of δ 2.2. The two signals integrating to 1H each at δ 6.0 and δ 6.55 are from two different amide N-H's, indicating a primary amide. The only structure of molecular formula $C_6H_{13}NO$ that is consistent with this spectrum is hexanamide.

$$\underset{0.9}{CH_3}\underset{1.4-22}{(CH_2)_4}\overset{\overset{\textstyle O}{\|}}{C}\underset{6.0,\,6.55}{NH_2}$$

Compound U

Problem 21.33 Propose a structural formula for the analgesic phenacetin, molecular formula $C_{10}H_{13}NO_2$, based on its ^1H-NMR spectrum.

This structure is not only consistent with the molecular formula, but also with the ^{1}H-NMR spectrum. The characteristic two doublets centered at δ 7.2 indicates the presence of a 1,4-disubstituted phenyl ring. The singlet at δ 9.65 integrating to 1H indicates a primary amide, and the singlet integrating to 3H at δ 2.02 indicates this is an acetamide. Finally, the typical ethyl splitting pattern for the signals at δ 1.32 and δ 3.96 indicates the presence of an ethyl group. The quartet of the ethyl group is shifted so far downfield that it must be part of an ethoxy group.

Problem 21.34 Propose a structural formula for compound V, an oily liquid of molecular formula $C_8H_9NO_2$. Compound V is insoluble in water and aqueous NaOH, but dissolves in 10% HCl. When its solution in HCl is neutralized with NaOH, compound V is recovered unchanged. The ^{1}H-NMR spectrum of compound V shows signals at δ 3.84 (s, 3H), 4.18 (s, 2H), 7.60 (d, 2H), and 8.70 (d, 2H).

Compound V is an amine based on its solubility in dilute HCl. The two doublets integrating to 2H each at δ 7.60 and δ 8.70 in the ^{1}H-NMR spectrum indicate there is a 1,4-disubstituted benzene ring. The singlet integrating to 3H at δ 3.84 indicates the presence of a methyl ester, and the signal integrating to 2H at δ 4.18 indicates the presence of a primary amine. The only structure consistent with the molecular formula of $C_8H_9NO_2$ and the ^{1}H-NMR spectrum is methyl 4-aminobenzoate.

$$\underset{\text{4.18}}{H_2N}-\underset{}{\text{C}_6\text{H}_4}-\overset{\overset{\displaystyle O}{\|}}{\underset{\text{3.84}}{C}}OCH_3$$

Compound V

Problem 21.35 Following is a ^{1}H NMR spectrum and a structural formula for anethole, $C_{10}H_{12}O$, a fragrant natural product obtained from anise. Using the line of integration, determine the number of protons giving rise to each signal. Show that this spectrum is consistent with the structure of anethole.

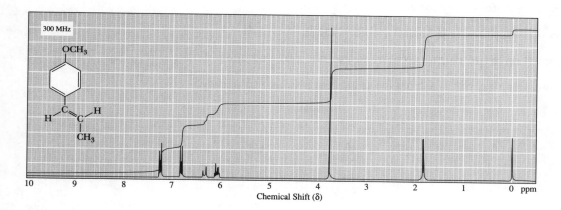

The chemical shift of the signals, the splitting patterns, and the number of protons corresponding to each signal are δ 1.81 (d, 3H), 3.77 (s, 3H), 5.95 (m, 1H), 6.33 (d, 1H), 6.80 (d, 1H), 7.23 (d, 2H). The aromatic ring of anethole contains a propenyl group which is para to a methoxy group. The 1,4-substitution pattern is responsible for the pair of doublets at δ 6.80 and 7.23. The vinylic proton nearest the aromatic ring will couple to only the hydrogen trans to it, giving the doublet at δ 6.33. The other vinylic proton couples to both the trans hydrogen and to the protons of the methyl group, giving the complex signal at δ 5.95. The methyl protons on the propenyl group are split into a doublet by the adjacent vinylic proton. Finally, the protons of the methoxy group are not coupled to anything and appear as a singlet at δ 3.77.

CHAPTER 22
Solutions to Problems

<u>Problem 22.1</u> A compound shows strong, very broad IR absorption in the region 3200-3500 cm^{-1} and strong absorption at 1715 cm^{-1}. What functional group accounts for both of these absorptions?

Both of these absorptions can be accounted for by a carboxylic acid. The absorption at 1715 cm-1 arises from the C=O bond while the broad absorption between 3200 and 3500 cm-1 arises from the O-H bond.

<u>Problem 22.2</u> Propanoic acid and methyl ethanoate are constitutional isomers. Show how to distinguish between them by IR spectroscopy.

$$\underset{\text{Propanoic acid}}{CH_3CH_2\overset{\overset{\displaystyle O}{\|}}{C}OH}$$
$$\underset{\substack{\text{Methyl ethanoate}\\\text{(Methyl acetate)}}}{CH_3\overset{\overset{\displaystyle O}{\|}}{C}OCH_3}$$

The propanoic acid will have a strong, broad absorption at 3200 to 3550 cm-1 due to the presence of the O-H bond of the carboxylic acid group. The IR spectrum of methyl ethanoate will not have an absorption in this region.

<u>Problem 22.3</u> Compound A, molecular formula, C_6H_{10}, reacts with H_2/Ni to give compound B, C_6H_{12}. See also the IR spectrum of compound A. From this information about compound A, tell
(a) Its index of hydrogen deficiency.

The index of hydrogen deficiency for compound A is two.

(b) The number of its rings and/or pi bonds.

Compound A must have two rings, two pi bonds, or one ring and one pi bond.

(c) What structural feature(s) would account for its index of hydrogen deficiency.

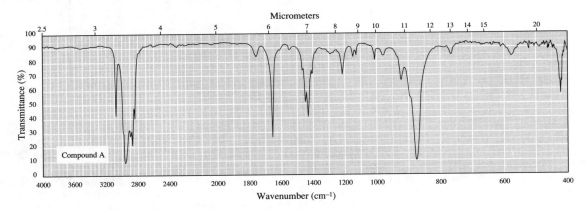

Compound A can only have one pi bond, because reaction with H2/Ni adds only 2 more hydrogen atoms. Thus, there must be one pi bond and one ring present in the structure. In addition, the double bond must be highly unsymmetrical, having a permanent dipole, to explain the prominent C=C stretching band at 1654 cm-1 seen in the IR spectrum. It would take more information than just the IR spectrum to unambiguously deduce the structure of compound A. For example, 1H-NMR and 13C-NMR would be helpful. The combined data would allow you to determine that compound A is methylenecyclopentane:

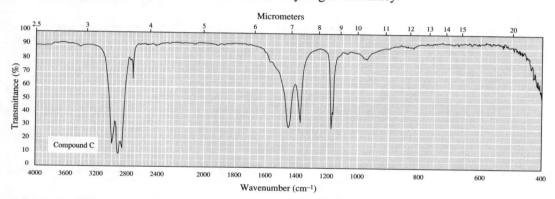

Methylenecyclopentane

<u>Problem 22.4</u> Compound C, molecular formula, C_6H_{12}, reacts with H_2/Ni to give compound D, C_6H_{14}. See also the IR spectrum of compound C. From this information about compound C, tell
(a) Its index of hydrogen deficiency.

The index of hydrogen deficiency of compound C is one.

(b) The number of its rings and/or pi bonds.

Compound C must have one ring or one pi bond.

(c) What structural feature(s) would account for its index of hydrogen deficiency.

Compound C must have one pi bond, since it can react with H_2/Ni and add two hydrogen atoms. However, the absence of any C=C stretching bands near 1650 cm^{-1} in the IR spectrum indicates that this must be an entirely symmetrical double bond.
It would take more information than just the IR spectrum to deduce unambiguously the detailed structure of compound C. For example, ^{1}H-NMR and ^{13}C-NMR would be helpful. The combined data would allow you to determine that the structure of compound C is 2,3-dimethyl-2-butene:

$$(CH_3)_2C =\!\!= C(CH_3)_2$$

2,3-Dimethyl-2-butene

<u>Problem 22.5</u> Following are infrared spectra of compounds E and F. One spectrum is of 1-hexanol and the other of nonane. Assign each compound its correct spectrum.

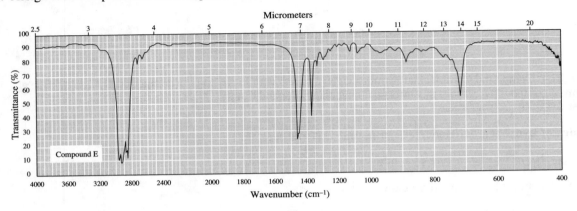

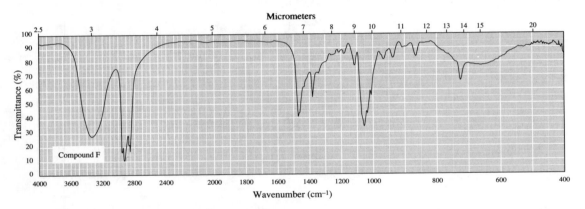

Both compounds have C-H bonds, so both spectra have C-H stretches and bends at 2900 cm^{-1} and 1450 cm^{-1}, respectively. On the other hand, the 1-hexanol has an OH group, that will give rise to an O-H and C-O stretching vibrations at 3340 cm^{-1} and 1050 cm^{-1}, respectively. These two features are in the second spectrum, so the second spectrum must correspond to 1-hexanol and the first spectrum must correspond to nonane.

<u>Problem 22.6</u> 2-Methyl-1-butanol and *tert*-butyl methyl ether are constitutional isomers of molecular formula C$_5$H$_{12}$O. Assign each compound its correct infrared spectrum, G or H.

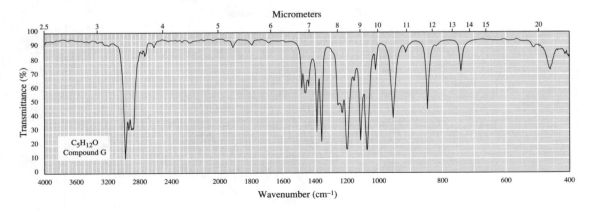

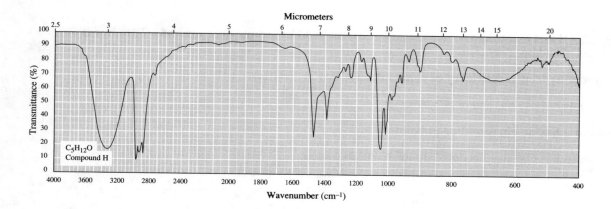

The molecules are similar except for the -OH group present in 2-methyl-1-butanol. Since the characteristic O-H stretch is present at 3625 cm^{-1} in the second spectrum, this verifies that the second spectrum corresponds to 2-methyl-1-butanol. Therefore, the first spectrum corresponds to *tert*-butyl methyl ether.

Problem 22.7 From examination of the molecular formula and IR spectrum of compound I C$_9$H$_{12}$O, tell
(a) Its index of hydrogen deficiency.

The index of hydrogen deficiency of compound I is four.

(b) The number of its rings and/or pi bonds.

Compound I has four rings and/or pi bonds.

(c) What one structural feature would account for this index of hydrogen deficiency.

A single benzene ring could account for an index of hydrogen deficiency of four.

(d) What oxygen-containing functional group it contains.

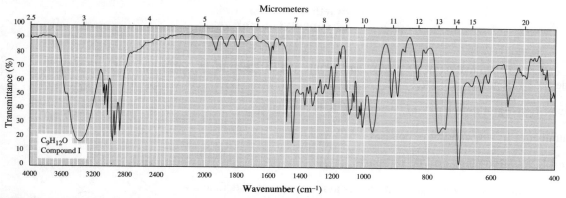

The strong broad absorption near 3400 cm^{-1} indicates the presence of an -OH group, so this must be the oxygen containing functional group. The spectrum is of 1-phenyl-1-propanol.

1-Phenyl-1-propanol

<u>Problem 22.8</u> From examination of the molecular formula and IR spectrum of compound J $C_5H_{13}N$, tell
(a) Its index of hydrogen deficiency.

The index of hydrogen deficiency of compound J is 0.

(b) The number of its rings and/or pi bonds.

Compound J has no rings and/or pi bonds.

(c) The nitrogen-containing functional group(s) it might contain.

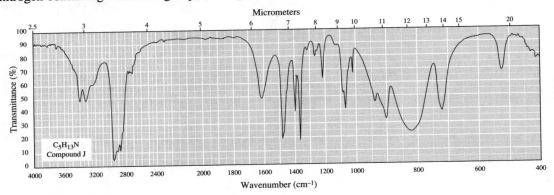

Compound J must contain an amine functional group, a prediction that is confirmed by the presence of the two broad absorptions at 3300 and 3400 cm^{-1}. The presence of the two bands indicates compound J is a primary amine. Compound J is 2,2-dimethyl-1-propanamine.

$$CH_3$$
$$|$$
$$CH_3CCH_2NH_2$$
$$|$$
$$CH_3$$

2,2-Dimethyl-1-propanamine
(Neopentylamine)

<u>Problem 22.9</u> From examination of the molecular formula and IR spectrum of compound K, $C_{10}H_{12}O$, tell
(a) Its index of hydrogen deficiency.

The index of hydrogen deficiency of compound K is five.

(b) The number of its rings and/or pi bonds.

Compound K has five rings and/or pi bonds.

(c) What structural features would account for this index of hydrogen deficiency.

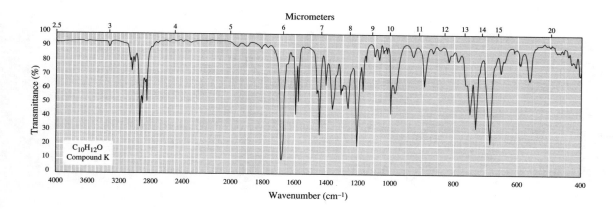

A benzene ring could account for part of this index of hydrogen deficiency, a prediction that is confirmed by the sharp absorptions at 1475 and 1600 cm^{-1}. In addition, compound K must contain a carbonyl group because of the strong absorption at 1690 cm^{-1}. Compound K is 1-phenyl-1-propanone.

$$\text{O}$$
$$\parallel$$
$$\text{—CCH}_2\text{CH}_3$$

1-Phenyl-1-propanone

Problem 22.10 From examination of the molecular formula and IR spectrum of compound L, $C_7H_{14}O_2$, tell
(a) Its index of hydrogen deficiency.

The index of hydrogen deficiency of compound L is one.

(b) The number of its rings and/or pi bonds.

Compound L has one ring or pi bond.

(c) The oxygen-containing functional group(s) it might contain.

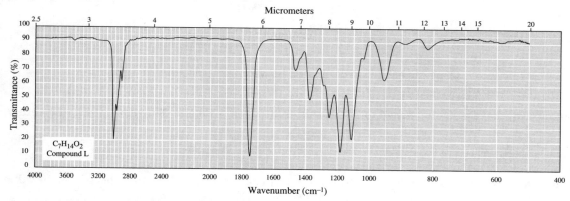

The strong absorption band at 1750 cm^{-1} indicates the presence of a carbonyl group in compound L. There is no broad -OH absorption between 2500 and 3300 cm^{-1} to indicate the presence of a carboxylic acid. Thus, it is likely that the oxygen containing functional group of compound L is an aldehyde, a ketone, or an ester. The strong C-O absorptions between 1100 and 1200 cm^{-1} indicate that compound L is an ester. Compound L is isopropyl butanoate.

$$\overset{\overset{\textstyle O}{\|}}{CH_3CH_2CH_2COCH(CH_3)_2}$$

Isopropyl butanoate

Problem 22.11 From examination of the molecular formula and IR spectrum of compound M, $C_6H_{13}NO$, tell
(a) Its index of hydrogen deficiency.

The index of hydrogen deficiency of compound M is one.

(b) The number of its rings and/or pi bonds.

Compound M has one ring or pi bond.

(c) The oxygen and nitrogen-containing functional group(s).

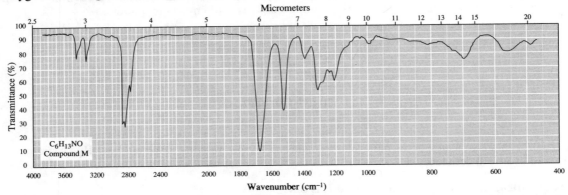

The presence of the carbonyl absorption at 1740 cm⁻¹ and the two broad absorptions at 3460 and 3590 cm⁻¹ indicate that the oxygen and nitrogen containing functional group in compound M is a primary amide. Compound M is hexanamide.

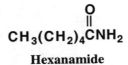

Hexanamide

22.12 Show how IR spectroscopy can be used to distinguish between the compounds in each set.
(a) 1-Butanol and diethyl ether

The 1-butanol will have a strong, broad O-H absorption between 3200 and 3400 cm⁻¹, the diethyl ether will not.

(b) Butanoic acid and 1-butanol

The butanoic acid will have a strong C=O absorption between 1700 and 1725 cm⁻¹, the 1-butanol will not.

(c) Butanoic acid and 2-butanone

The butanoic acid will have a strong, broad O-H absorption between 2400 and 3400 cm⁻¹, the 2-butanone will not.

(d) Butanal and 1-butene

The butanal will have a strong C=O absorption between 1720 and 1740 cm⁻¹, and the 1-butene will have a medium C=C absorption between 1475 and 1600 cm⁻¹ along with a strong vinylic C-H absorption near 3030 cm⁻¹.

(e) 2-Butanone and 2-butanol

2-Butanone will have a strong C=O absorption between 1705 and 1725 cm⁻¹, and the 2-butanol will have a strong, broad OH absorption between 3200 and 3400 cm⁻¹ along with a medium C-O absorption between 1050 and 1150 cm⁻¹.

(f) Butane and 2-butene

2-Butene will have a strong vinylic C-H absorption near 3030 cm⁻¹ and a medium C=C absorption between 1475 and 1600 cm⁻¹.

22.13 For each set of compounds, list one major feature that appears in the IR spectrum of one compound but not the other. Your answer should state what type of bond vibration is responsible for the spectral feature you list and its approximate position in the IR spectrum.

(a) 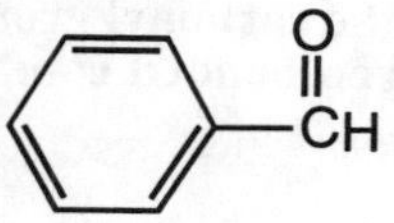   and

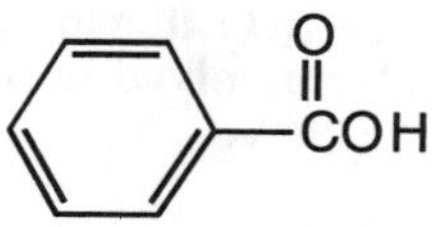

The benzoic acid on the right will have a strong, broad OH absorption between 2400 and 3400 cm⁻¹, the benzaldehyde will not.

(b)

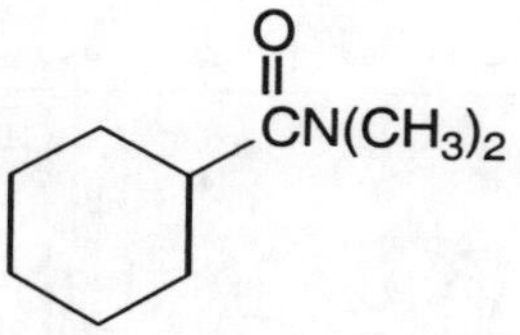

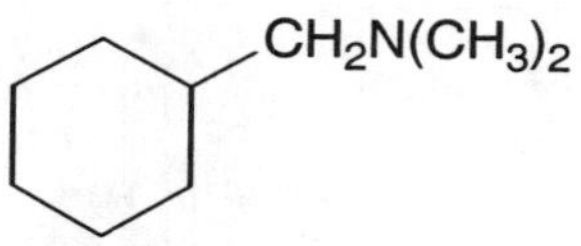

The amide on the left will have a strong C=O absorption between 1630 and 1680 cm⁻¹, the amine on the right will not.

(c) 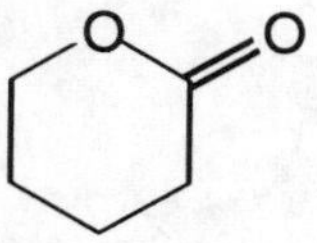   and   HO(CH₂)₄COH

5-Hydroxypentanoic acid on the right will have a strong, broad O-H absorption between 2400 and 3400 cm⁻¹ from the acid OH and a strong, broad OH absorption between 3200 and 3400 cm⁻¹ from the alcohol OH. The lactone will have neither of these absorptions.

(d) 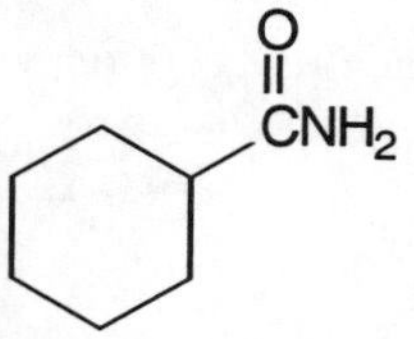   and

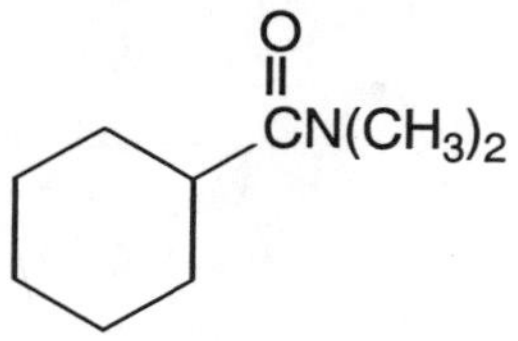

The primary amide on the left will have two broad N-H absorptions between 3200 and 3400 cm⁻¹, the *N,N*-dimethyl amide on the right will not.

Problem 22.14 Following is an infrared spectrum and a structural formula for methyl salicylate, the fragrant component of oil of wintergreen. The spectrum was recorded using the pure liquid between KBr plates. The dashed line between O of the C=O group and H of the -OH group indicates intramolecular hydrogen bonding. On this spectrum, locate the absorption peak(s) due to

(a) O-H stretching of the hydrogen-bonded -OH group (very broad and of medium intensity).

The O-H bond stretch is centered at 3190 cm^{-1}.

(b) C-H stretching of the aromatic ring (sharp and of weak intensity).

The aromatic C-H stretch is the sharp signal at 3075 cm^{-1}.

(c) C=O stretching of the ester group (sharp and of strong intensity).

The carbonyl group absorbance is centered at 1680 cm^{-1}. The carbonyl stretching frequencies of esters normally appear between 1735 - 1750 cm^{-1}. In the case of methyl salicylate, the fact that the carbonyl group is (1) conjugated with the pi bonds of the aromatic ring and (2) involved in internal hydrogen bonded combine to lower the stretching frequency to 1680 cm^{-1}.

(d) C=C stretching of the aromatic ring (sharp and of medium intensity).

Absorbances for the aromatic carbon-carbon bond stretch are at 1441 cm^{-1} and 1615 cm^{-1}.

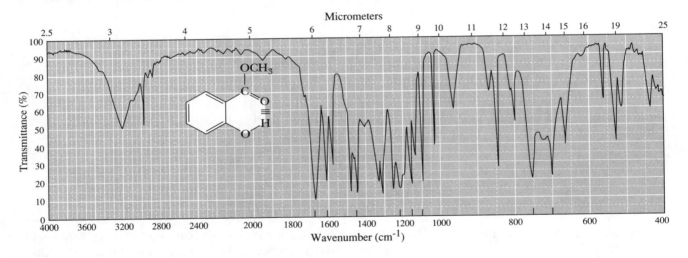